AF539868

PLANTS FOR NOVEL DRUG MOLECULES

About the Editors

Dr. Bikarma Singh born in India working as Senior Scientist in National Botanical Research Institute (NBRI), Lucknow one of the pioneer institute of Council of Scientific and Industrial Research known for largest research and development activities under Ministry of Science and Technology, Government of India. He is recognized as Assistant Professor in AcSIR (Ghaziabad), and registered as Ph.D. Supervisor in University of Jammu. He graduated up as an Honours and Gold Medalist in Botany from North-Eastern Hill University Shillong in 2005, and completed Doctorate Degree in Botany from Gauhati University Assam and Botanical Survey of India (Shillong) Meghalaya in 2012. Prior to joining CSIR-NBRI Lucknow in 2020, he was working as Senior Scientists in CSIR-Indian Institute of Integrative Medicine Jammu and involved in R&D related to Plant Sciences. He also served as Scientist-Ecologist in WAPCOS Limited Gurgaon.

Currently, Dr. Singh pursued research with passion and possesses 15 year's research experience in Plant Sciences, expertised in systematic and biotechnology, ethnobotany, value addition from essential oils, ecology, biodiversity germplasm conservation and EIA. He is serving as a reviewing member for National and International Scientific SCI and UGC listed journals, undertaken as lead leader for organizing many seminars and workshops. He received Prestigious recognition as 'Outstanding Scientist in Botany' in 2019 awarded by Venus International Foundation, Chennai and two of his Ph.D. Scholar received best research paper award, working in the area of Botany. He authored/co-authored 8 books and published 86 research papers/experimental findings in peer-reviewed International and National journals. His recent books are Botanical Leads for Drug Discovery, 'Plants for Human Survival and Medicine' and Plants of Commercial Values. He has 2 patents on medicinal aspects, and worked as PI/Co-PI/Team member for more than 40 research projects funded by Govt. and Private Organizations. He is supervising 5 Ph.D. students in Science to defend their doctorate degree. He is a member of several research committees, and delivered invited talk as key-note speaker/lecturer under various themes in several conferences, seminars and workshops.

Prof Yash Pal Sharma, born on January 01, 1969 in a remote village of District Udhampur, Jammu & Kashmir, is presently working as a Professor and Coordinator in the Department of Botany (UGC-SAP DRS II), University of Jammu, Jammu which is one of the India's leading academic and research Universities with NAAC accredited A+ grade. A post-graduate in Botany (1992) from the Department of Botany, University of Jammu, Prof. Sharma qualified UGC-CSIR NET-JRF in the same year and obtained his Ph.D Degree from the same organization in the year 1997. He began his career as an Assistant Professor in the Government College and later joined the Department of Botany, University of Jammu in the year 2000 and was elevated to the rank of Professor in the year 2011. At present, Prof. Sharma is actively involved in teaching varied subjects such as the diversity of lower plants and gymnosperms, mycology and plant pathology, plant metabolism, ethnobotany etc. to the postgraduate and research students in the subject of Botany. Besides his involvement in providing academic leadership to the University Department as Head of the Department of Botany (2015- 2017), he has so far successfully guided a total of 12 Ph.D & 14 M.Phil students and presently 8 research students are working under his supervision. Prof Sharma has published more than 106 research papers in peer-reviewed international and national journals and is also on the panel of reviewers of several journals of National and International repute. Prof Sharma has handled many research projects sanctioned by various funding agencies. In view of his research contributions, Prof Sharma was awarded Science Talent Promotion Scholarship in Botany from the Department of Science and Technology, J&K (1991), Dr. K.S. Krishnan DAE Research Fellowship as JRF in Life Sciences, from the Department of Atomic Energy, Bombay, Govt. of India (1993), Senior Research Fellowship from CSIR, New Delhi (1996), Young Scientist Award (2003) of J&K State Council for Science and Technology, Jammu and Kashmir, Prof. P.N. Mehra Young Scientist Award (2005), Prof. H.C. Dube Outstanding Young Scientist Award (2010) by the Indian Society of Mycology and Plant Pathology, Udaipur and Dr. S.K Shome Memorial Award by the Mycological Society of India, Chennai (2018). A life member of many professional/ scientific organizations of India, He has also organized several prestigious national scientific events at the University of Jammu, Jammu and is regularly involved in delivering popular lectures/invited talks in various academic and research events across the country and abroad.

PLANTS FOR NOVEL DRUG MOLECULES
Ethnobotany to Ethnopharmacology

Editors

Bikarma Singh
(B.Sc. Honours, M.Sc. Gold Medalist, Ph.D., NABET)
CSIR-National Botanical Research Institute
(Council of Scientific & Industrial Research)
Ministry of Science and Technology, Government of India
Rana Pratap Marg, Hazartganj, Lucknow-226001
Uttar Pradesh, India

Yash Pal Sharma
Department of Botany
University of Jammu
Jammu–180006, Jammu & Kashmir, India

NEW INDIA PUBLISHING AGENCY
New Delhi – 110 034

NEW INDIA PUBLISHING AGENCY

101, Vikas Surya Plaza, CU Block, LSC Market
Pitam Pura, New Delhi – 110 034, India
Email: info@nipabooks.com
Web: www.nipabooks.com

For customer assistance, please contact
Phone: + 91-11-27 34 17 17 Fax: + 91-11- 27 34 16 16
E-Mail: feedbacks@nipabooks.com

ISBN 978-93-89571-94-3

Composed and Designed by NIPA.

Preface

Scientific research on plants coupled with traditional knowledge for human health care used globally is supercilious. In developing countries, traditional medicine is often the only accessible and affordable treatment available for curing frequently occurred seasonal diseases. This book **'Plants for Novel Drug Molecules: Ethnobotany to Ethnopharmacology'** is the result of ongoing R&D on ethnobotanical and pharmacological aspects for discovering new drugs and nutraceuticals directly or indirectly helpful for human and animals health care. Validated scientific data gathered on plants always remain the basis for commercial medication and product formulations for the treatment of chronic diseases. Traditional herbal formulations and local home-made recipes have been used throughout history to prevent and cure diseases. People are coming using plants that were available within the geographical boundary addressing local health related issues. Actually the indigenous cultural tradition were exposed due to people immigration for trade and business, and gets often overwhelmed by modern scientific concepts and medications of various culture of different country.

Ethnobotany as an academic discipline surfaced in the nineteen century to disseminate scientific research ethics and describes the relationship between ethnic communities and plants, however, a similar term ethnobiology was also used sometime to signify the use of plants and animals by ancient communities. Even before the emergence of ethnobotany and ethnobiology as disciplines, various societies and individuals explored the relationship between human-plants-animals. The development of ethnobiology in fact started with the compilation of ancient medicinal knowledge in from Greece, Egypt and Asia. Actually, the old systems of medicine are believed to be the written compilation of contemporary folk knowledge, and thus folk knowledge considered as the precursor of all traditional medicinal systems.

As we enter the new decades of twenty-first century, medicines continue to constitute as one of the essential components of health care system in promoting health and preventing illness. Plants has been recognized as a rich source of novel drugs that form the ingredients in traditional system of medicine such as Ayurvedha, Siddha and Yoga. Drugs were manufactured and prepared by using plants as a main source of ingredient.

In recent decades, the discovery of several molecules such as berberine form the genus *Berberis* and *Coptis*, artimisinin from *Artemisia annua*, cannabidiol from *Cannabis sativa*, diosgenin from *Cheilocostus speciosus*, and several others bioactive molecules from ethnobotanical plants has a long history of use for human needs and wants, considered as the best examples of drug discovered for human from plants. The main idea to present this book is to provide the baseline scientific information on plant-people interaction and to get acquainted with current scientific discovery.

The present book is based on twenty five excellent scientific contributions of seventy two researchers from topmost research organizations. The book begin with plants used in Sowa-Rigpa system of food and medicine, followed by traditional uses of plants as medicine among Khasi tribe living in northeast India. This compilation contains several research techniques highlighting methods and analysis of documented data, and procedure for scientific validation, findings and results. Methods for assessing traditional knowledge of highly threatened plants such as *Hodgsonia heteroclita*, pharmacological applications of family asteraceae, ethnobotany of family apiaceae, plants used in managing leucorrhea, plants as animal care, phytochemistry of *Arisaema jacquemontii, Andrographis paniculata, Blumea lacera, Boerhaavia diffusa, Hemidesmus indicus, Pterocarpus santalinus, Rauwolfia serpentina, Rauwolfia tetraphylla,* and several other ethnobotanical and ethnopharmacological parameters used in studying current science is described here in this book. Besides, it is contain by several research topics focused to the clinical arena, plants used in relation to cancer, diabetes, skin disorders and many other aspects relates to animal and human health care. Today's food supplements derived from plants are of high demand, and this compilation also highlighted several plants used as nutraceuticals. It has been observed that herbs contain many bioactive compounds with powerful antioxidant properties as evidence from the scientific data, and few research data on lianas, lichens and role of allyl isothiocyanate as a bioprotective agent discussed added more value to this compilations. Focused theme such as ethnobotanical trends and techniques, phytochemistry, biological activities, ethnopharmacology and clinical studies is adding and contributing a lots value to this book in discovering leads for medicine formulations.

Editors and authors tried their best to convey the maximum knowledge through this book "**Plants for Novel Drug Molecules: Ethnobotany to Ethnopharmacology**" with respect to potential plants for human needs and medicines. We are sure and full confident that the readers will have a moral obligation to convey their suggestions on this book in near future for its better improvement.

Bikarma Singh
Yash Pal Sharma

Contents

Contributors

EDITORS

Bikarma Singh CSIR-National Botanical Research Institute (Council of Scientific & Industrial Research) Ministry of Science and Technology, Government of India Rana Pratap Marg, Hazartganj, Lucknow-226001, Uttar Pradesh, India

Yash Pal Sharma Department of Botany, University of Jammu, Jammu-180006, Jammu & Kashmir, India

CONTRIBUTORS

Bikarma Singh CSIR-National Botanical Research Institute (Council of Scientific & Industrial Research) Ministry of Science and Technology, Government of India Rana Pratap Marg, Hazartganj, Lucknow-226001, Uttar Pradesh, India

Yash Pal Sharma Department of Botany, University of Jammu, Jammu-180006, Jammu & Kashmir, India

Abhishek Dutta Department of Botany, University of Jammu, Jammu-180006, Jammu and Kashmir, India

Ajit Kumar Shasany CSIR-Central Institute of Medicinal and Aromatic Plants, Kukrail Picnic Spot Road, Lucknow-226015, Uttar Pradesh, India.

Amit K. Tripathi Center for Environmental Sciences & Engineering, School of Natural Sciences, Shiv Nadar University, Greater Noida-201314, Uttar Pradesh, India;

Anand Kishor Department of Botany, Veer Kunwar Singh University, Ara-802301, Bihar, India

Anita Gupta CSIR-National Botanical Research Institute, Lucknow-226001, Uttar Pradesh, India

Anjali Saini Plant Sciences (Biodiversity & Applied Botany Division), CSIR-Indian Institute of Integrative Medicine, Canal Road, Jammu-180001, Jammu and Kashmir, India

Anjina Devi Plant Cytology & Genetics Laboratory, Department of Botany, University of Jammu, Jammu-180006, Jammu and Kashmir, India

Anupam Changmai Medicinal Aromatic & Economic Plants Group, BSTD CSIR-North East Institute of Science and Technology, Jorhat-785006, Assam, India

Archana Ojha Department of Botany, North-Eastern Hill University, Shillong-793022, Meghalaya, India

Arun Chettri Department of Botany, Sikkim University, 6th Mile, Samdur, PO Tadong- 737102, Gangtok, Sikkim, India

Arushi Gupta Natural Product Chemistry Division, CSIR-Indian Institute of Integrative Medicine, Jammu-180001, Jammu & Kashmir, India.

Ashish Kumar CSIR-Central Institute of Medicinal and Aromatic Plants Research Centre, Boduppal, Hyderabad-500092, Telangana, India

Ashutosh K. Dash School of Pharmaceutical Sciences, Department of Pharmaceutical Chemistry, Shoolini University Solan-173229, Himachal Pradesh, India

Bashir Lone Natural Product Chemistry Division, CSIR-Indian Institute of Integrative Medicine, Jammu-180001, Jammu & Kashmir, India.

Bishander Singh Department of Botany, Veer Kunwar Singh University, Ara-802301, Bihar.

Debaraj Mukherjee Natural Product Chemistry Division, CSIR-Indian Institute of Integrative Medicine, Jammu, Jammu & Kashmir 180 001, India

Dipayan Ghosh CSIR-Central Institute of Medicinal and Aromatic Plants, Kukrail Picnic Spot Road, Lucknow-226015, Uttar Pradesh, India.

Eduardo Soares Calixto Post-Graduate Program in Entomology, Faculdade de Filosofia, Ciências e Letras, Universidade de São Paulo. 3900 Bandeirantes av., Zip-code-14040-901, Ribeirão Preto, SP, Brazil.

Gardinia Nongbri Department of Environmental Studies, North-Eastern Hill University, Shillong-793022, Meghalaya, India

Gargee Debnath Department of Environmental Studies, North-Eastern Hill University, Shillong-793022, Meghalaya, India

Harish Chander Dutt Ecological Engineering Laboratory, Department of Botany, University of Jammu, Jammu-180006, Jammu & Kashmir, India

Harpreet Bhatia Department of Botany, University of Jammu, Jammu-180001, Jammu and Kashmir, India

Harsh Singh Department of Botany, North-Eastern Hill University, Shillong-793022, Meghalaya, India

Jnanesha A CCSIR-Central Institute of Medicinal and Aromatic Plants Research Centre, Boduppal, Hyderabad-500092, India

Jyoti K Sharma Center for Environmental Sciences & Engineering, School of Natural Sciences, Shiv Nadar University, Greater Noida-201314, Uttar Pradesh, India

Kota Srinivas Plant Science Division, CSIR Indian Institute of Integrative Medicine, Canal Road, Jammu-180001, Jammu and Kashmir, India

Krishna Upadhaya Department of Basic Sciences & Social Sciences, North-Eastern Hill University, Shillong-793022, Meghalaya, India

Madhulika Bhagat School of Biotechnology, University of Jammu, Jammu-180006, Jammu & Kashmir, India

Mamta Bhat Department of Botany, School of Biosciences and Biotechnology, BGSB University, Rajouri-185234, Jammu and Kashmir, India.

Manoj Kumar Singh Kulbhaskar Ashram Post Graduate College, Prayagraj-211002, Uttar Pradesh, India.

Mohammed Asif Chowdhary Plant Sciences (Biodiversity & Applied Botany Division), CSIR-Indian Institute of Integrative Medicine, Canal Road, Jammu-180001, Jammu & Kashmir, India

Mohan Lal Medicinal Aromatic & Economic Plants Group, BSTD CSIR-North East Institute of Science and Technology, Jorhat-785006, Assam, India

Mohd Ahmad Center for Environmental Sciences & Engineering, School of Natural Sciences, Shiv Nadar University, Greater Noida-201314, Uttar Pradesh, India

Mohd Shahnawaz Plant Biotechnology Division, CSIR-Indian Institute of Integrative Medicine Canal Road, Jammu-180001, Jammu & Kashmir, India

Mudasir Nazir Bhat Plant Sciences (Biodiversity & Applied Botany Division), Academy of Scientific & Innovative Research, CSIR-Indian Institute of Integrative Medicine, Canal Road, Jammu-180001, Jammu and Kashmir, India

Namrata Sharma Department of Botany, University of Jammu, Jammu-180006, Jammu and Kashmir, India.

Narendra Kumar CSIR-Central Institute of Medicinal and Aromatic Plants, Kukrail Picnic Spot Road, Lucknow-226015, Uttar Pradesh, India

Nawang Tashi Ecological Engineering Laboratory, Department of Botany, University of Jammu, Jammu-180006, Jammu & Kashmir, India

Om Prakash Arya G.B. Pant National Institute of Himalayan Environment and Sustainable Development, North-East Regional Centre, Itanagar-791113, Arunachal Pradesh, India.

Opender Surmal Plant Sciences (Biodiversity & Applied Botany Division), Academy of Scientific & Innovative Research (AcSIR), CSIR-Indian Institute of Integrative Medicine, Canal Road, Jammu-180001, Jammu and Kashmir, India

Pooja Shukla CSIR-National Botanical research Institute, Lucknow-226001, Uttar Pradesh, India

Prasoon Gupta Natural Product Chemistry Division, CSIR-Indian Institute of Integrative Medicine, Jammu-180001, Jammu & Kashmir, India

Prem Prakash Singh Department of Botany, North-Eastern Hill University, Shillong-793022, Meghalaya, India

Rajendra Bhanwaria Genetic Resources & Agrotechnology Division, CSIR-Indian Institute of Integrative Medicine, Canal Road, Jammu-180001, Jammu & Kashmir, India

Rajendra Gochar Genetic Resources & Agrotechnology Division, CSIR-Indian Institute of Integrative Medicine, Canal Road, Jammu-180001, Jammu & Kashmir, India

Rasleen Sudan School of Biotechnology, University of Jammu, Jammu-180006, Jammu & Kashmir, India.

Rohit Arora Department of Biochemistry, Sri Guru Ram Das Institute of Medical Sciences and Research, Amritsar, Punjab, India

Sajan Thakur Ecological Engineering Laboratory, Department of Botany, University of Jammu, Jammu-180006, Jammu & Kashmir, India

Sakshi Bhushan Department of Botany, University of Jammu, Jammu-180006, Jammu & Kashmir, India

Saroj Arora Department of Botanical & Environmental Sciences, Guru Nanak Dev University, Amritsar-143005, Punjab, India.

Savita Sharma School of Biotechnology, Faculty of Sciences, Shri Mata Vaishno Devi University, Katra-182320, Jammu & Kashmir, India

Sharada Mallubhotla School of Biotechnology, Faculty of Sciences, Shri Mata Vaishno Devi University, Katra-182320, Jammu & Kashmir, India

Shiekh Marifatul Haq Department of Botany, University of Kashmir, Hazratbal, Srinagar-190006, Jammu & Kashmir, India

Shivali Verma Department of Botany, University of Jammu, Jammu-180006, Jammu & Kashmir, India.

Shreyans Kumar Jain Department of Pharmaceutical Engineering & Technology, Indian Institute of Technology, Banaras Hindu University, Varanasi-221005, Uttar Pradesh, India

Siva Hemalatha Department of Pharmaceutical Engineering & Technology, Indian Institute of Technology, Banaras Hindu University, Varanasi-221005, Uttar Pradesh, India

Sneha CSIR-Central Institute of Medicinal and Aromatic Plants, PO-CIMAP, Near Kukrail Picnic Spot Lucknow-226015, Uttar Pradesh, India

Sonam Tamchos Department of Botany, University of Jammu, Jammu-180006, Jammu & Kashmir, India

Sumit Singh Plant Sciences (Biodiversity & Applied Botany), Academy of Scientific and Innovative Research (AcSIR), CSIR-Indian Institute of Integrative Medicine, Jammu-180001, Jammu & Kashmir, India

Sunil Kumar School of Pharmaceutical Sciences, Department of Pharmaceutical Chemistry, Shoolini University, Solan- 173229, Himachal Pradesh, India

Susheel Verma Department of Botany, School of Biosciences and Biotechnology, BGSB University, Rajouri- 185234, Jammu & Kashmir, India

Tarkeshwar Dubey Department of Pharmaceutical Engineering and Technology, Indian Institute of Technology, Banaras Hindu University, Varanasi-221005, Uttar Pradesh, India.

Uma Bharti Department of Botany, University of Jammu, Jammu-180006, Jammu & Kashmir, India

Upasana Sharma Department of Applied Chemistry, Mahant Bachittar Singh College of Engineering and Technology, Babliana, Jammu-181101, Jammu & Kashmir, India

Vandolf M Kharbih Department of Botany, North-Eastern Hill University, Shillong, 793022, Meghalaya, India

Veenu Kaul Department of Botany, University of Jammu, Jammu-180006, Jammu & Kashmir, India

Venugopal Singamaneni Natural Product Chemistry Division, CSIR-Indian Institute of Integrative Medicine, Jammu-180001, Jammu & Kashmir, India

Wishfully Mylliemngap G.B. Pant National Institute of Himalayan Environment and Sustainable Development, North-East Regional Centre, Itanagar-791113, Arunachal Pradesh, India.

Zishan Ahmad Wani Conservation Ecology Lab, Department of Botany, Baba Ghulam Shah Badshah University, Rajouri-185 234, Jammu & Kashmir, India.

Glossary

Abscess: Localised collection of pus caused by suppuration in a tissue.

Acne: An inflammatory disease occurring in or around sebaceous glands.

Aflatoxin: Poisonous carcinogens that are produced by certain molds which grow in soil, decaying vegetation, hay and grains.

Agronomy: The science of crop production and soil management.

Allosteric Modulator: Drug that binds to a receptor at a site distinct from the active site. Induces a conformational change in the receptor, which alters the affinity of the receptor for the endogenous ligand. Positive allosteric modulators increase the affinity, whilst negative allosteric modulators decrease the affinity.

Amino acid: Amino acids are organic compounds containing amine and carboxyl functional groups, along with a side chain specific to each amino acid.

Amorphous: without a clearly defined shape or form.

Anaemia: Lack of enough blood in the body causing paleness.

Anaesthetic: Inducing loss of feeling or consciousness.

Analgesic: Relieving pain.

Analytic studies: Studies with control groups, namely case-control studies, cohort studies, and randomized clinical trials.

Annual: A type of flower or plant that lives for only one year.

Antagonist: Drug that attenuates the effect of an agonist. Can be competitive or non-competitive, each of which can be reversible or irreversible. A competitive antagonist binds to the same site as the agonist but does not activate it, thus blocks the agonist's action. A non-competitive antagonist binds to an allosteric (non-agonist) site on the receptor to prevent activation of the receptor. A reversible antagonist binds non-covalently to the receptor, therefore can be "washed out". An irreversible antagonist binds covalently to the receptor and cannot be displaced by either competing ligands or washing.

Antidiarrheal: Preventing or controlling diarrhea.

Antidote: An agent which neutralizes or opposes the action of a poison.

Apoptosis: Death of cells which occurs as a normal and controlled part of an organism's growth or development.

Ascocarp: An ascocarp is the fruiting body of an ascomycete phylum fungus. It consists of very tightly interwoven hyphae.

Autophagy: Autophagy (or autophagocytosis) is the natural, regulated mechanism of the cell that disassembles unnecessary or dysfunctional components.

Bioactivity: Biological activity describes the beneficial or adverse effects of a drug on living matter.

Biodiversity hotspot: Biogeographic region that is both a significant reservoir of biodiversity and is threatened with destruction. The term biodiversity hotspot specifically refers to 25 biologically rich areas around the world that have lost at least 70 percent of their original habitat.

Biodiversity: Reflects the number, variety and variability of living organisms.

Biological resources: Those components of biodiversity of direct, indirect, or potential use to humanity.

Biome: A major portion of the living environment of a particular region characterized by its distinctive vegetation and maintained by local climatic conditions.

Bovine: cattle.

Calibration: The action or process of calibrating something.

Carbohydrate: Biomolecule consisting of carbon, hydrogen and oxygen atoms, usually with hydrogen-oxygen atom ratio of 2:1.

Chromatin: Mass of genetic material composed of DNA and proteins that condense to form chromosomes during eukaryotic cell division.

Chromatography: Laboratory technique for the separation of mixture.

Clinical pharmacology: The study of the effects of drugs in humans.

Coniferous*:* The conifers are a division of vascular land plants containing gymnosperms, cone-bearing seed plants.

Crystallization: Chemical solid–liquid separation technique, in which mass transfer of a solute from the liquid solution to a pure solid crystalline phase occurs.

Cytotoxicity: Quality of being toxic to cells. Examples of toxic agents are an immune cell or some types of venom.

Deciduous: Shedding the leaves annually, as certain trees and shrubs.

Deciduous: A type of tree that sheds its leaves before the colder months.

Decoction: Concentrated liquor resulting from heating or boiling a substance, especially a medicinal preparation made from a plant.

Distillation: A process of evaporation and recondensation used for purifying liquids.

Diuretic: Diuretics, also called water pills, are medications designed to increase the amount of water and salt expelled from the body as urine.

Dose-response relationship: A relationship in which a change in amount, intensity, or duration of exposure is associated with a change in risk of a specified outcome.

Drip irrigation: The use of pipes to bring water into contact with the roots of plants.

Drug product: A finished dosage form, for eg., a tablet, capsule or solution that contains a drug substance

Drug substance: An active ingredient that is intended to furnish pharmacological activity or other direct effect in diagnosis, cure, mitigation, treatment or prevention of diseases or to effect the structure or any function of the human body

Drug: An agent that is used therapeutically to treat diseases. It may also be defined as any chemical agent and/or biological product or natural product that affects living processes

EC_{50}: Molar concentration of an agonist that produces 50% of the maximum possible response for that agonist.

Ecology: The study of the environment and how living things interact with it.

Ecosystem: Community of living and non-living things that interact by exchanging matter and energy.

Environment: Physical surroundings; all that is around you.

Enzymes: Proteins that start a chemical reaction.

Ecosystem: A dynamic complex of plant, animal and micro-organism communities and their non-living environment interacting as a functional unit.

ED_{50}: *In vitro* or *in vivo* dose of drug that produces 50% of its maximum response or effect.

Efficacy: Describes the way that agonists vary in the response they produce when they occupy the same number of receptors. High efficacy agonists produce their maximal response while occupying a relatively low proportion of the total receptor population. Lower efficacy agonists do not activate receptors to the same degree and may not be able to produce the maximal response.

Endemic: Distribution restricted to a particular area: used to describe a species or organism that is confined to a particular geographical region, for example, an island or river basin.

Essential Oil: Volatile perfumery material derived from a single source of vegetable or animal origin by a process, such as hydrodistillation, steam distillation, dry distillation or expression.

Ethanol: Form of natural gas that can be produced from corn.

Experimental studies: Studies in which the investigator controls the therapy that is received by each participant, generally using that control to randomly allocate patients among study groups.

Extract: A concentrate of dried, less volatile aromatic plant part obtained by solvent extraction with a polar solvent.

Extraction: The process of isolating essential oil with the help of a volatile solvent.

Fatty acid: A carboxylic acid consisting of a hydrocarbon chain and a terminal carboxyl group, especially any of those occurring as esters in fats and oils.

Flavonoid: Class of plant and fungus secondary metabolites.

Flavour: Refers to that characteristic quality of a material as affects the taste or perception.

Gastronomist: A connoisseur of good food; a gourmet.

Glutamic acid: An α-amino acid that is used by almost all living beings in the biosynthesis of proteins. It is non-essential in humans, meaning the body can synthesize it.

Half-life: Half-life ($t\frac{1}{2}$) is an important pharmacokinetic measurement. The metabolic half-life of a drug *in vivo* is the time taken for its concentration in plasma to decline to half its original level. Half-life refers to the duration of action of a drug and depends upon how quickly the drug is eliminated from the plasma. The clearance and distribution of a drug from the plasma are therefore important parameters for the determination of its half-life.

Heart palpitations: Abnormally rapid and irregular beating of the heart

Heterothallism: The term is applied particularly to distinguish heterothallic fungi, which requires two compatible partners to produce sexual spores from homothallic ones.

Humus: The organic component of soil, formed by the decomposition of leaves and other plant material by soil microorganisms.

Hydro-Distillation: Distillation of a substance carried out in direct contact with boiling water.

IC_{50}: In a functional assay, the molar concentration of an agonist or antagonist which produces 50% of its maximum possible inhibition. In a radioligand binding assay, the molar concentration of competing ligand which reduces the specific binding of a radioligand by 50%.

In vitro: Taking place in a test-tube, culture dish or elsewhere outside a living organism.

In vivo: Taking place in a living organism.

Incidence rate: Measure of the frequency of the disease or outcome. The number of new cases which develop over a defined time period in a defined population at risk, divided by the number of people in that population at risk.

Infusion: A process of treating a substance with water or organic solvent, with or without heating.

Insecticide: A type of chemical used to kill insects.

Liana: Any of various long-stemmed, woody vines that are rooted in the soil at ground level and use trees, as well as other means of vertical support, to climb up to the canopy to get access to well-lit areas of the forest.

Microorganisms: Tiny living things that can only be seen with a microscope.

Migraine: A periodic condition with localised headaches, frequently associated with vomiting and sensory disturbances

Monitoring: The performance and analysis of routine measurements aimed at detecting changes in the environment or health status of populations.

Morel: Name given to genus Morchella, an edible fungi closely related to the anatomically simpler cup fungi. These distinctive fungi have a honeycomb appearance due to the network of ridges with pits composing their cap.

Mushroom: A mushroom, or toadstool, is the fleshy, spore-bearing fruiting body of a fungus, typically produced above ground on soil or on its food source.

Mycelia: The vegetative part of a fungus, consisting of a network of fine white filaments (Hyphae).

Mycologist: The person deals with the study of fungi, including their genetic and biochemical properties, their taxonomy and their use to humans as a source for medicine, food, as well as their dangers, such as toxicity or infections.

Nutraceutical: A food stuff that is held to provide health or medicinal benefits in addition to its basic nutritional value also called functional food.

Odour: Property of a substance which stimulates and is perceived by the olfactory sense.

Organic acid: An organic compound with acidic properties.

Organic farming: Producing foods without the use of laboratory made fertilizers, growth subtances, or pesticides.

Organic matter: Dead plants, animals and manure converted by earthworms and bacteria into humus.

Osteoporotic: A disease where increased bone weakness increases the risk of a broken bone. It is the most common reason for a broken bone among the elderly.

Perfume: A suitably blended composition of various materials of synthetic and/or natural origin to give a desired odour effect. It is carried in a suitable medium to the extent of not more than 20 percent.

pH: A scale of measurement by which the acidity or alkalinity of soil or water is rated. A pH of 6 to 7.5 is considered "ideal" for most agricultural crops. Each plant, however, has its own "ideal" pH range.

Pharmacology: The study of the effects of drugs. The branch of biology concerned with the study of drug action, where a drug can be broadly defined as any man-made, natural, or endogenous (from within the body) molecule which exerts a biochemical or physiological effect on the cell, tissue, organ, or organism (sometimes the word pharmacon is used as a term to encompass these endogenous and exogenous bioactive species).

Phytochemistry: The study of phytochemicals, which are chemicals derived from plants.

Plant: Multicellular predominantly photosynthetic eukaryotes of the kingdom Plantae. Historically, plants were treated as one of two kingdoms including all living things that were not animals, and all algae and fungi were treated as plants.

Polymorphic: Occurrence of two or more clearly different morphs or forms, also referred to as alternative phenotypes, in the population of a species.

Polysaccharides: A carbohydrate (e.g. starch, cellulose, or glycogen) whose molecules consist of a number of sugar molecules bonded together.

Population: The number of living things that live together in the same place. In biology, a population is all the organisms of the same group or species, which live in a particular geographical area, and have the capability of interbreeding.

Potency: A measure of the concentrations of a drug at which it is effective.

Saprophyte: A plant, fungus, or microorganism that lives on dead or decaying organic matter.

Sclerotia: Compact mass of hardened fungal mycelium containing food reserves.

Screening: The presumptive identification of unrecognized disease or defect by the application of tests, examinations or other procedures which can be applied rapidly. Screening is an initial examination only and positive responders require a second diagnostic examination.

Side effect: Any unintended effect of a pharmaceutical product occurring at doses normally used in humans which is related to the pharmacological properties of the drug.

Signal: Reported information on a possible causal relationship between an adverse event and a drug, the relationship being unknown or incompletely documented previously. Usually more than a single report is required to generate a signal, depending upon the seriousness of the event and the quality of the information.

Solitary: Existing alone.

Species: A group of living organisms consisting of similar individuals capable of exchanging genes or interbreeding. The species is the principal natural taxonomic unit, ranking below a genus and denoted by a Latin binomial, e.g. *Homo sapiens*.

Specificity: The ability of a method, system or tool to correctly classify the proportion of persons who truly do not have a characteristic, as not having it.

Spectroscopy: The study of the interaction between matter and electromagnetic radiation.

Steam Distillation: Distillation of a substance by bubbling steam through it.

Stimulant: Making a body organ active.

Succulent: A plant (especially a xerophyte) having thick fleshy leaves or stems adapted to storing water.

Surveillance: Ongoing scrutiny, generally using methods distinguished by their practicability, uniformity, and rapidity, rather than by complete accuracy. Its main purpose is to detect changes in trends or distribution in order to initiate investigative or control measures.

Tannin: A yellowish or brown bitter tasting organic substance present in some galls , barks and other plant tissues.

Taxon (plural taxa): In biology, a taxon is a group of one or more populations of an organism or organisms seen by taxonomists to form a unit. Although neither is required, a taxon is usually known by a particular name and given a particular ranking, especially if and when it is accepted or becomes established.

Technology: Instruments, tools or inventions developed through research to increase efficiency.

Tincture: A cold alcoholic extract of natural fragrant material of vegetable or animal origin, the solvent being left in the extract as a diluent.

Topographic: Relating to the arrangement of the physical features of an area.

Traditional knowledge: The term traditional knowledge generally refers to knowledge systems embedded in the cultural traditions of regional, indigenous, or local community.

Urbanization: The growth of the city into rural areas.

Validation: Establishing documented evidence which provides a high degree of assurance that a specific process will consistently produce a product meeting its pre-determinant specifications and quality attributes.

Volatile: A material is said to be volatile when it has the property of evaporating at room temperature when exposed to atmosphere.

Yield: the amount of a crop produced in a given time or from a given place.

1

Folklore Plants Used in Tibetan Mountain Based Sowa-Rigpa System of Food and Medicine: A Close Look on Plant-People Perception to Herbal Cure

Bikarma Singh, Opender Surmal, Bishander Singh
Sumit Singh, Mudasir Nazir Bhat, Mohammed Asif Chowdhary
Sneha, Kota Srinivas and Mohd. Shahnawaz

Abstract

Since pre-historic times plants have tendered their services to mankind in the form of food, shelter and medicine. Due to the advancement of human civilization, various folklore systems of medicines were developed across the globe. Among the various folklore system of medicine, Sowa-Rigpa" is one of the oldest well recognised and widely accepted system of folk-lore medicine in the world. It is one of the commonly used of system of medicine in different parts of Tibet, Bhutan, China, former Soviet Union (some parts), Himalayan ranges, Magnolia

***Bikarma Singh*(✉)**
Botanical Garden, CSIR-National Botanical Research Institute, Lucknow-226001
Uttar Pradesh, India

Opender Surmal, Sumit Singh, Mudasir Nazir Bhat, Kota Srinivas and Mohammed Asif Chowdhary
Academy of Scientific and Innovative Research (AcSIR), Ghaziabad-201002, India; Plant Sciences (Biodiversity and Applied Botany Division), CSIR-Indian Institute of Integrative Medicine, Canal Road, Jammu-180001, Jammu and Kashmir, India

Bishander Singh
Department of Botany, Veer Kunwar Singh University, Ara-802301, Bihar, India

Sneha
CSIR-Central Institute of Medicinal and Aromatic Plants, PO-CIMAP, Near Kukrail Picnic Spot Lucknow-226015, UP, India

***Mohd. Shahnawaz* (✉)**
Plant Biotechnology Division, CSIR-Indian Institute of Integrative Medicine
Canal Road, Jammu-180001, Jammu and Kashmir, India
**All authors contributed equally and are first authors*

✉Corresponding author(s) email: drbikarma.singh@nbri.res.in; mskhakii@gmail.com

Plants for Novel Drug Molecules: Ethnobotany to Ethnopharmacology
Bikarma Singh & Yash Pal Sharma (eds.), (pp. 1-43)

Email: *info@nipabooks.com* Web: *www.nipabooks.com*

and, Nepal. The exact origin of Sowa-Rigpa system of medicine is still not known. In the present scenario due to the awareness among the people about the determental effects of the modern chemical drugs, people preferred to use herbal based system of medicine to treat diseases. So, it is needed to understand the floristic wealth of tradional system of medicine to identify the important plants with potential source of drug candidate molecules. In the present chapter an attempt has been made to study the folklore plants used in Tibetan Mountain based Sowa-Rigpa system of medicine and food. Total 332 plant species presented as the most important in Sowa-Rigpa system of medium.

Keywords: Tibetan plateau, Folk-lore Plants, Sowa-Rigpa, Medicinal Plants

1. Introduction

Since ancient times, people lived in different timezones across the globe had discovered plants for food (Powell, 1982) medicine (Paulsen 2010) and other benfits. Planetological studies revealed the usage of plant as a medince by humans traced back to around 60 thousands years (Shi 2010, Fabricant and Farnsworth 2001). Ötzi, the 5300 years old iceman discovered in 1991 from Itlay, is considered as one of the highly studied corpse (Zink et al. 2019), also suggests the usage of plants to treat diseases at that point of time in the human history (Learn 2018). Based on the ethinicity of the people, different folklore systems of food and medicines were developed, practiced and propogated from generation to generation in different parts of the world without proper documentation (Marett 1920). Yang (2000) reported around 2000 ethinic groups of folk-lore medicine in the world; each ethinic group has unique experience and traditional knowledge. The first written known manuscript (around 5000 years old) on the medicinal plants was scripted by sumarinas scholars and was reported from southern Mesopotamia (southern Iraq)(Petrovska, 2012). Among the various folklore system of medicine, "Sowa-Rigpa" is one of the oldest (2500 years old), well recognised and widely accepted systems of folk-lore medicine in the world (Gurmet 2004). In the present scenario, the inhabitants of the Tibet, Bhutan, China (some regions), former Soviet Union (some parts), India (Himalayan ranges), Magnolia and Nepal are reported to be using this sytem of medicine (Sen and Chakraborty 2017). The exact origin of Sowa-Rigpa system of medicine is still not clear, some researchers believe, Tibet as its place of origin, where as others belives its place of origin is in India or somewhere in China (Gurmet 2004, Kloos 2013, Pordié 2015, Kloos 2016). However, Sowa-Rigpa system of medicine shares significant similarity in both theory and practice with Ayurvedic system of medicine (Gerke 2011, Kloos 2016) due to the influence of Buddhism in Tibet from India during 3rd to 7th centuries. During that period of time, Bhuddist Scholars of India were getting invitation from Tibet to propogate Bhuddism, which leads to develop strong poltical cum religious relationship between the two countries (Gurmet 2004). As a result, numer of books on Indian culture, medicine, customs, rituals, religion, arts, and philosophy, etc., were written and translated into Tibetian language. So, the Tibetian literature of that era, reported to have around 25 books on medicine (Clifford 1994, Gurmet 2004). Thereafter the knowledge of folk-lore medicine of the Tibet was further influenced by other systems of medicines of surrounding nations, reults in the formation of various system of medicine in the Asian subcontinent e.g. "Sowa-Rigpa" (Science of healing) (Deane 2014, Salick et al. 2006). Due to advancement of science and technology, field of medicine also progressed and the modern system of medicine got world wide recognition. People armed with modern medicine technology challenged the herbal based traditional sytem of medicine due to lack of scientific evidence of the treatment even after their reports of effective treatments (Pan et al. 2014). However, things changed with time, due to various disadvantages of the modern chemical based drugs [side effect of the chemical based drugs (Huscher 2009; Minami 2010, Aronson 2015); unavailibility of the modern therpaies for various chronic diseases (Wagner 1998, Wagner 1999, Hankey 2005, Lee 2009), huge investment in the pharma industrial reseach and development (Moerman and Van-Der-Laan 2006, Schramm and Hu 2013, Artuso 2013), etc people have again started to revive the usage of herbal based system of medicine. In the present scenario, majority of people around the globe prefered the usage of herbal based system for the health purpose (Callahan 2001). Further, the idea to use plants as a potential source to rediscover the potential drug candidate molecules, also renewed the strategies to treat diseases to favour the notion of tradional system of medicine significantly (Seidl 2002, Li and Zhang 2007, Schmidt et al. 2007, Corson and Crews 2007).

It is very important to revive the floristic wealth and documentation of traditional system of medicine to overview the most important plant to be used to discover the drug candidate molecules using the integrative approaches. In the present chapter an attempt has been made to pool all the available literature on the folklore plants used in Tibetan Mountain based Sowa-Rigpa system under the following objectives: (1) to overview the topography and climatic conditions of Tibetan Mountain ranges; (2) to highlight floristic studies of medicinal plants of Tibetan Mountain ranges; (3) to enlist the plants used as vegetables, local chutney, spices, fruits, herbal formulations, beverages, narcotics and bakeries by the native people of the locality; and (4) to discuss the methods used to treat various diseases using herbal formulations.

2. Overview of Topography and Climatic Conditions of Tibetan Mountain Ranges

The Tibetan is largest plateau (1000 Km × 2500 Km) in the world (Powell and Conaghan 1975), with average elevation of 4500 m. It is located in central Asia and is perennially covered with snow and glaciers developed widely after decades. The Tibet Mountain can be divided into two types (i) east-west, and (ii) south-north mountain ranges. The Himalayan Mountains, Gangdise - Nyenchen Tanglha Mountains, Kara Kunlun- Tanggula Mountains and Kunlun Mountains are found in the east-west and Hengduan Mountains found in the south-north of the Tibet Mountain.

2.1. Topography

Tibet is the largest plateau on earth and continuously elevated area nearly 3500 by 1500 Km. It originated due to the collision of Indian and Asian continental plates (Powell and Conaghan 1975, Molnar and Tapponier 1975). The Tibetan plateau comprises 82% of the world's total surface area greater than 4 Km altitude and strongly affect climate by affecting atmospheric circulation over the entire northern hemisphere (Ruddiman et al. 1989). The climatic conditions of the Tibetian plateau ranging from arid to semi-arid zones over most of the area.

2.2. Climate

The mean altitude above 4000m from the sea level with area around 2.3 million Km^2 greatly influences the ecosystem of the Asian continent. The adequate information regarding the changing climates of the Tibetan plateau is still not available (Lin and Zhao 1996, Tang et al. 1998, Zheng and Zhu 2000, Zheng et al. 2000). However, the literature reveal strong evidence that the Tibetan plateau employs intense thermal and dyanamic impact on the climate (Manabe and Terpstra 1974). The area experiences uniform climate: freezing and dry windy air in winter and cool in summer. The climate shows considerable variations between day and night due to the effect of strong solar radiations that are characteristic of high altitudes.

2.2.1. Temperature

In summer Tibet experiences high temperatures, whereas in winters intense cold climate prevails. Low temperature is experienced through western part of the region. The annual average temperature ranges from 7.5°C to 7.6°C.

2.2.2. Rainfall

During the monsoon season (June to September), heavy rainfall took place during two month (July-August) per year with average rain fall of 120 to 125 mm per month.

2.3. People

The native people of this region are known as Tibetian people. They have also migrated to other parts of the world. They use their own language i.e. Tibetan languages. They mainly follow bhuddism and their ethinic folklore traditions.

2.4. Vegetation Composition

In the Tibetan mountain, during summers plants grows vegetativly and in the month of August most of the flora blooms. The vegetation found in these area are mainly of three types: (i) Temperate forests and species composition, (ii) Alpine forests and species composition and (iii) Alpine pastures and snow clad species composition.

2.4.1. Temperate forests and species composition

Temperate type of vegetation is found in the south-eastern margin of the plateau with 3000 m elevation. The climate is subtropical and monsoonal, annual temperature around 16.0 °C (maximum temperature: 21.5 °C; minimum temperature: 7.7 °C). The annual precipitation is reported around 850 to 1500 mm with highest rainfall during May and October. The vegetation comprises of *Pinus yunanensis, Tsuga chinensis, T. dumosa* and *Quercus* spp. Upto 2000 to 3000 m elevation, evergreen needle-leaved and deciduous broad-leaved mixed forests, dominated by *Pinus yunanensis, Tsuga chinensis, T. dumosa, Quercus* sp., were reported. Location with elevations higher than 3000-4000 m, evergreen conifer forests were reported to be dominated by *Abies fabri* with a *Rhododendron* shrub understorey.

2.4.2. Alpine forests and species composition

Alpine forests range between the altitudes of around 3400 to 4500 m above the sea level. These types of vegetation occur in the eastern part of the Plateau with very cold climate (average temperature of 0.9 °C). Alpine forests have recorded annual rainfall of around 560 and 860 mm with relative humidity in between 64% to 73%. In this type of vegetation Alpine meadows occur in flat-lying areas and alpine steppe on the mountain slopes (Chai et al. 1965). In this forest dominated families are Cyperaceae, Gramineae, Rosaceae, Asteraceae, Caryophyllaceae, Saxifragaceae and Ericaceae. The vegetation consists of ever green needle-leaved trees like *Picea purpurrea* and *Abies likiangensis* are growing on the wetter slopes of the mountains (Chai et al. 1965).

2.4.3. Alpine pastures and snow clad species composition

This type of vegetation occurs in the the southern Plateau, elevations around 4000 m above the sea level. The climate is very cold, average temperature around 1.5 °C, with annual rain fall between 400 to 550 mm (Zhou et al. 1976, Wang 1987). Below the elevation of 4400 m, the meadow vegetation was reported. The dominant families are Gramineae, Compositae, Cyperaceae, Ranunculaceae, Fabaceae, Liliaceae, Primulaceae, Caryophyllaceae, Labiatae, Saxifragaceae and Gentianaceae. Herbaceaous species like *Artimesia, A. wellbyi* and *A. stracheyi*. A shrub land occurs at an altitude of 4400 to 4700 m, dominated species are *Sabina pingii, Caragana versicolor* and *Potentilla* sp. At higher elevations above 5200 m, a sparse vegetation cover is found.

3. Floristic Studies of Medicinal Plants of Tibetan Mountain Range (Sowa-Rigpa System of Medicines)

The Sowa-Rigpa system of medicine is practiced since 'ancient' times in Ladakh. Most of the important medicinal plants are critically endangered. Documentation of the data on numerous facets of medicine, treatment methods, material used and socio cultural aspects is needed to ensure the long time survival.

3.1. Floristic Analysis of Medicinal Plants

A total of 332 plant species belonging to 205 genera and 65 families were documented as ethnomedicinal plants in Sowa-Rigpa system of medicines (Table 1; Fig. 1). Out of this 302 were herb species, shrub species were 14, tree species were 9 and 5 were climber species. The top ten dominant angiospermic plant families from the study area with number of species were Asteraceae (72) followed by Lamiaceae (20), Ranunculaceae (17), Apiaceae (15), Fabaceae (14), Gentianaceae (13), Polygonaceae (10), Boraginaceae (9), Papaveraceae (8) and Berberidaceae (7). Different plant was used in traditional system of medicines. Root (16%), stem (2%), leaves (16%), flowers (10%), Fruit (4%), tuber (1%) and rhizome (2%) of different plant species were applied as medicine in number of traditional ways to cure various diseases and ailments.

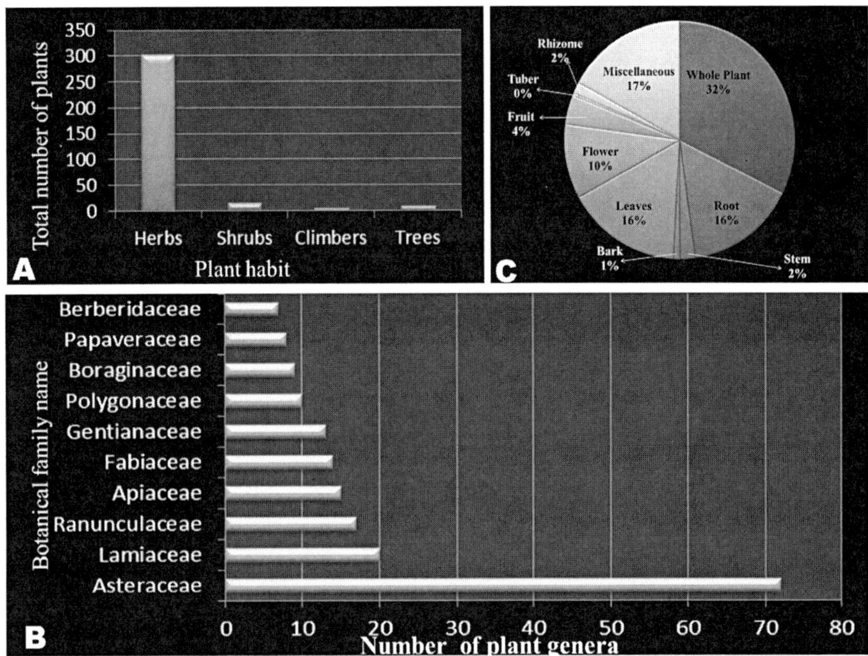

Fig. 1: Floristic details of the folklore plants growing along the mountains of Tibet: (A) Graphical representation of habits of the plants; (B) Ten most dominant families of the plants; (C) Percentage contribution of the different plant parts for medicinal usage.

3.2 Plants Used as a Food in Sowa-Rigpa System

The wild edible plants have proved an important source of natural food for humans since ages. Around 23 plant species belonging to 12 differenent botanical families were mainly used as food in Sowa-Rigpa system of medicine (Table 2). *Prunus armeniaca* (Khubani) is the main cultivated plant in Ladakh and many wild edible fruits along with khubani are dehydrated for prolonged winter use. Some common plant species used by the locals as various preparations in their hose holds are as *Artemisia gmelinii, Avenae fatua, Lathyrus sativus, Rumex hastatus, Tulipa clusiana* and several other plant species.

3.3 Plants Used as a Vegetable in Sowa-Rigpa System

Total 40 plant sepcies belongs to 20 families were reported to be used as a vegetable in Sowa-Rigpa system (Table 3) of medicine.

3.4 Plants Used in Sauce in Sowa-Rigpa System

Total 11 plant species as belongs to 5 families were reported to be used in the sauce preparation (Table 4).

Table 1: List of plants used as a medicibe in Sowa-Rigpa system of medicine

Sr. No.	Botanical Name	Sowa-Rigpa name	Family	Part Used	Sowa-Rigpa uses
1.	*Acantholimon lycopodioides* (Girard) Boiss.	Longze	Plumbaginaceae	Lv, F	Used in treatment of cardiac disorders.
2.	*Achillea millefolium* L.	NA	Asteraceae	WP	Shade dried plants boiled with common salt and herbal tea prepared to cure cold and cough.
3.	*Aconitum heterophyllum* Wall. ex Royle	Bong-dkar	Ranunculaceae	R	Helpful in arthritis, gout, swelling, pain, inflammation, body pain, lymph fluid diseases, intestinal worms and cardiac diseases.
4.	*Aconitum rotundifolium* Kar. & Kir.	Poma Karpo	Ranunculaceae	WP	Used as antidote for insect bites.
5.	*Aconitum violaceum* Jacquem. ex Stapf	NA	Ranunculaceae	R	Used as paste to cure high fever.
6.	*Actinocarya tibetica* Benth.	NA	Boraginaceae	Rt	Decoction prepared used to cure fever.
7.	*Ailanthus excelsa* Roxb.	Mahanumb	Simaroubaceae	Br	Used to cure asthma.
8.	*Ajania fruticulosa* (Ledeb.) Poljakov	NA	Asteraceae	F	Used against gastric disorders.
9.	*Ajania tibetica* (Hook.f. & Thomson) Tzvelev	Phematso	Asteraceae	Sh	Applied as antiseptic and used to cure fresh wounds and boils.
10.	*Alcea rosea* L.	MDog-ldan pho-lcam	Malvaceae	Fs	Taken to cure cough, asthma and throat infection.
11.	*Allardia tomentosa* Decne.	Lukmik sepo	Asteraceae	WP	Used to get relief from headache, fever and joint pain.
12.	*Allium carolinianum* DC.	Skotse	Amaryllidaceae	Br, Lv	Used to cure constipation.
13.	*Allium cepa* L.	Tsong	Amaryllidaceae	R, Lv	Used to get relief from vomiting
14.	*Allium przewalskianum* Regel	Skotche	Amarylidaceae	Lv	Decoction used to cure stomach pain.
15.	*Althaea officinalis* L.	Suzmool	Malvaceae	R	Used to cure asthma, cough and throat infection.

Contd.

16.	*Amaranthus spinosus* L.	NA	Amaranthaceae	Lv	Used to promote kidney function.
17.	*Anaphalis busua* (Buch.-Ham.) DC.	Shishing	Asteraceae	Lv	Taken to get relief from cold and cough.
18.	*Anaphalis nepalensis (Spreng.)* Hand.-Mazz.	Simula	Asteraceae	WP	Plant extract used to cure skin diseases.
19.	*Anaphalis triplinervis* (Sims) Sims ex C.B.Clarke	Ta-wa	Asteraceae	Lv, St, Fs, Fr	Taken as antidote to get relief from insect poisons, to stop bleeding and cure swelling.
20.	*Androsace aizoon* Duby	Zatikmukpo	Primulaceae	WP	Used to cure cough and indigestion.
21.	*Androsace mucronifolia* Watt	Zigsolo	Primulaceae	WP	Taken to cure abdominal pains.
22.	*Anemone falconeri* Thomson	Pharenge	Ranunculaceae	WP	Used against cold, fever and gastritis.
23.	*Eriocapitella rivularis* (Buch.-Ham. ex DC.) .Christenh. & Byng	Zukpa	Ranunculaceae	WP	Taken to cure cough, fever and cold.
24.	*Angelica glauca* Edgew.	NA	Apiaceae	R	Taken to cure cough and cold.
25.	*Anthriscus nemorosa* (M.Bieb.) Spreng.	Sunak	Apiaceae	WP	Used against inflammation and rheumatism.
26.	*Aquilegia moorcroftiana* Wall. ex Royle	Maneri	Ranunculaceae	F	Used to treat asthma.
27.	*Arabis recta* Vill.	Solo	Brassicaceae	R	Used as cooling agent against allergic and gastric disorders.
28.	*Arctium lappa* L.	Byi-bzyung	Asteraceae	R	Taken to cure tumors in uterus and also nerve disorders.
29.	*Eremogone festucoides (*Benth.*)* Pusalkar & D.K.Singh	Umarchua	Caryophyllaceae	R	Taken as tonic to increases lactation process in women and cows.
30.	*Arnebia euchroma* (Royle) I.M.Johnst.	NA	Boraginaceae	R, Lv	Used to control cough and improve hair growth.
31.	*Arnebia guttata* Bunge	NA	Boraginaceae	R	Used to cure cold and cough.
32.	*Artemisia absinthium* L.	Micah	Asteraceae	WP	Used to cure stomach pain.
33.	*Artemisia biennis* Willd.	Magrass	Asteraceae	S	Used to treat obesity
34.	*Artemisia brevifolia* Wall.	Mkhan-dkar	Asteraceae	Lv	Used against stomach complaints.

Contd.

35.	*Artemisia desertorum* Spreng.	NA	Asteraceae	WP	Used as antiseptic to cure fresh cut.
36.	*Artemisia dracunculus* L.	Tsar-bong	Asteraceae	Lv, Fs	Extract useful in stomach complaints.
37.	*Artemisia gmelinii* Weber ex Stechm.	Nurcha	Asteraceae	F	Used to treat cold and fever.
38.	*Artemisia laciniata* Willd.	Khampa	Asteraceae	R	Fresh juice applied on cuts and wounds.
39.	*Artemisia macrocephala* Jacquem. ex Besser	Khampa	Asteraceae	WP	Used against joint pain and to treats rheumatism.
40.	*Artemisia imponens* Pamp.	Burtse	Asteraceae	WP	Used to cure malaria fever.
41.	*Asperugo procumbens* L.	Til	Boraginaceae	S	Used to treat rheumatism.
42.	*Aster diplostephioides* (DC.) C.B.Clarke	Me-tog Lug-mig rigs-dang-po	Asteraceae	F	Used against cold, cough and fever.
43.	*Aster flaccidus* Bunge	Lukmik	Asteraceae	F, St	Used to treat eye treatment, liver disease and fever.
44.	*Astragalus annularis* Forssk.	Rungentso	Fabaceae	R	Extract used to purify blood.
45.	*Astragalus munroi* Bunge	NA	Fabaceae	R	Used to cure cold, cough and chronic bronchitis.
46.	*Phyllolobium tribulifolium* (Benth. ex Bunge) M.L.Zhang & Podlech	Yanglo	Fabaceae	WP	Used as diuretic agent in kidney disorders.
47.	*Astragalus zanskarensis* Bunge	Chisigma	Fabaceae	R	Used to eliminate intestinal worms.
48.	*Avena sativa* L.	Yupo	Poaceae	S	Used to treat burning sensation during urination.
49.	*Berberis brandisiana* Ahrendt	Skerpa	Berberidaceae	R, Br	Used to cure eye problem.
50.	*Berberis lycium* Royle	Daruhaldi	Berberidaceae	R	Extract taken as tonic to cure cold and cough.
51.	*Berberis pachyacantha* Bien. ex Koehne	NA	Berberidaceae	R	Used against fever.
52.	*Berberis ulicina* Hook.f. & Thomson	Shinnar	Berberidaceae	St	Used in arthritis, cough, fever and ring worm infections.
53.	*Berberis vulgaris* L.	Kirsing	Berberidaceae	R	Used to cure intestinal ulcers and lung infections.
54.	*Berberis zabeliana* (C.K.Schneid.) Jafri	Shinnar	Berberidaceae	R, Lv	Extract taken to control fever and

Contd.

					dysentery.
55.	*Bergenia stracheyii* (Hook.f. & Thomson) Engl.	Gatikpa	Saxifragaceae	Lv, R, Fs	Used to dissolve kindey stones, cuts and wounds.
56.	*Betula utilis* D.Don	Towa	Betulaceae	Br, R	Used against jaundice, burns and leprosy.
57.	*Bidens cernua* L.		Asteraceae	R	Used to cure fever and headache.
58.	*Bidens pilosa* L.	NA	Asteraceae	Lv	Used to cure cold and cough.
59.	*Biebersteinia odora* Stephan ex Fisch.	NA	Gerinaceae	R	Extract applied as antiseptic for infections and also taken as tonic for blood purification.
60.	*Elwendia persica* (Boiss.) Pimenov & Kljuykov	NA	Apiaceae	Fr	Used as remedy for abdominal pain.
61.	*Bupleurum longicaule* Wall. ex DC.	Sah-kukchak	Apiaceae	F	Applied against stomachache, and act as antidote to cure insect bites.
62.	*Bupleurum marginatum* Wall. ex DC.	Zeera Karpo	Apiaceae	R	Applied for solve liver problems.
63.	*Capparis spinosa* L.	Capra	Capparaceae	Lv, R	Used to treat paralysis, gout and act as tonic to cure toothache.
64.	*Capsella bursa-pastoris* (L.) Medik	NA	Brassicaceae	Lv	Acts as antivomit agent.
65.	*Caragana versicolor* Benth.	Trama	Fabaceae	St	Used to cure food poisoning, fever and throat infection.
66.	*Carum carvi* L.	Go-snyod	Apiaceae	Fr, R	Used to cure weak eye sight and also promotes one set of menstruation.
67.	*Centaurea depressa* M.Bieb.	NA	Asteraceae	WP	Used to get reflief from cold, cough and fever.
68.	*Elwendia persica* (Boiss.) Pimenov & Kljuykov	Spang-yan-karpo	Caryophyllaceae	Lv, Sh	Used to treat headache and bodyache.
69.	*Chaerophyllum reflexum* Aitch.	NA	Apiaceae	R	Used to promote urine discharge and removes painful urine.
70.	*Oxybasis glauca* (L.) S.Fuentes, Uotila & Borsch	Sanek	Chenopodiaceae	Lv	Act as purgative to remove stomach disorders.

Contd.

71.	*Chesneya cuneata* (Benth.) Ali	Taskya	Fabaceae	WP	Used to treat ulcers of gastro-intestinal tract.
72.	*Christolea crassifolia* Cambess.	Khahla	Brassicaceae	Rt	Act as purgative to treat stomach disorders.
73.	*Cicer microphyllum* Benth.	Sari	Fabaceae	Lv	Used to treat jaundice and sore throat.
74.	*Cicer songaricum* DC.	Merkhukhani	Fabaceae	Rt	Act as expectorant to cure cough and cold.
75.	*Cichorium intybus* L.	NA	Asteraceae	WP	Act as anti-rheumatic and also helful in relieving high fever.
76.	*Cirsium arvense* (L.) Scop.	Biangtser	Asteraceae	Lv	Extract taken as antivomit agent and also treats headache.
77.	*Clematis orientalis* L.	Emong	Ranunculaceae	Sh	Used to cure indigestion problems.
78.	*Clematis vernayi* (C.E.C. Fisch.) Grey -Wilson	E-mong Nakpo	Ranunculaceae	Lv, Fs, Fr	Used to treat indigestion.
79.	*Codonopsis ovata* Benth.	Klu-bdud-rdo-rje	Campanulaceae	Lv, T, F, Fr	Used to treat arthritis, gout, rheumatism, leprosy and joints pains.
80.	*Colchicum luteum* Baker	Tukpa	Liliaceae	Rh	Used to treat gout.
81.	*Comastoma tenellum* (Rottb.) Toyok.	Pangane Snungpo	Gentianaceae	F	Used to cure cough, fever and other respiratory disorders.
82.	*Coriandrum sativum* L.	NA	Apiaceae	Lv, S	Used to treat insomnia and jaundice.
83.	*Cortia depressa* (D.Don) C.Norman	Chakchoo	Apiaceae	WP	Taken to get relief from stomachache problems.
84.	*Corydalis falconeri* Hook.f. & Thomson	Ralchat	Papavaraceae	Sh	Used to treat cold and fever.
85.	*Corydalis flabellata* Edgew.	Tongil	Papavaraceae	WP	Helpful in controlling fever.
86.	*Corydalis govaniana* Wall.	NA	Papaveraceae	R	Decoction is given to cure fever.
87.	*Corydalis megacalyx* Ludlow & Stearn	Stongzil	Fumariaceae	WP	Used to cure pus and pain of boils.
88.	*Corydalis meifolia* Wall.	Tongrusilva	Papaveraceae	WP	Used to control fever.

Contd.

89.	*Corydalis rutifolia* (Sm.) DC.	Chimlo	Papaveraceae	WP	Applied to cure skin infections.
90.	*Cousinia thomsonii* C.B.Clarke	Biangtser nakpo	Asteraceae	Lv, R, F	Used to get relief from body pain, sprain and arthritis.
91.	*Cremanthodium arnicoides* (DC. ex Royle) R.D.Good	Reskunsemar	Asteraceae	F	Used to treat dysentery and peptic ulcers.
92.	*Cremanthodium arnicoides* (DC. ex Royle) R.D.Good	Nimma-gorgos	Asteraceae	F, Sh.	Used to treat peptic ulcer, dysentery and liver disorders.
93.	*Cremanthodium decaisnei* C.B.Clarke	Rhend	Asteraceae	WP	Used to cure bodyache.
94.	*Cremanthodium ellisii* (Hook.f.) Kitam.	Ming-chan-nagpo	Asteraceae	Lv, F, T, Fr	Used to cure inflammation of muscular tissue and act as antidote for poison caused by insects.
95.	*Crepis nicaeensis* Balb. ex Pers.	Gulthak	Asteraceae	WP	Applied to get relief from fever, backache, abdominal disorder and urinary problems.
96.	*Crocus sativus* L.	NA	Iridaceae	St	Used to treat gynecological disorders.
97.	*Crucihimalaya himalaica* (Edgew.) Al-Shehbaz	Sbiu-lapug	Brassicaceae	WP	Used to cure indigestion and also used as appetizer.
98.	*Crucihimalaya wallichii* (Hook.f & Thomson) Al-Shehbaz	Imatso	Brassicaceae	Lv	Decoction taken as appetizer.
99.	*Cuscuta approximata* Bab.	NA	Convolvulaceae	WP	Extract applied in treating warts and sores.
100.	*Cuscuta europaea* L.	Gulthak	Convolvulaceae	WP	Used against abdominal ailments, chest and lung infection.
101.	*Cynoglossum wallichii* G.Don	Nad-ma-Byar-ma	Boraginaceae	Lv, Fs, T, Fr	Used to get relief from bad smell and vomiting.
102.	*Dactylorhiza hatagirea* (D.Don) Soó	Wang-bolak-pa	Orchidaceae	R	Act as expectorant to treat cough and cold.
103.	*Dactylorhiza incarnata* (L.) Soó	Banglakh	Orchidaceae	R	Act as aphrodisiac to cure infertility.
104.	*Datura stramonium* L.	Dha-du-ra	Solanaceae	F, Fr	Used to treat sinusitis.
105.	*Delphinium cashmerianum* Royle	Cha-rkyang	Ranunculaceae	Lv, T, Fs, Fr	Treats dysentery and diarrheoa.

Contd.

106.	*Descurainia Sophia* (L.) Webb ex Prantl	Khampa	Brassicaceae	St, S	Used to cure chicken pox and bronchitis.
107.	*Dianthus anatolicus* Boiss.	Thangthorn	Caryophyllaceae	Lv	Used to cure stomach complaints, cough and cold.
108.	*Dontostemon glandulosus* (Kar. & Kir.) O.E.Schulz	Umnako	Brassicaceae	WP	Used to cure diarrhoea.
109.	*Dontostemon glandulosus* (Kar. & Kir.) O.E.Schulz	Umakno	Brassicaceae	Lv	Used to treat abdominal pain.
110.	*Dracocephalum heterophyllum* Benth.	NA	Lamiaceae	WP	Used to treat cold and cough.
111.	*Dracocephalum stamineum* Kar. & Kir.	Ghiromanko	Lamiaceae	WP	Applied against cough and headache.
112.	*Dysphania botrys* (L.) Mosyakin & Clemants	Snue	Amaranthaceae	F, Lv	Act as anthelmintic and as laxative to treat stomach complaints.
113.	*Dysphania botrys* (L.) Mosyakin & Clemants	Snue	Chenopodiaceae	F	Act as antihelmintic and laxative to cure stomach ailments.
114.	*Echinops cornigerus* DC.	Ekzima	Asteraceae	R, St, Fs	Used against food poisoning.
115.	*Hippophae rhamnoides* L.	Tsarmang	Elaeagnaceae	Fr	Used to treat pulmonary diseases, blood tumors, blood circulation and high altitude diseases.
116.	*Elsholtzia densa* Benth.	NA	Lamiaceae	WP	Used to treat gynecological disorders.
117.	*Elsholtzia eriostachya* (Benth.) Benth.	NA	Lamiaceae	WP	Helpful against various female disorders.
118.	*Elymus repens* (L.) Gould	NA	Poaceae	Rh	Used to treat the irritable condition of urinary bladder and promotes urination.
119.	*Elymus repens* (L.) Gould	Zamak	Poaceae	Rh	Used to promote urination.
120.	*Ephedra gerardiana* Wall.ex Stapf	NA	Ephedraceae	St	Used to treat fever.
121.	*Epilobium angustifolium* L.	Utpalwampo	Onagraceae	F, St	Used against abdominal pain and also against renal complaints.
122.	*Epilobium latifolium* L.	Utpalwampo	Onagraceae	F	Used to cure acidity.

Contd.

123.	*Epipactis helleborine* (L.) Crantz	NA	Orchidaceae	WP	Used to treat common cold.
124.	*Equisetum arvense* L.	Kihnn	Equisetaceae	WP	Act as diuretic to treat kidney problems.
125.	*Erigeron multiradiatus* (Lindl. ex DC.) Benth.& Hook.f.	Rayhanda	Asteraceae	WP	Taken as tonic for renal pain.
126.	*Erigeron poncinsii* (Franch.)Botsch.	Luchehugba	Asteraceae	WP	Used against worms infection and to treat wounds.
127.	*Psychrogeton poncinsii* (Franch.) Y.Ling & Y.L.Chen	Lukchung	Asteraceae	WP	Applied on wounds and cuts.
128.	*Neobrachyactis roylei* (DC.) Brouillet.	Sathi	Asteraceae	WP	Used to treat rheumatism.
129.	*Eriocaulon cinereum* R.Br.	NA	Eriocaulaceae	Lv	Applied to cure pimples.
130.	*Erodium tibetanum* Edgew. & Hook.f.	Zemma	Geraniaceae	St, F	Used to cure wounds and burns.
131.	*Euphorbia neriifolia* L.	Sehund	Euphorbiaceae	R	Used against asthma.
132.	*Euphorbia thomsoniana* Boiss.	Titri	Euphorbiaceae	WP	Used as purgative to treat stomach disorders.
133.	*Euphrasia himalayica* Wettst.	Skianglo	Orobanchaceae	Lv	Used to treat eye problems.
134.	*Eurycarpus lanuginosus* (Hook.f. & Thomson) Botsch.	Measlo	Brassicaceae	Lv	Used to stop haemorrhage of kidney.
135.	*Fagopyrum esculentum* Moench.	Bro	Polygonaceae	R	Applied to control painful urination.
136.	*Ferula jaeschkeana* Vatke	Thunak	Apiaceae	R, St, Lv	Applied to get relief from chest pain and wounds.
137.	*Filago arvensis* L.	Panksa	Asteraceae	WP	Used to treat gout and rheumatism.
138.	*Gagea gageoides* (Zucc.)Vved.	Gudgi	Liliaceae	Rh	Used to treat whooping cough.
139.	*Gagea serotina* (L.) Ker Gawl.	Kangkar	Liliaceae	B	Act as blood purifier.
140.	*Galium aparine* L.	NA	Rubiaceae	Lv	Act as sedative to treat nervous disorders and regulates urine discharge.
141.	*Galium hypocarpium* (L.) Endl.ex Griseb.	Rangche	Rubiaceae	Lv	Used against throat infection.
142.	*Galium serpylloides* Royle ex Hook.f.	Pemantso	Rubiaceae	Lv	Act as diuretic to treat kidney complaints and painful urination.

Contd.

143.	*Galium spurium* L.	NA	Rubiaceae	WP	Used against fever.
144.	*Gentiana algida* Pall.	Tikta	Gentianaceae	Fs	Used to treat epidemic fever.
145.	*Gentiana carinata* (D.Don) Griseb.	NA	Gentianaceae	WP	Used to heal injuries and controls stomach disorders.
146.	*Gentiana squarrosa* Ledeb.	Ziang	Gentianaceae	R	Act as sedative to control nerve disorders.
147.	*Gentiana tianschanica* Rupr. ex Kusn.	Pangane munpo	Gentianaceae	WP	Act as blood purifier.
148.	*Gentiana tubiflora* (G.Don) Griseb.	Spangyan	Gentianaceae	WP	Used to treat cough and stomach disorder.
149.	*Gentianella moorcroftiana* (Wall. ex Griseb.) Airy Shaw	NA	Gentianaceae	WP	Used to cure cold, cough and fever.
150.	*Gentianopsis detonsa* (Rottb.) Ma	Sheeti	Gentianaceae	F	Helpful to check nausea and cough.
151.	*Gentianopsis paludosa* (Hook.f.) Ma	Chubeething	Gentianaceae	F	Used to treat boils.
152.	*Geranium pratense* L.	NA	Geraniaceae	Lv	Used to treat fever.
153.	*Geranium sibiricum* L.	Eyamlomentok	Gerinaceae	Lv, R	Applied to cure dandruff.
154.	*Geranium wallichianum* D.Don ex Sweet	Le-gha-dur	Geraniaceae	R	Used to treat contagious fever, fever of lungs.
155.	*Gnaphalium stewartii* (Holub) C.B.Clarke ex Hook.f.	Gnatzemetz	Asteraceae	R	Act as antipyretic to treat high fever.
156.	*Halerpestes tricuspis* (Maxim.) Hand.-Mazz.	Chu-ruk-balak	Ranunculaceae	Lv, St, Fr, F	Decoction useful against diarrhoea. Also used in fever by amchis.
157.	*Halogeton glomeratus* (M.Bieb.) Ledeb.	Ischermang	Amaranthaceae	Lv	Used to cure itching.
158.	*Helianthus annuus* L.	NA	Asteraceae	S	Applied to controls swelling of kidney and burning sensation of urinary bladder and urine tract.
159.	*Heracleum pinnatum* C.B.Clarke	Tu-dkar	Apiaceae	R, Fr	Used to treat skin diseases and inflammation caused by vulnerable fever.
160.	*Herminium monorchis* (L.) R.Br.	NA	Orchidaceae	T	Used against fever and also used

Contd.

					in kidney disorders.
161.	*Hippolytia senecionis* (Jacquem. ex Besser) Poljakov ex Tzvelev	Purnak	Asteraceae	WP	Used to cure colic complaints.
162.	*Hippophae salicifolia* D.Don	Sermang	Elaegnaceae	Fr	Applied to cure cold, cough and lung complaints.
163.	*Hydrilla verticillata* (L.f.) Royle	NA	Hydrocharitaceae	WP	Used to treat boils and cuts.
164.	*Hyoscyamus niger* L.	Gyelamtag	Solanaceae	Lv	Act as narcotic and also used in nervous disorders.
165.	*Hyoscyamus pusillus* L.	Sastalula	Solanaceae	Lv	Used as antiseptic to treat infections.
166.	*Hypecoum leptocarpum* Hook.f. & Thomson	Parpata	Papaveraceae	R, Lv	Used to treat stomach disorders and acidity problems.
167.	*Impatiens glandulifera* Royle	Mava	Balsaminaceae	Lv	Used to treat wounds and burns.
168.	*Inula racemosa* Hook.f.	Manu	Asteraceae	R	Used against cold, cough and chest pain.
169.	*Inula rhizocephala* Schrenk	Riamko	Asteraceae	WP	Used in cold, cough and chest pain.
170.	*Iris hookeriana* Foster	NA	Iridaceae	WP	Act as sedative to control nerve disorders and regulates urine discharge.
171.	*Juglans regia* L.	Starga	Juglandaceae	Lv	Decoction is taken to cure to get relief from cough and cold.
172.	*Juncus articulates* L.	Imbuke	Juncaceae	Br	Used against bone dislocation and bone fracture.
173.	*Juniperus recurva* Buch.-Ham. ex D.Don	Sukpa	Cupressaceae	Lv	Decoction intakes to lower fever especially in children.
174.	*Jurinea ceratocarpa* (Decne) Benth. & Hook.f.	Turjit	Asteraceae	Lv	Extract used to cure wounds, joint pains and kidney disorders.
175.	*Koelpinia linearis* Pall.	Nodar	Asteraceae	WP	Used to cure rheumatism.
176.	*Krascheninnikovia ceratoides* (L.) Gueldenst.	Gabsan	Amaranthaceae	Lv	Used to control hyperacidity.
177.	'*Melanoseris lessertiana* (DC.) Decne.'	Tharnue	Asteraceae	Lv, Sh	Applied to cure rheumatism.
178.	*Lactuca tatarica* (L.) C.A.Mey.	Bshakha	Asteraceae	Lv	Used to cure headache, fever and

Contd.

					internal wounds.
179.	*Lagotis kunawurensis* Rupr.	NA	Plantaginaceae	WP	Decoction used to treat cold, cough and fever.
180.	*Lancea tibetica* Hook.f. & Thomson	Spa-yak-rtsa-ba	Phrymaceae	R Fs, Lv, Fr	Used to cure pulmonary diseases and lungs inflammation.
181.	*Lavatera kashmiriana* Mast.	NA	Malvaceae	WP	Used to control painful urination.
182.	*Leontopodium himalayanum* DC.	Tawa	Asteraceae	Lv	Used to cure headache.
183.	*Leontopodium nanum* (Hook.f. & Thomson) Hand.-Mazz.	Tzigma	Asteraceae	R	Used to treat dysentery.
184.	*Leontopodium nivale* (Cass.) Greuter	Tzima	Asteraceae	W	Act as antiseptic to cure wounds.
185.	*Lepidium latifolium* L.	Seoji	Brassicaceae	St, Lv	Used to cure rheumatism.
186.	*Lindelofia macrostyla* (Bunge) Popov	Shapur	Boraginaceae	Lv	Used to control hypertension.
187.	*Lomatogonium carinthiacum* (Wulfen) A.Braun	NA	Gentianaceae	WP	Used against various chest infections.
188.	*Lomatogonium rotatum* (L.) Fr. ex Fernald	NA	Gentianaceae	WP	Used for treatment of cold and cough.
189.	*Lycium ruthenicum* Murray	Umila	Solanaceae	Lv	Used as diuretic agent.
190.	*Lysimachia maritima* (L.) Galasso, Banfi & Soldano	Spengcha	Primulaceae	R	Act as antiseptic agent to cure infections.
191.	*Malaxis muscifera* (Lindl.) Kuntze	NA	Orchidaceae	R	Used to promote kidney function.
192.	*Malva parviflora* L.	Khubazi	Malvaceae	S	Used to treat cold, cough and fever.
193.	*Malva verticillata* L.	Cam-pa ma-ning	Malvaceae	S, Lv	Used as a remedy of cough.
194.	*Paracaryum thomsonii* C.B.Clarke	NA	Boraginaceae	Lv	Act as diuretic to cure kidney infection.
195.	*Meconopsis aculeata* Royle	Achatsarmum	Papaveraceae	Lv, F	Used against cardiac ailments.
196.	*Medicago lupulina* L.	Burahang	Fabaceae	WP	Used to treat jaundice and similar liver ailments.
197.	*Melica persica* Kunth	Awa	Poaceae	WP	Used to treat eye disorders.
198.	*Melilotus officinalis* (L.) Pall.	Ole	Fabaceae	WP	Decoction given to reduce fever.
199.	*Mentha longifolia* (L.) L.	Pho-lo-ling	Lamiaceae	Lv	Used to treat purgation phlegm diseases, swelling and indigestion.
200.	*Microula tibetica* Benth.	Demok	Boraginaceae	Rt	Used to cure lung diseases. *Contd.*

201.	*Sabulina kashmirica* (Edgew.) Dillenb. & Kadereit	Pangyankarpo	Caryophyllaceae	WP	Used to treat headache.
202.	*Pontederia vaginalis* Burm.f.	NA	Pontederiaceae	Lv	Used against cough.
203.	*Morina longifolia* Wall.ex DC.	Kim	Caprifoliaceae	WP	Used to cure indigestion.
204.	*Myrtama elegans* (Royle) Ovcz. & Kinzik	Umbu	Tamaricaceae	Lv	Act as blood purifier.
205.	*Myricaria germanica* (L.)Desv.	Umbo	Tamaricaceae	Lv	Used to control chronic bronchitis.
206.	*Najas marina* L.	NA	Hydrocharitaceae	Lv	Used to treat boils and goiter.
207.	*Nepeta annua* Pall.	Khamyu	Lamiaceae	WP	Used to cure arthritis.
208.	*Nepeta clarkei* Hook.f.	Che-ruk	Lamiaceae	Lv	Useful against septic wounds.
209.	*Nepeta coerulescens* Maxim.	Neimlo	Lamiaceae	WP	Used to cure dysentery and eye ailments.
210.	*Nepeta discolor* Royle ex Benth.	Nyomalo	Lamiaceae	Lv	Decoction given to treat cold, cough and fever.
211.	*Nepeta eriostachya* Benth.	Pebul-durse	Lamiaceae	WP	Used to cure eyesight and eye redness.
212.	*Nepeta floccosa* Benth.	NA	Lamiaceae	Lv	Decoction given against cold, cough and fever.
213.	*Nepeta glutinosa* Benth.	Gimaanko	Lamiaceae	Lv	Decoction given to lower diarrhoea, pneumonia and fever.
214.	*Nepeta leucolaena* Benth.	Bossthe	Lamiaceae	WP	Taken as cerebral tonic.
215.	*Nepeta longibracteata* Benth.	Piangku	Lamiaceae	WP	Used to cure stomach disorders.
216.	*Nymphoides peltata* (S.G.Gmel.) Kuntze		Menyanthaceae	WP	Act as cardiotonic.
217.	*Ocimum basilicum* L.		Lamiaceae	Lv	Act as antihelmintic and also against constipation.
218.	*Onosma hispida* Wall. ex G. Don	Demok	Boraginaceae	R	Applied against abdominal ulcers and bladder stone.
219.	*Orobanche hansii* A. Kern.		Orobanchaceae	WP	Used to stop haemorrhage of kidney.
220.	*Oxyria digyna* (L.) Hill	Chumtswa	Polygonaceae	Lv, Sh	Used to cure indigestion and gastric problems.
221.	*Oxytropis mollis* Benth.	Sheli-Dhambo-koya	Fabaceae	WP	Applied to cure wounds and cuts.

Contd.

222.	*Oxytropis tatarica* Baker	Sarkash	Fabaceae	WP	Act as diuretic agent to cure kidney problems.
223.	*Papaver nudicaule* L.	Tshersngonserpo	Papaveraceae	Lv	Act as analgesic and also cure cold and cough.
224.	*Paraquilegia microphylla* (Royle) J.R. Drumm. & Hutch.	Yumo deujin	Ranunculaceae	St, Lv, F	Used to cure gynaecological diseases and blood disorder.
225.	*Parnassia cabulica* Planch.ex C.B.Clarke	Dnyul-tik	Celastraceae	Lv, St, Fs	Used to treat local trigs-pa diseases.
226.	*Parnassia ovata* Ledeb.	Pangyan-Karpo	Celastraceae	WP	Used to cure cold and cough.
227.	*Pedicularis bicornuta* Klotzsch	Peyasang	Scrophulariaceae	Lv, Fs	Used to treat burns and helpful in rheumatism.
228.	*Pedicularis longiflora* Rudolph	Luguruk serpo	Scrophulariaceae	Lv, St	Act as diuretic agent to cure kidney problems and also cures liver problems.
229.	*Pedicularis oederi* Vahl	Luguruk serpo	Scrophulariaceae	WP	Used to treat headache, backache and food poisoning.
230.	*Pedicularis punctata* Decne.	Lugru-mug-po	Scrophulariaceae	Fs	Used to treat bacterial diseases caused by micro-organism.
231.	*Pedicularis pyramidata* Royle ex Benth.	Lungro-marpo	Orobanchaceae	WP	Used to cure headache.
232.	*Peganum harmala* L.	Sepan	Nitrariaceae	S, Lv	Used as a remedy for cold, cough and fever.
233.	*Salvia abrotanoides* (Kar.) Sytsma	Iskilling	Lamiaceae	WP	Used to control cough and headache.
234.	*Persicaria amplexicaulis* (D.Don) Ronse Decr.	Metcheran	Polygonaceae	R	Used to cure asthma and cold.
235.	*Bistorta vivipara* (L.) Delarbre	Langna	Polygonaceae	F, St	Used to treat abdominal pain and backache.
236.	*Phlomoides rotata* (Benth. ex Hook.f.) Mathiesen	Tapaq	Lamiaceae	WP	Used to treat bone fracture and joint swelling.
237.	*Alkekengi officinarum* Moench.	NA	Solanaceae	Fr	Taken to cure all kinds of kidney and urinary disorders.
238.	*Physochlaina praelata* (Decne.) Miers	langthang	Solanaceae	Lv	Used as vermifuge to treat insect

Contd.

					infections.
239.	*Picrorhiza kurrooa* Royle	NA	Scrophulariaceae	R	Used against cold, cough and fever.
240.	*Pisum sativum* L.	Sad-ma	Fabaceae	Fs, Fr	Used in treating irregular menstruation, nose bleeding and ruptured blood vessels.
241.	*Plantago depressa* Willd.	NA	Plantaginaceae	WP	Used to overcome cold, cough and fever.
242.	*Plantago himalaica* Pilg.	Tharum	Plantaginaceae	St	Used to cure diarrheoa.
243.	*Plantago major* L.	Tharum	Plantaginaceae	St	Used against dysentery and gastric problems.
244.	*Pleurospermum candollei* Benth. ex C.B.Clarke	Supka	Apiaceae	Fr	Used to cure renal pain.
245.	*Polygonatum multiflorum* (L.)All.	NA	Asparagaceae	Rh	Applied to promote urine discharge and removes painful urine.
246.	*Polygonatum verticillatum* (L.)All.	NA	Asparagaceae	Rh	Used to cure kidney disorders.
247.	*Polygonum aviculare* L.	NA	Polygonaceae	Lv	Used to cure inflammation and kidney pain.
248.	*Potentilla anserine* L.	Toma	Rosaceae	Lv	Used for treatment of diarrhea and stomach complaints.
249.	*Potentilla argyrophylla* Wall. ex Lehm.	Skyaldaypo	Rosaceae	R	Used to cure chest and throat pain.
250.	*Potentilla atrosanguinea* G.Lodd. ex D.Don	NA	Rosaceae	WP	Used to treat fever at an early stage.
251.	*Potentilla fulgens* Diels	NA	Rosaceae	Lv	Applied in curing stomachache and sore throat.
252.	*Potentilla multifida* L.	Thak-toma	Rosaceae	St	Used to treat insomonia
253.	*Prangos pabularia* Lindl.	Prangos	Apiaceae	Fr	Act as carminative agent to treat gas problem.
254.	*Primula denticulata* Sm.	Tarla-iching	Primulaceae	Sh	Used to treat cold, headache and gastric problems.
255.	*Primula macrophylla* D. Don	NA	Primulaceae	WP	Extract taken to treat cold and cough.

Contd.

256.	*Primula rosea* Royle	Troma	Primulaceae	R	Used to treat muscular pain.
257.	*Prunella vulgaris* L.	Keeche	Lamiaceae	F	Used cure brain disorders.
258.	*Pyrethrum pyrethroides* (Kar. & Kir.) B. Fedtsch. ex Krasch.	NA	Asteraceae	Fs	Used to cure fever.
259.	*Richteria pyrethroides* Kar. & Kir.	Serpan	Asteraceae	F	Used to cure fever.
260.	*Ranunculus brotherusii* Freyn	Chu-rugpa	Ranunculaceae	Lv, Fr, St	Used to cure indigestion.
261.	*Ranunculus hirtellus* Royle	Chepcha-mendok	Ranunculaceae	WP	Act as antihelmintic to treat various insect infections.
262.	*Ranunculus hyperboreus* Rottb.	Chechah	Ranunculaceae	S	Used to increase fertility and vitality.
263.	*Ranunculus trichophyllus* Chaix ex Vill.	Rengo	Ranunculaceae	WP	Used to treat diarrhoea.
264.	*Raphanus sativus* L.	NA	Brassicaceae	R	Act as diuretic agent and removes kidney stone.
265.	*Rhaponticum repens* (L.) Hidalgo	Ayu	Asteraceae	WP	Act as anti-pyretic agent to remove fever and also used as blood purifier.
266.	*Rheum australe* D. Don	Chun-tswa	Polygonaceae	R	Used to treat internal injuries and bone fracture.
267.	*Rheum spiciforme* Royle	Lacchu	Polygonaceae	R	Applied on swellings and wounds.
268.	*Rheum webbianum* Royle	Lacchu	Polygonaceae	Lv	Useful in control piles and chronic bronchitis.
269.	*Rhodiola heterodonta* (Hook.f. & Thomson) Boriss.	NA	Crassulaceae	R	Used against cold, cough and fever.
270.	*Rhodiola imbricata* Edgew.	NA	Crassulaceae	R	Used against cold, cough and fever.
271.	*Rhodiola sinuata* (Royle ex Edgew.) S.H.Fu	NA	Crassulaceae	F	Act as expectorant to treat cough and cold.
272.	*Rhododendron anthopogon* D.Don	Tali	Ericaceae	F, Lv	Used to treat cold, Cough and skin problems.
273.	*Rhododendron campanulatum* D.Don	Tama	Ericaceae	F	Used against cough, dysentery and fever.
274.	*Rosa webbiana* Wall. ex Royle	NA	Rosaceae	Fs, Fr	Used against fever due to food

Contd.

					poisoning.
275.	*Rubia cordifolia* L.	NA	Rubiaceae	R	Used to treat cold, cough and fever.
276.	*Rumex patientia* L.	Lhung-sho	Polygonaceae	WP	Used as febrifuge to reduce fevers and also heal wounds.
277.	*Sagina saginoides* (L.) H.Karst.	Gangachon	Caryophyllaceae	WP	Used to cure food poisoning and diarrhea.
278.	*Salix alba* L.	Mulching	Salicaceae	Lv	Used to treat fever.
279.	*Salix denticulata* Andersson	Langma	Salicaceae	Lv, St	Used to treat joint pain and menstrual disorders.
280.	*Santalum album* L.	NA	Santalaceae	W	Applied against gynaecological disorders
281.	*Sausssurea gnaphalodes* (Royle ex Royle) Sch.Bip.	Yuliang	Asteraceae	R	Used to treat pains especially arthritis.
282.	*Saussurea bracteata* Decne.	Phansi	Asteraceae	Fs	Used against cold and cough.
283.	*Aucklandia costus* Falc.	NA	Asteraceae	R	Used against cold, cough and fever.
284.	*Saussurea glacialis* Herder	Pangchee	Asteraceae	F	Applied for mental disorders and throat troubles.
285.	*Saussurea jacea* (Klotzsch) C.B.Clarke	Pashuk	Asteraceae	WP	Used to control painful urination.
286.	*Saussurea simpsoniana* (Fielding & Gardner) Lipsch.	Jogiphool	Asteraceae	WP	Used to cure asthma and boils.
287.	*Saussurea taraxacifolia* (Lindl.)Wall.ex DC.	Spangsea	Asteraceae	WP	Used to cure headache.
288.	*Saxifraga flagellaris* Willd.	Sumchutik	Saxifragaceae	Lv	Used to cure wounds and boils.
289.	*Saxifraga oppositifolia* L.	Sasomantso	Saxifragaceae	WP	Taken as tonic.
290.	*Saxifraga stenophylla* Royle	NA	Saxifragaceae	WP	Used to treat gynecological disorders.
291.	*Scorzonera divaricata* Turcz.	Tharnoo	Asteraceae	WP	Used to treat rheumatism.
292.	*Scorzonera virgata* DC.	NA	Asteraceae	WP	Used to cure jaundice.
293.	*Scrophularia koelzii* Pennell	Khirah	Scrophulariaceae	S	Used to treat rheumatic pain.
294.	*Hylotelephium ewersii* (Ledeb.) H.Ohba	Dachungpa	Crassulaceae	WP	Used against external injury and toothache.

Contd.

295.	*Senecio chrysanthemoides* DC.	Rgu-dus	Asteraceae	WP	Used to cure dysentery problems.
296.	*Senecio krascheninnikovii* Schischk.	Umarswah	Asteraceae	WP	Used to treat headache.
297.	*Seseli libanotis* (L.) W.D.J.Koch	Khumbak	Apiaceae	Sh	Act as stimulant.
298.	*Silene amoena* L.	Lug-sug	Caryophyllaceae	R	Used to treat nasal problems and hearing defects.
299.	*Sinopodophyllum hexandrum* (Royle) T.S.Ying	Olmose	Berberidaceae	Fr	Used to cure gynaecological diseases and also against skin problems.
300.	*Sisymbrium orientale* L.	Churukpa	Brassicaceae	Sh	Act as appetizer and carminative to remove gas disorder.
301.	*Sium latijugum* C.B.Clarke	NA	Apiaceae	WP	Used to cure boils, burns and other skin problems.
302.	*Solanum americanum* Mill.	Tsigma	Solanaceae	Fr	Applied to cure cold, cough and fever in early stages.
303.	*Sonchus oleraceus* (L.) L.	NA	Asteraceae	Lv	Used to treat jaundice and other liver disorders.
304.	*Sorghum bicolor* (L.) Moench	Toue	Poaceae	Gs	Used as appetizer and act as cooling agent.
305.	*Spinacea oleracea* (L.) Moench	NA	Chenopodiaceae	Fr	Used to reduce irritable condition of bladder and promotes urination.
306.	*Eriophyton tibeticum* (Vatke) Ryding	NA	Lamiaceae	WP	Decoction taken to treat fever.
307.	*Swertia petiolata* D.Don	Lchags-tig	Gentianaceae	Lv, St, Fs	Useful in fever and headache.
308.	*Swertia thomsonii* C.B.Clarke	Tikya	Gentianaceae	WP	Used to cure stomach disorder.
309.	*Hippolytia dolichophylla* (Kitam.) K.Bremer & Humphries	NA	Asteraceae	WP	Decoction is given against fever.
310.	*Tanacetum emodi* R.Khan	Phematso	Asteraceae	Sh	Act as antiseptic to treat various infections.
311.	*Tanacetum gracile* Hook.f. & Thomson		Asteraceae	WP	A decoction given against fever.
312.	*Taraxacum campylodes* G.E.Haglund	Khur-mong	Asteraceae	R, Fs, Lv, Fr.	Used as diuretic agent to cure kidney disorders.
313.	*Taraxacum sikkimense* Hand.-Mazz.	Han	Asteraceae	F, R	Used to treat fever and regulates urine discharge.

Contd.

314.	*Terminalia belirica* Wall.	NA	Combretaceae	Br	Used to treat urinary disorder.
315.	*Terminalia chebula* Retz.	NA	Combertaceae	Fr	Used to cure gynecological disorders.
316.	*Thalictrum foetidum* L.	Chak-choo	Ranunculaceae	R	Applied against eye ailments.
317.	*Thalictrum minus* L.	Chak-ackoo	Ranunculaceae	Sh, Lv	Used to treat gout and rheumatism.
318.	*Thermopsis inflata* Cambess.	Dugsrad	Fabaceae	WP	Used to cure swelling.
319.	*Thlaspi arvense* L.	Brega	Brassicaceae	WP	Used to cure indigestion and gastritis.
320.	*Thymus linearis* Benth.	Tumburu	Lamiaceae	WP	Applied to cure Stomach-ache and gastrointestinal problems.
321.	*Tragopogon gracilis* D.Don	Tharnoo	Asteraceae	F	Act as laxative to cure stomach problems.
322.	*Tribulus terrestris* L.		Zygophyllaceae	Fr	Used against cold and cough.
323.	*Triglochin palustris* L.	Marcha	Juncaginaceae	WP	Used to cure cuts and wounds.
324.	*Tulipa clusiana* DC.	Kapichung	Liliaceae	B	Taken as tonic to treat rheumatism.
325.	*Urtica hyperborea* Jacq. ex Wedd.	Zahchot	Utricaceae	WP	Used to cure cold and cough.
326.	*Valeriana jatamansi* Jones	NA	Caprifoliaceae		Used to cure abdominal pain and acts as anti-septic towards infection.
327.	*Verbascum thapsus* L.	Yug-pa-gser-pcha	Scrophulriaceae	S	Applied for treatment of asthma and chest pain.
328.	*Vincetoxicum canescens* (Willd.) Decne.	Tung-Me-Yung	Apocynaceae	WP	Used to cure dysentery.
329.	*Viola kunawurensis* Royle	Rholo	Violaceae	F	Used to cure heart diseases.
330.	*Vitis vinifera* L.		Vitaceae	Fr	Act as diuretic and removes kidney stone.
331.	*Waldheimia glabra* (Decne.) Regel	Thumba	Asteraceae	WP	Used to cure headache and fever.
332.	*Waldheimia stoliczkae* (C.B.Clarke) Cstenf.	NA	Asteraceae	WP	Decotion given to cure cold, cough and fever.

Abbreviations: WP-Whole plant, F-Flower, S-Seeds, B-Buds, Fr-Fruits, R-Roots, Br-Bark, Sh-Shoots, St-Stems, W-Wood, Bark-Br, G-Grain, Lv-Leaves, Rh-Rhizome, T-Trunk,

Table 2: List of plants used as food in Sowa-Rigpa system

Sr.No.	Plant name	Tibetan name	Family	Part Used	Method of Use
1.	*Avena fatua* L.	NA	Poaceae	Seeds	Grains grinded and used as flour to make roti
2.	*Artemisia gmelinii* Weber ex Stechm.	Makan pa	Asteraceae	Leaves and flowers	Mixed with floor of wheat to prepare ferments
3.	*Arnebia guttata*Bunge	Demokh	Boraginaceae	Roots	Used in making of stews and soups
4.	*Capsella bursa* Raf.	Shamsho	Brassicaeae	Leaves	Leaves used in various dishes
5.	*Chenopodium album*	Em	Chenopodiaceae	Tender leaves	Leaves used as vegetable and also cooked with other dishes
6.	*Convolvulus arvensis*	Ratcho	Convulvulaceae	Leaves	cooked as vegetable
7.	*Fagopyrum esculentum* Moench	Dyat	Polygonaceae	Seed	Seeds and leaves mixed with peas and potato and eaten as vegetable
8.	*Hordeum aegiceras* Nees ex Royle	Sherok	Poacea	Seed	Seeds used in vegetable
9.	*Hordeum vulgare* L.	Sherok	Poaceae	Seed	Grinded seeds used as food
10.	*Lathyrus humilis* (Ser.)	Kaown	Fabaceae	Seeds and pods	Directly eaten and some times cooked with meat and potata.
11.	*Lathyrus sativus* L.	Kaown	Fabaceae	Seeds and pods	Directly eaten and some times cooked with meat and potata.
12.	*Lactuca sativa* L.		Asteraceae	Young leaves	Eaten raw in salads and dried for wnters for cooking.
13.	*Rhodiola imbricata* Edgew.	Sholo	Crassulaceae	Tender,roots and leaves	Roots are boiled in water and mixed with floor and meat
14.	*Rumex hastatus* D.Don	Gugantse	Polygonaceae	Leaves	Cooked as flavouring agent
15.	*Rumex orientalis* Bernh. ex Schult. & Schult.f.	Shang tso	Polygonaceae	Leaves	Cooked as flavouring agent and with other

Contd.

16.	*Setaria italica* (L.) P.Beauv.		Poaceae	Seed	Seeds used in various cooking pur poses
17.	*Solanum nigrum* L.	Tsigma	Solanaceae	Leaves tubers	Mixed with peas and meat
18.	*Solanum tuberosum* L.	Alu	Solanaceae	Tubers	Mixed with meat
19.	*Sonchus oleraceus* (L.) L.	Khala	Asteraceae	Green leaves	eaten raw in salads and used as vegetable
20.	*Taraxacum campylodes* G.E.Haglund	Han	Asteraceae	Leaves	Leaves mixed with vegetable and then cooked
21.	*Thermopsis inflata* Cambess.	Kanjubopa	Fabaceae	Seeds	Used axs vegetable
22	*Triglochin palustris* L.	Makka	Juncaginaceae	Fruit	Eaten as bread after making flour
23	*Tulipa stellata* Hook.	Kapichong	Liliaceae	Bulbs	Locals eat raw bulbs.
23	*Urtica hyperborea* Jacq. ex.Wall.	Dzatsutt	Urticaceae	Tender leaves	Used in making of stews and soups

Table 3: List of plants used as a vegetablesin Sowa-Rigpa system

Sr.No.	Plant name	Local Name	Family	Part used	Using tecnique
1.	*Allium cepa* L.	Tson	Alliaceae	Bulb and leaves	Used in preparation of curry.
2.	*Arenaria holosteoides* (C.A.Mey.) Edgew.	Chikki	Caryophyllaceae	Shoot	Tender shoot cooked as vegetable.
3.	*Allium loratum* Baker	Skotche	Alliaceae	Whole	Cooked as a vegetable
4.	*Amaranthus spinosus* L.	Chulai	Amaranthaceae	Tender leaves	As avegetable
5.	*Anaphalis triplinervis* Sims ex C.B.Clarke	Yaktsu	Asteraceae	flower bud	Eaten raw as salads
6.	*Arnebia euchorma* (Royle) I.M.Johnst.	Troma	Boraginaceae	Roots	Roots fried in oil and eaten with ladakhi roti.
7.	*Beta vulgaris* L.	Mangol	Chenopodiaceae	Leaves roots	Used as food
8.	*Bidens pilosa* L.		Asteraceae	Leaves	Used in mixed vegetables
9.	*Brassica oleraceae* L.	Ful gobhi	Brassicaceae	Flower	Used as vegetable
10.	*Bunium persicum* (Boiss.) B.Fedtsch.	Kosnyot	Apiaceae	Seed, leaves	Used as a substitute of zeera
11.	*Arnebia guttata* Bunge	Demokh	Boraginaceae	Roots	Boiled in water and cooked as vegetable
12.	*Capparis spinosa* L.	Kabra	Caparidaceae	Tender leaves and unripe fruits	Used as vegetable and oil used for cookin purposes
13.	*Capsella bursa-pastoris* (L.) Medik.	Shamsho	Brassicaeae	Leaves	Consumed as vegetable
14.	*Capsella elliptica* C.A.Mey.	Shamsho	Brassicaeae	Leaves	Cosumed as a vegetable
15.	*Chenopodium album* L.	NA	Amaranthaceae	Leaves	Leaves boiled with water and fried with onion, garlic and consumed.
16.	*Chenopodium foliosum* Asch.	Sangsi	Amaranthaceae	Leaves	Leaves boiled with water and fried with onion, garlic and consumed
17.	*Christolea crassifolia* Cambess.	Solomarpo	Brassicaceae	Leaves	Leaves of the plant are cooked and eaten.
18.	*Cicer songaricumn* DC.	Seri-jriboo	Leguminosae	Unripe legumes	The plant is cooked and eaten with breads

19.	*Coriandrum sativum* L.	Osu	Apiaceae	Leaf and seed	Used as a vegetable and as a spice
20.	*Convolvulus arvensis* L.	Ratcho	Convolvulaceae	Seeds	Fried with onion n garlic.
21.	*Fagophyrum tatarcium* (L.) Gaertn.	Dyat	Polygonaceae	Leaves	Used as vegetable
22.	*Gentiana leucomelaena* Maxim.	Sakir	Gentianaceae	Leaves	Cooked along with other mixed vegetables
23.	*Lactuca dolichophylla* Kitam.	Kitam	Asteraceae	Leaves and shoots	Used as vegetable.
24.	*Lepidium latifolium* L.	Sousar	Brassicaceae	Leaves abd shoot	Consumed by locals in the form of vegetable.
25.	*Leucosphaera bainesii* (Hook.f.) Gilg	Sichi	Amaranthaceae	Leaves	Boiled leaves consumed as vegetable
26.	*Medicago sativa* L.	Hispit	Fabaceae	Leaves	As vegetable
27.	*Mentha arvensis* L.	NA	lamiaceae	Leaves	Leaves cooked and eaten along with butter.
28.	*Oxyria digyna* (L.) Hill	Suma	Polygonaceae	Leaves	Used as vegetable
29.	*Pisum sativum* L.	Sremma	Fabaceae	Fruit	Seeds consumed as green vegetable
30.	*Plantago asiatica* L.	Seri-jriboo	Plantaginaceae	Unripe legumes	Used as vegetable
31.	*Plantago himalaica* Pilg.	Karche	Plantaginaceae	Leaves	Used as vegetable.
32.	*Potentilla atrosanguinea* G.Lodd. ex D.Don	Skaildaepo	Rosaceae	Leaves	As flavouring agent and cooking with sarson.
33.	*Rheum emodi* Wall.	Karche	Polygonaceae	Leaves	As vegetable
34.	*Rhodiola heterodonta* (Hook. f. & Thomson) Boriss	Shrolo	Crassulaceae	Tender leaves and shoots	Leaves and shoots are boiled in water and then consumed along with curd and butter
35.	*Hylotelephium ewersii* (Ledeb.) H.Ohba (Ledeb.) H.Ohba	Gomni	Crassulaceae	Leaves	Used as vegetable.
36.	*Sedum tibeticum*Hook.f. & Thomson	Khumbak	Crassulaceae	Shoots	Used as vegetable.
37.	*Seseli libanotis* (L.) W.D.J.Koch	Lachu	Apiaceae	Tender leaves	Boiled leaves consumed as vegetable.
38.	*Solanum oleraceous* Ochoa	Khala	Solanaceae	Leaves	Used as vegetable.
39.	*Solanum tuberosum* L.	Alu	Solanaceae	Tubers	Tubers boiled and conusumed.
40.	*Urtica hyperborea* Jacq. ex Wedd.	Zwa	Urticaceae	Young shoots	Used to prepare soups

Table 4: List of plants used in sauce preparationin Sowa-Rigpa system

Sr.No.	Plant name	Local name	Family	Parts used
1.	*triplinervis Sims* ex C.B.Clarke	Yaktsu	Asteraceae	Flower buds used as salads and preparing chutneys
2.	*Bunium persicum* (Boiss.) B.Fedtsch.	Kosnyot	Apiaceae	Seed and leaves used as a substitute of zeera.
3.	*Elsholtzia densa* Benth.	Nag po	Lamiaceae	Whole plant used in chutney
4.	*Elsholtzia eriostachya* (Benth.) Benth.	Tsa tsa	Lamiacea	Leaves used for making chutney.
5.	*Gentiana acaulis* L.	Wangloo	Gentianaceae	Whole plants used in chutneys and salads
6.	*Mentha arvensis* L.	Pudina	Lamiaceae	For making chutney with onion and garlic
7.	*Mentha longifolia* (L.) L.	Jangli podina	Lamiaceae	Leaves used in preparing chutneys and as flavouring agents
8.	*Oxyria digyna* (L.) Hill	Suma	Polygonaceae	Leaves used in preparing chutney
9.	*Rumex patientia* L.	Shoma	Polygonaceae	Leaves used in chutneys and flavouring agents
10.	*Rumex hastatus* D.Don	Gugantse	polygonaceae	Leaves used in chutneys.
11.	*Rheum webbianum* Royle	Lachu	Polygonaceae	Leaves used for making chutneys.

3.5 Plants Used as Spices in Sowa-Rigpa System

Ten plant species belonging to 7 different botanical families are known to be used as spices in Sowa-Rigpa System (Table 5).

3.6 Plants Consumed as Direct Fruits in Sowa-Rigpa System

Among the various wild plants, total 24 plant species belongs to 11 different botanical families were reported to be used as a source of direct fruits (Table 6).

4. Medical Formulations in Sowa-Rigpa System of Medicine

The Amchis uses different types of ethno-medicinal preparations for curing different diseases. The plant material forms an important ingredient in different medical formulations. Besides plants, Amchis also uses animal based materials, minerals and salts in the drug formulations. The medicine is prepared from a single constituent or it consists of 30 to 40 different ingredients. The end products are used in either forms: powder (chema), tablet (rilvo), paste (degu), ointment (chukma) or decoction (Thang). In most of the cases the Amchis prescribe medicine mainly in the form of pills manufactured by local drug or pharmaceutical companies. The most of the medicine are purely plant based medicine, like A-gar 8, A-gar 12, A-gar 15, A-gar 31, A-gar 35 and many other medicine. The plant material is used in combination with different minerals and stones to cure different ailments, for example shilajet is the most important mineral locally called Tak-joon means Rock extract.

Different formulations are used by Amchis generally referred by the common names such as Koeu-Dingzor (A formulation of upto seven wild/ exotic plants with other material); Kurkum-Gyepta (A formulation s of nine wild/exotic plants with other materials); Zeu-get, Cheu-chick, Sungmel-chupka or Aru-chukpa (A formulation of 11, 13 or 15 wild/exotic plants with other materials. Olsee-Aerange (a formulation of 21 wild/exotic plants with other materials).

Apart from these formulations, people of the region made some indigenous preparations which are long used in ladakh.

4.1. Beverages

4.1.1. Phabs

The people of Trans Himalayan regim use barely based alcoholic beverages *chhang* and *aarak*. The fresh tablets of *phabs* are incubated with fresh twigs of shrub *Artemisia* sp. The *phabs* is used in hydrotherapy to treat arithrits and joint pain (Angmo and Bhalla 2014)

Table 5: List of plants used as spices in Sowa-Rigpa system

Sr. No.	Plant name	Local name	Family	Parts used
1.	*Allium macranthum* Baker	Byi'u sgog	Alliaceae	Bulbs
2.	*Allium carolianianum* DC	Koshuk	Amaryllidaceae	Dried leaves
3.	*Allium prattii* C.H.Wright	Rug sgog	Alliaceae	Bulbs
4.	*Allium roseum* L.	Dzim bu	Amaryllidaceae	Bulbs
5.	*Arnebia euchroma* (Royle ex Benth.) I.M. Johnston	Bri mog	Boraginaceae	Roots used as spice to cook meat.
6.	*Carum carvi L.*	Go snyod	Apiaceae	Seed used instead of zeer
7.	*Ligusticum pteridophyllum* Franchet	Bam po	Apiaceae	Root and Rhizome
8.	*Thymus linearis* Benth	Su lu	Lamiaceae	Leaves powdered with chilly used as condiment
9.	*Trigonella foenum-graecum* L.	Methi	Fabaceae	Leaves and seeds used as spices in preparing foods
10.	*Zanthoxylum tibetanum* C.C.Huang.	G.yer ma	Rutaceae	Fruit

Table 6: List of plants as a source of direct fruit in Sowa-Rigpa system

Sr. No.	Plant name	Tibetan name	Family	Part Used	Usages/uses
1.	*Berberis lycium* Mart. ex Schult. & Schult.f.	Daruhaldi	Berberidaceae	Fruits	Matured fruits eaten directly by local people
2.	*Berberis ulicina* Hook.f. & Thomson	Khizer	Berberidaceae	Ripen fruits	Rhizome eaten directly by the people
3.	*Capparis spinosa* L.	Jafri	Capparidaceae	Ripe fruits	Shepherds eat ripen fruits
4.	*Cicer microphyllum* Benth.	Srad-kar	Fabaceae	Unripe seeds	Used seeds eaten by local shepherds and children
5.	*Codonopsis ovtaa* Benth.	Lucut	Campanulaceae	Raw roots	Raw most consumed by local people in the fields
6.	*Elaeagnus angustifolius* (Rehder) C.Y.Chang	Chiolik	Elaeagnaceae	Ripe fruits	Children eat ripened fruits
7.	*Ephedra gerardiana* Wall. ex Stapf	Asmania	Ephedraceae	Ripe fruits	Ripe fruit sweet and eaten by shepherds and children
8.	*Ephedra intermedia* Schrenk & C.A.Mey.	Asmania	Ephedraceae	Ripe fruits	Ripe fruit sweet and eaten by shepherds and children
9.	*Hippophae rhamnoides* L.	Tsermang	Elaeagnaceae	Fruit	Strong aidic and eaten directly by shepherds
10.	*Hippophae tibetana* Schltdl.	Tsermang	Elaeagnaceae	Fruits	Fruits eaten and used in preprations of juices and squashs
11.	*Lathyrus humilis* (Ser.) Spreng	kacwn	Fabaceae	Seeds and Pods	directly eaten.
12.	*Lathyrus sativus* L.	kacwn	Fabaceae	Seeds and Pods	directly eaten
13.	*Malus pumila* Mill.	Kusu	Rosaceae	Fruit	eaten raw
14.	*Morus alba* L.	Oshe	Moraceae	Raw fruit	Matured fruits edible.
15.	*Sinopodophyllum hexandrum* (Royle) T.S.Ying	Papra	Berberidaceae	Fruits	Fruits are eaten by children in Zanskar areas
16.	*Prunus armeniaca* L	Khubani	Rosaceae	Fruits	Fruits consumed fresh and dehydrated for prolonged winter.

Contd.

17.	*Pyrus pyrifolia* (Burm.f.) Nakai	Naspati	Rosaceae	Fruit	Fruits edible.
18.	*Ribes alpestre* Wall. ex Decne.	Zasoot	Saxifragaceae	Ripe fruits	Fruits eaten by shepherds.
19.	*Ribes orientale* Desf.	Askuta	Saxifragaceae	Ripe fruits	Fruits eaten by shepherds
20.	*Rosa webbiana*Wall. ex Royle	Siah	Rosaceae	Ripe berrier	Fruits eaten by shepherds.
21.	*Solanum nigrum* L.	Tsigma	Solanaceae	Ripe fruits.	Children eat ripe fruits
22.	*Vitis vinifera* L.	NA	Vitaceae	Fruit	Ripen berries directly eaten
23.	*Hippophae rhamnoides* L.	Tsermang	Elaeagnaceae	Fruit	Fruit used to make juices and squashs.
24.	*Hippophae tibetana* Schltdl.	Tsermang	Elaeagnaceae	Fruit	Fruits eaten and used in making juices and squashs

4.1.2. *Balam*

Balam is catalyzing agent used for the production of fermented food in the tradional ethenic cltures. It is preapared by roasting flour of the sundried wheat pre-washed with water. Over the fire, the brown colour flour is mixed with spices viz. *Cinnamomum zeylanicum, Amomum subulatum,* and *Piper longum* and seeds of *Ficus religiosa* (Roy et al. 2004).

4.1.3. *Jann*

The traditional drinks of Bhotiyas have low contents of alcohol. Commonly prepared using *Oryza sativa, Triticum aestivum* and *Hordeum vulgare* (Roy et al. 2004).

4.1.4. *Sez*

This is the traditional food used by Bhotiyas. It is prepared from rice (*Oryza sativa*) and is mostly used in snacks. While preparation of rice Jan (chaul-ki-jann) sez is extracted (Roy et al. 2004).

4.1.5. *Saja or solja*

Saja or solja is the well known tea of cold desert of ladakh prepared from the freshly harvested leaves of *Bidens pilosa* L. the local inhabitants grow this plant in the kitchen garden (Angmo and Bhalla 2014).

4.1.6. *Chhang Chhang*

A local beer called Chhang Chhang is an important part of everyday life. It is made of chhang tablets and naked barley (*Hordeum vulgare*). Chhang tablets are also sold in markets and prepared at home (Targais et al. 2012).

4.1.7. *Gur gur chai*

The local salt tea is prepared by indigenous instrument known as "Gur chai". The tea is prepared from the tea leaves "Charil" mainly brought from ladakh

4.1.8. *Marcha*

It is the fermented food beverage prepared from rice (*Oryza sativa*) as outlined in Fig.2.

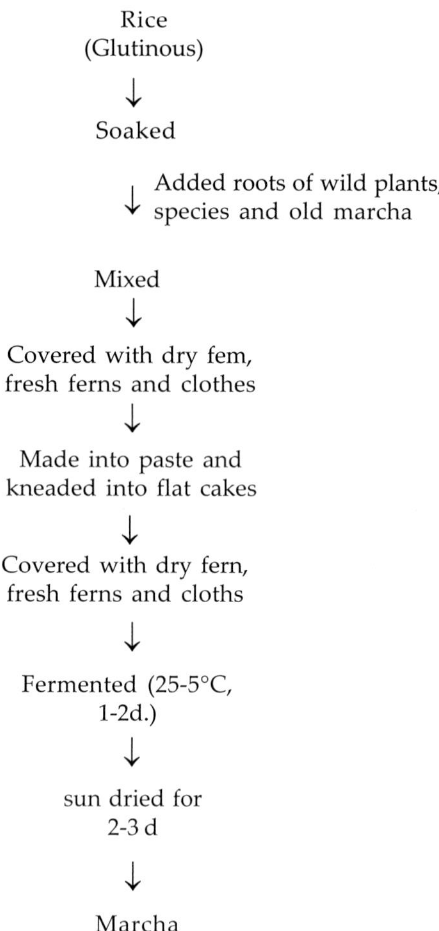

Fig.2: Flow chart of traditional method of marcha preparation (Tamang et al, 1996)

4.2. Narcotics

4.2.1. Berzeatsink

The rootsof *Lancea tibetica* are dried and roasted using fire and powdered after mixing with tobacoo. The powder is smoked using tobacco or consumed using milk.

4.2.2 Staspakchak

The sun dried (7-8 days) leaves of *Limmonium macrorhab* powdered. The powder is dissolved filled in bottle and left as such for one week.

4.2.3 Techepakchiatzen

The seed of *Peganum harmala* are roasted on a red-hot iron plate. Fine powder is obtained with the help of a sieve which is taken as such or smoked with tobacco.

4.2.4 Zimpating

The young twigs and immature fruits of *Tribulus terrestris* are dried and powdered. The powder is roasted and dissolved in milk. The mixture is taken after meals. It is dangerous if taken in excess.

4.2.5 Lingeatzish

Utricularia minor leaves are roasted on flat stone over fire. The roasted leaves are then agitated in water and then incubated for 13 to 14 days in a closed bottle and are usually consumed during winter.

4.3. Bakeries

4.3.1 Skyurchuk: it is the traditional bread used by the locals in Ladakh (Angchok et al, 2009)

4.3.2 Tagi buskhuruk: This is the type of unleavened bread prepared in a plate on a dung of the cow. (Angchok et al, 2009)

4.3.3 Tagi Thalkuruk: This is the unleavened bread prepared on the ashes (Angchok et al, 2009)

4.3.4 Mer-khour: Made by molten butter and well beaten egg white. It is prepared on a iron plate surrounded by hot ashes all side (Angchok et al, 2009)

4.4. Method of Treatment Techniques

4.4.1 Burning of moxa and puncturing of veins

The treatment of headache, paralysis, Inflammation of joints is done by moxibustion or by puncturing of veins by the needle. In moxibustion flower head or whole plant is used. Puncturing of veins is used for blood and skin disorders.

4.4.2 Cold and hot water immersion

The bath promotes energy. It prevents perspiration, laziness, fetid smell and itching of skin. Many hot water springs are found in the valley of ladakh Chumathang, Panamik, Chiling, markha, Serchan and pugga with large amount of sulphar and other minerals. Bathing in sulphar is beneficial for the person suffering with rheumatism.

5. Discussion

In the context of Indian system of medicine the medical pluralism is well known. The essential components of medical pluralism are ayurveda, allopathy, homeopathy and unani. In the context of ladakh the medical pluralism involves allopathy or biomedicine, shamanism (locally known as Lhawaism), Lamaism and amchi medicine (Ballabh and Chaurasia 2007). In trans-Himalayan region different form of medical practices had existed like Ihaba (shaman) and onpo (astrologer). The Amchi system of medicine under the influence of Buddhism overshadowed the other traditional medicinal practices in the region. Amchis (the people practicing amchi system of medicine) are the religious people had received more respect and faith. Ayurveda was introduced by *Bauddha Bhiksus* (monks) and propagated by their disciples in ladakh, lahaul-spiti, Tibet and other places wherever they preached. Over a period of time the budda was established to refer the medical practices being done by Amchis. Countries like Nepal, India, China and Mangolia has contributed in the evolution of Tibetan system of medicine. In ladakh the indigenous knowledge in the desolate area gave rise to the Tibetan system of medicine, the system of medicine was based on the availability of rich medicinal herbs in the valley (Kala Prakash 2005). In sowa rigpa system of medicine the health of the entire person is considered. The chronic diseases like arthritis, asthma, rheumatism, liver problem and diseases related to digestive and respiratory system are believed to be treated by sowa rigpa system of medicine (Wangchuk and Wangchuck 2006). Significant numbers of patients are treated every year by following the practices of sowa rigpa system of medicine (Anonymous 2006). The increasing number of patients visiting the sowa rigpa centres is a good indication. In the region people tend to attribute their well being to the modern medicine or other forms of local healings or traditional system of treatment such as spiritual or religious practices. One of the main reason for popularity of the sowa rigpa system of medicine is that the general perception that "herbal means safe" (Lhamo and Nebel 2011). The health care system of Bhutan integrates sowa rigpa and modern medicine offers an opportunity in health care decision making for patients (Lhamo and Nebel 2011). In india it is popularly practiced in ladakh, Himachal Pradesh, Arunachal Pradesh, Sikkim, Darjeeling and all over Tibetan settlements all over India. The recognition of sowa rigpa system of medicine has been achieved fast, the system of medicine was finally secured in the form of funds from the government, the research center in Leh was upgrade to India only National research institute for Sowa-Rigpa (Clifford 1994) Sowa-Rigpa originated from india is based on Jung-wa-lna (Panch Mahubuta/Five elements) and Nespa gsum (Tri-dosh/Three humours) theories (Gurmet,

2004). In Sowa-Rigpa system of medicine eight types of substances are used *Rinpoche –sman* (prized metal and stone), *Sa sman* (Drugs from mud and earth), Rdo sman (Drugs from stones), Shing smang (Drugs from trees), rtsi sman (essence and exudates medicines), *thang sman* (plant material for decoction/shrub), *sno sman* (herb), *srog chags sman* (Animal material) (Gurmet 2004). Due to poor infrastructure of modern medicine 60% of the the health care in ladakh is looked by Amchis whereas in Lahaul-Spiti it was up to 80%. About 55% of the Amchis are concentrated along the Indus valley followed by changthang plateau (19%) and Nubra valley (12%). It is interesting to note that amchis of the study area were apprehensive of their traditional practices due to fear of market forces which were making available all modern facilities available in remote corners. Now these amchis pass on their knowledge from one generation to next generation. Now the shift in socio-economic patterns and unwillingness of the new generation to make their career in amchi system of medicine is declining (Kala 2005). The Amchi system of medicine highly resembles with Ayurvedic practice. In amchis system of medicine, the local Amchis use minerals, hot water springs, branding with hot metal or other substances, vein puncturing, prayer, animal organs and herbs for the treatment of the disease. The Amchi has a long rectangular box, consists of drug preparation, in powdered form tied around the middle with a strip of leopards skin. This increases the efficacy and potency of the drugs. Several diseases like arthritis, skin disease, wound, sores and other diseases are cared as advised by Amchi by immersing the body of the patient in the water. In Tibetan system of medicine many high altitude plant species are used. Each plant species may be used in single or several Amchi formulations. The medicines made by local Amchis consists fresh plant species and their parts. The medicinal preparations are made up of a single constituent or combination of 30 to 40 constituents (Tsarong 1986, Phuntsog 2006).

6. Conclusion

Sowa Rigpa system of medicine is rich mixture of art, culture, philosophy and religion. It is the holistic understanding of body/soul. Our findings of the present study showed that the Sowa-Rigpa system of medicine has held a house hold name in the culture of ladakh. The local medical practioners i.e. Amchis play an important role in the traditional health care system of tribal population. The local Amchis are highly skilled, they use wild medicinal plants combined with medicinal stones, minerals and certain exotic herbs for the preparation of herbal remedies. While knowing about the Sowa-Rigpa system of medicine system of medicine we documented around 500 medicinal plants in our study and their various aspects used in Sowa Rigpa system off medicine. Most of the documented species belong to herbs. The treatment from the sowarigpa system of medicine is sought by all age groups of population. The integration of sowarigpa system of medicine with modern medicine not only adds value to the health care system but provides patients with choices of health care system and different options of treatment. The relationship of Indian system of medicine with Sowa-Rigpa begins with the wider context of medical pluralism. However showing such an influence young generation still shows least interest in adopting the culture of sowa-rigpa system of medicine. Integrating the system with modern health care system and more government support may prove effective to enhance its declining conditions. Sowa-rigpa practitioners and institutions must now rapidly adapt 21st century pressures under the growing of integrative medicine paradigm. This system of medicine can complement with western medical institutions through mutual trust between health practitioners. Despite having many hard working Tibetan doctors and interesting students, there still exist many challenges like government recognition and support and translation of texts and ability of the students to gain valuable clinical knowledge.

Conflict of Interest

All author declares no conflict of interest for this manuscript.

Acknowledgements

All authors are indebted to Director, CSIR- Indian Institute of Integrative Medicine for providing the necessary support to carry out this work. The contribution of all the participants who put their full hearted support to accomplish this work are dully acknowledged.

References

Angchok D, Dwivedi SK, Ahmed Z (2009) Traditional foods and beverages of Ladakh. Indian Journal of Traditional Knowledge 8(4):551-558.

Angmo K, Bhalla T C (2014). Preperation of *Phabs*-an indigenous starter culture for production of traditional aalcholic beverage *chhang* in ladakh. Indian journal of Traditional Knowledge 13(2):347-351.

Anonymous (2006) Ministry of Health Bhutan: Traditional health services. *Annual Health*

Aronson JK (2015) Meyler's side effects of drugs: the international encyclopedia of adverse drug reactions and interactions. Elsevier.

Artuso A (2013) Drugs of natural origin: Economic and policy aspects of discovery, development, and marketing. Routledge.

Ballabh B Chaurasia (2007) Traditional medicinal plants of cold desert Ladakh. used in treatment of cold cough & fever. Journal of Ethnopharmacology 112(2): 341-349. DOI: 10.1016/J.Jep.2007.03.020

Callahan D (2001) The role of complementary and alternative medicine: Accommodating pluralism. Georgetown University Press.

Chai X, Lang HQ, Jin SR (1965) Swamps in Zoigê Plateau.Science Press, Beijing [in Chinese. Web of Science, pp. 1– 72.

Clifford T (1994) Tibetan Buddhist medicine and psychiatry: The diamond healing. Motilal Banarsidass Publ.

Corson TW, Crews CM (2007) Molecular understanding and modern application of traditional medicines: triumphs and trials. Cell 130(5), 769-774.

Deane S (2014) Sowa Rigpa, spirits and biomedicine: lay Tibetan perspectives on mental illness and its healing in a medically-pluralistic context in Darjeeling, Northeast India (Doctoral dissertation, Cardiff University).

Fabricant DS (2001). The value of plants used in traditional medicine for Drug Discovery. *Environmental Health Perspect*ive 109:69–75.

Gerke B (2011) Correlating biomedical and Tibetan medical terms in amchi medical practice. Medicine between Science and Religion: Explorations on Tibetan Grounds, 127-152.

Gurmet P (2004) Sowa-Rigpa: Himalayan art of healing. Indian Journal of Traditional Knowledge 3(2): 2012-2018.

Hankey A (2005) CAM Modalities can stimulate advances in theoretical biology. Evidance Based Complement Alternative Medicine. 2:5–12.

Huscher D, Thiele K, Gromnica-Ihle E, Hein G, Demary W, Dreher R, Zink A, Buttgereit F, (2009) Dose-related patterns of glucocorticoid-induced side effects. Annals of the Rheumatic Diseases, 68(7):1119-1124.

Kala CP (2005) Health traditions of Buddhist community and role of amchis in trans-Himalayan region of India. Current Science 25:1331-1338.

Kloos S (2013) How Tibetan medicine in exile became a "Medical system". East Asian Science, Technology and Society: An International Journal 7(3):381-395.

Kloos S (2016) The recognition of sowa rigpa in India: How Tibetan medicine became an indian medical system. Medicine, Anthropology, Theory 3(2):19-49.

Learn JR (2018) 5000-year-old 'Iceman' may have benefited from a sophisticated health care system. Archaeology. doi:10.1126/science.aav3516

Lee BJ, Forbes K (2009) The role of specialists in managing the health of populations with chronic illness: the example of chronic kidney disease. BMJ 339:b2395.

Lhamo N, Nebel S (2011) Perceptions and attitudes of Bhutanese people on Sowa Rigpa, traditional Bhutanese medicine: a preliminary study from Thimphu. Journal of Ethnobiology and Ethnomedicine, 7(1):3.

Li XJ, Zhang HY (2008) Western-medicine-validated anti-tumor agents and traditional Chinese medicine. Trends in Molecular Medicine 14(1):1-2.

Lin Z, Zhao X (1996) Spatial characteristics of changes in temperature and precipitation of the Qinghai-Xizang (Tibet) Plateau. Sci. China, Ser. D 39:442-448.

Manabe S, Terpstra TB (1974) The effects of mountains on the general circulation of the atmosphere as identified by numerical experiments. Journal of Atmospheric Sciences 31(1)3-42

Marett RR (1920) *Psychology and Folk-lore*. London, Methuen

Minami M, Matsumoto S, Horiuchi H (2010) Cardiovascular side-effects of modern cancer therapy. Circulation Journal, 1008100855-1008100855.

Moerman L, Van-Der-Laan S (2006) TRIPS and the pharmaceutical industry: Prescription for profit?. Critical Perspectives on Accounting 17(8):1089-1106.

Molnar P, Tapponnier P (1975) Cenozoic tectonics of Asia: Effects of a continental collision: Science 189:419-426.

Pan SY, Litscher G, Gao SH, Zhou SF, Yu ZL, Chen HQ, Zhan, SF, Tang MK, Sun JN, Ko KM (2014) Historical perspective of traditional indigenous medical practices: the current renaissance and conservation of herbal resources. Evidence-Based Complementary and Alternative Medicine DOI: http://dx.doi.org/10.1155/2014/525340

Paulsen BS (2010) Highlights through the history of plant medicine, In: Bioactive compounds in plants – benefits and risks for man and animals, Bernhoft A (Editor), Oslo: The Norwegian Academy of Science and Letters,Norway

Petrovska BB (2012) Historical review of medicinal plants' usage. Pharmacognosy reviews, 6(11), 1-5. doi: 10.4103/0973-7847.95849

Phuntsog ST (2006) Ancient Materia Medica: Sowa-Rigpa (Tibetan Science of Healing). Paljor Publications Ltd., New Delhi.

Pordié L (2015) Genealogy and ambivalence of a therapeutic heterodoxy. Islam and Tibetan medicine in North-western India. Modern Asian Studies 49(6): 1772-1807.

Powell CM, Conaghan PJ (1975) Tectonic models of the Tibetan plateau. Geology 3(12):727-731.

Powel CM, Conoghan PJ (1973) Plate tectonics and the Himalayan. Earth and Planetary Source Setters 20(1), 1-12.

Powell JM (1982) The history of plant use and man's impact on the vegetation. In: Biogeography and ecology of New Guinea (pp. 207-227). Springer, Dordrecht.

Roy B, Kala CP, Farooquee NA, Majila BS (2004) Indigenous fermented food and beverages: a potential for economic development of the high altitude societies in Uttaranchal. Journal of Human Ecology. 15(1):45-49.

Ruddiman WF, Kutzbach JE (1989) Forcing of late Cenozoic northern hemisphere climate by plateau uplift in southern Asia and the American West. Journal of Geophysical Research: Atmospheres, 94(D15):18409-18427.

Salick J, Byg A, Amend A, Gunn B, Law W, Schmidt H (2006) Tibetan Medicine Plurality. Economic Botany, 60(3): 227-253.

Schmidt BM, Ribnicky DM, Lipsky PE, Raskin I (2007) Revisiting the ancient concept of botanical therapeutics. Nature Chemical Biology 3(7): 360.

Schramm ME, Hu MY (2013) Perspective: The evolution of R&D conduct in the pharmaceutical industry. Journal of Product Innovation Management 30:203-213.

Seidl PR (2002) Pharmaceuticals from natural products: current trends. Anais da Academia Brasileira de Ciências 74(1):145-150.

Sen S, Chakraborty R (2017) Revival, modernization and integration of Indian traditional herbal medicine in clinical practice: Importance, challenges and future. Journal of Traditional and Complementary Medicine, 7(2): 234-244.

Shi QW, Li LG, Huo CH,. Zhang ML, Wang YF (2010) Study on natural medicinal chemistry and new drug development. Chinese Traditional and Herbal Drugs 41: 1583–1589.

Tamang JP, Thapa S, Tamang N, Rai B (1996) Indigenous Fermented Food Beverages Of Darjeeling Hills And Sikkim: Process and Product Characterization. Journal of Hill Research 9(2): 401-411.

Tang M, Cheng G, Lin Z (Eds.) (1998) Contemporary Climatic Variations over Qinghai-Xizang (Tibetan) Plateau and their Influences on Environments (in Chinese). Guangdong Science and Technology Press, Guangzhou.

Targais K, Stobdan T, Mundra S, Ali Z, Yadav A, Korekar G, Singh SB (2012) Chhang-A barley based alcoholic beverage of Ladakh, India. Indian Journal of Traditional Medicine 11(1): 190-193

Tsarong TJ (Ed.) (1986) Handbook of traditional Tibetan drugs: their nomenclature, composition, use, and dosage.Tibetan Medical Publications.
Wagner EH (2004) Chronic disease care. BMJ 328:177–178.
Wagner EH (1998) Chronic disease management: what will it take to improve care for chronic illness? Effective Clinical Practice 1:2–4.
Wagner EH (1999) Managing chronic disease. BMJ 318:1090.
Wang MH (1987) Pollen assemblages at peat bogs in Zoigê Plateau and its palaeovegetation and palaeoclimate. Scientia Geographica Sinica 7: 147– 154.
Wangchuck AD, Wangchuck QA (2006) Treasures of the thunder dragon: A portrait of Bhutan. Penguin Books India.
Yang JS (2000) Tibetan medicine. In Ethnobotany and Sustainable Utilization of Plant Resources. Xu JC. Kunming: Science and Technology Press of Yunnan, pp. 65-75.
Zheng D, Zhang Q, Wu S (Eds.) (2000) Mountain Genecology and Sustainable Development of the Tibetan Plateau. Kluwer Academic Publishing, Dordrecht
Zheng D, Zhu L (Eds.) (2000) Formation and Environmental Changes and Sustainable Development on the Tibetan Plateau: Proceedings of International Symposium on the Qinghai-Tibetan Plateau, Xining, China, 21 – 24 July 1998. Academy Press, Beijing.
Zhou KS, Chen SM, Ye YY, Liang XL (1976) Study of some issues of Quaternary palaeogeography in the Everest area based on palynological analysis. The Scientific Expedition Report of Everest Area, Science Press, Beijing. pp. 79– 92
Zink A, Samadelli M, Gostner P, Piombino-Mascali D (2019) Possible evidence for care and treatment in the Tyrolean Iceman. International Journal of Paleopathology 25:110-117.

2

Traditional Uses of Plants as Medicine Among Khasi Tribe of Meghalaya, Northeast India

Prem Prakash Singh, Gargee Debnath, Archana Ojha Gardinia Nongbri, Anita Gupta, Vandolf M Kharbih, Harsh Singh and Krishna Upadhaya

Abstract

The Khasi tribe of Meghalaya depends on wild medicinal plants for curing a number of ailments since ages. The traditional knowledge systems which evolved through trial and error method are an integral part of their culture. The objectives of the present study were to highlight the medicinal plants used by the Khasi tribe to treat various diseases, provide scientific validation of the medicinal plants by pharmacological screening using secondary literature and suggest strategies for its sustainable management. A total of 363 plants distributed in 283 genera and 129 families were documented. The life forms of these plants were in the order of herb > tree > shrub > climber > parasite > epiphyte. The herbal remedies for treating various ailments are prepared by using the whole plant either individually or in

Prem Prakash Singh, Archana Ojha, Anita Gupta, Vandolf M Kharbih, Harsh Singh
Department of Botany, North-Eastern Hill University, Shillong, 793022, India

Gargee Debnath and Gardinia Nongbri
Department of Environmental Studies, North-Eastern Hill University, Shillong, 793022, India

Anita Gupta
CSIR-National Botanical Research Instititue, Lucknow-226001, Uttar Pradesh, India

Krishna Upadhaya (✉)
Department of Basic Sciences and Social Sciences, North-Eastern Hill University, Shillong, 793022 India
**All Authors Contributed Equally*

✉*Corresponding author email: upkri@yahoo.com*

Plants for Novel Drug Molecules: Ethnobotany to Ethnopharmacology
Bikarma Singh & Yash Pal Sharma (eds.), (pp. 45-83)

Email: *info@nipabooks.com* Web: *www.nipabooks.com*

combination with other plant parts. Cross validation with the help of secondary literature showed the presence of active compounds in these plants. Therefore, it is suggested that traditional knowledge should be properly documented and scientifically explored with more emphasis on species specific phytochemicals and their mode of action.

Keywords: Khasi, Life form, Traditional knowledge system, Medicinal plant Pharmacology.

1. Introduction

The use of bioresources for human welfare can be traced back to prehistoric times when primitive man and women recognized the importance of natural products (plants, animals and microorganisms) which they use as food, cloths, shelter and medicines (Bandaranayake 2006, Yuan et al. 2016). Herbal medicine is the oldest form of health care used to alleviate and treat diseases by traditional know-how since ancient times across all cultures throughout history (Bandaranayake 2006, Alves and Rosa 2007, Dias et al. 2012, Yuan et al. 2016). The knowledge and use of herbal traditional medicines also known as complementary and alternative medicine (CAM) was a trial and error method by early humans. In ancient times they often consumed poisonous plants which led to vomiting, digestive problem, or other toxic reaction and even death (Yuan et al. 2016). Led by their natural instinct, taste, and experience, these early humans treated illness by using plants, animal parts, and minerals that were not part of their usual diet (Bandaranayake 2006). All these knowledge have also increased their ability to distinguish useful plants with beneficial effects from those that were toxic. Over a period of time various combinations of plants were prepared to be used as medicine to cure various ailments (Kayang et al. 2005). This knowledge was passed on from one generation to another, thus laying the foundation for many systems of traditional medicine all over the world (IARC and WHO 2002, Bandaranayake 2006, Dias et al. 2012). Today the herbal traditional medicines are of great importance to the welfare of entire humanity. Subsequently, various forms of traditional medicine systems evolved which uses naturally available products from nature *viz.* Indian system of traditional medicine (Ayurveda), traditional Chinese medicine (TCM), Japanese traditional medicine (Kampo), traditional Korean medicine (TKM) and Unani systems of medicines (Fabricant and Farnsworth 2001, Yuan et al. 2016). According to World Health Organization (WHO 2004) it is estimated that about 80% of the people living in developing countries still rely on herbal medicinal products as a primary source of healthcare (Farnsworth et al. 1985) and is viewed as an integral part of the culture in those communities (Bodeker et al. 2005, Bandaranayake 2006).

The herbal medicine which is considered as essential part of the traditional health care system have witnessed a tremendous surge in acceptance and public interest in the past few decades. The natural therapies are once again receiving scientific attention to develop new drugs that are non-toxic and inexpensive (Couzinier and Mamatas 1986). In contrast to the conventional treatment (western medicine) which is often associated with some degree of side-effects, people are inclined more towards the use of herbal medicine treatment (Welz et al. 2018). Although herbal medicines have promising potential and are widely used, there is also a growing concern over the scientific validation of the traditional herbal medicine with respect to their potential adverse effects, mode of use and dosages (Ekor 2014).

Meghalaya is a hilly state in North-east India and a part of Indo-Burma biodiversity hotspot. It is floristically one of the richest states of India. The rich diversity of the state is due to large variation in altitude, climatic condition, edaphic factors and the existence of different vegetation types (Haridasan and Rao 1985-1987, Upadhaya 2015). Due to its diverse cultural and traditional practices and high richness in plants having ethnobotanical value, the state has immense scope of ethnobotanical studies (Kayang et al. 2005). The state is inhabited by Khasis, Jaintias and Garos and identified as Scheduled Tribes (Ministry of Tribal Affairs 2013). Among these three tribes, Khasi are the largest tribes and are closely associated with nature, and possess immense ethno-botanical knowledge about the plants available around them. Generally, a few individuals (traditional healers or locally called as *Nong ai dawai kynbat* or *Kabiraij*) from each village, possesses good knowledge on administering wild medicinal plants for treatment of various ailments. This traditional knowledge is passed on from one generation to another orally (Upadhaya et al. 2004).

During the last three decades a number of studies have been carried out to document the medicinal plants usage among the Khasis (Rao 1981a and b, Joseph and Kharkongar 1981, Kharkongar and Joseph 1981, Neogi et al. 1989, Maikhuri and Gangwar 1993, Laloo et al. 2006, Ahmed and Borthakur 2005, Kayang et al. 2005, Lakadong 2009, Kayang 2007, Mao et al. 2009, Medhi and Chakraborti 2009, Hynniewta 2010, Hynniewta and Kumar 2008, 2010, Singh and Borthakur 2011, Mir et al. 2014, Sen et al. 2016, Kharjana and De 2015). However, there is lack of scientific validation of the medicinal plants used by the Khasi tribe. Thus it is important and interesting to scrutinize that traditional uses are supported by definite pharmacological property (Jain and Tarafder 1970, Mauri and Pietta 2000). The pharmacological screening of plants for medicinal compounds provide a scientific basis for the continued use of the plants, thereby validating their historical utilization by traditional healers and herbalists, and also provide society with sources of new, effective and safe drugs (Taylor et al. 2001). Therefore, the present paper aims to: i) highlight the medicinal plants used by the Khasi tribe of Meghalaya to treat various diseases ii) provide scientific validation of the medicinal plants by pharmacological screening using secondary literature and iii) suggest strategies for the sustainable management of medicinal plant resources.

2. Material and Methods

2.1. Study Area

Meghalaya has a total geographical area of 22,429 sq. km and is situated between 25°05′ N and 26°10′ N latitudes and 89°47′ E and 92°47′ E longitude. It is bordered on the northwest, north and east by Assam and south and southwest by Bangladesh. The state is a collection of undulating hills with an east west orientation. The climate of the state is influenced by its topography and is controlled by seasonalwinds like the south-west monsoon and the north-east winter-winds. The altitude of the state ranges from 50-1990 m a.s.l. The lower elevation experiences fairly high temperature whereas the high elevation has moderate temperature. The climate of the area is monsoonal with distinct wet (May-October) and dry (November-March) seasons (Upadhaya 2015). However, to the south, Cherrapunjee and Mawsynram, records the highest rainfall (~11,000-12,000 mm) in the World. The vegetation type ranges from tropical, subtropical -broad leaved, -pine forest to lower temperate type. In addition to this there are also bamboo forest and grassland spread throughout the state (Haridasan and Rao 1985-1987, Rao and Hajra 1986, Upadhaya et al. 2012). The forest cover of the state is about 76.8% (FSI 2017).

According to 2011 census, the population of Meghalaya is 29.7 lakhs. The overall average density of population in the state is 132 persons/ sq.km (Census of India 2011) and around 80% of the population of the state lives in rural areas. The tribal population makes up about 86.1% of the population. In terms of tribal composition the state has three distinct regions namely, Khasi Hills, Jaintia Hills and Garo Hills. The total populations of Khasis is about 14.1 lakhs and are dominant in East Khasi-South west-, West Khasi- and Ribhoi- districts of the state.

2.2. Ethno-Botanical Surveys

In the present study, systematic and extensive ethno-botanical surveys were carried out in the Khasi dominated areas across the state during the period (2014-2018). During the field survey, the informants were asked about their traditional knowledge, plant use, part/s used and disease treated. The plant specimens were identified with the help of local flora

(Kanjilal et al. 1934-1940, Balakrishnan 1981-1983, Haridasan and Rao 1985-1987). The herbaria at Botanical Survey of India, Eastern Regional Centre, Shillong were also consulted for correct identification. The data collected during the present study was supplemented with earlier data existing for the Khasi tribe. The scientific validation of the traditional medicinal plant was established by the presence of active substances, or through pharmacological findings based on the related articles collected from different sources viz. Google Scholar, Researchgate, academia.edu, shodhganga, Science Direct, inflibnet.

3. Results and Discussion

3.1. Species Diversity

A total of 363 plants distributed in 283 genera and 129 families were recorded in the present study which is commonly used in traditional health care by the Khasi tribe. Of these, 345 were angiosperms, 2 gymnosperms and 10 pteridophytes (Appendix 1). Among the life forms, herbs with 152 species (41.9%) were the dominant component followed by 107 species of trees (29.4%), 67 shrubs (18.4%), 33 (9.1%) climbers, three parasites (0.8%) and one epiphyte (0.3%) (Fig. 1, Appendix I).

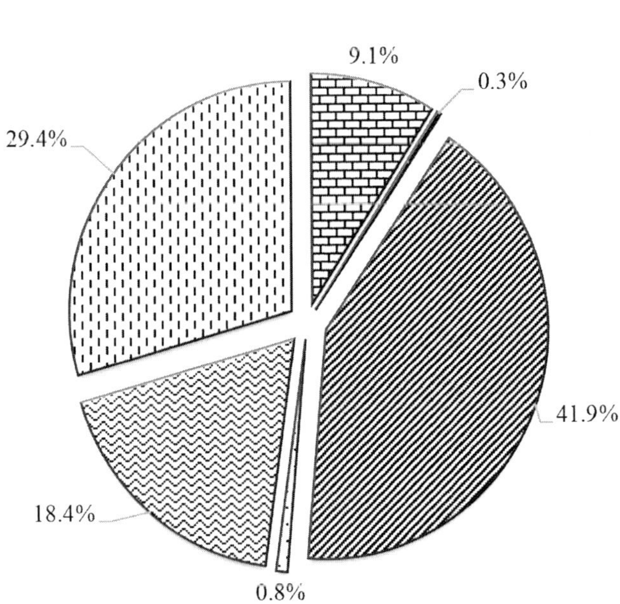

Fig. 1: Percentage of different life forms in the ethno-medicinal flora of Khasi tribe

In term of family richness, Asteraceae with 33 species was the dominant family followed by Lauraceae (14 species), Zingiberaceae (13 species), Euphorbiaceae and Lamiaceae (11 species each), Rosaceae, Rubiaceae and Verbanaceae with 10 species each. There were 27 families that were represented by two species each whereas, 64 families were monospecific (Appendix I).

3.2. Part Used

The study revealed that the herbal remedies for treating various ailments is prepared by using the whole plant either individually or in combination with other plant parts [leaf, inflorescence, underground parts (bulbs, rhizomes, roots, tubers, corms, pseudobulbs), bark, fruit, stem, flower (inflorescence), young shoot (fronds) and wood] or the individual parts. The use of whole plant was mainly for herbaceous species followed by shrubs (Table 1). In majority of the cases, leaves (159 species) are used for the treatment of various diseases followed by underground parts (95 species), whole plants (50 species), barks (46), fruits (36), stems (16 species), seeds (8 species), flowers (7 species), young shoots (6) and woods (2 species) as shown in Fig. 2.

The remedies from these plants were often utilized in the form of extracts, juice, paste and powder. Pastes made from leaves, fruits, bark, seeds and stem were applied on cuts, wounds, boils and skin diseases. Other methods include chewing the raw plant and inhaling smoke or vapor generated by burning. Some plants were boiled, while others were applied directly in fresh form or topically.

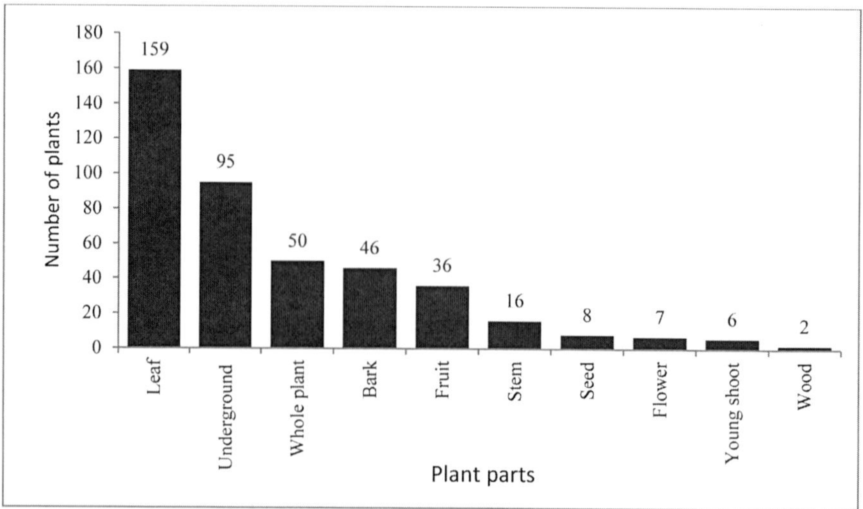

Fig. 2: Number of species and different plant parts used for treatment of various ailments

Table 1: Life form wise use of plant parts in Khasi traditional medicines

Plant parts	Life forms					
	Climber	Epiphyte	Herb	Parasite	Shrub	Tree
Bark	1	-	-	1	6	38
Bulb	-	-	4	-	-	-
Corm	-	-	1	-	-	-
Flower	1	-	2	-	1	2
Frond	-	-	4	-	-	-
Fruit	5	-	1	-	9	21
Inflorescence	-	-	-	-	-	1
Leaf	12	1	68	-	37	41
Pseudobulbs	-	-	1	-	1	-
Rhizome	3	-	21	-	-	-
Root	5	-	25	-	11	14
Seed	2	-	2	-	1	3
Stem	4	-	1	-	4	7
Tuber	2	-	7	-	-	-
Twig	-	-	-	-	-	1
Whole plant	5	-	28	2	9	6
Wood	-	-	-	-	-	2
Young shoot	-	-	-	-	-	1

Coix lacryma-jobi *Ilex khasiana* *Houttuynia cordata*

Nepenthes khasiana *Erythrina stricta* *Rubia cordifolia*

Ilex embelioides *Pouzolzia hirta* *Schima khasiana*

Swertia angustifolia *Prunella vulgaris* *Mimosa pudica*

Photo Plate. Medicinal plants of Khasi Hills

Appendix – I: List of medicinal plants used by the Khasi tribe of Meghalaya

Species	Family	Local name	Habit	Part use(s)*	Ailments / Diseases	Active compound#	Sources**
Acer laevigatum Wall.	Sapindaceae	*Dieng-than*	T	L	Sprain	-	19
Achyranthes aspera L.	Amaranthaceae	*Sohbyrthit*	H	L	Wounds, dental problems	+	4,13
Acmella paniculata (Wall. ex DC.) R.K.Jansen	Asteraceae	*Jasat*	H	Fl	Tooth ache	+	13,8,14
Acorus calamus L.	Araceae	*Bet, Ryniaw*	H	L/Rh	Rheumatism	+	7,8,15,16
Adenia trilobata (Roxb.) Engl.	Passifloraceae	*Sohmynthar*	C	L	Body ache	-	14
Adenostemma lavenia (L.) Kuntze	Asteraceae	*Soh byrthit*	H	L	Cuts, wounds, insect bites	+	12
Adiantum lunulatum Burm. f.	Adiantaceae	*Tyrkhang khyllai*	H	L	Temperature	+	15, 18
Aegle marmelos (L.) Corrêa	Rutaceae		T	F	Dysentery, dioarrhoea	+	6
Aesculus assamica Griff.	Sapindaceae	*Dieng-sangkenrop*	T	B/L	Fever	+	3
Agapetes mannii Hemsl.	Ericaceae	*Dieng-Sohlapydong*	S	B	Fractured bones	-	19
Agapetes odontocera (Wight) Benth. & Hook.f.	Ericaceae	*Dieng-sohlmut, Jalamut*	S	B	Fractured bones	-	19
Ageratina riparia (Regel) R.M.King & H.Rob.	Asteraceae	*Kyrbet jewslew*	H	WP	Stomach problem	+	7
Ageratum conyzoides (L.) L.	Asteraceae	*Ksangd agiem, Kyrbat myngai*	H	L	Cuts, Snake bites, insect sting, swellings and tumor	+	13, 4,8,15,16
Alangium chinense (Lour.) Harms	Cornaceae	*Diengsohkympel*	T	R	Astringent, anthelmintic	+	3
Albizia chinensis (Osbeck) Merr.	Mimosaceae	*Dieng phallut*	T	B	Insect bites	+	18, 19
Aleuritopteris subdimorpha (C.B.Clarke & Baker) Fraser-Jenk.	Sinopteridaceae	*Tyrkhang lieh*	H	Fr	Kidney stones	-	14
Allium sativum L.	Liliaceae	*Ryrsun*	H	Bu	Febrifuge	+	8

Contd.

Allium tuberosum Rottler ex Spreng.	Liliaceae	*Jyllang*	H	WP	Mouth sores, cough	+	5,8
Aloe vera (L.) Burm.f.	Aloaceae	*Jatyngkieh*	H	L	Leprosy, child delivery, malaria, Snakebite, gastric disorders, stomach aches	+	14
Alstonia scholaris (L.) R.Br.	Apocynaceae	*Dieng-ryteng*	T	B	Headache, fever	+	19
Alysicarpus monilifer (L.) DC.	Fabaceae		H	WP	Snake bites	+	16,15
Amaranthus tricolor L.	Amaranthaceae	*Jada basaw*	H	R	Irregular menses	+	14
Amomum aromaticum Roxb.	Zingiberaceaae	*Ilashi saw*	H	Rh	Skin diseases	+	16
Ampelocissus barbata (Wall.) Planch.	Vitaceae		C	Rh	Boils	-	16
Anaphalis adnata Wall. ex DC. Planch.	Asteraceae	*Tiew riem*	H	L	Boils, urinary tract infection, kidney stones, eye drop	+	14
Andrographis paniculata (Burm.f.) Nees	Acanthaceae		H	R	Rheumatic pains, liver disorders	+	4
Anemone rivularis Buch.-Ham. ex DC.	Ranunculaceae		H	L	Cure sinus	+	10
Antidesma acidum Retz.	Euphorbiaceae	*Dieng-japew*	T	L	Diphoretic	+	3
Antidesma bunius (L.) Spreng.	Euphorbiaceae	*Soh-syllai*	T	L	Joint pains	+	18
Antidesma puncticulatum Miq.	Euphorbiaceae		T	L	Joint pains	+	16
Ardisia paniculata Roxb.	Myrsinaceae	*Dieng-Soh-botul*	S	R	Haemorhoea, jaundice	+	3, 19
Areca catechu L.	Arecaceae	*Kwai*	T	F	Boils	+	8
Argemone maxicana L.	Papaveraceaea		H	WP	Laxative	+	4
Argyreia hookeri C.B.Clarke	Convolvulaceae	*Dieng Langsying*	C	F	Boils	-	14
Arisaema consanguineum Schott	Araceae	*Soh-bsein*	H	R	Paralysis, epilepsy, Stomach problems, evil eye, snake bites, insect stings, malaria	-	19
Aristolochia saccata Wall.	Aristolochiaceae		C	Rh	Cough, fever, malaria, chicken pox, tonic,	+	17

Contd.

					pneumonia, vermifuge, headache, astringent		
Artemisia indica Willd.	Asteraceae	*Jaiaw*	H	L	Sinusitis	+	14
Artemisia nilagirica (C.B.Clarke) Pamp.	Asteraceae	*Khel bijak*	S	L	Asthma, brain disease, sores	+	4
Asparagus filicinus Buch.-Ham. ex D.Don	Liliaceae	*Bah niang soh pet*	C	R	Hypertension, wounds, urinary troubles	+	6,7,19
Asparagus racemosus Willd.	Liliaceae	*Bah niang soh pet*	C	L/R	Diarrhoea, stomach	+	6,7,14
Astilbe rivularis Buch.-Ham. ex D.Don	Saxifragaceae	*Pdah*	S	L	Tooth ache, blood purification	+	12
Averrhoa carambola L	Averrhoaceae	*Sohpyrshong*	T	F	Jaundice	+	18
Azadirachta indica A. Juss.	Meliaceae	*Dieng nim*	T	L	Diarrhoea, dysentery, tuberculosis, heart diseases	+	16,17
Bambusa tulda Roxb.	Poaceae	*Lungsieh*	T	St	Nerve pains	+	19
Bauhinia variegata L.	Caesalpiniaceae	*Dieng tharlong*	T	Fl	Piles, dysentery	+	1
Begonia josephi A.DC.	Begoniacea	*Jajew*	H	Bu	Stomach pain, indigestion	-	
Begonia palmata D.Don	Begoniacea		H	WP	Allergies	-	16
Begonia roxburghii A.DC.	Begoniacea	*Jajew jylwang*	H	L/Tu	Measles, diarrhoea and dysentery, fever, body temperature	+	16,8,17
Begonia rubronervia Hort. ex Klotzsch	Begoniacea		H	R	Dysentery	-	16
Berberis wallichiana DC.	Berberidaceae	*Dieng-mat-shynrang*	S	B/R	General and unspecified	-	3
Bergenia ciliata (Haw.) Sternb.	Saxifragaceae	*Pymon/ Tiew japang*	H	L	Paralysis, rheumatism, Injuries, bone fractures	+	14
Betula alnoides Buch.-Ham. ex D.Don	Betulaceae	*Dienglieng*	T	R	Indigestion, flatulence	+	18
Bidens pilosa L.	Asteraceae	*Soh byrthit*	H	L	Gastric disorders	+	13

Contd.

Bidens biternata (Lour.) Merr. & Sherff	Asteraceae		H	L	Snake bite	+	16
Biophytum sensitivum (L.) DC.	Oxalidaceae		H	L	Headache, giddiness, fever	+	16
Blumea balsamifera (L.) DC.	Asteraceae		H	WP	Expectorants, colds	+	4
Boehmeria macrophylla Hornem.	Urticaceae	*Dieng soh khasim*	S	L/St	Blood purifier	+	6
Bombax ceiba L.	Bombacaceae	*Dieng-syrah*	T	B	Snake bite	+	19
Bonnaya reptans (Roxb.) Spreng.	Scrophulariaceae	*Neinglik*	H	L/R	Snake bite, urinary ailment	-	9,16
Brucea javanica (L.) Merr.	Anacardiaceae	*Dieng-sohma*	T	Se	Stomach trouble, skin trouble, papyloma, rheumatism, intestinal worm	+	3,8,9,16
Bryophyllum pinnatum (Lam.) Oken	Crassulaceae	*Bat rben/ syntiew thylliej*	H	L	Dysentery, skin diseases	+	14
Buddleja asiatica Lour.	Loganiaceae	*Duma khynnah*	S	WP	Mentally disturbed patients	+	14, 19
Buddleja macrostachya Benth.	Buddlejaceae	*Jalong krem*	S	L	Venereal disease	+	18
Calendula officinalis L.	Asteraceae	*Tiew monger*	H	Fl/L	Wounds, ear ache	+	14
Callicarpa arborea Roxb.	Verbenaceae	*Dieng-lakhiot*	T	WP	Fever, gastric diseases, giddiness, headache, skin diseases, scorpion sting, carminative, headache, giddiness, fever	+	8, 13, 16
Camellia kissii Wall.	Theaceae	*Dieng-tyrnem*	T	L	Tumors	-	1,2,3
Canna indica L.	Cannaceae		H	Rh	Lower body temperature	+	3
Cannabis sativa L.	Cannabinaceae		S	L	Stomach problem, hemorrhage, disorders of spleen	+	12
Capsicum annuum L.	Solanaceae	*Sohmynken*	H	F	Cat bite	+	14

Contd.

Careya arborea Roxb.	Barringtoniaceae	*Soh-kundur*	T	F/Fl	Cold and cough	+	19
Casearia vareca Roxb.	Flacourtiaceae	*Dieng-soh-rang*	T	L	Anthelmintic, ear ache, anticancer, anti-vermicidal properties	+	3,16
Cassia fistula L.	Ceasalpinaceae	*Dieng laroo*	T	F	Stomach problem	+	7
Celtis tetranda Roxb.	Ulmaceae		T	F	Febrifuge, sores, anemia, tonic	+	3
Celtis timorensis Span.	Ulmaceae		T	Wo	Amenorrhea, colic diseases	+	3
Centella asiatica (L.) Urb.	Apiaceae	*Batmonia*	H	Wp	Boils, tumors, dysentery, diarrhea, cough, blood purification	+	5,6,8,13,16
Centranthera grandiflora Benth.	Orobanchaceae		S	Wp	Stomachache, eye sores, urinary troubles, cataract, night blindness, leprosy	+	16
Cephalostachyum latifolium Munro	Poaceae	*Sylli*	T	St	Skin bleeding	-	7
Cheilanthes albomarginata C.B.Clarke	Cheilanthaceae	*Tyrkhang lieh*	H	Fr	Burns	+	8
Cheilocostus speciosus (J.Koenig) C.D.Specht	Zingiberaceaae	*Sla pangmat*	H	Rh	Nausea, vomiting	+	17
Chenopodium ambrosioides L.	Chenopodiaceae	*Sla-sam*	S	Wp	Nerve tension	+	4
Chromolaena odorata (L.) R.M.King & H.Rob.	Asteraceae	*Latna-iong*	S	L	Dysentery, cuts, wounds	+	15,4,16
Cinnamomum pauciflorum Nees	Lauraceae	*Dieng-lorthia*	T	B	Expectorant, deodorant, diuretic, carminative, colic pain, bronchitis, asthma, diarrhea, nausea	+	1,3

Contd.

Cinnamomum bejolghota (Buch.-Ham.) Sweet	Lauraceae	*Lakhyrdop*	T	St/B	Body swellings, bodyaches, bone fractures	+	14
Cinnamomum glaucescens (Nees) Hand.-Mazz.	Lauraceae	*Dieng Pingwait*	T	L	Sinusitis in adults, nose blockage	+	14
Cinnamomum tamala (Buch.-Ham.) T.Nees & Eberm.	Lauraceae	*Dieng-sia-sia*	T	B/L	Diarrhea, carminative, cold, cough	+	16,3
Cinnamomum verum J.Presl	Lauraceae	*Dalshini*	T	L	Cough	+	16
Cirsium verutum (D.Don) Spreng.	Asteraceae	*Soh-chiia*	H	L	Gastro-intestinal disorder	+	19
Citrus medica L.	Rutaceae	*Sohkwit*	T	L/R	Laxative, anthelmintic, astringent, stimulant, anthelmintic, digestive, carminative, dysentery, cold	+	1,2,3,8
Citrus latipes (Swingle) Yu.Tanaka	Rutaceae	*Sohkymphor*	T	F/L	Appetizer, gout, rheumatism, rashes and ringworm	+	1,2
Clematis gouriana Roxb. ex DC.	Ranunculaceae	*Jyrmi bteng*	C	St	Injuries, fractured bone.	+	14
Cleome viscosa L.	Capparaceaea		H	L	Ear troubles	+	4
Clerodendrum infortunatum L.	Verbenaceae	*Dieng-kylasla*	S	R	Astringent, gonorrhea	+	3
Clerodendrum paniculatum L.	Verbenaceae	*Jaiong jarem*	S	L	Anemia, blood purification	+	13
Clerodendrum bracteatum Wall. ex Walp.	Verbenaceae	*Jarem saw*	S	L	Antihemorrhagic, headaches	-	14
Clerodendrum glandulosum Lindl.	Verbenaceae	*Sla jarem*	S	L	Antihemorrhagic, headaches	+	16,8
Coelogyne stricta (D.Don) Schltr.	Orchidaceae		S	Pb	Blood dysentery	+	16
Coix lacryma-jobi L.	Poaceae	*Sohriew*	S	L	Stomach problem	+	8
Colocasia esculenta (L.) Schott	Araceae	*La wang*	H	Co/L	Skin diseases	+	13

Contd.

Combretum pilosum Roxb. ex G.Don	Combretaceae		C	St	Stem medicinal	-	3
Cordia fragrantissima Kurz	Boraginaceae	*Bahari*	T	B/F	Astringent, Fever, anthelmintic, diarrhea, skin diseases, diuretic, expectorant, lung and spleen diseases	-	3
Corylopsis himalayana Griff.	Hemamelidaceae	*Ja-iur*	S	WP	Diarrhea, dysentery, gastro-intestinal disorder	-	19
Crassocephalum crepidioides (Benth.) S.Moore	Asteraceae	*Jathymmai, Sla-aeroplane*	H	L	Cuts,constipation, wounds, stomach disorder	+	8, 12, 13
Crinum amoenum Ker Gawl. ex Roxb.	Amaryllidaceae	*Rynsun Blei*	H	Bu	Astringent, expectorant, cough, skin diseases, clearing hair	-	14
Croton caudatus Geisel.	Euphorbiaceae	*Soh-lambrang*	T	Wp	Malaria, cholera	+	3
Cryptocarya amygdalina Nees	Lauraceae	*Dalgappa*	T	B/L	General and unspecified	-	3
Curcuma aeruginosa Roxb.	Zingiberaceaae	*Shyrmit iong / Syiang Blei*	H	Rh	Anemia, bronchitis, inflammation, jaundice rheumatism, white leprosy	+	14
Curcuma amada Roxb.	Zingiberaceaae	*Ban kitril dawai sohpieng*	H	Rh	Body swellings, injuries, bone fractures, skin diseases (cattle and chicken), toothaches, dental caries, poisoning	+	14
Curcuma angustifolia Roxb.	Zingiberaceaae	*Niang-sohpet*	H	Rh/L	Stomach problems	+	18
Curcuma longa L.	Zingiberaceaae	*Shynrai Shyrmit stem*	H	Rh	Children for gripe	+	5, 14

Contd.

Curcuma zedoaria (Christm.) Roscoe	Zingiberaceaae		H	Rh	Wounds, injuries, sprains, nail fall off, anti-bacterial, blood dysentery, urinary system, anti- allergic	+	17
Cuscuta reflexa Roxb.	Convolvulaceae		C	Wp	Bodyache, rheumatism	+	16
Cymbidium aloifolium (L.) Sw.	Orchidaceae		H	Se	Headache, fever	+	11
Cynodon dactylon (L.) Pers.	Poaceae	*Semtyrpai*	H	Wp	Diarrhea,dysentery, fever, small pox, tonic	+	14
Cyperus rotundus L.	Cyperaceae	*Phlang*	H	Wp	Cuts	+	17,13
Dendrobium chrysanthum Wall. ex Lindl.	Orchidaceae	*Tiew lyngskw*	H	St	Healing wounds	+	8
Dendrobium moschatum (Buch.-Ham.) Sw.	Orchidaceae	*Tiew dieng*	H	L	External Injuries, fractured bone	-	12,11
Desmodium griffithianum Benth.	Fabaceae	*Sohbyrthit*	H	Wp	Fractured bone	-	14
Desmodium triflorum (L.) DC.	Fabaceae	*Kynbat shria shria*	S	L	Dysentery, diarrhea	+	14
Dichrocephala integrifolia (L.f.) Kuntze	Asteraceae	*Kynbat snianglbat sohphie*	H	L	Wounds, injuries, throat problems	+	14
Dicliptera chinensis (L.) Juss	Acanthaceae		H	R	Tonic	+	4
Dicranopteris linearis (Burm.f.) Underw.	Gleicheniaceae		H	Wp	Tonic	+	16
Dioscorea deltoidea Wall. ex Griseb.	Dioscoreaceae	*Phan ka khlam*	C	Tu	Dysentery, jaundice, delerium	+	14
Dioscorea pubera Blume	Dioscoreaceae	*Dawai suhjyndong*	C	Tu	Kidney stones	+	14
Diospyros pilosiuscula G.Don	Ebenaceae		T	Wp	Stomach disorder, piles, kidney stone, diarrhea, dysentery	-	1,3
Diplopterygium volubile (Jungh.) Nakai	Gleicheniaceae	*Tyrkhang nar*	H	Fr	Epilepsy	-	14
Dipsacus asper Wall. ex C.B.Clarke	Dipsacaceae	*Tiew-stem*	H	L	Skin disease	+	4

Contd.

Dischidia bengalensis Colebr.	Asclepiadaceae	*Dawai bsein*	C	Wp	Snake bites	-	14, 19
Dischidia nummularia R.Br.	Asclepiadaceae	*Kynbat kudam*	C	L	Cuts, wounds, injuries, bone fractures	-	8
Disporum cantoniense (Lour.) Merr.	Liliaceae		S	St	Cure dysentery, stomach problems	+	16
Docynia indica (Wall.) Decne.	Rosaceae	*Sok-pho-btet*	T	F/L	Cold, cough, dysentery, diarrhea	+	19
Drymaria cordata (L.) Willd. ex Schult.	Caryophyllaceae	*Bat-nongrim*	H	L	Burns, skin diseases, snake bites, leprosy	+	4,8,12,15,16,17
Dysoxylum excelsum Blume	Meliaceae	*Sla-luchai*	T	L/Tw	Malarial fever, blood pressure	-	19
Dysoxylum gotadhora (Buch.-Ham.) Mabb.	Meliaceae	*Bol-narang*	T	Se/Wo	General and unspecified	+	3
Ecbolium ligustrinum (Vahl) Vollesen	Acanthaceae		H	R	Jaundice	+	4
Eclipta prostrata (L.) L.	Asteraceae		H	L	Liver problems, spleen enlargement, headache, brain tonic, cataracts	+	4
Elephantopus scaber L.	Asteraceae	*Kynbat-skur*	H	L/R	Diarrhea, urinary problem, pimples, insect stings, contraceptive	+	4,9,13
Eleusine indica (L.) Gaertn.	Poaceae	*Lang krai*	H	R	Chewed for dental cavities, tooth aches	+	1
Elsholtzia blanda (Benth.) Benth.	Lamiaceae	*Bat-skain*	H	L	Repellant, cuts and injuries	+	15,12,16
Elsholtzia pilosa (Benth.) Benth.	Lamiaceae		H	L	Skin diseases	+	16
Embelia ribes Burm.f.	Myrsinaceae	*Bakul lata*	C	Wp	Astringent, nervous debility, skin diseases and leprosy	+	3

Contd.

Emilia sonchifolia (L.) DC. ex DC.	Asteraceae	*Soh-byshet*	H	L	Eye problem	+	4
Engelhardtia spicata Lechen ex Blume	Juglandaceae	*Dieng Iyba*	T	I/L	Scabies, skin diseases	+	1, 3
Entada rheedii Spreng.	Fabaceae		C	Se	Dandruff	+	16
Erigeron trilobus (Decne.) Boiss.	Asteraceae		H	L	Wounds, skin disease	-	4
Eriosema himalaicum H.Ohashi	Fabaceae	*Sohpen*	H	Tu	Dysentery, bad breath	-	8
Eryngium foetidum L	Apiaceae	*Dhonia Bhoi, Dhonia khlaw*	H	L	Epilepsy	+	8
Erythrina arborescens Roxb.	Papilionaceae	*Dieng-song*	T	St	Tooth ache, prevent dental carries	+	8,14,18
Erythrina stricta Roxb.	Fabaceae	*Dieng-songdkhar*	T	B	Sedative, carminative, digestive, anthelmintic, expectorant, diuretic, stomach ache, itching, rheumatism, burning, fever, fainting, asthma, leprosy	+	3
Erythroxylum kunthianum A.St.-Hil.	Erythroxylaceae	*Dieng-painkhar*	T	B	Stimulant	-	1,2,3
Eulophia speciosa (R.Br.) Bolus	Orchidaceae		H	L	Ear pain	-	11
Eulophia spectabilis (Dennst.) Suresh	Orchidaceae	*Bat piat*	H	Tu	Blood clotting	+	14
Euonymus lawsonii C.B.Clarke ex Prain	Celastraceae		T	B	Syphilis, indigestion, liver disorder, lice	-	1,2,3
Eupatorium cannabinum L.	Asteraceae		S	L	Visceral obstruction, fever, swelling of legs and scrotum	+	4
Eupatorium adenophorum Hort.Berol. ex Kunth	Asteraceae	*Bat iong*	S	L	Injuries	+	5,6,15
Eurya acuminata DC.	Theaceae	*Dieng-shit*	T	L	Astringent, digestive, carminative, diuretic	+	3

Contd.

Eurya japonica Thunb.	Theaceae	*Dieng-pyrshit*	T	L	General and unspecified	+	3
Fagopyrum acutatum (Lehm.) Mansf. ex K.Hammer	Polygonaceae	*Jarain*	H	L	Blood purification	+	5
Ficus hispida L.f.	Moraceae	*Dieng-lapong*	T	B/F	Emetic, anaemia, jaundice, fever, ulcer	+	3
Ficus heterophylla L.f.	Moraceae	*Soh eitblang shynrang*	S	F	Boils	-	14
Ficus hirta Vahl	Moraceae	*Soh eitblang kynthei*	S	F	Boils	+	14
Flemingia procumbens Roxb.	Fabaceae	*Sohphlang*	H	Tu	Deworming, anthelmintic	+	5, 8
Galinsoga parviflora Cav.	Asteraceae	*Tiew-lien*	H	L/R	Bleeding, snake bite, insect stings	+	16,15
Galium asperifolium Wall.	Rubiaceae	*Kynbat mynsawl Kynbat laiproh*	H	WP	Injuries, body sores with maggot attack, removal of corns	+	14
Garcinia cowa Roxb. ex Choisy	Clusiaceae	*Rengran*	T	F/L	Cathartic, dysentery, stomach trouble	+	2,3
Garuga pinnata Roxb.	Burseraceae	*Dieng khiang*	T	F/L/St	Indigestion, conjunctivitis, asthma	+	18
Gaultheria fragrantissima Wall.	Ericaceae	*Jirkap, Lathynrait*	S	L/R	Rheumatism, paralysis, migraines, pneumonia, wormicidal, menstrual disorder	+	3,5,8
Geranium nepalense Sweet	Geraniaceae	*Batlmieng*	H	L	Tooth ache, bleeding gums	+	8
Glochidion lanceolarium (Roxb) Voight	Euphorbiaceae		T	B	Stomach ailments	-	3
Gomphostemma parviflorum Wall. ex Benth.	Lamiaceae		S	L	Forehead	+	16
Grewia multiflora Juss.	Tiliaceae	*Dieng-tyrbhong*	T	B	Cuts, wounds, flatulence, kill intestinal worms (cattle)	-	3

Contd.

Gynocardia odorata R.Br.	Flacourtiaceae	*Sohliang*	T	Se	Leprosy, nausea, rheumatism	+	18
Hedychium villosum Wall.	Zingiberaceaae	*Syein-kloo*	H	Rh	Jaundice	+	19
Hedyotis scandens Roxb.	Rubiaceae	*Mo-shoh shu*	C	L	Gastric troubles, cough and cold.	+	9
Hedyotis uncinella Hook. & Arn.	Rubiaceae	*Bat iong*	H	L	Insect stings, skin diseases	+	16,8
Helicia excelsa (Roxb.) Blume	Proteaceae		T	B	Insect stings	+	3
Heliotropium indicum L.	Boraginaceae	*Mationg-blag*	H	Wp	Stomach ache	+	4
Helixanthera ligustrina (Wall.) Danser	Loranthaceae	*Mang-kariang*	P	B	Gastric, other stomach problem	-	19
Helixanthera parasitica Lour.	Loranthaceae	*Mangkaring heh sla*	P	Wp	Bodyache (analgesic)	+	14
Hibiscus rosa-sinensis L.	Malvaceae	*Tiew ot*	S	L	Boils	+	14
Hodgsonia heteroclita (Roxb.) Hook.f. & Thomson	Curcubitaceae	*Soh risa*	C	R	Fever	+	
Hodgsonia macrocarpa (Blume) Cogn.	Curcubitaceae	*Tyr-khang*	C	Wp	Gastric ulcer, malarial fever	+	19
Houttuynia cordata Thunb.	Saururaceae	*Jamyrdoh*	H	Wp	Iron deficiency, blood purification, blood Sugar, sores and boils.	+	5, 8,12
Hypericum japonicum Thunb.	Hypericacaeae	*Bat saw rit*	S	Wp	Antidote for snakebite	+	8
Hypocharis radicata	Asteraceae	*bat jhur kthang*	H	L	Stomach upset	+	18
Ilex khasiana Purkay.	Aquifoliaceae		T	B/R	Cold, cough, tuberculosis	-	1,3
Ilex embelioides Hook.f.	Aquifoliaceae		T	B/R	Cold, cough, tuberculosis	-	1,2,3
Impatiens benthamii Steenis	Balsalminaceae	*Sohpohloh*	H	L	Boils	-	14
Impatiens racemosa DC.	Balsalminaceae		H	L/R	Rheumatic pain	-	4
Iresine herbstii Hook.	Amaranthaceae	*Syntiew saw*	H	L	Rheumatism, paralysis, anaemia	+	14
Iris domestica (L.) Goldblatt & Mabb.	Iridaceae	*Kynbat phlang phiang*	H	Wp	Urinary tract infection, kidney stones	+	14

Contd.

Isodon wightii (Benth.) H.Hara	Lamiaceae		S	L	Corns	+	16
Itea chinensis Hook. & Arn.	Saxifragaceae	*Dieng*	T	L	Skin diseases	-	9
Ixora acuminata Roxb.	Rubiaceae		S	L/R	General and unspecified	-	3
Justicia gendarussa Burm.f.	Acanthaceae		S	L	Headache, influenza, gall stones	+	17
Justicia adhatoda L.	Acanthaceae		S	L	Abortifacient, heart trouble, bronchitis, leprosy, jaundice, blood purification,	+	15, 4
Kaempferia galanga L	Zingiberaceaae	*Sying khmoh, Sying shmoh, Ingsmoh*	H	Rh	Gastric ulcer, asthma, skin disease	+	8, 17
Kydia calycina Roxb.	Malvaceae	*Dieng-misiri*	T	B/L/R	Rheumatism, lumbago, febrifuge	+	3
Lactuca laevigata C.B.Clarke	Asteraceae		H	L	High blood pressure, diabetes, skin infection	-	8
Laggera pterodonta (DC.) Sch.Bip. ex Oliv.	Asteraceae	*Bat nongrim*	H	WP	Rheumatism	+	14
Lantana camara L.	Verbenaceae	*Sohpang-khleh*	S	L/R	High blood pressure, malaria, liver troubles, rheumatism	+	13,5,14
Launaea asplenifelia Hook.f.	Asteraceae	*jalyngkdeh*	H	L	Blood purification	+	5
Leucas ciliata Benth.	Lamiaceae	*Bat nianglynur*	H	Wp	Antidote for snakebite	+	8
Leucosceptrum canum Sm.	Lamiaceae	*Soh Kjit*	T	R	Malaria	+	8
Lindenbergia griffithii Hook.f.	Scrophulariaceae		H	Wp	Bronchitis	-	4
Lindenbergia muraria (Roxb. ex D.Don) Brühl	Scrophulariaceae	*Bat saw stem syntiew*	H	Wp	Snakebites	-	14
Lindera latifolia Hook.f.	Lauraceae	*Dieng-jalang*	T	B/L	Skin diseases	-	1,3
Lindera pulcherrima (Nees) Hook.f.	Lauraceae	*Dieng-jaburit*	T	B	Cold, cough, worm, rheumatism, wounds	+	3,12

Contd.

Lindernia anagallis (Burm.f.) Pennell	Linderniaceae		H	L	Burns, injuries, bone fractures, gall bladder, eye problems	+	16
Litsea cubeba (Lour.) Pers.	Lauraceae	*Dieng-sying*	T	F	Deodorant, insect repellent	+	3
Litsea khasyana Meisn.	Lauraceae	*Dieng mosu*	T	R	Chronic bronchitis	-	18
Litsea salicifolia (J. Roxb. ex Nees) Hook.f.	Lauraceae	*Dieng-lali*	T	B/L	General and unspecified	+	3
Lonicera japonica Thunb.	Caprifoliaceae	*Jermei-ren*	C	L	Astringent, pulmonary diseases	+	3
Lyonia ovalifolia (Wall.) Drude	Verbenaceae	*Diengla-samiang*	T	R	Whooping cough treatment, skin diseases	+	3
Lysimachia racemosa Lam.	Primulaceae		H	L	Deworming	-	10
Magnolia champaca (L.) Baill. ex Pierre	Magnoliaceae	*Shap*	T	B	Stimulant, purgative	+	3
Mahonia nepalensis DC. ex Dippel	Berberidaceae	*Dieng-niangmat*	S	B	Eye diseases	+	3,12
Mahonia pycnophylla (Fedde) Takeda	Berberidaceae	*Dieng chandan*	S	F	Eye diseases, tapeworm	-	2
Mallotus philippensis (Lam.) Müll.Arg.	Euphorbiaceae	*Dieng chandan*	T	F	Tapeworm	+	18
Mazus pumilus (Burm.f.) Steenis	Scrophulariaceae	*Bat iong*	H	Wp	Injuries, body sores, skin diseases	+	14
Melasma avense (Benth.) Pennell	Scrophulariaceae	*Bat khlamsyiar*	H	Wp	Chicken in diseases	-	14
Melastoma malabathricum L.	Melastomaceae	*Dieng-sohkhling*	S	L	Stop bleeding	+	4,7,13
Melia azedarach L.	Meliaceae	*Dieng ja rasang*	T	B	Anthelmintic, malarial fever, poultice on skin eruption.	+	
Mentha arvensis L.	Lamiaceae	*Pudina*	H	L	Stomach problem	+	5
Mesua ferrea L.	Calophyllaceae		T	B	Cough, anaemia	+	17, 19
Micromelum pubescens Blume	Rutaceae	*Dieng-sohsat*	T	R	Cough	+	3
Mikania micrantha Kunth	Asteraceae	*Bat refugee*	C	L	Diarrhea	+	13,6,14

Contd.

Mimosa pudica L.	Mimosaceae		H	R	Snake bites, diabetes	+	4,5,13,14,16
Monochoria hastata (L.) Solms	Pontederiaceae	*Sla-tiewlily*	H	L	Skin diseases, leprosy	-	5
Morus australis Poir.	Moraceae	*Sohlyngdkhur*	S	F	Anaemia	+	14
Myrica esculenta Buch.-Ham. ex D.Don	Myricaceae	*Sohphie*	T	Wp	Astringent, carminative, antiseptic, asthma, chronic bronchitis, fever, lung infections, dysentery, toothache, worms, jaundice	+	1,3,5,6
Myrica integrifolia Roxb.	Myricaceae	*Siohlia*	T	B	Dysentery	+	8
Nasturtium officinale R.Br.	Brassicaceae	*Tyrso-um*	H	Wp	Pneumonia, pulmonary ailments.	+	
Neanotis wightiana (Wall. ex Wight & Arn.) W.H.Lewis	Rubiaceae	*Shkor-maina*	H	L	Wounds, bleeding, snakebite	+	4,8
Nepenthes khasiana Hook.f.	Nepenthaceae	*Kserphare*	C	Fl	Headache	+	1,2,15,17
Nephrolepis cordifolia (L.) C. Presl	Polypodiaceae	*Tyrkhang*	H	Tu	Migraine, dizziness	+	14
Inula cappa (Buch.-Ham. ex D.Don) DC.	Asteraceae	*Ja raikhoh shynrang*	H	R	Anti-allergic	+	14
Ocimum tenuiflorum L.	Lamiaceae		H	L	Cold, cough	+	4
Oenanthe javanica (Blume) DC.	Apiaceae	*Jatira*	H	L	Anaemia	+	14
Olax acuminata Wall. ex Benth.	Olacaceae		T	L	Cathartic	+	3
Oldenlandia verticillata L.	Rubiaceae		H	L	Fever	-	16
Olea dioica Roxb.	Olacaceae	*Poreng*	T	B	Febrifuge	+	3
Osbeckia capitata Benth. ex Naudin	Melastomaceae	*Soh Pythem*	H	Wp	Snake bite, muscle swelling	-	2
Osbeckia crinita Benth. ex C.B. Clarke	Melastomaceae	*Syntiew-khra*	S	R	Diarrhoea, wounds, against Snake bites	-	5,7,12
Osbeckia stellata Buch.-Ham. ex Ker Gawl.	Melastomaceae	*Soh-lyngkthut*	S	L	Wounds, snake bites, nose bleeding, dysentery	+	16

Contd.

Oxalis corniculata L.	Oxalidaceae	*Jabuit*	H	L	Diarrhea, cough, snake bites, fever	+	4,5,6,813, 14,15,17
Oxalis debilis var. *corymbosa* (DC.) Lourteig	Oxalidaceae	*Dike-mesing*	H	Wp	Stomach ache	+	4
Paederia foetida L.	Rubiaceae	*Kynbat-iwtuing, Jyrmismaiwtung*	C	L/Wp	Dysentery, snakebite	+	6
Panax bipinnatifidus var. angustifolius (Burkill) J.Wen	Araliaceae		H	Rh	Bleeding, pain swelling, tonic	+	10,1
Pandanus tectorius Parkinson ex Du Roi	Pandanaceae		T	L	Leprosy, paralysis, skin diseases	+	12
Parochetus communis D.Don	Papilionaceae	*Khia knoi*	H	Wp	Stomach disorders	+	12
Passiflora edulis Sims	Passifloraceae	*Sohbrap*	C	L	Blood dysentery, stomach problem	+	5, 6,8
Peliosanthes teta andrews	Asparagaceae		H	Tu	Wounds, fresh Cuts, burns, eczema and emollient, antibacterial, menstrual flow, tumors, enlargement of spleen and liver, jaundice, good laxative	-	16
Peperomia heterophylla Miq.	Piperaceae	*Bat kudam*	E	L	Wounds, injuries	-	14
Persea bombycina (King ex Hk.f.) Koster.	Lauraceae	*Som*	T	L	Hair	+	3
Persea gamblei (King ex Hook. f.) Kosterm.	Lauraceae	*Omgthat*	T	B/L	Muscular pain	+	3
Persicaria capitata (Buch.-Ham. ex D.Don) H.Gross	Polygonaceae		H	Wp	Insect stings	+	16
Persicaria chinensis (L.) H. Gross	Polygonaceae	*Jasch*	H	R	Urinary disorders	+	9
Persicaria nepalensis (Meisn.) Miyabe	Polygonaceae	*Jakyrphuh*	H	Rh	Leprosy, paralysis, skin diseases	+	8

Contd.

Phlogacanthus thyrsiflorus Nees	Acanthaceae		S	L/St	Diarrhea, dysentery, fever	+	16
Phoebe attenuata (Nees) Nees	Lauraceae	*Bonsum*	T	F	Sores	-	3
Pholidota imbricata Lindl.	Orchidaceae		H	Pb	Rheumatism, arthritis, paralysis, migraines, pneumonia	-	16
Phrynium pubinerve Blume	Marantaceae	*Sla met*	H	Rh	Tumour	+	8
Phyllanthus retusus Dennst.	Euphorbiaceae		S	B/L	Diuretic, astringent, diuretic	+	3
Phyllanthus clarkei Hook.f.	Euphorbiaceae		S	L	Headache, giddiness, fever	-	16
Phyllanthus glaucus Wall. ex Müll.Arg.	Euphorbiaceae	*Sohrynniaw*	S	L	External sores	+	14
Phyllanthus parvifolius Buch.-Ham. ex D.Don	Euphorbiaceae		T	Wp	Astringent, diuretic, jaundice, diarrhea, dysentery, fever	-	3
Picrasma javanica Blume	Simaraubaceae	*Bor-jagreng*	T	B/L	Stomach problem	+	1,3
Pinus kesiya Royle ex Gordon	Pinaceae	*Kseh khasi*	T	YS	Boils	-	1,8
Piper betle L.	Piperaceae	*Tympew/ pathi*	C	L	Cuts, wounds, cough	+	14
Piper griffithii C.DC.	Piperaceae	*Mrit khlaw*	C	Se	Cough, stomach trouble, diarrhea, dysentery	-	1,2
Piper mullesua Buch.-Ham. ex D.Don	Piperaceae		C	F/L	Cough	-	16
Piper peepuloides Wall.	Piperaceae	*Sohmrit khlaw*	C	F	Whooping cough	+	14
Plantago asiatica subsp. *erosa* (Wall.) Z.Yu Li	Plantaginaceae	*Shkor blang*	H	L	Wounds	+	
Plantago major L.	Plantaginaceae	*Kyrblang*	H	L	Burns	+	12
Plumbago zeylanica L.	Plumbaginaceae		S	R	Piles, diarrhea, dyspepsia	+	4
Pogostemon purpurascens Dalzell	Lamiaceae	*Sal-smalwking*	H	L	Burns, wounds	+	4

Contd.

Polyalthia longifolia (Sonn.) Thwaites	Annonaceae	*Diengther*	T	B	Febrifuge	+	3
Polygonatum oppositifolium (Wall.) Royle	Liliaceae	*Syiog maw*	H	Rh	Rheumatism	-	14
Polystichum pseudotsussimense Ching	Aspidiaceae	*Tyrkhang nar Bhoi*	H	Fr	Boils	-	14
Pongamia glabra Vent.	Fabaceae		T	L/R	Cough, leprosy	+	3
Potentilla fulgens Diels	Rosaceae	*Lynniang*	H	R	Gums and teeth, diarrhea, anti-diabetic, stomach problem, common cold, high blood pressure	+	5,8
Pouzolzia hirta Blume ex Hassk.	Urticaceae	*Memsleh*	H	R	Dysentery	+	5,10,12
Pouzolzia zeylanica (L.) Benn.	Urticaceae	*Miensa-miyo*	H	L/R	Hair tonic, boils, bone fracture	+	9
Prunella vulgaris L.	Lamiaceae	*Jahynwet*	H	L	Antispasmodic, cuts, expectorant,wounds, piles	+	4, 8
Prunus napaulensis (Ser.) Steud.	Rosaceae	*Soh-iong*	T	Wp	Astringent, diuretic, dropsy, hardwood astringent, acrid, refrigerant	+	3
Prunus persica (L.) Batsch	Rosaceae	*Soh phareng*	T	L	Diarrhea	+	14
Psidium guajava L.	Myrtaceae	*Soh priam*	T	L	Diarrhea, blood and chronic dysentery	+	8,14, 18
Pteridium aquilinum (L.) Kuhn	Polypodiaceae	*Tyrkhang shatri*	H	Rh	Urinary tract infection	+	8
Pteris wallichiana J. Agardh	Pteridaceae	*Tyrkhang shatri*	H	Rh	Leprosy, paralytic, skin diseases boils, gall stones	+	14
Pyrus pashia Buch.-Ham. ex D.Don	Rosaceae	*Soh-shur*	T	L	General and unspecified	+	3

Contd.

Randia longiflora Lam.	Rubiaceae	*Jyrnishiah-tiewkrot*	T	F	Insecticidal and insect repellant	-	3
Ranunculus diffusus DC.	Ranunculaceae	*Kynbat pang bniat*	H	L	Toothaches	-	14
Remusatia vivipara (Roxb.) Schott	Araceae	*Shriew maw/ Kynbat tah thung*	H	Tu	Fever, rickets disease	+	14
Rhododendron arboreum Sm.	Ericaceae	*Tiewsaw*	T	B	Headache, diarrhea, blood dysentery	+	3,6
Rhus succedaenia L.	Anacardiaceae	*Dieng-khlaw*	T	F	Phthisis, astringent, tonic, expectorant, stimulant, diarrhea, dysentery	+	3
Rhynchotechum ellipticum (Wall. ex D. Dietr.) A.DC.	Gesneriaceae	*Jarmehek*	H	L	Fractured bones	+	14
Ricinus communis L.	Euphorbiaceae		T	L	Rheumatism, swelling	+	16
Rosa indica L.	Rosaceae	*Tiew kulab saw*	S	Fl	Skin, pigmentation on the face	+	14
Rotala rotundifolia (Buch.-Ham. ex Roxb.) Koehne	Lythraceae	*Bat dohkoid*	H	L	Boils	+	8,14
Rotheca serrata (L.) Steane & Mabb.	Verbenaceae +	*Rilong-phlang* 19	H	R	Rheumatism, sprain, scabies, eczema, cut-wounds		
Rubia cordifolia L.	Rubiaceae	*Soh-misem*	C	R/St	Stomach ache, eye, ear trouble, insects stings	+	4, 7, 12,15
Rubus barberi H.E.Weber	Rosaceae	*Sohnepbah*	C	F/St/B	Cough, mouth ulcers	+	14
Rubus ellipticus Sm.	Rosaceae	*Soh-shiah*	S	WP	Dysentery	+	16,6, 5,7,12
Rubus micropetalus Gardner	Rosaceae	*sohepbah*	S	F/St	Cough, mouth ulcers	-	8
Rubus moluccanus Aiton	Rosaceae	*Soh nybbah*	S	R	Cuts, blood clotting, swelling	+	9

Contd.

Salix tetrasperma Roxb.	Salicaceae	*Jamynrei*	T	B	Febrifuge, rheumatism, swellings, epilepsy, piles, stones in bladder	+	1,3
Sansevieria trifasciata Prain	Liliaceae	*Waitlam*	H	R	Cancer, leprosy, paralytic patients	+	14
Sapindus attenuatus Wall.	Sapindaceae		T	F	Pimples	-	3
Saprosma ternatum (Wall.) Hook.f.	Rubiaceae		T	L	Flatulence, stomach ache, indigestion	+	3
Sarcandra glabra (Thunb.) Nakai	Chloranthaceae	*Tiew krismas*	S	L/R	Wounds, irregular menstrual bleeding, cure blood deficiency	+	16,8
Schima khasiana Dyer	Theaceae		T	B	Poulticing skin eruption	-	1,2,3
Schima wallichii Choisy	Theaceae		T	L/St	Skin, anthelmintic	+	16,7
Schisandra elongata (Blume) Baill.	Schizandraceae	*Mesaw*	C	F/L	Stomach disorder	-	12
Scutellaria discolor Colebr.	Lamiaceae	*Lahi*	H	L	Stomach ache, skin disease, snake bite	+	15,4,16
Senna alata (L.) Roxb.	Caesalpiniaceae	*Dawai khniang*	S	L	Skin diseases	+	14
Shorea robusta Gaertn.	Dipterocarpaceae		T	B	Diarrhea, dysentery.	+	16
Sida rhombifolia L.	Malvaceae	*Synsar dwar*	S	L	Cuts, wounds	+	14
Sigesbeckia orientalis L.	Asteraceae	*Soh-barthudip*	S	L	Wounds	+	4
Skimmia laureola Franch.	Rutaceae		S	L	General and unspecified	+	3
Smilax ferox Wall. ex Kunth	Smilacaceae	*Shiah krot*	C	R	Boil, cancer, eye problems	+	8
Smilax glabra Roxb.	Smilacaceae	*Khong*	C	L	Gall bladder stones	+	12
Solanum myriacanthum Dunal	Solanaceae		S	F	Tooth infection	+	
Solanum torvum Sw.	Solanaceae	*Soh pdok*	S	Se	Tooth ache, insect stings, snake bite	+	14,8,15,16
Solidago virgaurea L.	Asteraceae		H	Wp	Dropsy	+	4
Sonerila maculata Roxb.	Melastomaceae		H	R	Stomach ailment	-	16
Sphaeranthus indicus L.	Asteraceae		H	R/Se	Anthelmintic, stomach ache, piles	+	4

Contd.

Spilanthes acmella (L.) L.	Asteraceae	*Byshit-iong*	H	L	Tooth ache, mouth boil	+	13,4
Spondias pinnata (L.f.) Kurz	Anacardiaceae	*Dieng-sohpier*	T	B	Dysentery, rheumatism, dyspepsia	+	3
Stachytarpheta dichotoma (Ruiz & Pav.) Vahl	Verbenaceae	*Jaiong khlaw*	S	L	Malaria fever	+	14
Sterculia villosa Roxb.	Sterculiaceae	*Tluh*	T	B/St	Allergic rashes (Anti-allergic)	+	14
Strobilanthes cusia (Nees) Kuntze	Acanthaceae	*Lamuh iong*	S	L	Rheumatic patient	+	14
Swertia angustifolia Buch.-Ham. ex D.Don	Gentianaceae	*Charita*	H	L/R	Malaria	+	14
Swertia chirayita (Roxb.) Buch.-Ham. ex C.B.Clarke	Gentianaceae	*Charita*	H	L/R	Malaria	+	8
Symplocos racemosa Roxb.	Symplocaceae	*Bolimitap*	T	B	Diarrhoea, menstrual disorder, indigestion, ulcer, eye diseases, tonic	+	3
Symplocos cochinchinensis var. *laurina* (Retz.) Noot.	Symplocaceae	*Dieng Sohkpei*	T	L	High fever	+	14
Symplocos lucida (Thunb.) Siebold & Zucc.	Symplocaceae	*Dieng-pei*	T	L	Tumors	-	3
Symplocos paniculata (Thunb.) Miq.	Symplocaceae	*Diengiong, Jajew ding*	T	B/R	Kidney stones, constipation	+	14
Tabernaemontana divaricata (L.) R.Br. ex Roem. & Schult.	Apocynaceae	*Syntiew khlaw*	S	Wp	Anthelmintic, skin diseases, insect bites, headache, fever	+	8
Taxus baccata L.	Taxaceae	*Dieng seh-blei*	T	L	Cough	+	18
Terminalia chebula Retz.	Combretaceae	*Soh salukah*	T	F/R	Diuretic, conjunctivitis.	+	18
Tetrastigma serrulatum (Roxb.) Planch.	Vitaceae		C	Rh	Boils	-	16
Thalictrum foliolosum DC.	Ranunculaceae		H	R	Disease, fever, jaundice	+	4

Contd.

Thysanolaena latifolia (Roxb. ex Hornem.) Honda	Poaceae	*Synsar*	S	Wp	Jaundice	+	8,16
Trevesia palmata (Roxb. ex Lindl.) Vis.	Araliaceae	*Dieng-lakor*	T	L/R	Stomach ache	+	3
Tricyrtis maculata (D.Don) J.F.Macbr.	Liliaceae	*Sohkhia Blei*	H	L	Insect bites, boils	+	14
Uraria crinita (L.) DC.	Papilionaceae	*Dieng-kharia*	S	R	fever	+	4
Uraria rufescens (DC.) Schindl.	Papilionaceae		S	L	Febrifuge	+	4
Valeriana hardwickii Wall.	Valerianaceae		H	L	Skin diseases	+	16
Valeriana jatamansii Wall.	Valerianaceae	*Jatung*	H	Wp	Fractured bone, nail fall	+	8
Viburnum foetidum Wall.	Caprifoliaceae	*Dieng-sohlang*	S	F/L	Astringent, emetic, hemorrhage	+	3,5
Viola arcuata Blume	Violaceae	*Jamaiang*	H	L	Urinary, spleen disorders	-	8
Viscum articulatum Burm.f.	Loranthaceae	*Mangkaring*	P	Wp	Stomach problem.	+	8
Vitex negundo L.	Verbenaceae	*Pasutia*	S	Wp	Toothaches	+	3
Wedelia wallichi L.	Asteraceae	*Sok-so*	H	R	Cholera	-	4
Zanthoxylum acanthopodium DC.	Rutaceae	*Jaiur khlaw*	S	F/L	Stomach disorders, vermicide, fever, cold, cough, skin disease	+	8
Zanthoxylum armatum DC.	Rutaceae	*Jaiur*	T	F	Toothaches	+	14
Zanthoxylum khasianum Hook. f.	Rutaceae	*Soh-tiewshiah*	S	B/F	Carminative, stomachic, anthelmintic	-	1,3
Zephyranthes rosea Lindl.	Amaryllidaceae	*Tiew piat*	H	Bu	Skin diseases, eczema, scurvy	+	14
Zingiber montanum (J.Koenig) Link ex A.Dietr.	Zingiberaceaae	*Syiog Blei/Syiog shmoh shynrang*	H	Rh	Poisoning, mouth sores, tongue blisters, cataract, night blindness	+	14
Zingiber officinale Roscoe	Zingiberaceaae	*sying*	H	Rh	Chest trouble, anti-hemorrhagic,	+	5

Contd.

					urinary tract infection, evil eye emmenagogue, white discharge, rheumatism, gynecological problems, stomachache		
Zingiber rubens Roxb.	Zingiberaceaae	*Sying makhir*	H	Rh	Fever, cough, cold	+	14
Zingiber zerumbet (L.) Roscoe ex Sm.	Zingiberaceaae	*Sying khlaw*	H	Rh	Toothaches	+	14, 18
Ziziphus jujuba Mill.	Rhamnaceae	*Sok-broi*	T	F	Carminative	+	16

*B = Bark, Bu= Bulb, Co= Corm, Fl=Flower, Fr = Frond, F= Fruit I= Inflorescence, L= Leaf, Pb = Pseudobulb, Rh=Rhizome, R= Root Se= Seed, St= Stem, Tu= Tuber,Tw= Twig, Wp= Whole plant=Wp,Wo =Wood,Ys= Young shoot, T=Tree, S= Shrub, H=Herb, C= Climber, E=Epiphyte, P= Parasite**1=Mir et al. 2014, 2= Lakadong, 2009, 3= Laloo et al. 2006, 4= Neogi et al. 1989, 5= Kharjana and De, 2015, 6= Ahmed and Borthakur, 2005, 7= Hynniewta and Kumar, 2010, 8= Hynniewta and Kumar, 2008, 9= Maikhuri and Gangwar, 1993, 10= Mao et al, 2009, 11= Medhi and Chakraborti, 2009, 12= Rao 1981, 13= Sen et al. 2016, 14= Hynniewta, 2010, 15= Joseph and Kharkongar, 1981, 16= Kharkongar and Joseph, 1981, 17=Rao, 1981, 18= Kayang et al. 2005, 19= Singh and Borthakur, 2011.# '+' indicates present and '–' indicates absent

Among the plant parts used by the Khasi tribes in their traditional health care practices, herbs (38.8%) parts are widely used to treat various ailments followed by trees (32.2%), shrubs (18.6%), climbers (9.4%), parasitic plants (0.7%) and the least by epiphytes (0.2%). From various life forms,leaves (37.4 %) constitutes the most important ingredient in the Khasi medicines followed by roots (12.9%), whole plants (11.6%), barks (10.8%) and the least by corms, inflorescences, pseudobulbs, twigs, woods and young shoot (<0.5%) (Table 1).

3.3. Ailments Treated

A wide range of ailments were treated using these plants. The majority of the species (52.3%) had multiple-therapeutic uses, while 47.3% were used for treating specific diseases. *Callicarpa arborea, Cordia fragrantissima, Aloe vera, Myrica esculenta* and *Erythrina stricta* were used for more than 10 different ailments (Fig. 3). Some plant species were used for treatment of more than one ailments eg. *Diospyros pilosiuscula, Tabernaemontana divaricata, Grewia multiflora, Elephantopus scaber, Lyonia ovalifolia, Prunella vulgaris, Bryophyllum pinnatum, Brucea javanica, Coix lacryma-jobi, Ficus hispida, Lindera pulcherrima, Pteridium aquilinum, Houttuynia cordata, Eclipta prostrata, Eulophia spectabilis, Citrus latipes, Ageratum conyzoides* and *Kaempferia galanga* were used for treatment of five ailmens. Similarly, *Justicia adhatoda, Centella asiatica, Phyllanthus parvifolius, Cheilocostus speciosus, Nepenthes khasiana, Prunus napaulensis, Zanthoxylum acanthopodium, Salix tetrasperma, Zingiber zerumbet, Symplocos racemose* and *Schima wallichii* were used for treating six different ailments, *Rhus succedaenia, Gaultheria fragrantissima, Curcuma aeruginosa* and *Potentilla fulgens* for seven ailments, *Cinnamomum pauciflorum, Curcuma longa*

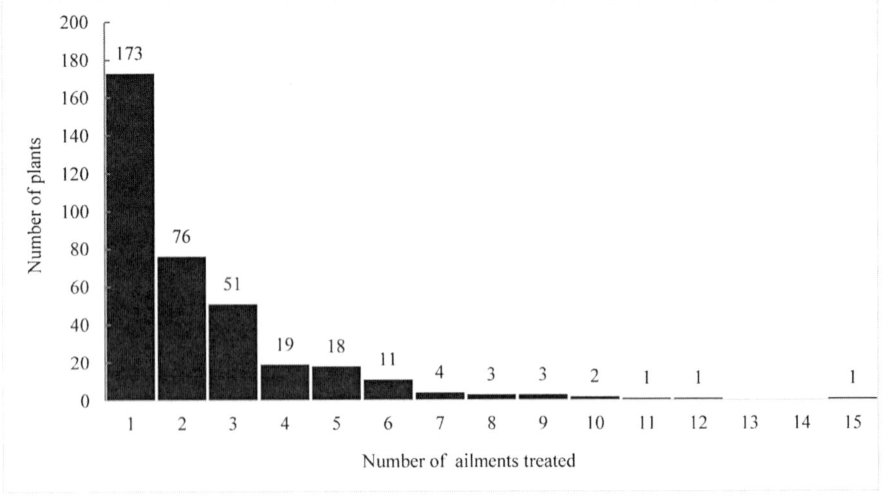

Fig. 3: Number of plants used to treat various ailments

and *Acorus calamus* for eight ailments. *Citrus medica, Vitex negundo* and *Zingiber montanum* for nine ailments, *Callicarpa arborea* and *Cordia fragrantissima* for ten ailments, *Aloe vera* for eleven ailments, *Myrica esculenta* for twelve ailments and, *Erythrina stricta* for 15 different ailments.

Among the ailment categories, the gastro-digestive disorders were treated with the highest number of plant species (233), followed by external injuries, bites and dermatological problems (229 species), head, thermoregulatory nervous system problems (110), oro-dental and respiratory problems (93), cancer and musculoskeletal problems (72), urogenital, gynecological and venereal problems (43), circulatory problems (35) (Table 2).

Table 2: Number of plant species used against various ailment category

Ailment category	Ailment sub-category	Number of plants
Digestive system and gastrointestinal problems	Indigestion, vomiting, nausea, spasms, constipation, intestinal worms, gastric ulcers, stomach pain, liver problems, spleen problems, dysentery and diarrhea.	233
External injuries, bites and dermatological problems	Scabies, ringworms, leprosy, rashes, pimples, acne, itching, dandruff, burns, insect bites, snake bites, caterpillar stings, poison consuming, cuts and wounds.	229
Head, thermoregulatory and nervous system problems	Malaria, hypertension, measles, cold, stress and tension, insomnia, anxiety, restlessness and fever.	110
Oro-dental and respiratory problems	Bleeding gums and nose, toothache, mouth sores, tongue blisters, nasal congestion, dental caries, influenza, bronchitis, pneumonia, cough, pulmonary infections, asthma and tuberculosis.	93
Cancer and musculo skeletal problems	Cancer, paralysis, muscular sprain and pain, bone dislocation, inflammation, rheumatism, obesity and weakness.	72
Urogenital, gynecological and venereal problems	Hydrocoel, diabetes, polyurea, gonorrhea, kidney stone, amenorrhea, pregnancy related problems, Abortion	43

Contd.

Circulatory problems	Heart trouble, hematoma, high blood pressure and anemia.	35
Ophthalmological and odological problems	Eye injury, conjunctivitis, eye sight problems, ear pain and eye redness.	25
General	General and unspecified	10
veterinary	Animal disease	3

3.4. Pharmacological Aspects

In the present analysis, active compounds were present in 78.5 % of the plant species. For instance, *Solanum torvum* contains minerals, vitamins, antioxidants, nutrients and biomolecules such as alkaloids, flavonoids, saponins, tannins, glycosides and phenolic compound. These phytochemicals have been found to have antioxidant-, antifungal-, antibacterial-, antiulcer, antihypertensive- and metabolic-correction, nephroprotective-, cardioprotective-, antidiabetic-, analgesic- and anti-inflammatory-, immunomodulatory- and erythropoietic- activity (Jaiswal 2012, Yousaf et al. 2013). Similarly, a comparative studies on the anti-proliferative potentials of *Melia azedarach* and *Azadirachta indica* on cancer cell lines showed that flavonols was abundant in the leaves of *Melia azedarach* (Jaferi et al. 2013). *Eupatorium adenophorum* is commonly used for cuts and wounds by the locals.The essential oil extracted from the inflorescence (esquiterpenoid) and root (monoterpenes) of this plant displayed moderate to strong antioxidant activity and have potential as antifungal and antibacterial agents (Ahluwalia et al. 2014).

Viburnum foetidum and *Houttuynia cordata* have been reported to contain gallic acid, protocatechuic acid, pvanillic acid, p-coumaric acid, sinapic acid, and ellagic acid in varying amounts and are valuable antioxidant components (Seal et al. 2016). The extracts of leaves of *Achyranthes aspera* exhibited good antioxidant effect used extensively for cuts by the local people (Edwin et al. 2008). In *Zingiber officinale* an important medicinal plant, as many as 63 active compounds have been identified (Jolad et al. 2004). The main compounds include immuno-modulatory, anti-tumorigenic, anti-inflammatory, anti-apoptotic, anti-hyperglycemic, anti-lipidemic and anti-emetic actions. Ginger is a strong anti-oxidant substance and may either mitigate or prevent generation of free radicals (Ali et al. 2008). *Colocasia esculenta* have been tested for anti-cancer, anti-diarrheal, astringent, nervine tonic, and hypolipidemic activity (Prajapati et al. 2011). A mention may be made about the pharmacological studies on *Psidium guajava* that have demonstrated the antioxidant, hepatoprotection, anti-allergy, antimicrobial, antigenotoxic, antiplasmodial, cytotoxic, antispasmodic, cardioactive, anticough, antidiabetic, antiinflamatory and antinociceptive activities properties. The species also has a wide range of clinical applications for the treatment of infantile rotaviral enteritis, diarrhoea and diabetes (Gutiérrez et al. 2008). Some of the plants marked as absent (-) (21.5%) in the Appendix-I do not necessary mean absence of active compound, but suggest lack of pharmacological study. However, literature survey clearly indicates that at the genus level the active compounds are present.

4. Conclusion

In the study it was observed that the tribal people were not aware of the active principles present in the plants which help in treating a particular disease. However, the presence of biological active compounds in these tried and tested wild plants revealed that they had in-depth knowledge about the medicinal plants. Such vast knowledge should be properly documented and scientifically explored as it may have potential for development of new drugs. Therefore, it is suggested that more emphasis should be given to evaluate species specific phytochemicals and their mode of action. There is immense scope of ethanobotanical studies in Meghalaya state due to its high natural forest cover which hold huge reserve of wild medicinal plants. The state also has high potential for plantation and commercialization of medicinal plants because of its favorable agro-climatic and natural landscape (Hynniewta and Kumar 2008). In our present study we enumerated 363 wild medicinal plants used locally by Khasi traditional health practitioners. The rural areas in the state lacks proper road connectivity and medical facilities and people rely heavily on local health practitioner and wild herbal plants to cure their ailments. However, the huge traditional knowledge on medicinal plants is on the verge of extinction due to erosion of tribal culture and lack of people's interest in taking up this practice as a profession (Kayang et al. 2005). Besides, the increasing popularity of the western medicines and ease of finding medicine have also added to its further decline. Inspite of all these constrains there are few people who are still practicing this system of medicine. Uncontrolled harvesting of wild medicinal plants has pushed some of the vital medicinal plants on a verge of extinction that is futher aggravated by anthropogenic disturbances such as deforestation, mining and landuse changes.Therefore, it is suggested that local people should be encouraged for plantation of wild medicinal plants in homegardens to ensure its conservation.

Abbreviation Used

CAM: Complementary and Alternative Medicine; TCM: Traditional Chinese medicine; TKM: Traditional Korean Medicine; WHO: World Health Organization.

Conflict of Interest

The authors declare no conflict of interest.

Acknowledgements

The authors are thankful to the Headman, traditional healers and the local people for providing us the detailed information on medicinal plants. Financial assistance received from Department of Science and Technology, Government of India (DST-Traditional Knowledge Systems, PAC-SES-DST-12141219-863) in the form of a research project is also acknowledged.

References

Ahluwalia V, Sisodia R, Walia S, Sati OP, Kumar J &Kundu A (2014). Chemical analysis of essential oils of *Eupatorium adenophorum* and their antimicrobial, antioxidant and phytotoxic properties.*Journal of Pest Science* 87:341–349.

Ahmed AA& Borthakur SK (2005). *Ethnobotanical wisdom of Khasis (Hynniewtreps) of Meghalaya*.Published byBishen Singh and Mahendra Pal Singh, Dehradun, India. 305 pages.

Ali BH, Blunden G, Tanira MO & Nemmar A (2008). Some phytochemical, pharmacological and toxicological properties of ginger (*Zingiber officinale* Roscoe): A review of recent research. *Food and Chemical Toxicology* 46(2): 409-420. doi.org/10.1016/j.fct.2007.09.085.

Alves RR & Rosa IM (2007). Biodiversity, traditional medicine and public health: Where do they meet? *Journal of Ethnobiology and Ethnomedicine* 3:14.doi:10.1186/1746-4269-3-14

Balakrishnan NP (1981-1983).*Flora of Jowai, Meghalaya*, Vol.I & II.Published by Botanical Survey of India, Howarh, 666pages.

Bandaranayake WM (2006). Quality control, screening, toxicity, and regulation of herbal drugs,pp.25–57. In: Ahmad I, Aqil F & Owais M (Eds.) Modern Phytomedicine: *Turning Medicinal Plants into Drugs*. doi:10.1002/9783527609987.ch2.

Bodeker C, Bodeker G, Ong CK, Grundy CK, Burford G & Shein K (2005). *WHO Global Atlas of Traditional, Complementary and Alternative Medicine*. Published by World Health Organization. Geneva, Switzerland, 347 pages.

Census of India (2011). Provisional population totals-India data sheet. Published by Office of the Registrar General Census Commissioner, India, Indian Census Bureau.

Couzinier JP &Mamatas S (1986). Basic and applied research in the pharmaceutical industry into natural substances,pp 57–61. In: Barton D and Ollis WD (Eds.), Advances in Medical Phytochemistry. *Proceedings of the International Symposium on Medicinal Phytochemistry*, John Libbey & Co, Morocco. London.

Dias DA, Urban S & Roessner, U (2012). A historical overview of natural products in drug discovery. *Metabolites* 2(2): 303-336.

Edwin S, Edwin JE, Deb L, Jain A, Kinger H, Dutt KR & Amal RA (2008). Wound healing and antioxidant activity of *Achyranthes aspera*, *Pharmaceutical Biology* 46:824-828.

Ekor M (2014). The growing use of herbal medicines: issues relating to adverse reactions and challenges in monitoring safety. *Frontiers in Pharmacology* 4:177. doi.org/10.3389/fphar.2013.00177

Fabricant DS & Farnsworth NR (2001). The value of plants used in traditional medicine for drug discovery. *Environmental Health Perspectives* 109: 69–75.

Farnsworth NR, Akerele RO, Bingel AS, Soejarto DD & Guo Z (1985).Medicinal Plants in Therapy.*Bulletin of the WHO* 63: 965–981.

FSI (2017). Forest Survey of India.India State of Forest Survey Report, 2017 (Available online at http://fsi.nic.in/isfr2017/meghalaya-isfr-2017.pdf).

Gutiérrez RM, Mitchell S & Solis RV (2008). *Psidium guajava*: a review of its traditional uses, phytochemistry and pharmacology. *Journal of Ethnopharmacology* 117:1-27.doi: 10.1016/j.jep.2008.01.025.

Haridasan K & Rao RR (1985-1987). *Forest Flora of Meghalaya*.Vol.I & II.Bishen Singh & Mahendra Pal Singh. 937 pages.

Hynniewta SR (2010). Ethnobotanical studies in Khasi hills, Meghalaya.D. Phil. Thesis, Department of Botany, North-Eastern Hill University, Shillong (Unpublished).

Hynniewta SR & Kumar Y (2008). Herbal Remedies among the Khasi Traditional healers and village folks in Meghalaya. *Indian Journal of Traditional Knowledge* 7:581-586.

Hynniewta SR & Kumar Y (2010). The lesser-known medicine Ka Dawai Ñiangsohpet of the Khasis in Meghalaya, Northeast India.*Indian Journal of Traditional Knowledge* 9(3):475-479.

Jafari S, Saeidnia S, Hajimehdipoor H, Ardekani MRS, Faramarzi MA, Hadjiakhoondi A& Khanavi M (2013). Cytotoxic evaluation of Melia azedarach in comparison with, Azadirachta indica and its phytochemical investigation. *DARU Journal of Pharmaceutical Sciences*, 21:37.doi: 10.1186/2008-2231-21-37.

Jain SK & Tarafder CR (1970). Medicinal plant-lore of the santals.*Economic Botany* 24: 241-278

Jaiswal BS (2012). *Solanum torvum*: a review of its traditional uses, phytochemistry and pharmacology. *International Journal of Pharma and Bio Sciences* 3: 104 – 111.

Jolad SD, Lantz RC, Solyom AM, Chen GJ, Bates RB & Timmermann BN (2004). Fresh organically grown ginger (*Zingiber officinale*): composition and effects on LPS-induced PGE2 production. *Photochemistry*. 65:1937-1954.

Kharkongor P & Joseph J (1981).Folklore medicobotany of rural Khasi and Jaintia tribes in Meghalaya,pp. 124-136. In: Jain SK (Ed) *Glimpses of Indian Ethnobotany*, Oxford and IBH New Delhi.

Kanjilal UN, Kanjilal PC, Das A, De RN& Bor NL (1934-1940). *Flora of Assam*. Vols 1-5. Government Press,Shillong.

Kayang H, Kharbuli B, Myrboh B & Syiem D (2005). Medicinal plants of Khasi hills of Meghalaya, India, pp. 75-80. In: III WOCMAP Congress on Medicinal and Aromatic Plants-Volume 1: *Bioprospecting and Ethnopharmacology*.10.17660/ActaHortic.2005.675.9.

Kayang H (2007).Tribal knowledge on wild edible plants of Meghalaya, Northeast India. *IndianJournal of Traditional Knowledge* 6:177-181

Kharjana PJ& De A (2015). A study of indigenous knowledge on conservation and use of some medicinal plants in Umthlong village, West Khasi Hills District,Meghalaya, India.*Journal of Science and Environment Today* 1:57-64

Joseph J& Kharkongor P (1981). A preliminary ethnobotanical survey in the Khasi and Jaintia hills, Meghalaya,pp. 115-123. In: Jain SK (Ed) *Glimpses of Indian Ethnobotany*, Oxford and IBH New Delhi.

Lakadong NJ (2009). Assessment of endemism, rarity and conservation status of a few medicinal plant species of Meghalaya.D. Phil. Thesis, Department of Botany, North-Eastern Hill University, Shillong.

Laloo RC, Kharlukhi L, Jeeva S & Mishra BP (2006). Status of medicinal plants in the disturbed and the undisturbed sacred forests of Meghalaya, northeast India: population structure and regeneration efficacy of some important species. *Current Science* 90: 225-232.

Maikhuri RK & Gangwar AK (1993). Ethnobiological notes on the Khasi and Garo tribes of Meghalaya, Northeast India. *Economic Botany* 47:345. https://doi.org/10.1007/BF02907348

Mao AA, Hynniewta TM & Sanjappa M (2009). Plant wealth of Northeast India with reference to ethnobotany.*Indian Journal of Traditional Knowledge* 8:96-103

Mauri P & Pietta P (2000). Electrospray characterization of selected medicinal plant extracts. *Journal of Pharmaceutical and Biomedical Analysis* 23:61–8.

Medhi RP& Chakrabarti S (2009). Traditional knowledge of NE people on conservation of wild orchids. *Indian Journal of Traditional Knowledge* 8:11-16.

Ministry of Tribal Affairs (2013). Demographic Status of Scheduled Tribe Population of India.Government of India. https://tribal.nic.in/Statistics.aspx.

Mir AH, Upadhaya K& Choudhury H (2014). Diversity of endemic and threatened ethnomedicinal plant species in Meghalaya, North-East India. *International Research Journal of Environmental Sciences* 3: 64-78.

Neogi B, Prasad MNV & Rao RR (1989). Ethnobotany of Some Weeds of Khasi and Garo Hills, Meghalaya, Northeastern India.*Economic Botany* 43:471-479

Prajapati R, Kalariya M, Umbarkar R, Parmar S & Sheth N (2011).*Colocasia esculenta*: A potent indigenous plant. *International Journal of Nutrition, Pharmacology, Neurological Diseases* 1: 90-96.

Rao RR (1981a). Ethnobotanical studies on the flora of Meghalaya-Some interesting reports of herbal medicines,pp. 137-148.In: JainSK (Ed). *Glimpses of Indian Ethnobotany*, Oxford and IBH New Delhi.

Rao RR (1981b). Ethnobotany of Meghalaya: medicinal plants used by Khasi and Garo tribes. *Economic Botany* 35: 4-9.

Rao RR & Hajra PK (1986). Floristic diversity of eastern Himalaya-in a conservation perspective.*Proceedings of the National Academy of Sciences* (*Animal sciences/plant sciences*) 103-125.

Seal T, Pillai B & Chaudhuri K (2016). Identification and quantification of phenolic acids by HPLC, in three wild edible plants viz. *Viburnum foetidum*, *Houttuynia cordata* and *Perilla ocimoides* collected from North-Eastern Region in India. *International Journal of Current Pharmaceutical Review and Research* 7: 267-274

Sen S, Pathak SK & Suiam ML (2016). Weed flora of tea plantations of Ri-Bhoi District of Meghalaya, India with a glimpse on its ethnobiological value. *World Scientific News* 56: 82-96.

Singh B & Borthakur SK (2011). Wild medicinal plants used by tribal communities of Meghalaya. *Journal of Economic and Taxonomy Botany* 35: 331-339

Taylor JLS, Rabe T, McGaw LJ, Jäger AK & Van Staden J(2001). Towards the scientific validation of traditional medicinal plants. *Plant Growth Regulation* 34: 23-37.

Upadhaya K, Pandey HN, Law PS & Tripathi RS (2004). Plants of ethnobotanical importance in the sacred groves of Jaintia hills of Meghalaya. *Indian Forester* 131: 819-828.

Upadhaya K, Choudhury H & Odyuo N (2012). Floristic diversity of Northeast India and its conservation,pp 61-71.In: Mandal R.K. (Ed.), *International Environmental Economics: Biodiversity and Ecology*. Discovery Publishing House Pvt. Ltd. New Delhi.

Upadhaya K (2015). Structure and floristic composition of subtropical broad-leaved humid forest of Cherapunjee in Meghalaya, Northeast India. *Journal of Biodiversity Management and Forestry* 4:2-8.

Welz A N, Emberger-Klein A & Menrad K (2018). Why people use herbal medicine: insights from a focus-group study in Germany. *BMC Complementary and Alternative Medicine* 18(1): 92.doi.org/10.1186/s12906-018-2160-6

WHO (2004). *Guidelines on Good Agricultural and Collection Practices (GACP) for Medicinal Plants.* World Health Organization, Geneva.

IARC (International Agency for Research on Cancer) & WHO (World Health Organization), Working Group on the Evaluation of Carcinogenic Risks to Humans, (2002). Some traditional herbal medicines, some mycotoxins, naphthalene and styrene (No. 82).World Health Organization.

Yousaf Z, Wang Y & Elias B (2013). Phytochemistry and pharmacological studies on *Solanum torvum* Swartz. *Journal of Applied Pharmaceutical Science* 3(04):152-160.

Yuan H, Ma Q, Ye L & Piao G (2016). The traditional medicine and modern medicine from natural products. *Molecules* 21(5): pp559.doi.org/10.3390/molecules 21050559.

3

Assessing Ethnobotanical Value and Threat Status of *Hodgsonia heteroclita* (Roxb.) Hook.f. & Thomson, A Lesser Known Liana Species of Sikkim Himalaya

Arun Chettri

Abstract

The paper deals with taxonomy, ethnobotany and threat status of lesser known liana species Hodgsonia heteroclita *(Roxb.) Hook.f. & Thomson. The local people use in terms of foods and rites and its role in ecosystem have been reported. Local people used the seeds of liana as food because of high nutritive value. Threat assessment has been following IUCN version 3.1 to evaluate the threat status of the species in Sikkim Himalaya. Out of the five criteria, data on two criteria could be collected and was used to evaluate the threat category of the species. The criteria used for threat assessment were (a) reduction in population and (b) geographical range in the form of either B1 (extent of occurrence) or B2 (area of occupancy) or both. Considering the population status and geographical range of the species, it has been classified as "Least Concern" (A1acd; B2ab(ii,iv)).*

Keywords: Ethnic uses, *Hodgsonia heteroclita*, Sikkim Himalaya, Ecosystem

Arun Chettri (✉)
Department of Botany, Sikkim University, 6th Mile, Samdur, PO Tadong- 737102 Gangtok, Sikkim India

✉*Corresponding author email: achettri01@cus.ac.in*

Plants for Novel Drug Molecules: Ethnobotany to Ethnopharmacology
Bikarma Singh & Yash Pal Sharma (eds.), (pp. 85-90)

Email: *info@nipabooks.com* Web: *www.nipabooks.com*

1. Introduction

Hodgsonia heteroclita Hook.f. & Thomson belonging to family Cucurbitaceae is a wild edible liana that grows in the tropical and subtropical forest between 250-1500 m above sea level in the Asian subcontinent. The species is locally known by the name of Kanan champa lahara, Gheuphal, Ghiuphul (*Nepali*) and Kurehio (*Lepcha*). This species was identified as a promising oil yielding cucurbit by Chinese scientists in the late 1960s due to high oil contents in their seeds. The dried leaves have medicinal properties. The decoction from boiled leaves are taken orally during nose complaint and fevers (Semwal et al. 2014) The species is believed to have its origin in north-east India, China (Western Yunan) and Malaysia and is found along southern temperate Asia to tropical Asia (Hooker 1855, Zeven and Zhukovsky 1975). Although, species has been playing important role in the life of local people of Sikkim, no record has been done about its ethnic uses and associated traditional knowledge. The paper in essence presents the taxonomy, threat status and ethnic uses of the species in the Sikkim region.

2. Diversity and Plant Enumeration

Cucurbitaceae is a moderately large family of flowering plant with nearly 100 genera and more than 750 species (Ajuru & Okoli 2013). *Hodgsonia heteroclita* (Fig.1.1) is a perennial liana, woody large climbers. Stems 20-30m in girth, angular and branches stout, leaves coriaceous, suborbicular in outline, 20-32 cm long, 25-42 cm broad, lobes oblong, apiculate, base cordate glabrous on both surfaces, lobed to middle. Tendrils robust, 2–5-fid, plants dioecious. Male flowers minutely brown tomentose, calyx tube 7-10 cm, lobes 5mm, calyx and bracts densely studded with glands; corolla 5cm, yellow outside, white within, very long fimbriate upto 10 cm, stamens 3; filaments inconspicuous; anthers connate, linear, 1-celled, 2-celled; anther cells conduplicate. Female flowers on peduncle 3-5 cm, calyx and corolla as in male flowers; ovary globose, 1-locular; ovules 12, horizontal; placenta 3, parietal with pairs of ovules attached on each side; style long; stigma 3-lobed; lobes 2-fid, exserted. Fruit large, 7-15 cm long, 7-15 cm diameter, brown tomentosse, 10-12 grooved (Chen 1995, Shiu-Ying 1964). Perfect seeds usually 6, each having a rudimentary or barren seed attached to its side and seeds is large, flat, ellipsoid.

Phenology: Flowering: February-March, Fruiting: September-December.

Geographical distribution: Indian Himalaya: Assam, Arunachal Pradesh, Meghalaya, Manipur, Mizoram, Nagaland, Tripura, West Bengal (Darjeeling) and Sikkim; 9th mile, Namcheybong, Rhenock, 32 No.(East Sikkim.), Dzongu (North Sikkim). Global- Bangladesh, China, India, Malaysia and Nepal.

Specimens examined: Sikkim, 9th mile, Radang on 16/07/2014, A.R Subba (SU).

Forest type in which the species occur in Sikkim Himalayas: Subtropical forest in sloppy rocky land.

Habitat characteristics: A common woody canopy forming liana in the subtropical zone of Sikkim Himalaya

Successional status: Mid-successional species.

Functional group: Liana.

Associated species: It is usually found in association with tree species such as *Schima wallichi* (DC.) Korth**.,** *Castonopsis indica* (Roxb.) Miq., *Duabanga grandiflora* (Roxb. ex DC.) Walp., *Ostodes paniculatus* Blume, *Terminalia myriocarpa* Van Heurck & Müll. Arg., *Albizia lebbeck* (L.) Benth., *Ficus religiosa* L., and bamboo species.

Threat status: Least Concern

*Ethnobotanical Importance***:** Seeds are edible to humans, and fruits are edible by monkey, squirrel, rodents and birds.

Fig. 1.1 a-c Plant with fruits in its natural habitat **d)** Leaves, **e)** Flowers and **f)** Seed

Less known ethnic uses: Lepcha tribe of Sikkim used it as rituals plant for performing their traditional rites. An important issues in the management of this species is the processing of the fruits and seeds, respectively, for its utilization as food. The seed are embedded in a stone structure "pyrene" which make extraction of the seeds laborious, time consuming and thus make it expensive (De Wilde WJJO & Duyfjes 2001, Shiu-Ying 1964, Senwal et al. 2014). It is usually grown in the dry sloppy rocky areas.

3. Study Area and Methods

The study was conducted in east, north (Dzongu) and south districts of Sikkim Himalaya during the year 2015-2016. The species was found in subtropical forest at an elevation of 250-1500m above sea level. The information about its ethnic uses and its importance was collected through direct interviewing the traditional folk healers, local people of villages around the areas of occurrence of species. The main focus of present study was to document the local existing knowledge associated with the species and to record local name of plant and plant parts used for various purposes. The herbarium and seed specimens are deposited in the Dept. of Botany, Sikkim University.

4. Results and Discussion

4.1. Habitat and Associated Species

The preferred habitats of species are dry rocky sloppy areas in the subtropical region of Sikkim Himalaya. It usually forms association with other tree species like *Castanopsis indica, Schima wallichi*, etc. It is locally known as "Kurehio" in *Lepcha* which mean favourite food for birds and "Gheuphal" (*Nepali*) meaning butter fruit, because it tastes as butter. The population of the species is declining day by day due to seeds collection and habitat fragmentation by human interference like slope cutting, road construction etc. Therefore it needs immediate conservation of its habitat.

4.2. Ethnobotanical Uses

Seeds are used as wild edible food for making pickle, curry or taking it directly as dry fruits as it contains lots of oil. Wild animals like monkey, squirrels, other rodents and birds are also dependant on this species for their food. The fruit have potential to work as important food during starvation and natural disasters. Local tribal people, especially *Lepcha* used it for religious ritual. People collects seeds and fruits from forest areas only and they don't cultivate in their private land because *Hodgsonia heteroclita* is woody climber and it needs extra support of tree as host. Moreover, the land is limited per individuals and lianas are not good for under growing shrubs, herbs and crop plant.

4.3. Threat Assessment

IUCN (2001, 2012), version 3.1 has been followed to evaluate the threat status of the species in Sikkim Himalaya (downloaded from www.iucn.org/knowledge/publications doc/publications/). Out of the five criteria, data on two criteria could be collected and was used to evaluate the threat category of the species. The two criteria used for threat assessment are (1) reduction in population, and (2) geographical range in the form of either B1 (extent of occurrence) or B2 (area of occupancy) or both. Considering the population status and geographical range of the species (Table 1) it has been classified as "Least Concern" (A1acd; B2ab(ii,iv).

Table 1: Population data of *Hodgsonia heteroclita* used for classification of threatened category under IUCN, version 3.1

A. Population Reduction	A1. <50% over 3 generation	a. Direct observation : many occurrencec, quality of habitat: anthropogenic disturbances actual or potential levels of exploitation: extraction of seeds, leaves for ritual purpose
B. Geographic range	B2. Area of occupancy (>2,000 km^2)	a. Known to exist at more than 10 locations b. Continue decline (ii) area of occupancy >2,000 km2 (iv) number of locations or subpopulation >10 locations

5. Future Direction

It is a prospective liana species for mass cultivation in the wasteland, rocky terrain and regenerating forests habitat. Being a lesser known species, its nutritional potential has not been exploited at par with other cultivated crops. Therefore, in future local people in the region may be encouraged to cultivate this plant with desirable application of agro-technique. Since, whole shoot of the plant are edible or medicinal, this may be considered as prospective species for cultivation to supplement the nutritional requirement of the local people.

6. Conclusion

Hodgsonia heteroclita being a locally important species in Sikkim Himalaya needs to be planted in large scale. It has potential to be high oil yielding underutilized species in the region so that socio-economic condition of local people could be uplifted by addition of scientific value. Valued for its edible seeds (as nuts) and important as ethnic uses, species may be considered as a prospective species for cultivation. It also support lots of rodents, monkes and birds in terms of food, so it needs to be cultivated and planted more in the forest areas for making balance in ecosystem in terms of food chain. It is being noticed that the population *Hodgsonia heteroclita* is reducing largely in the state due to seeds collection and habitat fragmentation by human interference like slope cutting, road construction etc, therefore needs immediate conservation of its habitat .

Abbreviation Used

IUCN-International Union for Conservation of Nature; SU-Sikkim University.

Conflict of Interest

Author does not have conflict of interest.

Acknowledgements

Author would like to thank Department of Botany, Sikkim University for providing necessary help in the study of the species. The authors are local people for sharing their valuable knowledge on the use and importance of species.

References

Ajuru M G & Okoli BE (2013).The Morphological Characterization of the Melon Species in the Family Cucurbitaceae Juss., and their Utilization in Nigeria. International Journal of Modern Botany 3(2): 15-19 DOI: 10.5923/j.ijmb.20130302.01

Chen SK (1995) *Hodgsonia.* In: C.Y. Wu, C. Chen & S.K. Chen (eds.). Flora Yunnanica. 6: 376–378. Science Press, Beijing.

De Wilde WJJO & BEE Duyfjes (2001). Taxonomy of *Hodgsonia* (Cucurbitaceae), with a note on the ovules and seeds. *Blumea* 46: 169–179.

Grierson AJG & Long DG (1991). Flora of Bhutan 2, 1: 127–128. Royal Botanic Garden, Edinburgh.

Hara H (1963). Spring flora of Sikkim Himalaya: 1– 169. Hoikusha Publishing Co, Ltd, Osaka.

Hooker JD (1855). Illustrations of Himalayan plants: t. 1–3. L. Reeve, London.

Hu Shiu-Ying (1964). The economic botany of *Hodgsonia. Economic Botany* 18: 167-179.

Semwal DP, Bhatt KC, Bhandari DC, Panwar NS (2014). Anote on the distribution, ethnobotany and economic potential of Hodgsonia heteroclite (Roxb.) Hook. f. & Thoms. in North-eastern India. *Indian Journal of Natural Products and Resources* 5 (1): 88-91.

Senwal DP, Bhatt KC, Bhandare DC & Panwar N S (2014). A note on distribution, Ethnobotany and economic potential of *Hodgsonia heteroclita* (Roxb.) Hook.f. & Thoms in North-Eastern India. *Indian Journal of Natural Products and Resources* 5 (1): 88-91.

Zeven AC & Zhukovsky PM (1975). Dictionary of cultivated plants and their centres of diversity, centres for agricultural publishing and documentation, Wageningen, 76.

4

Tribal Knowledge on Ethnobotanical Plants of Uttarakhand Himalaya

Anjali Saini and Bikarma Singh

Abstract

Ethnobotany is the area of study which include direct interaction of human and plants. This communication compiles and evaluates 301 ethnobotanically important medicinal plants used by ethnic tribes for the treatment of different ailments living in Uttarakhand Himalaya. The state is a home of five major tribes, viz., *Bhoxa, Bhotiya, Tharu, Raji, and Jaunsari. These tribal people have age-old relations with plants, and their indigenous knowledge passed from one generation to another. In this communication, a total of 90 plant families were reviewed and recorded that Asteraceae, Lamiaceae, Rosaceae, Fabaceae, Amaranthaceae Apiaceae, Combretaceae, Euphorbiaceae, Anacardiaceae and Malvaceae members are frequently used as medicinal plants. Total 68 roots of different species, followed by leaves of 55 species, whole plants of 29 species, fruits of 26 species and 28 types of seeds of different plants families used by locals for the treatment of various diseases such as fever, cold, cough, diabetes, skin problems, digestive problems and other associated ailments. The mode of application of herbal formulations varies from tribe to tribe. Herbaceous communities belonging to 160 species were the frequently used plant groups followed by trees, shrubs, and lianas. The decoction was recorded as the most frequently used method of preparation. Medicinal plants sold in the market was not primary for local people rather herbal plants locally*

Anjali Saini
Plant Sciences (Biodiversity and Applied Botany Division), CSIR-Indian Institute of Integrative Medicine, Jammu 180001, Jammu and Kashmir, India

Bikarma Singh(✉)
Botanical Garden, CSIR-National Botanical Research Institute, Lucknow-226001 Uttar Pradesh, India

**Academy of Scientific and Innovative Research (AcSIR), Ghaziabad-201002, Uttar Pradesh India*

✉*Corresponding author email: drbikarma.singh@nbri.res.in*

Plants for Novel Drug Molecules: Ethnobotany to Ethnopharmacology
Bikarma Singh & Yash Pal Sharma (eds.), (pp. 91-140)

Email: *info@nipabooks.com* Web: *www.nipabooks.com*

available used as food, spice and beverages preparations. Therefore, there is a need for proper documentation and conservation planting of these unique medicinal plants growing in the Himalayas.

Keywords: Medicinal plants, Tribal communities, Treatment of diseases, Uttarakhand Himalaya, Conservation

1. Introduction

Ethnobotany refers to the study of relation between human and plants (Harshburger 1896). The indigenous dependency of human beings on plants for their livelihoods initially started by the old generation sometimes 10,000 years before (Bhatt et al. 2009). Ethnobotanical research investigate the knowledge on cultural interaction of people with plants, and the ethnobiological information for bringing new areas in the field of research. Ecological monitoring has received enormous interest in resource management (Berkes et al. 2000, Huntingto 2000). Ethnobiological knowledge also finds the way local people use flora for different purposes and the people incorporate plants into their tradition and religion. In many developing countries, most of the population depends on ethnomedicinal knowledge to cure diseases due to poor staffing and much expensiveness of drugs (Joshi and Pant 2012). Many agencies like World Wildlife Fund (WWF) and UNESCO as a part of their people and plant initiatives have enhanced the research on ethnobotanical knowledge and merged perceptions and practices of people in resource management at the local level (Cunningham 2001). Ethnomedicine is used for maintaining health and curing diseases based on people's faith, skill, traditional knowledge, methods and practices (Phondani et al. 2010). Due to rapid socio-economic, environmental, and technological changes the knowledge of ethnomedicinal plants are vanishing (Sharma et al. 2012, Gunderson and Holling 2002). Therefore, the knowledge of ethnomedicinal plants must be documented and preserved through systematic studies before extinction. The main attention are how used, managed and perceived for mankind including food, medicine, rituals and social life (Sharma et al. 2011).

The relationship between human and plants is not limited to the use of food, clothing and shelter but it also used for religious ceremonies, ornamentation and much important for healthcare. The connection of people with plants in earlier times evidenced as they created representations of plants, drawing them on stone or molding them in clay. Such pictures grant modern ethnobotanists with evidence related plant origins and functional as tangible indicators of the importance of these people attached to plants. From old times, humankind has used plants in attempting to cure diseases and related physical suffering. Ancient people have some information about medicinal plants, gained as a result of assessment and error. India is habitat to many languages, cultures, and values which inturns contributed to the enormous diversity of traditional information and practices of the people, which include the use of medicinal plants to cure diseases and their possible causative agents. Hence, plants have been used both in the prevention and treatment of a variety of diseases of humans and their animals from time immemorial. Abebe and Ayhehu (1993) reported that 80% of the Ethiopian inhabitants rely on traditional medicine for their well being. More than 95% of medicines are made from plants like many other developing countries. Most of the knowledge that traditional healers and local people had either vanished or transferred to generation by word of mouth. Limited research works have been done on medicinal plants used for curing animals with specific ailments.

The tribal and indigenous people of India have learned to live in most unfriendly environmental circumstances (Jain 1991). The most interesting feature regarding these people is that they live in areas that are enormously wealthy in biodiversity. India has one of the major concentrations of tribal communities in the world, almost 68 million tribal people belonging to 227 ethnic groups and 573 tribal communities living in different geographic locations within the country. According to the Indian constitution, *'Tribe'* means a group with the traditional territory, specific name, familiar language, strong family relations, association with tribe structure, tribal authority and rigid preference to religion and faith. Functional independence, homogeneity, an ancient mean of exploiting resources, economic backwardness, rich culture, ritual and slightest desire to alter are some of the other features dominated in tribes. Scheduled tribes (*Tharus, Buxas, Rajis* and *Bhotias*) are living in the Central Himalaya (Gaur et al. 2010). These tribal communities present an important degree of cultural. The northern part of India has an immense diversity of medicinal plants because of its distinct geography and ecological marginal conditions (Kumar et al. 2011). Therefore, present documentation mainly focused these tribes. This review paper focused at compiling and evaluating of published literature on traditional medicinal plants and our own collection from field.

2. Study Area and Data Collection

Several research studies with prominence on ethnobotany, ethnomedicine, and diversity of medicinal and aromatic plants conducted between 1979 to 2019 were consulted for the present study coupled with our field collection. Uttarakhand (latitude 28°43'N to 31°27'N; longitude 77°34' E to 81°02' E) is one of the floristically rich Himalayan state. The state is bounded on the northwest by Himachal Pradesh, on the southwest Uttar Pradesh state, on the east Nepal country, and on the north Tibaten hills, covering an area of 53,483 Km^2. The entire state is covered by 34,662 km^2 of forest area, which constitutes 64.79% of its geographic area. By lawful status, reserved forests constitute 71.08%, protected forest 28.51%, and unclassed forests 0.41% of the total area. Major forest types occurring in the state are classified as tropical moist deciduous, tropical dry deciduous, and sub-tropical pine, Himalayan moist temperate, Himalayan dry temperate, sub alpine, and alpine forest. Forests are largely distributed throughout the State with conifers and sal are the main forest formations. The current study was carried out in 13 districts in the Uttarakhand state of India. Tharu, Bhoxa, Bhotiya, Van-Raji and Jaunsari are ethnic minority tribe of western Himalaya (Gour and Sharma 2011). They are custodian of traditional knowledge especially about wild flora and fauna which had helped them through generations to sustain themselves in deep forests (Sharma et al. 2013).

The dialectical relationship between original information and practices shape the ecosystem and affects the essential plant populations. By introducing local knowledge and use in the process of scientific study, new hypotheses for the sustainable preservation of the resources can be developed. Local knowledge and use have to be studied to extend suitable management measures that construct on both scientific and indigenous knowledge. Due to varying opinions of the local people and rising influence of global commercialization and socio-economic conversion, native knowledge on plant resource use is constantly thinning. Many useful plant resources are diminishing at an alarming rate because of many factors such as lack of organized sustainable and scientifically monitored cultivation, proper harvesting techniques, proper management techniques, and lack of knowledge of social factors, Moreover, the local information on the use of less known plants are also rapidly decreasing. This study, therefore, reviews the local knowledge and use of plant resources of Uttarakhand Himalaya along altitudinal and longitudinal gradient.

3. People and Culture

The present study paying attention to the habitat, culture, economy and ethnomedicinal uses of plants used by the tribals of Uttarakhand. Even though, a lot of studies have been done on the ethnomedicinal uses of the plants described from the various parts of Uttarakhand (Gupta et al. 1995, Singh et al. 1997, Gaur 1999, Kala 2005, Joshi et al. 2010). The northernmost part of Uttarakhand (Kumaun and Garhwal) is known as *"Bhot"* region, consisting of sub-alpine and alpine zones bordering Tibet. It is supposed that *'Bhotias'* are a trans human society of semi-mongoloid people of Tibetan origin (Fuchs 1982, Kala 2002, Ratha et al. 2015). They show similar racial and cultural resemblance to the tibetans. The Bhotia region is called *Bod or Bhot* which is considered as a corrupt form of *Bod*, and it means *"Follower of Buddhism"*. These people in the Dharchula area of Pithoragarh district are known as *Shauka*. The eight major Bhotia groups in the state (*Johari, Juthora, Darmi, Chudans, Byansi, Marccha, Tolcha* and *Jad*) are distributed over eight key river valleys known as Johar, Darma, Byans, Chaudans (Pithoragarh District); Mana, Niti (Chamoli District); Nilang and Jadung (Uttarkashi District). The Bhoxa tribe is one of the main communities spread in different districts of the Sub-Himalayan belt of Uttarakhand (Sharma and Painuli 2011, Gairola et al. 2013). The mountainous Jaunsar-Bawar region of Dehradun district in Uttarakhand state is the abode of the Jaunsari people. They are particularly concentrated through chakrata tehsil with smaller numbers in the surrounding tehsil. The population of Jaunsari are approximately 96,995 (Census 2011). Among the local societies of the Central Himalaya, the Raji is one of the less developed and smallest tribal society (1,300 persons in Uttarakhand) inhabiting forested pockets in Champawat, Pithoragarh and Udham Singh Nagar districts of Uttarakhand. Eastern Himalaya is named as 'Dooars'. It is an ethnobotanically rich region due to the presence of 'Tharu' tribal communities since long back. About 1, 69,209 people of Tharu communities living in the Uttarakhand.

4. Results and Discussion

4.1. Medicinal Plant Diversity

Bhotiya, Bhoxa, Jaunsari, Raji and Tharu community largely depends on traditional healthcare practices for their well-being. A total of 301 plant species belonging to 90 families were documented to be used by these tribe community in various healthcare remedies for different diseases of human care (Table 1). This includes 12 species of lianas, 65 shrubs, 160 herbs and 45 species of trees (Fig.1). Maximum species belong to family

Asteraceae (19 species), Lamiaceae (16 species), Euphorbiaceae (15 species), Ranunculaceae (11 species), Polygonaceae (10 species), Fabaceae, Liliaceae, Rosaceae (9 species each), Poaceae (8 species), Combretaceae (7 species), Amaranthaceae, Zingi beraceae, Anacardiaceae, Moraceae (6 species each), Acanthaceae, Araceae, Caesalpiniaceae, Solanaceae, Convolvulaceae (5 species each) followed by family Ericaceae, Mimosaceae, Rutaceae, Asparagaceae, Saxifragaceae, Valerianaceae (4 species each), while leftover 61 families were represented by 1-3 species. However, thirty-six percent of documented medicinal plants are in a stable stage and are easily available in the region.

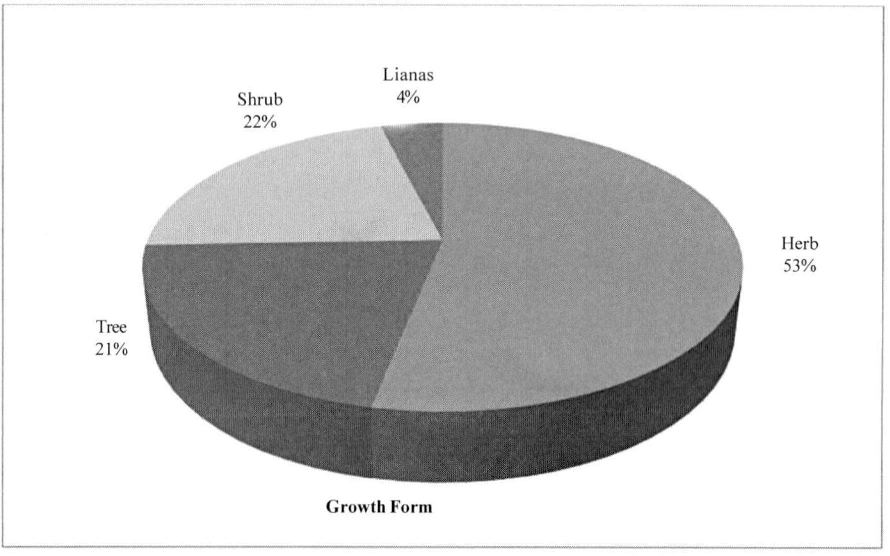

Fig. 1: The growth-form of plants used as ethnomedicine

4.2. Plant Parts Used and Disease Cure

The local communities use underground parts such as roots, rhizomes, and tubers of about 30% of documented species. In terms of above-ground parts, the people use leaves including twigs of 18% species, flower 5% species, bark including stem outgrowth 6.5% species, seeds 10% species, fruits 10% species and whole plant 10% species in different herbal preparations as medicine. A single plant is used in the preparation of more than one remedy and treatment of many ailments. In terms of plant species used for a particular ailment, a maximum of recorded species are used in relieving pain including bodyache, joint pain, abdominal pain and headache followed by species used in cuts and wounds, and indigestion respectively. There are several species used in treatment of diabetes, cold and cough, dysentery and diarrhea, fever and jaundice. In terms of traditional knowledge on medicinal plant use among the different age groups, the elderly respondents have more information in comparison to the younger generation. Moreover, the respondent having limited education (upto primary) or illiterate category are on top in comparison to the individual having higher education and also married respondents possess more knowledge in comparison to unmarried individuals.

4.3. Tribes and Uses of Medicinal Plants

The present investigation reveals 301 plant species used as medicine in Uttarakhand. Table 1 presented with botanical name, local name, family, tribe using the plant, disease cured and made of usage. The plant species are arranged alphabetically with their scientific name and their use. The present review shows that traditional medicines are quite common among all tribes in Uttarakhand Himalaya. This study indicated tribal people from different parts of Uttarakhand still depend on medicinal plants for their primary healthcare. The knowledge of these medicinal plants acquire from the tribal communities is valuable for researches in the field of pharmacology, clinical and toxicological studies. It is important to recognize the role of traditional health care practices for future sustainable development. Ethnomedicinal study of tribal people can fetch out many more competent herbal drugs. Therefore efforts should be made for the documentation of such precious knowledge for future perspectives. At present, many of the vital plant species are on the edge of reduction, therefore, such type of studies will be essential green future perspective. Inspection of the literature shows part of the Central Himalayan tract inhabited by all five tribes are still less studied. Therefore, tireless hard work has been made to record t knowledge of plants used as herbal remedies for the treatment of diverse diseases in the state of Uttarakhand.

Table 1: Ethnomedicinal plants of Uttarakhand Himalaya used as herbal medicine

Botanical Name	Local Name	Family	Tribe	Used Part	Disease	Mode of use
Abrus precatorius L.	Ratti	Fabaceae	Jaunsari	Leaf, Seed, Root	Fever, Asthma, Chest pain	Leaves, seed and root decoction drunk in fever, asthma, chest pain.
Acacia catechu Senegalia catechu (L.f.) P.J.H.Hurter & Mabb	Khair	Mimosaceae	Bhoxa, Bhotiya	Leaf, Bark	Blood dysentery, Diarrhoea, Indigestion, Fatigue	Leaf juice (1-2 teaspoon), is taken with milk two times a day as remedy for blood dysentery. Decoction of bark (2 teaspoon) given for the cure of diarrhoea. Bark extract used in indigestion and fatigue.
Acacia nilotica (L.) Delile	Babur	Mimosaceae	Bhoxa	Leaf, Bark	Diarrhoea, Dysentery	Leaves juice is mixed with 4-5 spoon of curd and taken orally to cure diarrhoea. 2-3g powder of bark taken to get relief from dysentery.
Achyranthes aspera L.	Latjiri, Apamarg	Amaranthaceae	Bhoxa, Bhotiya, Jaunsari	Root, Leaf, Whole plant	Ring worm, Eczema, Muscular Cramps, Malarial Fever, Dysentery	Root paste applied on ring worm, 100g leave powder mixed with 20g ghee used on eczema. Root decoction applied on muscular cramps. Root extract taken in malaria fever. Whole plant decoction used to care dysentery.
Achyranthus bidentata Blume	Golda, Apamarga	Amaranthaceae	Jaunsari, Raji	Root,Leaf	Snakebite, Dogbites, Mouth blisters	Root past applied on wounds and snake bite, leaf powder tablet (5g) with jiggery is given during dog bites for 3 days and root paste is also used on mouth blisters.

Aconitum lethale Griff.	Mithabish, Mitha	Ranunculaceae	Bhotiya	Root	Leprosy, Arthritis, Neuralgia, Hair Fall, Rheumatism, Musculoskeletal	Root past used to cure on leprosy. Rhizome oil applied to get cured from arthritis, sciatica and neuralgia, Tuber used on skin for rheumatism, musculoskeletal disorder
Aconitum heterophyllum Wall. ex. Royle	Atis	Ranunculaceae	Bhotiya	Root	Headache, Dyspepsia, Stomachache, Bronchitis, Fever, Arthritis, Wound	Root paste applied on headache, 50g tuber boil with 400ml water given in dysentry, 5g root powder mixed with 1 cup water in stomachache,1-2g root powder given in bronchitis, fever, arthritis,
Aconitum ferox (Buehli) Mukherjee	Meethabish	Ranunculaceae	Jaunsari	Rhizome	Rheumatism, Neuralgia, Paralysis, Puerperal Fever	Rhizome paste used in rheumatism, neuralgia, Paralysis, Puerperral fever.
Acorus calamus L.	Bach	Acoraceae	Bhotiya, Jaunsari	Root, Leaf, Flower	Cough, Sore Throat, Sprain region, Fever coryza	Dried root powder with honey used in cough and sore throat. Ground dry roots boiled with mustard oil is used on sprain region. Leaves and flower decoction is given to cured cough and fever.
Adhantoda vasia Justicia adhatoda L.	Basinga	Acanthaceae	Jaunsari,	Leaf, Raji	Whooping cough, Flower,	Leaf juice and root paste is Skin disease, given to get relief from headache, dysentery.
				Root	Headache, Dysentery, Cold, Bronchitis	Mixture of flower and leaves is given in common cold and bronchitis.

Adhantoda zeylanica Medik	Bansa, Banshoi	Acanthaceae	Bhoxa, Jaunsari	Leaf	Diarrhoea, Dysentery	The mixture of leaves 1-2 teaspoonful is given to cure dysentery and diarrhaea.
Adiantum sp.	Bansa, Banshoi	Acanthaceae	Bhotiya	Leaf	Burn Scar	Fresh 50g leaf paste used on effected part for 3-4 days on a regular basis to cure burn scar.
Aegle marmelos Correa	Bel	Rutaceae	Jaunsari	Leaf	Diarrhoea	Leaf paste and fruit decoction given in diarrhoea.
Aesculus indica (Wall. ex Cambess.) Hook.	Pangar	Hippocas-tanaceae	Bhotiya, Raji	Fruit	Dermatitis, Rheumatic pain	Matured fruit coat is crushed and made into a paste for dermatits. Warm paste of fruit is used on joints during rheumatic pain.
Ageratum conyzoids L.	Jangli pudina	Asteraceae	Raji	Leaves	Wound	Leaves juice is used to avoid bleeding and pus formation.
Aesculus aptera DC.	Karu	Asteraceae	Jaunsari	Root	Inflammation, Anthelmintic	Root powdered is taken by mouth for treatment of body inflammation and as anthelmintic.
Ajuga integrifolia Buch.-Ham. ex D.Don	Neelkan-thior, Nilkhanti	Lamiaceae	Jaunsari, Raji	Leaves, Root	Ear problem, Malaria, Snake bite, Menstrual cycle, Acidity, Indigestion	Leaves juice is dropped into the ear cavity (4-5 drops 2 time a day for 3 days). Root extract is given in malaria, astringent, febrifuge. Leaf juice is applied on skin disease. Root extract is used as antidote of snake bite. Leaves paste is given in acidity and indigestion.
Ajuga brachystemon Maxim	Neelkanti	Lamiaceae	Jaunsari	Whole plant	Malaria	Plant extract given to cure malaria.
Albizia lebbeck (L.) Benth	Saris	Mimosaceae	Bhoxa	Bark	Diarrhoea	Powder of bark, about 3g with sugar is taken orally 2 time a day in diarrhoea.

Allium cepa L.	Pyaz	Liliaceae	Bhotiya	Bulb	Diarrhoea, Snake bite, Wound, Fever, bodyache	One bulb mixed with 5g salt with water drunk to cure for Diarrhoea. 30ml bulb juice is mixed with 30 ml mustard oil apply on snake bite and scorpion sting. Tuber paste with warm water given to cure fever and bodyache.
Allium humile Kunth	Pharan, Jambu, Koch	Liliaceae	Bhotiya	Leaf and Bulb	Asthma, Pectoral complaints	5g leaves and 10g kuth root powder fried in 2 spoon of ghee taken to cure asthma.
Allium sativum L.	Lahsun	Liliaceae	Bhotiya	Bulb	Paralysis, Carbuncles, Mascular pain, Earache	5 bulbs fried in 50 ml mustard oil applied on paralytic infected part before sleeping. Bulbs are fried in 30ml of mustard oil apply on muscular pain and earache.
Allium wallichii Kunth	Ladum, Jangli lahsun	Liliaceae	Bhotiya	Leaf	Indigestion, Wound, body warming	Dried leaves are given during indgestion. Leaf juice applied on wound. Dried leaves are consumed for body warming during winters
Allium stracheyi Baker.	Jambu	Liliaceae	Jaunsari	Leaf	Wound	Leaf juice mixture which used on wound.
Amomum subulatum Roxb.	Badi Elaichii	Zingrberaceae	Bhotiya	Seed	Stomachache, Gastric problems, Swelling	Raw crushed seeds are given in case of stomach ache. Paste of seeds is used to cure swelling.

Amaranthus cruentus L.	Choulai	Amaranthaceae	Jaunsari	Stem	Fungus infection	Stem used to cure fungal affected part of body during rainy season.
Angelica glauca Edgew	Choru, Taskar, Gandrayani	Apiaceae	Bhotiya	Root	Cold, Indigetion, Dyspepsia, Colic, Stomach pain	1g root and 1g leaf powder of basil mixed in a cup of tea for cold. 50g roots boiled in 200ml water and strain used for indigestion. Dried root chewed for dyspepsis. 1-3g root powder given along with luke warm water given for abdomal pain, vomiting. Root decoction is given in colic.
Anacyclus pyrethrum (L.) Lag	Akarkara	Asteraceae	Bhotiya	Root	Toothache, Cough	Root paste beneficial in toothache. Dried root powder with honey taken to cure cough.
Anemone obtusiloba D.Don	Ratanjot	Ranunculaceae	Jaunsari	Seed	Food poisoning	Seed decotion is given in food poisoning.
Anemonastrum polyanthes (D.Don) Holub	Kanchpool	Ranunculaceae	Jaunsari	Root	Diarrhea	Root decoction is given in diarrhea.
Anemone vitifolia Buch.-Ham.	Mudeela, Dudhia mohora	Ranunculaceae	Jaunsari	Root, Leaf	Ring worm, Eczema, Lactation problem	Leaf or root paste applied on ring worm and eczema effected part of body. Powder of root is mixed with wheat flour aken to increas lactation in women.
Anogeissus latifolia (Rox. ex DC.) Wall. ex Bedd.	Dhaudi	Combretaceae	Bhoxa	Stem bark	Dysentery, Diarrhoea	The mixture of stem bark is taken by mouth 2 times a day for dysentery and diarrhoea.
Argemone mexicana L.	Pili Katili	Papaveraceae	Bhoxa,	Seed oil, Leaf	Leprosy, Rheumatism, Dysentery	Seed oil is used externally on effected area of rheumatism. Leaf paste with turmeric applies in leprosy. Leaves powder 2g is taken orally 3 times a day with unboiled milk to get relief from dysentery.

Arnebia benthamii (Wall. ex G.Don) I.M. Johnst.	Latiari	Boraginaceae	Bhotiya	Leaf	Baldness, Wound healing, Blood Pressure, Skin disease	5g root crushed and mixed with 50ml mustard oil rubbed on head for baldness. Fresh leaves paste apply on wound. 1g root powder with warm water given for high blood presure. Dry root paste with mustard oil is used on skin for skin disease.
Arisaema flavum (Forssk.) Schott	Bang	Araceae	Raji	Rhizome	Wounds, Snake or Scorpion bite	Decoction of rhizome used to cure wound. Rhizome paste with water snake or Scorpion stung.
Arisaema jacquemontii Blume	Khaprya, Saperi mausi, Khyan bank	Araceae	Bhotiya, Jaunsari	Root, Fruit	Worm infection, Snake bite	Root powder is taken to cure worm infection. Fruit decoction given as snake bite antidote.
Arisaema propinquum Schoot	Meen	Araceae	Jaunsari	Root	Erysipelas, Scabies	Root paste is used externally on scabies.
Arisaema speciosum Mart	Bankh, Cober lily	Araceae	Bhotiya	Seed	Snake bite	Seed is rubbed on skin to make numbness of skin area prior to incision in snake bite.
Arisaema tortuosum (Wall.) Schott	Bankh, Chham-bosa, Bag-Mungri	Araceae	Bhotiya, Jaunsari	Fruit, Rhizome	Piles, Snake bite, Rheumatism, Breathing disorders	Fruit paste is prepared and applied on piles. Rhizome paste with water is applied on snake or scorpion bite place. Corm extract is taken as antidote to snake bite. Tuber powder is used in rheumatism and breathing disorders.

Artemisia nilgerica (C.B. Clarke) Pamp.	Pati, Kharia, Chhamboor, Chhamra	Asteraceae	Bhotiya, Jaunsari, Raji	Root, Leaf	Intestinal worm, Cuts, Wound, Malarial Fever,	100g fresh roots or leaves are kept overnight in cold water and taken water 5-6 day at morning empty stomach to cure intestinal worm (Wormicide). Leaves juice is appied on wound and cuts. Leaf paste applied in malaria fever, headache. Leaves decoction is given 3 times a day for 7 days in malaria fever.
Artemisia roxburghaina Wall. ex Besser	Chhambus	Asteraceae	Jaunsari	Root, Leaf	Haedache, Stomachache, Boils, Blisters	Paste of leaves is used to cure headache and cuts. Root extract is taken by in stomachache. Root paste is applied to boils blisters 2 times in a day for 5-7 days.
Asclepias curassavica L.	Kahwa	Asclepiadaceae	Jaunsari	Root	Constipation	Powdered root is taken during constipation 2 times a day for 5-7 days.
Asparagus adscendens Roxb.	Jhirni	Asparagaceae	Bhotiya, Jaunsari, Raji	Seed	Muscular pain, Sexual debility, Urinogenital disorder	Seeds are eaten and oil obtained used to cure muscular pain. Powder of dried tuberous root gives for sexual debility and urinogenital disordes.
Asparagus filicinus Buch-Hamvex D.Don	Kerua, Kureda	Asparagaceae	Bhotiya	Tuber	Polyurea, Galactogogue, Bone fracture	Tuber powder 3-6gm taken with luke worm milk in polyurea. 3-6 gm powder taken with milk as an effective galactogogue. Paste of tuber with maize powder is applied as plaster at fracture bone.
Asparagus racemosus Willd.	Sataver, Jhirana	Asparagaceae	Bhotiya	Root	Eplilepsy	Root powder is mixed with ghee to cure epilepsy.

Astragalus condolleanus Royle	Hlachhya, Rudanti	Fabaceae	Bhotiya, Jaunsari	Root	Joint pain, Skin disease, Tuberculosis	Root powder useful in joint pain and also used to purify blood Root decoction used to cure tuberculosis.
Barleria cristata L.	Kalabansa	Acanthaceae	Jaunsari, Raji	Whole plant	Cough, Bronchitis, Dermatitis	Plant ash with honey is given 3 times a day for cough and bronchitis. Leaf paste is applied externally in dermatitis.
Bauhivia vahlii Wight and Arnott	Malu	Caesalpiniaceae	Raji	Stem	Skin irruption	Stem bark paste is used to cure skin irruption.
Berberis aristata DC.	Chotr, Kingod, Kilmora	Berberidaceae	Bhotiya, Raji	Root	Eye infection, Indigestion	Root mixture made from 100gm of root with 250ml of water used to cure eye infection. Extract of fruit is given in indigestion and tiredness.
Berberis chitria Chotar	Kingore, Lindl.	Berberidaceae	Jaunsari	Fruit, Bark, Root	Jaundice, Eye disorder	
Berberis lycium Royle	Kasmal, Chatroi, Kirmor	Berberidaceae	Bhotiya, Jaunsari	Root, Fruit	Diabetes, Piles, Skin disease, Stomach pain, Eye inflamation	5gm of root is crushed, steeped in 250ml of water overnight and filterate is used to cure diabetes. 50g of root powder is mixed with 50ml of water to treat piles. Root decoction used to cure skin diseases. Fruit paste given 2 times a day for 3-4 days in case of stomach pain and constipation.
Betula alnoids Buch.-Ham. ex D.Don	Saur	Betulaceae	Bhotiya	Bark	Eye intection	Bark ash 100g is blend with 2 spoonfuls of ghee to make a paste used on the eye lid to cure eye disease.

Betula utillis D.Don	Bhojpatra, Bhuj, Bhojyuda	Betulaceae	Bhotiya	Resin	Cold, Cough, Asthma, Miscarriage	50gm resin of bhojpatra boiled in 250ml water and added 2 spoonful of ghee and half spoon of salt and drunk to cure cold and cough. 50gm resin of bhojpatra is crushed with 20 gm leaf powder of Chandra and fried in 2 spoonfuls apricot (*Prunus armeniaca* L.) oil and taken to treat asthma. 100gm resin and 50gm of kirol seed kernels combine make a paste mixed with honey and mixture given to pregnant women to prevent the miscarriage.
Bergenia ciliata Sternb.	Silphori, Dhyupati	Saxifragaceae	Bhotiya, Jaunsari	Root, Leaf	Paralysis, Piles, Dyspepsia, Urolithiasis, Kidney stone	Boiled leaves dried and used in paralysis. Roots are chewed for piles. 20ml root decoction in empty stomach used to cure dyspepsia. Root powder 1-3 g is used in urolithiasis. Leaf powder is used as tea against kidney stone.
Bergenia ligulata (Wall) Engl.	Pasher Bhed, Bhotiya Chai	Saxifragaceae	Bhotiya	Rhizome	Kidney stone, Cough, Cold, Indigestion, Urinary disorder	Dried rhizome is chewed to get rid of kidney stone. Leaf powder is used to cure cough, cold, indigestion, urinary disorder.
Bergenia stracheyi (Hook.f.& Thomson) Engler	Silphora, Gneepati	Saxifragaceae	Bhotiya	Root	Urinary and Kidney Trouble, Urinary calculi	100g fresh roots are washed and chewed to cure the urinary and kidney trouble. Root powder is useful in urinary calculi.
Bidens bipinnata L.	Arka-jhar	Asteraceae	Raji	Leaf	Skin irritation	Leaves crused and juice rubbed on itching feet to get relief from skin irritation.

Bistrota amplexicaulis Greene	Banamunda	Polygonaceae	Bhotiya	Root		
Boerhavia diffusa L.	Punarnava, Santi Ghass	Nyctaginaceae	Bhoxa, Raji, Jaunsai	Root, Leaf	Eye problem, Bronchitis, Asthma, Jaundice, Dysentery, Diarrhoea	Root decoction is used in jaundice, asthma, bronchitis. Leaf juice mixed with goat's milk is dropped in eyes to cure cataract problem. Root powder is taken against dysentery and diarrhoea.
Bombex ceiba L.	Semal, Sembar	Bombacaceae	Bhoxa, Raji	Calyx, Gum	Stomachache, Dysentery, Diarrhoea	Powder of calyx is given by mouth for stomachache and dysentery. Gum is used to cure diarrhoea.
Brassica compestris L.	Sarso	Brassicaceae	Bhotiya	oil	Measles	Oil used to cure measles.
Brassica juncea L.	Rai	Brassicaceae	Bhotiya	Vegetable	Anemia	Vegetable used for anemia
Bridelia retusa (L.) Spreng.	Khaja	Euphorbiaceae	Bhoxa	Stem bark	Rheumatism	Stem bark used externally to get ride of rheumatism.
Butea frondosa Roxb.	Dark	Fabaceae	Bhotiya	Flower	Urine infection	Flower is cooked and mixed with half of glass water to cure urine infection.
Butea monosperma (Lam.) Taub.	Dhak	Fabaceae	Bhotiya, Bhoxa	Seed, Flower, Stem bark	Constipation, Backache, Dysentery	Seed are burnt and given with milk to stop constipation. Decoction of flower is given to treat backache. Stem back juice is administered orally to cure dysentery.

Buchanania cochinchinensis (Lour.) Almeida	Chirongi	Anacardiaceae	Tharu	Seed oil	Skin disorders	Seed oil is used in treatment of skin disease.
Callicarpa macrophylla Vahl	Daya, Daiya	Verbenaceae	Jaunsari, Raji	Fruits, Leaves,	Rheumatism, Joint pain, Urinary trouble, Bilsters	Fruits and leaves are used in rheumatic pain. Leaves are heated and tied over affected joints and rheumatic pain. Fruits are eaten during urinary trouble. Fruit paste mixed with yoghurt and taken orally to cure mouth blisters.
Calotropis procera Br.	Aak	Asclepiadaceae	Jaunsari, Raji, Bhotiya	Leaf, Flower, Roots, Latex	Migraine, Headache, Fever, Cholera, Asthma, Abortifacient	Powder of 5g dried leaves mixed with jaggery given orally for 5 days to treat migraine. Leaves paste is applied to cure headache. Flower bud paste is given in fever. Latex, roots bark and flower is used in treatment of asthma. Powder of dried leaves mixed with jaggery taken orally before sunrise for 5 days to cure migraine. Juice of whole plant is taken orally for abortifacient.
Cannabis sativa L.	Bhang	Cannabaceae	Bhotiya	Leaf	Piles, Stomach pain, Stomach worms	Leaf paste is applied on piles. Crushed leaves are given to cure stomach pain.
Carduus nutans L.	Bushiand	Asteraceae	Jaunsari	Whole plant	Sprain, Inflamation, Boils	Plant paste is applied externally in sprain, inflamation and on boils for 3-7 days.
Capsicum annuum L.	Mirch	Solanaceae	Bhotiya	Fruit	Rabies, Snake and Scorpion Sting	Two fruits milled and powder used to cure rabies. Fruit milled and made powder to cure snake and scorpion bite.

Carum carvi L.	Kalajeera, Thoya, Bhotiya jeera	Apiaceae	Bhotiya, Jaunsari	Seed	Stomachache, Flatulence, Indigestion, Stone	20gm of Kalajeera seeds are mixed with 5gm of *Plantago ovata* Forssk. Crushed with 5gm black salt and 10 gm kuth root for stomachache. 1-2gm of fruit powder given orally with lukewarm water in abdominal pain dyspepsia, flatulence, indigetion and worm infection. Seeds are rosted ground and salt is mixed in them which is an effective formula for stone and stomach pain.
Cassia fistula L.	Amaltas	Fabaceae	Bhoxa, Raji	Leaf, Stem bark, Fruit	Rheumatism, Stomach worm, Dysentery, Diarrhoea	Paste of leaves applied externally on rheumatic organs. Stem bark extract given to cure stomach pain. Fruit pulp is used for the treatment of dysentery and diarrhoea.
Senna occidentalis L.	Chakunda	Fabaceae	Jaunsari	Leaf, Root, Fruit	Skin diseases, Cuts, Wounds, Bone fracture	A homogeneous ethanolic extract is prepared from the stem of *Cassia occidentalis* and used on wounds, cuts, skin itching, swelling and on fracture.
Senna tora (L.) Roxb.	Chakunda	Fabaceae	Jaunsari	Leaf, Root, Fruit	Cuts, Wounds, Bone fracture	Seeds and leaves firstly crushed and then used to treat skin disease such as leprosy, ringworm, itching and psoriasis and also for snakebites. Seeds are crushed into powdered form and taken orally with water for the constipation

Cedrus deodara (Roxb. ex D.Don) G.Don	Devdar	Pinaceae	Bhotiya	Bark oil, Wood oil	Stomach worm, Itching, Dermatitis, Boils, Blisters, Wound	Oil is used to remove stomach worm. Bark oil is apply on itching. Oil is mixed with *Rubus paniculatus* Sm. leaf powder and used externally to treat dermatitis. Wood oil is externally applied to bils and blisters 3 times a day for 3-7 days. The essential oil is used in joint pain, dendruff and used on wound filling of livestock.
Celosia argentea L.	Salera	Amaranthaceae	Bhoxa	Seed, Leaves	Diarrhoea, Dysentery,	Seed powder is given for diarrhoea. Leaves juice is taken orally for dysentery, 2 times a day for 2 days.
Centella asiatica (L.) Vrb.	Brahmi	Apiaceae	Botiya	Whole plant	Stomach worm, Leucorrhoea, Mental disorder, Epilepsy, Skin disease	Powder of whole plant is used in 1 cup of water for stomach worm. Juice is made by mixing 10gm of fresh leaves, 5gm root powder of kuth and 5gm of atis root powder, for leucorrhoea. Leaf powder is mixed with cow ghee against epilepsy. Leaf paste is prepared for application for mental disorder and given for skin disease.
Melanoseris macrorhiza (Royle) N.Kilian	Karatu	Asteraceae	Bhotiya	Leaf	Headache	Leaves juice is drunk to treat headache.
Cirsium verutum (D.Don) Spreng	Biskanara	Asteraceae	Bhotiya	Rhizome,	Tuberculosis, Rheumatism	100gm root boiled with 500ml water is prepared as a 50ml decoction mixed with deodar oil externally applied to treat rheumatism. 100gm dried root is milled with 20gm kuth root and 20gm of dried chandra

						leaves, this mixture dipped into 500ml water for 2-3 hours and then taken for tuberculosis.
Cirsium wallichii DC.	Nache kulpu, Mancheu	Asteraceae	Bhotiya	Root, Fruit	Boils, Headache	Root paste applied externally on boils. Infusion of mancheu and *Pyrus pashia* Buch. & Ham to cure headache.
Cissampelos pareira L.	Jaljiamini	Menisperma-ceae	Bhoxa, Jaunsari, Raji	Leaf, Root	Rheumatism, Diabetes	Leaves paste is used externally on rheumatic pain. Aqueous extract of leaves is given to treat dysentery. Root powder with water is given once a day for 40 days in diabetes.
Clematis orientalis L.	Latta	Ranunculaceae	Raji	Leaf	Migraine, Headache, Fever, Cholera, Asthma, Abortifacient	1-2 drops of leaf juice is dropped into ear to get relief from migraine.
Clinopodium umbrosum O.Ktze	Barikura	Lamiaceae	Jaunsari	Leaf	Cuts	Leaf juice applied on cuts 2 time a day for 2-5 days.
Cocculus hirsutus (L.) W. Theob.	Jaljamuni	Menisperma-ceae	Bhoxa	Leaf	Dysentery, Diarrhoea	Leaves powder is taken orally in dysentery and diarrhoea.
Coleus barbatus (Andrews) Benth. ex G.Don	Chhyanglang Jari	Lamiaceae	Bhotiya	Root	Internal ulcer, Cancer, Tumour, Skin disease	Dried root is chewed to cure intestinal ulcer and cancer. Root paste is used externally on wound and tumer; it is also effective for skin problem.
Colebrookea oppositifolia Sm.	Bursong	Lamiaceae	Raji	Leaf	Cough	Leaf juice is swallowed in cough.
Combretum roxburghii Spreng.	Patajar	Combretaceae	Jaunsari	Root	Snake bite, Cuts	Root paste is given orally with cow's urine as an antidote to snake bite. Leaf juice is used on cuts.
Corydalis cornuta Royle	Chitra jhar	Fumariaceae	Bhotiya	Root	Leprosy	Root paste is applied to treat leprosy.

Cordyceps sinensis (Berk.) Sacc.	Nabu, Keedu	Clavicipitaceae	Bhotiya	Whole plant	Stomach problems	Whole plant dips into tradition green tea and taken to cure stomach problem.
Hellenia speciosa (J.Koenig) S.R. Dutta J.E.Smith	Keu	Zingiberaceae	Bhoxa	Root	Rheumatism	Root powder is mixed with mustard oil paste and used externally for rheumatism.
Crataeva adansonii Jacobs	Baana	Capparaceae	Bhoxa	Leaf	Rheumatism	Leaf paste used to treat rheumatism.
Cryptrolepis buchananii Roemer	Dhudi	Asclepiadaceae	Bhoxa	Bark	Rheumatism	Bark mixture is given orally in rheumatic pain and leaf apply externally on rheumatic place.
Cucumis sativus L.	Kakree	Cucurbitaceae	Bhotiya	Seed	Urinary disorder	Seeds are grounded and mixed with water for urinary disorder.
Cuminum cyminum L.	Zeera	Apicaceae	Jaunsari	Seed	Cold, Cough, Asthma, Miscarriage	Tea made up with zeeea and ajwan is given for cold.
Curculigo orchioides Gaerth.	Samusli	Hypoxidaceae	Bhoxa	Root	Dysentery, Diarrhoea	Infusion of root is taken orally in dysentery.
Curcuma longa L.	Haldi	Zingiberaceae	Bhotiya	Rhizome	Boils	Rhizome paste is used on boils.
Cuscuta europaea L.	Akash lagui	Cassalpiniaceae	Jaunsari	Whole plant	Skin disease	
Cuscuta reflexa Roxb.	Akesh bel	Convolvulaceae	Bhotiya	Whole plant	Ear disease, Itching	Whole plant is grounded, filtered and drops used in ear disease. Whole plant is grounded and mixed with *Rumex dentatus* L. leaf powder and mustard oil paste for itching.
Cyathula tomentosa (Roth) Moq.	Sirli	Amaranthaceae	Jaunsari	Root	Boils	Root paste applied on boils.
Cymbopogan martini (Roxb.)	Mirchya ghass	Poaceae	Bhotiya	Leaf	Rabies, Ear disease,	Fresh leaves are milled and paste made is used to cure rabies.

Watson					Ring worm infection, Dermatitis	Fresh leaves are crushed then filtered drops used ear disease. Leaves powder is mixed with mustard oil rubbed in the ringworm infection part. Leaf powder is mixed with mustard oil rubbed on dermatitis.
Cynodon dactylon (L.) Persoon	Dubghass, Dubra	Poaceae	Bhoxa, Raji	Whole plant	Rheumatism, Dysentery, Nasal bleeding	Plant paste applies externally in rheumatism. 100gm of root powder together with 5gm of *Valeriana jatamansi* Jones powder and 5gm of sugar taken with 1 glass of water twice a day. Whole plant crushed with water, 2-3 drops of water extract used for nasal bleeding.
Dactylorhiza hatgirea D.Don	Hatajari, Panja	Orchidaceae	Bhotiya	Root	Cut, Menstruation cycle, Wounds, Chronic fever, Diarrhea, Diabetes	Root paste is applied on cut and wound. 50gm fresh root crushed in cold water then filtered and drunk for the regulation of menstruation cycle for 15 days twice a day. Tuber powder is taken orally for chronic fever, diarrhea. The dry root is taken with water for diabetes.
Dalbergia sissoo DC.	Seesam	Fabaceae	Bhoxa	leaf	Dysentery, Diarrhoea	Leaves paste is mixed with curd is taken to cure dysentery and diarrhoea.
Daphne papyracea Wall. ex. Stued	Chambua	Thymelaeaceae	Jaunsari	Root, Stem bark	Boil, Eczema	Root paste is applied to boils. Ash of stem bark applied externally on eczema
Datura	Dhatura	Solanaceae	Bhotiya	Seed	Rheumatism,	Seed paste is used to cure

stramonium L.						rheumatism.
Delphinium brunonianum Royle	Kasturi Kamal	Ranunculaceae	Bhotiya	Leaf	Burn, Boil	A leaf decoction is applied on burn, cut and boil
Delphinium denudatum Wall. ex Hook. f. & Thomson	Nirbishi	Ranunculaceae	Bhotiya	Root	Snake and Scorpion bite, Boil	Root paste applie to snake and scorpion bite. The root decoction is used as a blood purifier. Extract of the root is used in intestinal pain amd poisoning.
Dicliptera bupeuroides Nees	Kuthi	Acanthaceae	Raji	Seed	Dysentery	Seed crushed and taken with water for the treatment of dysentery.
Dioscorra bulbifera L.	Geth, Gainthi	Dioscoreaceae	Bhoxa, Bhotiya, Raji, Jaunsari	Tuber	Cough, Cut, Boil, Pimples, Dysentery	The tuber is roasted in hot ash and taken with black salt to cure old cough. Dried rhizome with water applies on cuts, boils and pimples. The root decoction is used for dysentery.
Dioscorra deltoidea Wall	Tairu	Dioscoreaceae	Jaunsari	Rhizome	Urinogenital disorder	Rhizome decoction is given to urinogenital disorder
Echinochloa frumentacea Link	Jhangora	Poaceae	Bhotiya	Fruit	Javndice	Taken to cure Javndice
Eleusine. coracana L	Kodhu, Manduva	Poaceae	Jaunsari	Leaf	Typhoid	Crushed leaves mixed with water and jaggery to cure typhoid.
Elsholtzia fruticosa (D.Don) Rehd.	Padadu	Lamiaceae	Jaunsari	Leaf	Intestinal worm	Leaf juice is used as an anthelmintic and given to expel worm in children.
Phyllanthus emblica L.	Amla	Euphorbiaceae	Jaunsari	Fruit	Stomach problems	Fruit extract given to cure stomach problem.
Epherdra gerardiana Wall.	Ain	Ephedraceae	Bhotiya	Shoot	Fracture	Shoot paste used to treat fracture.
Euphorbia geniaculata Orteg.	Dudhlaee	Euphorbiaceae	Jaunsari	Latex	Toothache	Latex of plant is used for toothache.

aspera L.						
Euphorbia hirta L	Dudhibari, Laldudhi	Euphorbiaceae	Jaunsari, Raji, Bhoxa	Latex, Leaf	Piles, Wart, Bronchial infection, Asthma, Toothache	Latex with curd given for piles, wart, bronchial infection, asthma. Latex of plant is dropped on toothache.
Euphorbia ligularia Roxb.	Syon	Euphorbiaceae	Raji	Latex	Earache	Luke warm latex is dropped in ear during earache.
Euphorbia thymifolia L.	Choti-dudhi	Euphorbiaceae	Raji	Whole plant	Diarrhoea, Cholera	Whole plant is crushed with water and taken in diarrhoea and cholera.
Evolvulus alsinoides L.	Sankhpushpi	Convolvulaceae	Jaunsari	Flower	Cough, Cold, Asthma, Bronchitis	Flower extract is given in cough, asthma and bronchitis.
Excoecaria acerifolia Didr.	Dudhlu	Euphorbiaceae	Bhotiya	Root	Stomach worm	Roots paste applied externally to expel stomach worm.
Fagopyrum dibotrys D.Don	Phapar ogal, Kanyala		Bhotiya	Leaf	Fever, Urolithiasis	Leaves are consumed as vegetable for extremities, fever and urolithiasis.
Ficus benghalensis L.	Bar, Bargad	Moraceae	Raji	Root	Metrorrhagia	Crushed root used metrorrhagia.
Ficus palmata Forssk.	Beru	Moraceae	Raji	Latex	Boil, Cuts, Wounds	Milky latex is used to cure boils, cuts and wounds.
Ficus racemosa L.	Gular	Moraceae	Bhoxa	Latex	Diarrhoea, Dysentery	Latex is internally used for the treatment of diarrhoea and dysentery.
Ficus religiosa L.	Peepal	Moraceae	Raji	Bark	Cut, Wound, Skin disease	Bark mixed with turmeric powder is applied externally on cuts, wounds and skin disease.
Ficus semicordata Buch.-Ham. ex Sm.	Khaina	Moraceae	Bhotiya	Latex	Baldness	Milky latex is applied on baldness.
Flacourtia indica (Houtt.) Merr.	Kangou	Flacourtiaceae	Jaunsari	Stem bark	Eye inflamation, Constipation	Stem bark decoction is used for treatment of eye inflamation two

						times a day for 7 days. Fruit is taken for constipation.
Fraxinus micrantha Lingels-Heim	Ango	Oleaceae	Bhotiya	Inner bark	Liver disease	Root of *Rubus foliolosus D. Don* (Kala Hansyalu) infusion with inner bark of Ango is used for treatment of liver disease.
Galium aparine L.	Less Kuri	Rubiaceae	Bhotiya, Jaunsari	Whole plant	Cut	Plant juice is applied on cuts.
Gentiana kurroo Royle	Kaudi	Gentianaceae	Bhotiya	Rhizome	Indigestion, Stomach	Paste of rhizome given with warm water in indigestion, stomach pain.
Geranium ocellatum Jacquem ex Cambess	Kyaphulya	Geraniaceae	Bhotiya	Whole plant	Liver disorder, Fever	Decoction of whole plant is given liver disorder and fever.
Girardinia palmata (Forsk.) Gaud.	Koru Khuska	Urticaceae	Jaunsari	Root	Sprain	Warm root paste is applied externally on sarain 3 time a day for 3-7 days.
Gloriosa superba L.	Agnisika, Karihari	Liliaceae	Bhoxa, Jaunsari	Root	Rheumatism	Crushed roots are fried in mustard oil and paste applied on rheumatism externally. Tuber powder given in painful delivery and suppressed urination.
Glycine max (L.) Merr.	Kala bhatt	Fabaceae	Bhotiya	Seed	Kidney stone	Seed boiled in water to make a hot liquid soup to cure kidney stone.
Grewia optiva J.R.Drumm. ex Burret	Bimal	Tiliaceae	Jaunsari	Bark	Bone fracture	Turmaric and wheat flour are fried in mustard oil and put on the bone fracture part and than bimal bar is placed on it and bandaged.
Hordeum vulgare L.	Jaun	Poaceae	Bhotiya	Seed	Headache	Seed paste is applied to get relief from headache.
Hartmania rosea Smith	Massillo	Onagraceae	Jaunsari	Leaf	Nose bleeding	Leaf paste used on forehead to stop nose bleeding.

Hedera nepalensis K.Koch	Mithiyar	Araliaceae	Jaunsari	Leaf	Anthelmintic	Leaf juice is given twice a day for 7 days in anthelmintic.
Hedychium coronarium Koen.	Kachari	Zingiberaceae	Jaunsari	Rhizome	Dysentery	Rhizome paste is given in dysentery two times a day for 3-5 days.
Hedychium spicatum Bach.-Ham.	Kachoor, Sathi	Zingiberaceae	Jaunsari, Raji	Bark, Dry Fruit, Rhizome	Stomachache, Rheumatism, Arthritis, Neuro-muscular disorder	Bark decoction is mixed with jaggery is drunk to cure stomachache, rheumatism and arthritis. Dry fruit is powdered and mixed with jaggery is taken to cure body swelling. Rhizome powder taken orally for neuro-muscular disorder.
Helicteres isora L.	Marora	Sterculiaceae	Bhoxa	Fruit	Diarrhoea, Dysentery	Fruit decoction of with sugar given orally for diarrhoea and dysentery.
Euploca strigosa (Willd.) Diane & Hilger	Gorakhpan	Boraginaceae	Bhotiya	Leaf	Burning sensation of urination	Leaf extract is given to get relieve from burning sensation during urination.
Hibiscus rosa-sinensis L.	Gudhal	Malvaceae	Bhotiya	Flower	Abortion	One mature flower is fried with 10ml ghee and taken for abortion.
Hibiscus sabdariffa L.	Patson	Malvaceae	Bhoxa	Seed	Dysentery	Seed powder is given orally in dysentery.
Hippophae rhamnoides L.	Amesh	Elaegnaceae	Bhotiya	Fruit	Cold, Cough, Cancer	100ml mixture of fruit juice is mixed with 1 spoonful of sugar and 30-40gm of finger millet (*Eleusine coracana* (L.) Gaetn.) and taken to treat cold and cough. Juice of mature fruit with 250 gm of sugar is used in 1 liter juice to cure cancer.

Hippophae salicifolia D.Don	Bogdiaric-huk, Amil	Elaegnaceae	Bhotiya	Fruit	Cut, Wound, Diarrhoea, Dehydration	Fruit juice is used to give strength to pthiasis patient. Seed oil is applied externally on cut and wound. Fruit juice or extract is given for diarrhoea and dysentery.
Holoptelea integrifolia (Roxb.) Planchon	Papri	Ulmaceae	Bhoxa	Oil	Rheumatism	Oil of seed is used to get relief from rheumatism.
Holarrhena pubescens Wall.	Kuda	Apocynaceae	Bhoxa, Bhotiya	Bark	Dysentery	Bark powder is taken during dysentery.
Hordeum vulgare L.	Jau	Poaceae	Bhotiya	Leaf	Ear disease	Fresh leaves are crushed filtered with drops being applied on ear disease.
Hypericum dyerii Wall.	Pingona	Hypericaceae	Jaunsari	Leaf	Cut, Wound	Leaf paste applied on cuts and wounds.
Inula grandiflora Willd.	Phulente	Asteraceae	Jaunsari	Root	Boils, Blisters	Root paste is applied on boils and blisters.
Ipomea carnea Jacquin	Behaya	Convolvulaceae	Bhoxa	Leaf	Rheumatism	Leaf paste applied externally in rheumatic area.
Ipomoea nil (L.) Roth.	Bharar	Convolvulaceae	Raji	Seed	Abortifacient	Crushed seeds are considered as abortifacient when taken in heavy doses.
Ipomoea purpurea (L.) Roth		Convolvulaceae	Raji	Whoel plant	Syphilis	Whole plant is grounded and poultice is used to treat syphilis.
Jatropa curcas L.	Banarandi	Euphorbiaceae	Bhoxa	Seed oil	Rheumatism	Seed oil is applied externally in rheumatism.
Juglans regia L.	Akhrot	Juglandaceae	Bhotiya	Seed oil, Fruit, Twig	Swollen legs	Seed oil rubbed on swollen legs of pregnant women. Paste of fruit pulp is applied on lesion of alopecia. Twig is used as tooth brush to prevent dental carries.

Juniperus communis L.	Billa	Cupressaceae	Bhotiya	Fruit, Flower	Liver disorders	Fruit juice and flower bud extract is given for liver disorder.
Juniperus himalaica R.R.Stewart	Budhkesh	Asteraceae	Bhotiya	Root	Skin infection	Root paste applied to cure boils
Juniperus recurva Buch. Ham. D.Don	Padamas, Juniper	Cupressaceae	Bhotiya	Leaf	Insect repellent	Fumigation of leaves used as insect repellent.
Kalanchoe pinnata Pers.	Nanno	Crassulaceae	Jaunsari	Whole plant	Burns	Paste of whole plant is used on burns.
Lannea coromandelica (Houtt.) Merr.	Jhingan	Anacardiaceae	Tharu	Bark	Cuts, Burns, Toothache, Skin eruption, Ulcers	Bark paste used in burns, cuts, toothache, skin eruptions and ulcers.
Leptodermis lanceolata Wall.	Chirar	Rubiaceae	Raji	Leaf, Flower	Fever	Leaves and flower are boiled with water and extract is given to cure fever.
Leucas cephalotes Spreng.	Dronpushpi	Lamiaceae	Jaunsari	Whole plant	Diaphoretic, Snake bite, Anthelmintic	Plant decoction used for diaphoretic, snake bite and anthelmintic.
Lilium oxypetalum (D.Don) Baker	Sur	Liliaceae	Bhotiya	Bulb	Swollen limbs	Bulbs are eaten and paste applied on the swelling in the limbs.
Litsea glutinosa (Lour.) Robins	Maida-lakri	Lauraceae	Jaunsari	Wood	Bone fracture	Wood paste is applied on bone fracture.
Litsea umbrosa Nees	Surud	Lauraceae	Jaunsari	Seed oil	Skin itching, Insecticide	Seed oil applied to cure skin itching and used as insecticide.
Lyonia ovalifolia (Wall.) Drude	Anyar	Ericaceae	Bhotiya	Bud	Boldness	10gm of buds are crushed with 5gm of akhrot bark, mixed with 10 ml mustard oil and applied paste for boldness.
Macrotyloma uniflorum (Lam.)	Ganheth	Fabaceae	Bhotiya	Seed	Kidney stone	250gm of seeds are boiled with 1L water and extracts used

Verdc.						continuosly for 7 days to cure kidney stone.
Madhuca longifolia (J.Konig ex L.) J.F. Macbr	Mahua	Sapotaceae	Bhoxa	Flower	Diarrhoea	Flower decoction is mixed with sugar and drunk cure of diarrhoea.
Maianthemum purpureum (Wall.) LaFrankie Frankie	Puyanu	Asparagaceae	Bhotiya	Seed	Dyspepsia	50gm dried leaves are fried in 4-5 spoonful of *P. persica* oil for dyspepsia.
Mallotus philippensis (Lam.) Müll.Arg. Muell.-Arg.	Kamil	Euphorbiaceae	Jaunsari	Fruit	Boils, Blisters	Fruit powder is mixed with mustard oil and applied to cure boils and blisters.
Malva verticillata L.	Tankaghas, Yantoul	Malvaceae	Bhotiya	Whole plant	Urine problem, Skin infection	Whole plant paste is used on hypogastrium to induce urine flow in retention. Whole plant paste is used in skin infection.
Megacarpaea polyandra Benth. ex Madden	Barmoa, Ruki	Brassicaceae	Bhotiya	Root, Leaf	Asthma, Stomachache, Abdominal pain, Diarrhea, Fever	100ml root mixture is mixed with 1 spoonful of sugar and 4-5 drops of apricot (*P. armeniaca*) oil and used for asthma. 100gm of dried leaves are mixed with 10gm of thuner bark powder and fried in 2-3 spoonful of ghee for stomachache. Root powder is given in abdominal pain and diarrhoea. Leaf juice taken during fever.
Mesua ferrea L.	Nagkesar	Clusiaceae	Raji	Root, Stem flower, Seed oil	Menorrhagia, Sores, Wounds, Rheumatism, Snake bite	Root paste is taken as an antidote against snake bites. Stamen of flower is taken before 5 days of menstruation to control menorrhagia. Seed oil is applied for healing sores, wounds and

						rheumatism.
Mentha arvensis L.	Podina	Lamiaceae	Bhotiya	Leaf	Diarrhoea	Fresh leaves are milled with lashum to make a chatni and taken during diarrhea.
Mimosa pudica L.	Sharmili	Mimosaceae	Bhoxa	Whole plant	Dysentery	Whole plant decoction is taken orally in dysentery twice a day.
Momordica charantia L.	Karela	Cucurbitaceae	Bhoxa	Fruit, Leaf	Rheumatism	Fruit infusion (2-3 teaspoonfuls) is used internally for rheumatism. Leaf juice is used for intestinal worm.
Morus alba L.	Shahtoot	Moraceae	Bhotiya	Fruit	Cough, Cold	Fruit juice is taken in cure of cough and cold.
Morina longifolia Wall.	Biskandru	Caprifoliaceae	Bhotiya	Leaf	Boils, Wounds	Juice of leaves is mixed with 3-4 drops of mustard oil and applied externally on the boils and wounds.
Moringa oleifera Lam.	Sondi	Moringaceae	Jaunsari	Leaf	Snake bite	Root paste is used to treat snake bites.
Murraya koenigii (L.) Spreng	Gandhela	Rutaceae	Bhoxa, Jaunsari	Stem bark, Leaf	Boils, Dysentery, Diarrhoea	Stem bark paste is used externally on boils. Leaf juice is given orally in dysentery and diarrhoea.
Musa balbisiana Colla	Kela	Musacaceae	Bhoxa	Root	Diarrhoea	Root aqueous extract is given to cure diarrhoea.
Myrica esculenta Buch.-Ham. ex D.Don	Kaphal	Myricaceae	Bhotiya, Jaunsari	Fruit, Leaf	Cough, Cold, Headache	Threshed dry outer layer of fruit is taken before sleeping to cure of cough, cold and headache. Leaf paste is applied externally to get relief from headache.
Myristica fragrans Houtt.	Jayphal	Myristicaceae	Bhotiya	Fruit	Cough	Fruit paste is applied on the neck and chest to get relief from cough.
Nardostachys grandiflora DC.	Masi, Masidhoop,	Valerianaceae	Bhotiya	Root	Rheumatism	10gm of root paste mixed with 50 gm ghee, mildly heated for 5 minutes

	Jatamansi					and rubbed on joints to treat rheumatism.
Nardostachys jatamansi DC.	Masi, Jatamasi	Valerianaceae	Bhotiya, Jaunsari	Rhizome	Hysteria	Rhizome used to cure epilespy and hysteria.
Nepeta discolor Royle ex Benth.	Khirku	Lamiaceae	Bhotiya	Leaf	Tuberculosis	150ml of dried leaves decoction is mixed with 2-3 spoonful of honey given for tuberculosis.
Nicotiana tabacum L.	Tabacco	Solanaceae	Jaunsari	Leaf	Scabies	Leaves juice is applied to cure scabies affected part.
Ocimum tenuiflorum L.	Tulsi	Lamiaceae	Bhoxa	Whole plant	Diarrhoea	Whole plant decoction is taken in diarrhoea.
Orchis habenarioides King & Pant	Salam Mishri	Orchidaceae	Bhotiya	Tuber	Stop excessive bleeding during child birth	200ml decoction of tuber mixed with milk and 1 tablespoon of honey given to stop excessive bleeding during child birth.
Origonum vulgare L.	Bantulsi	Lamiaceae	Jaunsari	Leaf	Bronchitis, Whooping cough, Diarrhoea, Colic	Leaves extract given for bronchitis, whooping cough, diarrhoea and colic.
Oroxylum indicum (L.) Kurz.	Arula	Bignoniaceae	Bhoxa	Seed	Dysentery	Seed powder is taken in dysentery.
Ougeinia oojeinensis(Roxb.) Hochr.	Sandan	Fabaceae	Bhoxa	Gum	Dysentery	Gum is used in treatment of dysentery.
Paeonia emodi Wall. ex Royle	Chandra	Paeoniaceae	Bhotiya	Leaf	Dyspepsia, Dysentery	50gm leaves are crushed with 10gm dried barmoa root for dyspepsia. Leaves are crushed and made into a juice for dysentery.
Panicum miliaceum L.	Cheena	Poaceae	Bhotiya	Seed	Measles	Seed used to cure measles.
Paris polyphyllum Sm.	Satuwa	Liliaceae	Bhotiya	Tuber	Fever, Headache, Poisoning, Wound	Tuber powder given with warm water for fever, headache, poisioning and wound healing.

Parnassia nubicola Wall. ex Royle	Nirbisi	Celastraceae	Bhotiya	Root	Wound, Cut, Insect bite	Root paste applied on wounds, cuts and insect bits.
Pedalium murex L.	Gokhru	Pedaliaceae	Bhotiya	Berries	Gential problem	Paste of berries is made to provide benefits in gential problem.
Perilla frutescens (L.) Britt.	Bhangiri	Lamiaceae	Jaunsari	Seed oil	Arthritis	Seed oil is massaged 2 times a day in arthritis for long time.
Phyllanthus amarus Schumach. & Thonn.	Buiamla	Euphorbiaceae	Bhoxa	Leaf	Dysentery	Leaf infusion is administered in dysentery.
Phyllanthus emblica L.	Amla	Euphorbiaceae	Bhotiya	Fruit	Digestive problem	Dried fruit is mixed with fruit powder of harad and bahera is given for indigestive problems.
Picrorhiza kurrooa Royle ex Benth.	Kutki, Kadvi	Plantaginaceae	Bhotiya	Root	Fever, Jaundice, Stomachache, Diabetes	50gm dried root milled with 2 spoonfuls of sugar and drunk with water for fever. 50gm dried root with 3 spoonful sugar socked over night and drunk in morning to jaundice. 50g root mixture boiled mildly for 5-10 mint and mixed with honey for stomachache, constipation and gastric disorder. 30 gm root powder with 5gm black piper mixed with honey is given for high fever. 5gm of root is milled, steeped in a copper pot with 250ml water overnight, then filter and drunk in morning for diabetes.
Pinus roxburghii Sargent	Khaida	Pinaceae	Jaunsari	Gum, Root	Sprain, Tuberculosis	Fresh gum is applied on sprain affected part for 10-15 days. Plant root is mixed with root of *Quercus leucotrichophora*, made a paste and

						mixed with cow's milk and given orally for tuberculos is 2 times a day for 40 days.
Pinu wallichiana A.B.Jacks.	Kail	Pinaceae	Bhotiya	Resin	Sprain, Fracture	Plant resin is normally heated and used to plaster on fracture part and covered with *B. vtilis* bark immediately on sprain and fracture.
Piper longum L.	Peepla	Piperaceae	Bhotiya	Root	Diabetes	Dried root powder is mixed with grounded stems of chiraita, seeds of ajwain and roots of kuth to make a powder for diabetes.
Plantago depressa Willd.	Luhurya, Isabgol	Plantaginaceae	Jaunsari	Leaf, Seed	Cut, Wound, Piles, Stomach aliments	Leaf extract and seed are eaten during cuts and wounds, piles, stomach aliments
Plantago lanceolata L.	Luhurya, Isabgol	Plantaginaceae	Jaunsari	Leaf, Seed	Cut, Wound, Piles, Stomach aliments	Leaf extract and seeds are eaten to cure cuts and wounds.
Isodon japonicus (Burm.f.) H.Hara	Chhirchhida	Lamiaceae	Jaunsari	Leaf	Earache	Leaf juice is used earache twice a day for 2-7 days.
Pleurospermum angelicoides (Wall. ex DC.) Bent.	Chhippi	Apiaceae	Bhotiya	Root	Fever, Dyspesia	50gm root, 10gm of cumin seed and 7-8 grains of black pepper are mixed together and boiled with 200ml water on fair flame for up to 5-10 minutes and kept for cooling, this liquid is drunk to cure fever. 50gm root paste is mixed with 20gm of dried leaves of *Allium* sp. and fried with 3 spoonful of ghee for dyspesia.
Plumbago	Chita	Plumbaginaceae	Jaunsari	Root	Rheumatism	Root paste with buttermilk is used

zeylanica L.						in rheumantism and gout twice a day for 2 weeks.
Sinopodophyllum hexandrum (Royle) Ying	Bankakri, Dhyuchu rapu	Berberidaceae	Bhotiya	Root	Cancer, Cuts, Wound, Fever, Jaundice, Skin disease	Root paste is used to cure cancer, cuts and wound. Root soaked in water for 2-3 hours and taken during fever and jaundice. Root paste is used on skin disease.
Polygonum capitatum Buch.-Ham.	Dhadhura	Polygonaceae	Jaunsari	Root	Boils	Root paste is applied to cure boils.
Polygonum cirrhifolium Royle	Mahameda	Polygonaceae	Bhotiya	Leaf, Root	Joint pain, Cuts, Wound	Young leaves paste is used during joint pain. Root powder paste is used on cuts and wound.
Polygonum nepalensis Mesin.	Bhotia chai	Polygonaceae	Bhotiya	Bark, Root	Pregnancy, Pain, Hoarseness of throat	100ml decoction of bark is given to women for normal child birth (pregnancy). 10g dried roots are chewed to cure pain and harshness of throat.
Polygonum rumicifolium Royle.	Khyakjari	Polygonaceae	Bhotiya	Tuber	High fever	Tuber powder is given to get relief from high fever.
Polygonatum verticillatum (L.) All.	Salam misri, Khyol, Amrita, Hddivaila	Liliaceae	Bhotiya	Rhizome, Leaf	Leucorrhoea, Oligospermia, Joint pain, Bone fracture	1gm root powder is blend with one glassful of water to cure leucorrhoea. 2-4gm rhizome powder with milk taken during general weakness oligospermia. Leaves paste is used on joint pain. Rhizome paste applies on bone fracture as plaster.
Potentilla fulgens Hook.	Bajradanti	Rosaceae	Jaunsari	Fruit	Stomatitis	Fruit juice used in stomatitis.

Potentilla lineata Trevir.	Bjradanti	Rosaceae	Bhotiya	Root	Toothache	Root paste is used to cure toothache.
Potentilla nepalensis Hook.	Jakjari	Rosaceae	Jaunsari	Whole plant	Worms excel	Whole plant juice is taken orally twice times a day for 7 days to expel worms in children.
Potentilla sundaica (Blume) Kuntz.	Golliusu	Rosaceae	Bhotiya	Whole plant	Throat cough	Whole plant dried and pounded along with *Coleus forkohlii* (Willd.) Briq. (Coleus) and *Eugenia caryophyllus* (Spreng.) Bullock & Harrison (Clove), made into small tablet to be taken for throat cough.
Portulaca oleracea L.	Jark	Portulacaceae	Bhotiya	Leaf	Jaundice	Leaves juice is used to cure jaundice.
Prinsepia utilis Royle	Danthali	Rosaceae	Bhotiya	Seed oil	High blood pressure, Rheumatism, Joint pain, Boils, blisters	Seed oil is useful in high blood pressur and cholestrol. Oil massage gives relief from rheumatism and joint pain. Paste of oil cake is used externally to cure boils and blisters.
Prunus armeniaca L.	Chullu	Rosaceae	Jaunsari	Seed oil	After pregnancy	Seed oil is given orally after child birth for strenght and also massaged on the body in case of weakness for 2-3 weeks.
Prunus cerasoides D.Don	Panya	Rosaceae	Jaunsari	Bark, Leaf	Body swelling	Bark and leaf paste used as psychomedicine and body swelling.
Prunus persica (L.) Batsch	Aaru	Rosaceae	Bhoxa	Seed oil, Leaf	Rheumatism, Dysentery	Seed oil is externally applied in rheumatism. Leaves infusion is administered in dysentery.
Psidium guajava L.	Amrood	Myrtaceae	Bhotiya	Leaf	Blisters	Leaves are chewed to get relief from blisters.
Punica granatum L.	Anar	Lythraceae	Jaunsari	Leaf	Epilepsy	Leaves are boiled in water with 10

						rose leaves, reduced to half of its volume, filtered and used for curing epilepsy.
Pupalia lappacea (L.) Juss.	Sirali	Amaranthaceae	Jaunsari	Root	Boils	Root paste is used on boils.
Pyrus pashia Buch.- Ham.	Kainth	Rosaceae	Jaunsari	Fruit	Fever	Liquid extract of fruit is taken 3 times a day for 2 days in fever.
Phyllanthus niruri L.	Bhumiamla	Euphorbiaceae	Bhotiya	Whole plant	Diabetes	Extract of dried plant in hot water is made and given to diabetic patients.
Quercus leucotrichophora A.Camus	Banj	Fagaceae	Bhotiya, Jaunsari	Seed, Stem bark	Snake and scorpion bite, Dysentery	Seed paste applied on snake bite or scorpion sting. Ash of stem bark mixed with mustard oil is used on sprains. Bark decoction is taken in dysentery 2 times a day for 2-4 days.
Rauvolfia serpentina Benth.	Sarpag-andha	Apocynaceae	Jaunsari	Root	Fever, Anxiety, Epilepsy Nervous disorders	Root powder is used in fever, anxiety, epilepsy, intestinal and nervous disorders.
Rheum australe D.Don	Dolu, Tantari	Polygonaceae	Bhotiya	Root	Swelling, Pain, Fracture, Boils	20g roots heated gently used to plastered on fractured part covered with a bandage, it reduces the swelling, pain and fracture. Root paste with hot water applied over fractured supported by splint made up of bam. Root paste can be applied to treat boils.
Rheum emodi Wall. ex Meissn.	Dolu	Polygonaceae	Bhotiya, Jaunsari	Root	Wound, Boils, Pimples, Bone	Dried root paste with water is used on wound, boils and pimples.

					ache, Muscular pain, Bwise	Rhizome paste cooked with turmeric and ghee is used on bone ache, muscular pain and bwise.
Rheum webbianum Royle	Archa,	Polygonaceae	Bhotiya	Leaf, Root Tatri, Tanturi	Boils, Wound, Stomachache	100gm of dried leaf powder is mixed with *Rumex hastatus* D.Don leaf powder with 2 spoonful of mustard oil and used externally on boils and wounds. Root powder is given with milk to get relief from stomachache.
Rhododendron anthopogon D.Don	Awon, Palu	Ericaceae	Bhotiya	Leaf	Headache, Ringworm	100gm leaves powder mixed with sugar and 200ml water added and boiled for a few minutes, after cooling juice is taken to cure headache. Leaf powder mixed with mustard oil, rubbed to ring worm infected part.
Rhododendron arboreum Sm.	Burans	Ericaceae	Bhotiya	Flower	Cardiac disorder	Flower juice is prepared and applied to get relief from cardiac disorder.
Rhododendron campanulatum D.Don	Ratpa, Chimura	Ericaceae	Bhotiya			
Rhus parviflora Roxb.	Samak dana	Anacardiaceae	Jaunsari	Bark, Leaf	Cholera, Stomachache	Decoction of bark and leaves is used to cure cholera and stomachache.
Ricinus communis L.	Arandi	Euphorbiaceae	Bhoxa	Root	Rheumatism	Root infusion is taken with sugar to cure rheumatism. Seeds oil is used for massaging in rheumatism.
Rosa webbiana Wall. exhoyle	Dhwada, Syanapala, Jagligubb	Rosaceae	Bhotiya	Leaf	Boils, Ulcers, Urine infection	Leaves paste is used on boils and ulcers. Leaves juice of red rose is taken against urine infection.

Roylea cinerea (D.Don) Ball.	Kadiya	Lamiaceae	Jaunsari	Whole plant	Malaria fever	Aqueous extract of whole plant is taken 2 times a day for 2-4 days to get relief from malarial fever.
Rubia cordifolia L.	Manjitha	Rubiaceae	Bhotiya, Jaunsari	Whole plant	Acne, Boils, Bilsters	Whole plant pulp rubbed with honey is recommended as a cure for acne and dark spots on face. Plant paste is applied to boils and blisters 2 time day for 2-5 days.
Rumex hastatus D.Don	Almora	Polygonaceae	Jaunsari	Root	Cut, Wound, Boils, Snake bite, Gastric	Root extracts used on cuts, wound. Root paste is taken 2 times a day for 2-5 days. Root extract is given for snake bite. Root is chewed in gastric.
Rumex nepalensis Spreng.	Kharas, Jangli palak	Polygonaceae	Jaunsari, Raji	Root, Leaves	Boils, Blisters, bite insect	Warm root paste is used on boils and blisters 2 times a day for 2-7 days. Young leaves crushed and juice applies on insect bites.
Saccharum spontaneum L.	Kusha	Poaceae	Jaunsari	Leaves	Pus affected part	Pulp of crused leaves is used on the pus affected part for two weeks.
Sapindus saponaria L.	Reetha	Sapindaceae	Bhotiya	Seed	Snake and scorpion sting	Seeds are milled and drunk with water for snake/ scorpion sting.
Saussurea costus (Falc.) Lipsch.	Kuth, Pachhak	Asteraceae	Bhotiya	Root	Snake and scorpion sting, Jaundice, Toothache, Gastric trouble, Intestinal colic, Edema Fever, Cold	Root paste is applied during snake biting/ scorpion sting. 100ml of kuth tube decoction is mixed with 4-5 drops of apricot oil (*P. armeniaca*) and 1/2 spoonful salt, few small drops of this fusion placed on toothache. Root paste is used on itching. Root decoction is taken in empty stomach against gastric troubles and intestinal colic. Root powder mixed with rose water

						applied on edema of feet. Root powder is given for fever and cold. Root powder is taken with water pain.
Saussurea obvallata (DC.) Edgew.	Brahm kamal, Kaulkappu	Asteraceae	Bhotiya	Leaf, Root, Seed, Flower	Boil, Cut, Wound, Cardiac disorder, Mental disorder, Urine tract infection	100ml dried leaves decoction is mixed with 1/2 spoonful of salt and few drops of this is used on boils, cuts and wounds. 200ml mixture of root or leaves is mixed with 2-3 spoonful of deodar oil and used externally to cure cardiac disorder. Seeds are crushed into a powder; this powder is steeped in water overnight then filtered taken for mental disorder. Flower is cooked with taga misri and taken to cure urine tract infection.
Sarcococca saligna Muell.-Arg.	Tilhari	Buxaceae	Jaunsari	Root	Boils, Bilsters, Diuretic	Liquid extract of root is taken by mouth to cure boils and blisters and for diuretic purpose two times a day for 2-3 days.
Selenium vaginatum (Edgew) C.B.Clarke	Bhutkesh	Apiaceae	Bhotiya	Root	Joint pain, Menstrual cycle, Hysteria	Root paste is applied on joint pain. Root decoction taken orally with ginger and pepper powder to regulate menstrual cycle. Fumigation of root is given to restore consciousness in hysteria.
Semecarpus anacardium L.f.	Bhilao	Anacardiaceae	Bhoxa, Jaunsari	Seed oil	Rheumatism	Seed oil is applied externally in rheumatism.
Sesbania sesban (L.) Merr.	Jayantee	Fabaceae	Bhoxa	Seed	Dysentery	Seed powder is recommended for dysentery.

Shorea robusta Gaerth.	Sakhu	Dipterocar-paceae	Bhoxa	Gum, Bark	Dysentery	Gum with curd is given orally in dysentery. Bark decoction is given internally for diarrhoea.
Sida cordifolia L.	Bala	Malvaceae	Raji	Stem bark, Root	General debility	Stem bark or root powder is taken in general debility.
Sida rhombifolia L.	Khareti	Malvaceae	Raji	Root bark	Urinary trouble, Leucorrhoea	Root bark powder with milk and sugar is taken for urinary infections and leucorrhoea.
Silene kumaonensis Williams	Khushu	Caryophyllaceae	Bhotiya	Root	Hair problem	Root powder used for washing hair to get relief from hair problem.
Smilax aspera L.	Kubardara	Smilacaceae	Jaunsari	Root	Diuretic, Diaphoretic, Rheumatism, Arthritis	Root paste is given as diuretic, diaphoretic, rheumatism and arthritis.
Solanum viarum Dunal	Khalar kawaoe, Khem, Khari	Solanaceae	Jaunsari	Fruit	Jaundice, Rheumatism, Gout	Fruit aqueous extract is given to treat jaundice two times a day for 7 days. Fruit paste is used to cure rheumatism and gout.
Sphaeranthus indicus L.	Garakh-mundi	Asteraceae	Bhotiya	Leaves, Flower	Allergy, Itching, Toothache	Leaves paste is used in itching. Raw flower are eaten for toothache.
Sphaeranthus senegalensis DC.	Mundi	Asteraceae	Bhoxa	Fruit	Diarrhoea	Fruit powder is taken in diarrhoea.
Spondias pinnata L.	Amra	Anacardiaceae	Jaunsari, Raji	Bark gum, Fruit	Stoach problem, Diarrhoea, Dysentery	Fruit extract and bark gum is used for stomach and ear problem. Fruit decoction is taken in diarrhoea and dysentery.
Stephania elegans Hook. f. & Thomson	Gangeri	Menispermaceae	Bhotiya	Rhizome	Lung disease	Rhizome is cooked and taken with food to relief lung disease.
Sterculia villosa Roxb.	Udal	Sterculiaceae	Bhoxa	Gum	Dysentery	Gum is given with curd to cure dysentery.

Swertia angustifolia Buch-Ham. ex D.Don	Chirotu	Gentianaceae	Jaunsari	Whole plant	Blood disease, Malaria	Whole plant extract is used for blood disease and malaria.
Swertia chirata (Wall.) C.B.Clarke	Cheraita	Gentianaceae	Bhotiya	Leaf	Diabetes	5gm of leaves and stem are crushed, steeped in 250ml water over night, then filter and drunk for diabetes.
Taxus baccata L.	Thuner, Luet	Taxaceae	Bhotiya	Bark	High Blood Pressure, Bone fracture, Cancer	1gm dry bark is mixed with 1gm salt, 1 spoon of ghee and a cup of warm water for high blood pressure and cancer. Yolk of egg is mixed with thuner bark and made a paste used to plaster in fracture part with help of bark for 28-35 days in bone fracture.
Taraxacum officinale Weber ex. Wiggers	Kanphul	Asteraceae	Bhotiya, Raji	Root, Latex	Pimples, Digestive disorder, Fever, Sinusitis, Blisters, Skin irruption	Root powder 1-2 taken in ache pimples and digestive disorder. Infusion of root is given for fever and sinusitis. Decoction of inflorescence is given orally for blisters. Latex is used externally in skin irruption.
Terminalia arjuna (Roxb. ex DC.) Wight & Arn.	Arjuna	Combretaceae	Bhoxa, Bhotiya	Bark	Bone fracture, Dysentery	Bark is crushed and made a paste, used for plaster on fracture part then covered bark 28-35 days for bone fracture. Bark powder is taken during dysentery.
Terminalia bellirica (Gaerth.) Roxb.	Bahera	Combretaceae	Bhoxa, Bhotiya, Jaunsari	Fruit	Digestive problems	Dried fruit powder without seed used for digestive problem.

Terminalia chebula Retz.	Harad	Combretaceae	Bhoxa, Bhotiya, Jaunsari	Fruit	Digestive problems	Dried fruit powder without seed used for digestive problem.
Thalictrum foliolosum DC.	Mamiri	Ranunculaceae	Jaunsari	Root	Eye inflamation, Ophthalmia, Coic fever	Root decoction is used to cure eyes inflamation, ophthalmia, and colic fever.
Thalictrum javanicum Blume	Mamiri, Pilij	Ranunculaceae	Bhotiya	Root	Diabetes, Eye problem, Abdominal disease	5gm root is boiled with 250ml water and make decoction for diabetes. Root paste used as collyrium in eye disease and to improve better eye sight. Root juice is given for abdominal disease.
Thymus linearis Benth ex Wall.	Balmajhad	Lamiaceae	Bhotiya	Whole plant	Flatulence, Sour belching	Whole plant powder soaked in water for whole day is given in flatulence and sour belching.
Thymus serpyllum L.	Jowan	Lamiaceae	Jaunsari	Whole plant	Headache, Dysentery, Vomiting	Plant decoction is given 2 times for 2-5 days as herbal tea to cure headache, dysentery and vomiting.
Tinospora cordifolia (Wild.) Miers	Giloe	Menispermaceae	Bhoxa, Bhotiya	Whole plant	Gouts, Uric acid accumulations, Dysentery	Whole plant crushed and soaked into water overnight, next mornning water is removed and collected the left powdery layer dried and used to get relief from gouts and uric acid. Root infusion is used for dysentery.
Tinospora sinensis (Cour.) Merrill	Giloe	Combretaceae	Jaunsari	Stem, Leaf	Debility, Leprosy, Urinary trouble, Colic fever	Stem and leaf juice is used for debility, leprosy, urinary trouble and malaria fever.
Trachyspermum ammi (L.) Sprague	Ajuwan	Apiaceae	Jaunsari	Seed	Stomachache	Seed juice is given in stomachache 3 times a day.
Trachyspermum copticum (L.) Link	Ajwain	Apiaceae	Bhotiya	Seed	Gastric disorders, Menstrual cramps	Rosted seeds are given for delivery time to relieve pain, also help in gastric disorder. Tea of

						ajwain drunk at menstrual periods to relieve cramps.
Mallotus nudiflorus (L.) Kulju & Welz1	Gutel	Euphorbiaceae	Bhoxa	Root	Rheumatism	Root decoction is give internally for rheumatism (4 teaspoonfuls).
Tridax procumbens (L.) L.	Phulli	Asteraceae	Bhoxa	Leaf	Dysentery	Leaves infusion is taken to treat dysentery.
Triumfetta rhomboidea Jacq.	Leswa-kura	Tiliaceae	Raji	Leaf	Cut, Wound	Leaves are pulverized with curd and used to cure cuts and wounds.
Ulmus wallichiana Planch.	Chamme-rmwa	Ulmaceae	Bhotiya	Bark	Bone fracture	Bark paste apply on bone fracture.
Urtica dioica L.	Kandali	Urticaceae	Bhotiya, Jaunsari	Leaf, Root,	Rabies, Sprain, Earache	5g fresh leaves crushed and mixed with 1gm chili powder, made a paste without water and applied to cure rabies.Warm root paste is used externally on sprain. Leaf juice is droped in ear for earache.
Valeriana jatamansi Jones	Sameula	Dipsacaceae	Jaunsari	Root	Aphrodisiac, Mental disorder	Roots powder is given for aphrodisiac, mental disorder.
Valeriana wallichii DC.	Tagar	Valerianaceae	Bhotiya	Root, Leaf	Urinary disorder, Headache	Root extract is made with water against urinary disorder. Leaves are crused and rubbed on head to get relief from headache.
Verbascum thapsus L.	Kukalenga	Scrophulariaceae	Jaunsari, Raji	Root, Leaf, Flower	Snake bite, Boils, Eyes problem	Root paste is used as antidote to snake bites. Flower powder mixed with mustard oil is used on boils. Leaf juice is used as eye drop to cure eye problems.
Vernonia cinerea (L.) Less.	Sehdevi	Asteraceae	Bhoxa	Whole plant	Dysentery	Plant decoction is given during dysentery.
Vigna mungo	Kali dal	Fabaceae	Bhotiya	Seed	Bone fracture	250gm seeds are mixed with 250ml

(L.) Hepper						water, and then used as plaster with bhojpatra bark on fracture for 28-35 days during bone fracture.
Viola betcnicifolia J.E. Sm.	Banafsa	Violaceae	Jaunsari	Flower	Respiratory problem	Decoction of flowers along with cinnamon, clove, and fennel is used to treat respiratory tract problems.
Vitex nigundo L.	Sambhalu, Sinwali	Verbenaceae	Bhoxa, Jaunsari, Raji	Root, Leaf, Fruit	Rheumatism, Arthritis, Anthelmintic, Sprains, Eye inflamation	Root decoction is given to cure rheumatism (2-3 teapoonful). Leaf decoction used for rheumatism. Fruit decoction used as anthelmintic. Warm leaf paste is used on sprains. Leaf juice is dropped into eye inflamation.
Withania somnifera Dunal	Ashwag-andha	Solanaceae	Jaunsari	Leaf, Root	Urinary disorder, Fever, Insomnia	Leaf juice and root powder used for urinary disorder, fever and insomnia.
Woodfordia fruticosa (L.) Kurz.	Dhaudi	Lythraceae	Bhoxa, Jaunsari	Roots	Rheumatism, Gout, Dysentery	Root infusion used internally to treat rheumatism. Flower paste applied to cure gout. Juice of flower is given internally for dysentery.
Zanthoxylum alatum Roxb.	Timoor	Rutaceae	Bhotiya, Raji	Seed, Stem twig	Gastric, Cold, Cough, Dysentery, Pyorrhea	Seeds are chewed to get relief from toothache. Seed epicarp soup with salt is given for gastric, common cold and cough. Cooked seed is taken for dysentery. Stem bark paste is used on pyorrhea.
Zanthoxylum armatum DC.	Timru, Timoor, Tyamoor	Rutaceae	Bhotiya, Jaunsari	Bark, Leaf	Toothache, Headache	Paste of bark is used on toothache. Leaves are boiled in water and bath to treat headache.
Zingiber officinale	Adrak	Zingiberaceae	Bhotiya,	Rhizome	Paralysis,	10gm rhizome is mixed with 10ml

Roscoe			Jaunsari		Epilepsy	honey and eaten in paralysis. Rhizomes are crushed and mixed with cow ghee to cure epilepsy.
Ziziphus mauritiana Lam.	Ber	Rhamnaceae	Bhoxa	Bark	Diarrhoea	Bark decoction given to get relief from diarrhoea.
Ziziphus nummularia (Burn.f.) Wight & Arn.	Jharber	Rhamnaceae	Bhoxa	Root	Dysentery, Diarrhoea	Root powder given for dysentery and diarrhoea.

5. Wild Fruits and Their Ethnobotanical Uses

Uttarakhand is one of those places which is appropriate for wild edible fruiting plants due to their unique geographic and climatic conditions and overwhelming taste of fruits which is fascinated by people residing in the state as a rich resource of nutrition. Some of the edible fruits along with traditional use are given below:

- *Amelanchier spicata* (Lam.) K.Koch (local name-Saskatoon, family-Rosaceae): Fruits used to eat fresh, to made jam, jelly, and sauce, and also makes a fine beverage.
- *Ficus palmata* Forssk. (local name-Pheru, Bedu, family-Moraceae): Fruit contains 45% of juice, and edible source of minerals, phosphorus and a small amount of vitamin C.
- *Morus alba* L. (local name- Mulberries, family- Moraceae): Fruits used to make from making jams or jellies. The ripe fruit is edible and is widely used in pies, tarts, wines, cordials, and herbal tea.
- *Myrica esculenta* Buch.-Ham. ex D. Don (local name- Kafal or Kaphal, family-Myricaceae): Mature fruits edile raw and princkle in also made from fruit.
- *Prunus armeniaca* L. (local name- Khubani, family-Rosaceae): Fruit rich in carotene and vitamin C, provides a precious source of food, eaten fresh, as jams, dried or cooked in meat dishes. Fruit kernels can also be eaten, pressed to make almond oil or used medicinally.
- *Prunus persica* (L.) Batsch (local name- Plum; family- Rosaceae): Used widely in the preparation of jellies, jams, and desserts. People use dry fruit as edible nutrient source.
- *Punica granatum* L. (local name-Darim, family- Lythraceae): Fruit juice is known as a delicacy and is made into excellent sherbet, which has diuretic and cooling effect rich in glucose, fructose, tannins, oxalic acid, and reduces thirst in cases of fevers, supplies the required minerals and helps the liver to preserve vitamin A.
- *Pyracantha crenulata* (Roxb. ex D.Don) M.Roem. (Local name-Ghigharu, Family-Rosaceae): Fruits can be made into a preservative and leaves are used in the preparation of herbal tea, sunburn creams and many facial creams.
- *Pyrus pashia* Buch.-Ham. ex D.Don (local name- Mahal, family-Rosaceae): Ripe fruit is taken as raw food.
- *Pyrus pyrifolia* (Burm.f.) Nakai (local name-Shiara, family-Rosaceae): Fruits are juicy and edible raw.

- *Ribes nigrum* L. (local name- Blackcurrant, family-Grossulariaceae): Fruits useful as an antioxidant source and in treating rheumatoid arthritis and night and fatigue-related visual impairment.
- *Rosa canina* L. (local name- Dog Rose, family- Rosaceae): Used to make syrup, tea, and marmalade. Fruits have been used internally as tea for the treatment of viral infections.
- *Rubus ellipticus* Sm. (local name- Hisalu, Ashilo; family-Rosaceae): Used as a plant providing free energy packets for the people who are traveling mountains as they are more locally available.
- *Rubus niveus* Thunb. (local name-Kala Hinsalu, family-Rosaceae): Fruits are enjoyed fresh.
- *Rubus occidentalis* L. (local name- thimble berry, family-Rosaceae): Fresh leaves used in herbal teas.
- *Viburnum opulus* L. (local name: Guelder rose, family- Adoxaceae): During cooking, it is used as a cranberry substitute when making preserves and jellies. It can be eaten either raw or cooked.
- *Vitis vulpina* L. (local name-Jungli angur, family- Vitaceae): Fruits are raw edible.

6. Conclusion

There are several research studies available that have been carried out for the evaluation and validation particularly on the parameters of ethical knowledge of plants and modern science. Under the studies, it would be an immense scope to find out lesser-known medicinal plants which are utilized for folk medicine and also have a scientific basis and finally isolation, characterization, and development of newer and lead phytomedicine components. Indigenous system of medicine is a leading healthcare provider approximately in the globe mainly in the rural and remote areas. People in mass numbers depend on the indigenous system of medicine for their primary healthcare mostly in under developed or developing countries. Ethical knowledge of plants has a very great history of their effectiveness; modern research also accepted the significance of such medicine. Folk medicinal plants are considered as a crucial source of new drugs and mainstreaming of such medicine is vital for the peoples. Many measures have been taken in India to uplift such medicine and to incorporate them into clinical practice. Evidence-based incorporation of folk medicine in clinical practice will help to provide quality healthcare to all.

Abbreviation Used

No abbreviations

Conflict of Interest

Authors declare no conflict of interest for this manuscript.

Acknowledgements

Authors would like to thank Director IIIM for facilities and encouragement.

References

Abebe D & Ayehu A (1993). Medicinal plants and Enigmatic Health practices of Northern Ethiopia. Published by BSPE, University of California, California.

Berkes F, Colding J, Folke C (2000). Rediscover of traditional ecological knowledge as adaptive management. *Ecological Application* 10: 1251–1262.

Bhatt D, Joshi GC & Tiwari LM (2009). Culture, Habitat and Ethno-Medicinal practices by Bhotia Tribe people of Dharchula Region of Pithoragarh District in Kumaun Himalaya, Uttarakhand. *Ethnobotanical Leaflets* 13: 975-983.

Cunningham AB (2001). Applied Ethnobotany. People, Wild Plant Use and Conservation. Earthscan, London.

Gairola S, Sharma J, Gaur RD, Siddiqi TO & Painuli RM (2013). Plants used for treatment of dysentery and diarrhoea by the Bhoxa community of district Dehradun, India. *Journal of Ethnopharmacology* 150(3): 989-1006.

Gaur RD (1999). Flora of the District Garhwal North West Himalaya with Ethnobotnaical notes. Tran media Publication, Srinagar Garhwal.

Gaur RD, Sharma J & Painuli RM (2010). Plants used in traditional healthcare of livestocks by Gujjar community of Sub-Himalayan tracts, Uttarakhand, India. *Indian Journal of Natural Products and Resources* 1(2): 243-248.

Gour RD & Sharma J (2011). Indigenous knowledge on the utilization of medicinal plants diversity in Shiwalik region of Garhwal Himalaya, Uttarakhand. *Journal of Forest Science* 27(1): 23-31.

Gunderson LH & Holling CS (2002). Panarchy: understanding transformations in human and natural systems. Island Press, Washington, D.C., USA.

Gupta MP, Corea MD, Soils PN, Jones A Galdames C (1995). Medicinal Plants inventory of Kuna Indians. *Journal Ethnopharmacology* 40: 77–109.

Harshburger JW (1896). Purpose of Ethnobotany. *Botanical Gazette* 21:146–154.

Huntingto HP (2000). Using traditional ecological knowledge in science: methods and application. *Ecological Applications* 10: 1270–1274.

Jain SK (1991). Dictionary of Indian Folk Medicine and Ethnobotany. Deep publications, Paschim Vihar, New Delhi, India.

Joshi B & Pant SC (2012). Ethnobotanical study of some common plants used among the tribal communities of Kashipur, Uttarakhand. *Indian Journal of Natural Products and Resources* 3(2): 262-266.

Joshi M, Kumar M & Bussmann RW (2010). Ethnomedicinal Uses of Plant Resources of the Haigad Watershed in Kumaun Himalaya. *Journal of Medicinal and Aromatic Plant Science and Biotechnology* 4(1): 43–46.

Kala CP (2005). Current status of Medicinal plants used by traditional Vaidys in Uttaranchal Sate of India. *Ethnobotany Research and Application* 267–278.

Kala CP (2002). Indigenous Knowledge of Bhotia tribal community on wool dying and its present status in the Garhwal Himalaya, India. *Current Science* 83: 814-817.

Kumar M, Sheikh MA & Bussmann RW (2011). Ethnomedicinal and ecological status of plants in Garhwal Himalaya, India. *Journal of Ethnobiology and Ethnomedicine* 7: 32, doi: 10.1186/1746-4269-7-32

Phondani PC, Maikhuri RK, Rawat LS, Farooquee NA, Kala CP, Vishvakaram SCR, Rao KS & Saxena KG (2010). Ethnobotanical useds of plants among the Bhotiya tribal communities of Niti valley in central Himalayan, India. *Ethnobotany Research & Applications* 8: 233–244.

Ratha KK, Joshi G C, Rungsung W & Hazra J. (2015). Use pattern of high altitude medicinal plants by bhotiya tribe of niti valley, Uttarakhand. *World Journal of Pharmacy and Pharmaceutical Sciences* 4(6): 1042-1061.

Sharma J & Painuli RM (2011). Plants used for the treatment of rheumatism by the Bhoxa tribe of district Dehradun, Uttarakhand, India. *International Journal of Medicinal and Aromatic Plants* 1(1): 28-32.

Sharma J, Gairola S, Gaur RD & Painuli RM (2011). Medicinal plants used for primary healthcare by Tharu Tribe of Udham Singh Nagar Uttarakhand, India. *International Journal of Medicinal and Aromatic plants* 1(3): 228-233.

Sharma J, Gairola S, Gaur RD & Painuli RM. (2012). Forest utilization pattern and socio-economic status of Van-Gujjar tribe in sub-Himalayan tracts of Uttarakhand, India. A case study. *Forestry Studies in China* 14(1): 36-46.

Sharma J, Gaur RD & Painuli RM. (2011). Folk herbal medicines used by the Gujjar tribe of sub-Himlayan tract, Uttarakhand. *Journal of Economic and Taxonomic Botany* 35(1):224-230.

Sharma J, Gaur RD, Gairola S, Painuli RM & Siddiqi TO (2013). Traditional herbal medicines used for the skin disorders by the Gujjar tribe of Sub-Himalayan tract, Uttarakhand. Indian *Journal of Traditional Knowledge* 12(4): 736-746.

Singh VK, Ali ZA Siddiqui MK (1997). Folk medicinal plants of Garhwal and Kumaon forest of Uttar Pradesh, India. *Hamdard Medicus* 40: 35–47.

5

Pharmacological Potential of Ethnomedicinal Plants of Asteraceae Family from Arunachal Pradesh Northeast India

Wishfully Mylliemngap and Om Prakash Arya

Abstract

Asteraceae is one of the largest family of flowering plants and have been used by mankind for a variety of purposes including food, medicine, cooking oils, sweeteners and tea infusions. In India, Asteraceae is 4th largest family of angiosperms and ranked third among dominant families of angiosperms in Arunachal Pradesh. The indigenous communities of Arunachal Pradesh possess a rich knowledge on utilization of the plant resources around them for food, fodder, medicine and other purposes. Due to the remoteness and inaccessibility, the indigenous communities are mostly depended on their traditional medicines using plant resources around them for treating minor and major ailments. The effectiveness of these plants in traditional medicine has also been confirmed after isolation and characterization of the bioactive compounds responsible for their pharmacological activities. The present study is a review on ethnomedicinal plants belonging to Asteraceae family that are being used by different tribes of the state. The tribal communities have been using 36 medicinal plant species from this family as recorded from published literature and field study. Maximum number of species have been found to exhibit antimicrobial, antioxidant, antihelminthic, analgesic and anti-malarial properties. Among the recorded species, Ageratum conyzoides, Spilanthes acmella, Artemisia nilagirica, A. maritima, Blumea balsamifera, B. lacera, Eclipta prostrata, Vernonia cinerea and Chrysanthemum indicum were already being used in Ayurveda, Unani, Siddha,

Wishfully Mylliemngap(✉) and Om Prakash Arya

G.B. Pant National Institute of Himalayan Environment and Sustainable Development North-East Regional Centre, Itanagar, Arunachal Pradesh-791113, India

✉*Corresponding author: wishm2015@gmail.com*

Plants for Novel Drug Molecules: Ethnobotany to Ethnopharmacology
Bikarma Singh & Yash Pal Sharma (eds.), (pp. 141-168)

Email: *info@nipabooks.com* Web: *www.nipabooks.com*

Homeopathy, Folk and Sowa-Rigpa systems of Indian Medicine. Therefore, the study highlighted the pharmacological potential of ethnomedicinal plantsfrom Asteraceae family which would provide opportunities for future drugdiscovery.

Keywords: Traditional medicine, indigenous communities, pharmacological activity, conservation.

1. Introduction

The state of Arunachal Pradesh, which covers a major portion of the eastern Himalaya, with a geographical area of 83, 743 sq. km, is known for its rich diversity of flora and fauna. It comprises a part of the Indo-Burman biodiversity hotspots of the world (Myers et al. 2000). Arunachal Pradesh is home to as many as 26 major tribal communities and over 110 sub-tribes each having unique cultural and traditional values, thus exhibiting a rich cultural diversity. The use of plants as medicine for treating minor and major ailments has been an age-old history and practice of the different indigenous communities, particularly those living in remote areas. Due to the remoteness of these areas, this practice remained the most convenient, economical and easily accessible mean of medical treatment. Several studies conducted on ethnomedicinal plants used by different tribes of the Arunachal Pradesh have revealed that they have a vast wealth of knowledge in traditional healthcare practices using a number of wild and semi-domesticated plant species. These plants have been reportedly used for treating a variety of diseases and ailments ranging from cuts and wounds, skin diseases, gastrointestinal problems to jaundice, malaria, diabetes and cancer. A number of these species were found to be widely used in Ayurveda, Siddha, Unani, Homeopathy and other systems of Indian medicine (http://www.medicinalplants.in/). Therefore, the state has a great potential in developing the medicinal plant sector for further pharmaceutical research and validation of medicinal properties of these locally available plants.

Asteraceae or Compositae is the one of the largest family of flowering plants with cosmopolitan distribution, except in Antarctica. Till date,1,911 genera and 32,913 accepted species nameshas been reported in the Asteraceae family (http://www.theplantlist.org/1.1/browse/A/Compositae).

The Asteraceae family is distributed worldwide except for Antarctica but diverse in the tropical and subtropical regions of North America, the Andes, eastern Brazil, southern Africa, the Mediterranean region, central Asia, and southwestern China. Majority of the species of Asteraceae are herbaceous, though shrubs or even trees form a considerable component of the family occurring primarily in the tropical regions of North and South America, Africa and Madagascar and on isolated islands in the Atlantic and Pacific Oceans.

In India, Asteraceae is the fourth largest family of angiosperms with 167 genera and 950 species after Poaceae, Orchidaceae and Leguminosae. It ranks third among the dominant families of angiosperms in Arunachal Pradesh with 68 genera and 186 species (http://arunachalforests.gov.in/forest %20statistics.html). This family includes a great diversity of species, including annuals, perennials, stem succulents, vines, shrubs and trees. Asteraceae is the largest family of medicinal plants in northeast India (Saklani & Jain 1994).

Plants of the Asteraceae family are characterised by the special type of inflorescence called as 'head' or 'capitulum'. The capitulum (plural: capitula) is a specialized indeterminate inflorescence that can contain 1 to hundreds of individual flowers (florets). The flowering sequence in the capitulum is nearly always from the outside to the center, that is, centripetal. The head comprises of two types of florets, the outer florets are strap-shaped called as 'ray florets' and the inner circular shaped florets called 'disc florets'. The calyx is modified into pappus or absent or scale-like. At the base of the head, and surrounding the flowers before opening, is a bundle of sepal-like bracts or scales called 'phyllaries', which together form the involucre that protects the individual flowers in the head before opening. The flowers of a head may be all hermaphrodite (*Ageratum*), or ray-florets are female or neuter and inner ones hermaphrodite, or male; rarely the complete head bears unisexual flowers.

Species of Asteraceae family have been used by mankind for a variety of purposes including food, medicine, cooking oils, sweeteners and tea infusions. Many species such as Marigold, Dahlia, Zinnia and Chrysanthemum have been used as ornamental plants. Phytochemistry of this family revealed the presence of polyfructanes (especially inulin) as storage carbohydrates. They produce iso/chlorogenic acid, isoflavonoids, pentacyclic triterpene alcohols, terpenoid essential oils, various alkaloids, a variety of fatty acids in the seeds, tannins and no iridoids (Stevens 2001 onwards). Some taxa,e.g., *Chrysanthemum parthenium* and *Arnica montana* accumulate sesquiterpene lactones (e.g. Parthenolides) which are important natural products responsible for the pharmacological activities of many botanical drugs. Review of existing literature revealed that a number of medicinal plants from this family were already being used by different indigenous communities of Northeast India and Arunachal Pradesh in particular, in their traditional healthcare practices. These species have been found to possess a variety of medicinal properties ranging from antioxidants, anti-inflammatory, analgesic to anticancer, antidiabetic and antimalarial. Therefore, this study was carried out to draw attention to the potential of Asteraceae family as a source of raw materials in pharmacological and phytochemical studies for future drug development.

In the present review, a list of plants belonging to Asteraceae family which were being used by different indigenous communities of Arunachal Pradesh in their traditional medicine have been compiled from various published sources such as journals, books, theses, dissertations and other scientific publications. Pharmacological activities of the recorded species have been reviewed from global published literature. Further, the traditional use categories of the recorded species were matched with its reported pharmacological activities as an attempt to support its effective use in traditional medicine for a particular ailment.

2. Traditional Medicinal Uses

A total of 36 species belonging to 23 genera have been documented, which were being used in traditional medicine by different tribes of Arunachal Pradesh (Table 1). These plants were used by the local people and traditional medical practitioners (TMPs) to treat a variety of minor and major ailments including cuts and wounds, cold, fever, cough, gastrointestinal problems, pain reliever, antidote for snake and insect bites, diabetes, malaria, cancer, jaundice and other diseases. The diseases treated were categorised into 9 different groups (Table 2). Maximum number of species were being used for treatment of gastrointestinal disorders (24%) followed by general health (23%), as pain reliever (18%) and dermatological disorders (15%) (Fig. 1). Among the recorded species, *Ageratum conyzoides, Spilanthes acmella, Artemisia nilagirica, A. maritima, Blumea balsamifera, B. lacera. Eclipta prostrata, Vernonia cinerea* and *Chrysanthemum indicum* were reported to be widely used in Ayurveda, Unani, Siddha, Homeopathy, Folk and Sowa-Rigpa systems of Indian Medicine (http://www.medicinalplants.in).

Table 1: Traditional uses and pharmacological properties of ethnomedicinal plants of Asteraceae family from Arunachal Pradesh, Northeast India

Botanical name	Vernacular name*	Traditional medicinal uses	Pharmacological properties
Ageratum conyzoides L.	Pakku (A); Eww Namya (G); Eh gaar (T); Passo payo (N); Namninyng (Tangsa, Padam); Maon/ Ngonamshu (Mon); Pasong (Aka); Chengache (Lisu); Gundhuaɔon (Ch); Padribha (Kt)	Check bleeding and wound healing (Das et al. 2013,Baruah et al. 2013, Bam et al. 2015); conjunctivitis, dysentry and diarrhoea (Kagyung et al. 2010, Srivastava & Adi Community 2009, Srivastava & Nyishi Community 2010); root juice taken as antihelminthic (Perme et al. 2015, Das 2003); Khamptis used in fire burnt skin after mixing with *Ficus villosa* resin (Tag et al. 2007)	Anticancer, antioxidant (Adebayo et al. 2010); antiprotozoal (Nour et al. 2010); analgesic, antipyretic and anti-inflammatory (Okunade 2002); Anthelmintic (Sharma et al. 1979); Antidiabetic (Nyunai et al. 2009); antimicrobial (Singh et al. 2016); antibacterial (Odeleye et al. 2014)
Artemisia indica Willd.	Laglin (A); Tapin (N)	Bodyache, asthma, skin diseases and allergy, nose block, cure redness of eyes, headache, (Kala 2005, Srivastava & Adi Community 2009, Srivastava & Nyishi Community 2010); believed to be effective in breast cancer (Perme et al. 2015)	Anti-inflammatory and analgesic properties (Sagar et al. 2010)
Artemisia maritima L.	-	Blood purification (Kala 2005)	Antimalarial (Valecha et al. 1994) hepatoprotective (Janbaj & Gilani 1995); antimicrobial (Mitscher et al. 1972); antibacterial (Yashphe et al. 1979); anthelminthic (Sareen et al. 1961); Antidiabetic (Twaij & Al-Badr 1988); cytotoxic and antitumors (Hartwell 1971)
Artemisia nilagirica	Illumu (I), Titapati	Wound and nose bleeding (Ghosh et al.	Antimicrobial (Shafi et al, 2004);

Contd.

(C.B.Clarke) Pamp	(Nep); Merangma (Mon); Syowum (Aka); Tape nain (HM); Makampi (Kt)	2014); Cough, asthma, fever (Nimachow et al. 2011, Shankar et al. 2003); headache, stomach pain (Kohli 2003); asthma, sores, scabies and inflammations (Kala 2005, Namsa et al. 2011, Perme et al. 2015); acts as disinfectant (Tag & Das 2004); Fresh leaf paste applied on swelling, boil, eruption and old wound to prevent microbial infection (Tag et al. 2007)	antifungal (Chowdhury et al. 2003); antibacterial (Rao et al. 2006); antiulcer (Sagar et al. 2008); anticancer and antioxidant (Devmurari & Jivani 2011); anti-asthmatic (Aparna & Duraiswamy 2009)
Artemisia parviflora Buch. Ham. ex Roxb.	Taping roming (N)	Older people use to carry bundles on their back for 4-6 hours per day to get relieve from backache (Srivastava & Nyishi Community 2010)	Antifungal (Ahuja et al. 2011)
Artemisia vulgaris L.	Khampa/Khanme (Mon)	Leaves used as leech repellent (Bam et al. 2015); root used as tonic; plant used as antihelminthic (Shankar & Rawat 2008)	Antitumor (Abdelhamed et al. 2013); antimicrobial (Poiatã et al. 2009); antiviral (Meneses et al. 2009); antioxidant (Temraz & El-Tantawy 2008); antiinflammatory (Tigno & Gumila, 2000); analgesic (Pires et al. 2009)
Bidens biternata (Lour.) Merr. & Sherff	Tagam Nyeinyam/ Takam Pechi (N)	Eye, ear drops and asthma (Ghosh et al. 2014)	Antimicrobial (Reddy et al. 2015); anticancer (Pradeesh 2018); antioxidant (Zahara et al. 2015); antibacterial and antifungal activities (Ahmed et. al.2016)
Bidens pilosa L.	Hou bok (Ap); Robashing (Mon); Cheae (Lisu); Buki (HM)	Wound healing, ulcers, ear and eye problems (Khongsai et al. 2011); skin inflammation (Namsa et al. 2011); hepatitis, urinary tract infection, malaria (Perme et al. 2015)	Antidiabetic (Chien et al. 2009); antitumoral (Kviecinski et al. 2008); antimicrobial (Deba et al. 2008); hepatoprotective (Yuan, et al, 2008; Kviecinski et al. 2011); and

Contd.

			antioxidant (Chiang et al. 2004, Krishnaiah et al. 2011); Antimycobacterial (Newton et al. 2000)
Blumea balsamifera DC.	Yanang (Kt); Tangloti (A)	Diabetes, body pain (Das & Tag 2006); Leaves are used in menstrual problems (Payum et al. 2015); Helminthiasis in children (Das 2003)	Anti-microbial and anti-inflammatory (Li et al. 2008, Ragasa et al. 2005); antiplasmodial effects (Noor et al. 2007); antitumor (Li et al. 2008, Saewan et al. 2011); hepatoprotective (Pu et al. 2000); antioxidant (Nguyen et al. 2004)
Blumea fistulosa Kurz	Rumbdum (A); Yamang hak (Kt)	Diarrhoea (Ali & Ghosh 2006, Payum et al 2015, Srivastava & Adi Community 2009); Fresh juice of leaves applied to reduce swelling from sting of wasps (Tag et al. 2007)	NA
Blumea lacera DC.	Sidirebaisak (Ch)	Relieve pain and swelling (Sarmah et al. 2008)	Anti-inflammatory, analgesic, antipyretic (Khair et al. 2014); antiviral (Chiang et al. 2004), cytotoxic activities against breast cancer cells (Uddin et al. 2011)
Carthamus tinctorius L.		Fruit juice used to cure jaundice (Shankar et al. 2012)	Antidiabetic, antitumour, antiinflammatory and analgesic, anticoagulant, antioxidant, hepatoprotective (Dehariya and Dixit 2015,Gautam et al. 2014)
Chromolaena odorata L.	Daglin (G); Telimbabo (A); Wo pohing (No)	To stop bleeding from leech bite sites (Bam et al. 2015); Cuts and wounds (Gupta 2006, Shankar et al. 2003); Leaf juice in used in bleeding and as pain reliever; headache, fever	Anthelminthic (Mishra et al. 2010); antimalarial (Ongkana 2003); analgesic (Jena and Chakraborty et al. 2010); anti-Inflammatory, Antipyretic and

Contd.

		(Kala 2005, Payum et al. 2015, Srivastava & Adi Community 2009, Tangjang et al. 2011); anti- gonorrhoeal, skin disease (Perme et al. 2015)	antispasmodic (Taiwo et al. 2000), anti-oxidant (Phan et al. 2001); antimicrobial activity (Chomnawang et al. 2005)
Chrysanthemum indicum L.	Hahkak rangkak (Wancho)	Pain relief (Wanjen et al. 2011); Chest pain, prostate cancer, diabetes, stomach ache, fever, dysentery, cold, swelling (Primary data)	Anti-inflammatory (Xiao-Li Wu et al. 2013); Hepatoprotective (Jeong et al. 2013)
Cirsium lepskyi Petral.	-	Indigestion (Kala 2005)	NA
Crassocephalum crepidiodes Moore.	Gianda (N); Harang nimang 'Babuk' (Bangnis); Jakpangon (Mon)	Indigestion and diarrhea (Doley et al. 2010); antiseptic for cuts and wounds, stomach disorder, gastric trouble, headache; (Gupta 2006, Kala 2005, Namsa et al 2011, Srivastava & Adi Community 2009); Anti-malarial (Perme et al. 2015)	Antidiabetic (Bahar et al. 2017); anti-cancer (Thakur et al 2019)
Dichrocephala bicolor (Roth) Sch.		Digestive problems (Kala 2005)	Antihelminthic (Wabo et al. 2013); antimicrobial and antidiarrheal (Kamgang et al. 2015); anti-inflammatory and cytotoxic (Lee et al. 2015)
Eclipta prostrata (L.) L.	Keharaj (A); Bhumiraj (Ch)	Antiseptic in fresh cuts and wounds (Kala 2005); Decoction of leaf is given in dysentery with vomiting (Yongam 2007); Leaves used as hair conditioner as well as tonic (Payum et al. 2015); jaundice and fever (Shankar et al. 2012, Sarmah et al. 2008)	Anthelminthic (Bhinge et al. 2010); Hepatoprotective (Chandra et al. 1987, Ma-Ma et al. 1978); antidiabetic (Jaiswal et al. 2012); antimicrobial (Panghal et al. 2011); antioxidant (Sinha and Raghuwanshi 2016)
Erigeron bonariensis L.		Nose block (Kala 2005); sinus problem (Perme et al. 2015, Srivastava & Adi Community 2009)	hepatoprotective (Saleem et al. 2014); antimicrobial (Shah et al. 2013)

Contd.

Gerbera piloselloides (L.) Cass.	Daglentado (A)	Rheumatic pain (Srivastava & Adi Community 2009); cold, fever, acute conjunctivitis (Kala 2005, Perme et al. 2015)	Antioxidant (Wang et al. 2014)
Gnaphalium affine D.Don	Pangnesir (A)	Treatment of common cold, gout (Perme et al. 2015)	Anti-inflammatory (Huang et al. 2015); antioxidant activity (Zeng et al. 2013)
Gynura bicolor (Roxb. ex Willd.) DC.		Intestinal worms (Kala 2005)	Antioxidant, cytotoxic (Teoh et al. 2013)
Gynura cusimbua (D. Don) Moore	Buli	Leaf or shoot decoction is applied in fresh cut (Primary data)	Antiangiogenic activity (Ma et al. 2017)
Gynura nepalensis DC.	-	Indigestion (Kala 2005)	Antioxidant and α-glucosidase inhibitory activities (Quiming et al. 2016)
Inula cappa (Buch-Ham ex D.Don) DC	Chengache (Lisu)	Leaves crushed with those of *Plantago asiatica* and *Lobelia angulata* and juice taken for jaundice (Shankar et al. 2012)	Anti-inflammatory activities (Wu et al. 2015); antibacterial (Priydarshi et al. 2015)
Laggera pterodonta (DC.) Sch.-Bip. ex Oliver	Dindo eh (HM)	Leaf paste used for treating inflammation and swelling, anthelminthic (Perme et al. 2015, Tag & Das 2004)	Antiviral (Li et al. 2004); antiinflammatory (Wu et al. 2006); hepatoprotective and anti-oxidant (Wu et al. 2007); antitumor (Liu et al. 2008); antibacterial (Egharevba et al. 2010)
Mikania micrantha H.B.K.	Japani lota (A); Tare (N)	Dysentry and diarrhoea (Kagyung et al. 2010); Itching, skin diseases, headache (Kala 2005); wound healing; Warmed leaves kept on eyes repeatedly to cure any type of eye trouble; snake bite and scorpion sting (Das et al. 2013, Srivastava & Nyishi Community 2010).	Antibacterial, antitumor, cytotoxic, analgesic (Zhuang et al. 2010, Li et al. 2013); anti-inflammatory (Jyothilakshmi et al. 2016); anticancer (Dasgupta et al. 2014); antioxidant (Chethan et al. 2012).

Contd.

Mikania scandens Willd.	Chakpan (Ts); Hipini (No); Namleriyong (A)	Crushed leaves applied to cuts and wounds to stop bleeding (Khongsai et al. 2011,Tangjang et al. 2011); Diarrhoea (Tangjang et al. 2011); insect bites and sting, gastric ulcer (Perme et al. 2015)	Analgesic and antioxidant effect (Hasan et al. 2009); anti inflammatory (Banerjee et al. 2014); hepatoprotective (Maity et al. 2012)
Solidago virgaurea L.	Siyuji (Lisu)	Leaves paste used for treatment of swelling (Sarmah 2010)	Anticancer (Gross et al. 2002); Anti-inflammatory (El- Ghazaly 1992)
Sonchus arvensis L.	-	Stomach-ache, gastritis (Kala 2005), Root juice in water used to cure jaundice (Shankar et al. 2012)	Anti-inflammatory (Poudel et al. 2015); hypoglycaemic and analgesic (Sah et al. 2015)
Sonchus asper (L.) Hill	-	Indigestion (Kala 2005)	antioxidant and antibacterial properties (Jimoh et al. 2011); Hepatoprotective activity (Aftab-Ullah et al. 2015)
Sonchus wightianus DC.	Balakhar (Mon)	Dysentry (Kar & Borthakur 2008)	NA
Spilanthes acmella L.	Namlang marching (M); Marcha (Ap)	Crushed roots and flowers used for toothache (Khongsai et al. 2011); malaria, fever (Perme et al. 2015)	Antifungal (Rani and Murty, 2006); antipyretic and local anaesthetic (Chakraborty et al. 2010); antiinflammatory/analgesic (Peiris et al. 2001, Chakraborty et al. 2004); pancreatic lipase inhibitor (Ekanem et al. 2007); antinociception (Whittle 1964, Ahmed et al. 2004); diuretic (Ratnasooriya et al. 2004); vasorelaxant and antioxidant (Wongsawatkul et al. 2008,Hossain et al. 2012); immunomodulatory (Wu et al. 2008)
Spilanthes paniculata Clarke	Marsha/Marshang (A); Yamoglun (Kt);	Stomach problem (Ghosh et al. 2014, Gupta 2006); constipation (Bhuyan 2015,	Hepatoprotective and antioxidant activity (Ali et al. 2012);

Contd.

	Byadhi (N); Soeloe (No)	Kala 2005); toothache and tongue infection (Ali & Ghosh, 2006, Baruah et al. 2013, Tangjang et al. 2011); cough, fever and cold (Gupta 2006); bodyache (Khongsai et al. 2011); leaves mixed with *Piper nigrum* used for killing intestinal worms (Srivastava & Nyishi Community 2010)	antinociceptive (Hossain et al. 2012); antimicrobial (Morshed et al. 2011)
Taraxacum officinale Wigg.		Used as liver tonic, jaundice (Shankar et al. 2012)	Antitumor, cytotoxic, antioxidant (Chun & Kitts 2003, Sigstedt et al. 2008); immunostimulant (Lee et al. 2012); antimicrobial (Yarnell & Abascal 2009); anti-inflammatory (Jeon et al. 2008); hypoglycemic (Önal et al. 2005); depurative (Modaresi & Resalatpour 2012); hepatoprotective (Mahesh et al. 2010); hypolipidemic (Choi et al. 2010).
Vernonia cinerea (L.) Less	Reacheaful (Ch)	Indigestion (Kala 2005); leaf and stem paste applied on forehead for headache (Sarmah et al. 2008)	antidiarrhoeal (Nagaraj & Venkateswarlu 2013); Antibacterial (Rizvi et al. 2011); antimicrobial (Arumugam et al. 2010, Latha et al. 2009), nephroprotective (Sreedevi et al 2011); anti- inflammatory (Mazumder et al 2003); analgesic and antipyretic activity (Iwalewa et al. 2003,); Antitumor (Sangeetha and Venkatarathinakumar 2011); Antioxidant (Chethan et al. 2012)

Abbreviations: * A- Adi; I- Idu mishmi; K- Khamba; Adi.M- Adi-Minyong; M- Memba; Mon- Monpa; G- Galo; N- Nyishi; As- Assamese; Nep- Nepali; Ap- Apatani; Kt- Khampti; T- Tagin; Ch-Chakma; No-Nocte; Ts – Tangsa; NA – not found in literature

Photo Plates 1: *Ageratum conyzoides* 2. *Crassocephalum crepidioides* 3. *Chromolaena odorata* 4. *Spilanthes paniculata*; 5. *Spilanthes acmella*; 6. *Artemisia* sp.

Table 2: Traditional use categories of diseases/disorders

Traditional use category	Diseases/disorders
General health	Health tonic, blood purifier, blood coagulant, common cold, cough, fever, asthma, eye infection, ENT problems, menstrual disorder, hepatitis, chest pain
Gastro-intestinal disorder	intestinal parasites, diarrhoea, dysentry, gastritis, indigestion, gastric ulcer, stomachache
Dermatological	Cut and wound, burns and scalds, boils, rashes, skin inflammation, dandruff
Urological	diabetes, urinary tract infection, gonorrhoea
Pain relief	headache, toothache, bone fracture, arthritis, gout, bodyache
Snake/insect bite	snake bite, insect bites/sting
Cancer	Cancer of different organs
Malaria	-
Jaundice	-

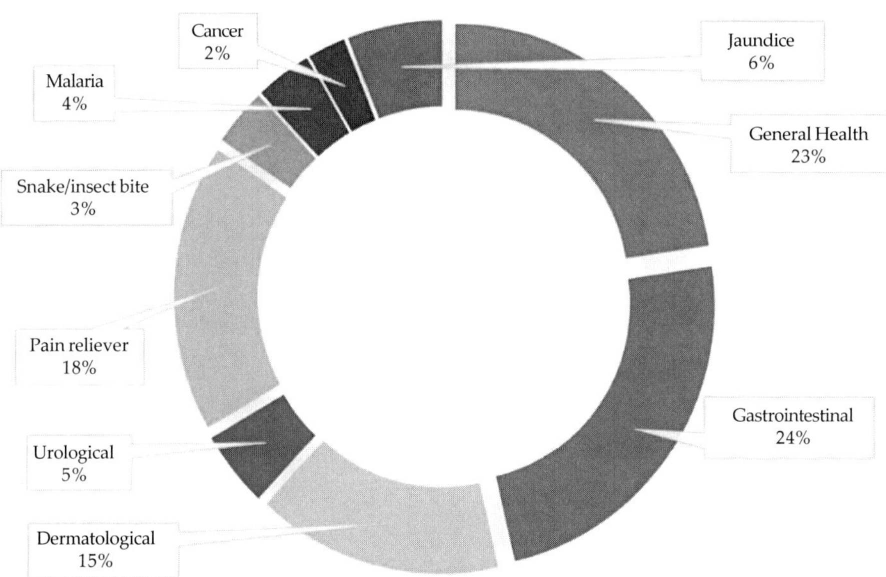

Fig. 1: Percentage of species used for treating different ailments

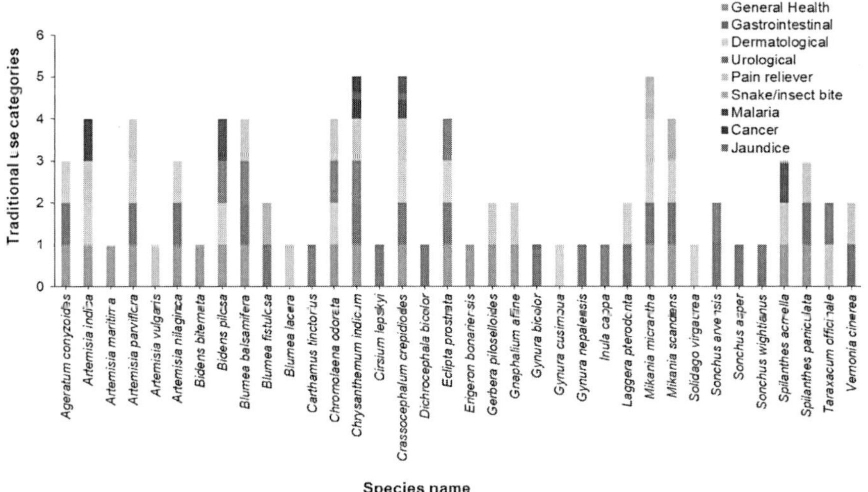

Fig. 2: Traditional use categories of the recorded species

3. Pharmacological Activity

In order to comprehend the medicinal values of the recorded plant species and to support their utilization in traditional healthcare systems, reported pharmacological activities exhibited by each species were compiled from global literature of published works in journals, books, theses, dissertations and other scientific reports. Pharmacological properties exhibited by the species were categorised into 10 broad groups, viz., antimicrobial, antioxidant, anti-inflammatory, analgesic, antipyretic, antihelminthic, hepatoprotective, anticancer, antidiabetic and antimalarial. Pharmacological studies of only 33 species out of the 36 recorded species were available in literature, related studies of the remaining 3 species viz., *Blumea fistulosa, Cirsium lepskyi* and *Sonchus wightianus* were not found in literature. The maximum number of species were found to exhibit antimicrobial and antioxidant properties (21 species each), followed by antihelminthic (18 species), analgesic (17) and anti-malarial (12). The minimum number of species exhibiting a particular pharmacological property was three 3 indicating that all the recorded species possess multiple pharmacological properties (Fig. 3).

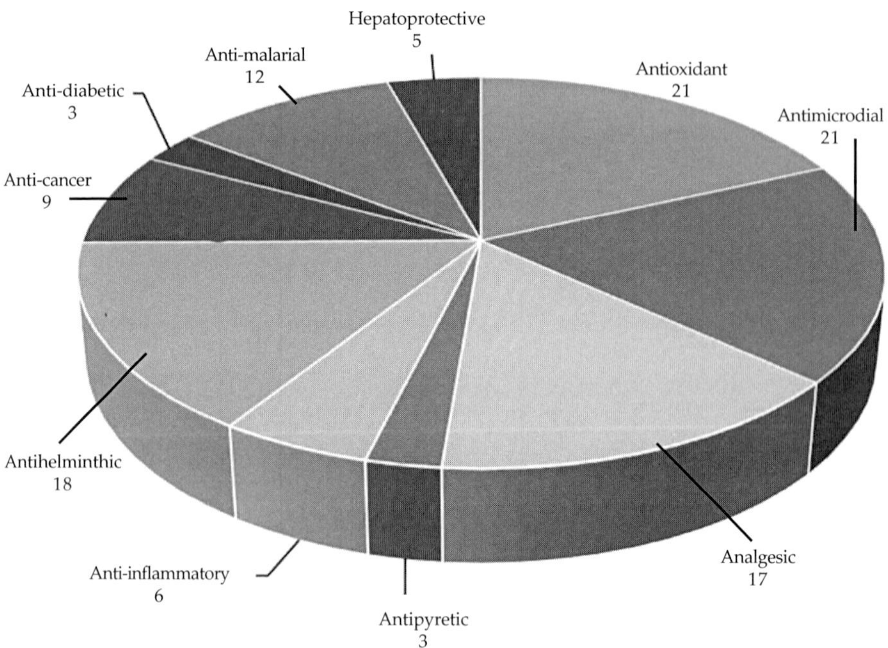

Fig. 3: Number of species exhibiting a particular pharmacological property

More than 50% of the species exhibited at least 4 out of the ten (10) pharmacological properties considered in the present study. *Ageratum conyzoides* ranked the highest with 8 pharmacological properties followed by *Chromolaena odorata* with 7 properties and *Artemisia maritima, Blumea balsamifera, Carthamus tinctorius, Taraxacum officinale* and *Vernonia cinerea* with 6 properties each. Seven species, viz., *Artemisia nilagirica, Bidens pilosa, Blumea lacera, Eclipta procera, Laggera pterodonta, Mikania micrantha* and *Spilanthes acmella* were reported to exhibit up to 5 pharmacological properties. The remaining species exhibited 4 or lesser number of pharmacological properties (Fig. 4)

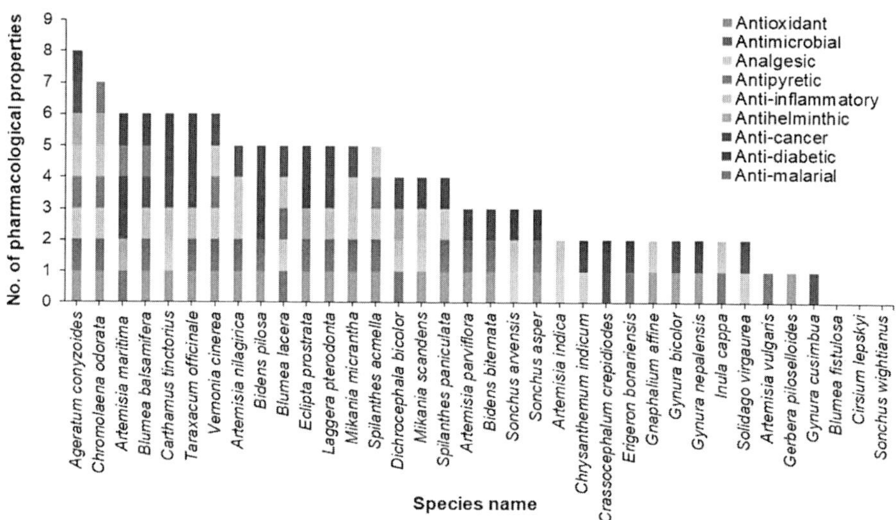

Fig. 4: Ranking of species in descending order based on number of pharmacological properties

4. Validation of Traditional Knowledge through Pharmacological Activities

Traditional knowledge in medicinal plant usage is an integrated part of the indigenous communities of Arunachal Pradesh. The knowledge is developed through trial and error and has been constantly evolving through decades of association with nature. A closer look into the traditional uses of the recorded species and comparison with scientific studies on pharmacological properties revealed the existence of one or more pharmacological properties which more or less justified their claimed in curing certain ailments. This approach was followed here to compare the traditional uses of each recorded species with the reported pharmacological properties in order to validate its effectiveness in traditional medicine.The use of these species for one or more diseases/disorder was being justified, at least partially, by comparing the traditional uses with the reported pharmacological activities.

Out of the 33 species whose pharmacological properties have been determined and reported by different authors, 18 species were found to possess at least one traditional use that could be validated by one or more of these pharmacological properties. The traditional uses of the remaining 15 species could not be validated from their reported pharmacological properties. Nevertheless, these 15 species have wide traditional uses and were found to exhibit multiple pharmacological properties. This calls for further pharmacological investigation of these species for scientific validation of each traditional usage in order to realize the potential of each species for future drug discovery.

The traditional use of *Ageratum conyzoides* to stop bleeding of wounds and treatment of burn injury may be substantiated by its anticoagulant, antimicrobial and anti-inflammatory activities (Okunade 2002). Traditional uses of *Artemisia indica* in treatment of bodyache, headache, skin diseases, allergy and redness of eyes could be validated by its reported analgesic and anti-inflammatory properties (Sagar et al. 2010). *Artemisia nilagirica,* traditionally used for treatment of wounds, sores, scabies and as disinfectant has been found to possess antimicrobial properties (Shafi et al. 2004). Moreover, it has been reported to possess anti-asthmatic property (Aparna & Duraiswamy 2009) which validated its traditional use in treatment of asthma. *Artemisia vulgaris* root, traditionally used as tonic was reported to exhibit antioxidant properties (Temraz & El-Tantawy 2008). Traditional treatment of hepatitis by using *Bidens pilosa* could be linked to its reported hepatoprotective activity (Yuan et al. 2008, Kviecinski et al. 2011) while its effective uses for wound healing and urinary tract infections may be due to its reported antimicrobial properties (Deba et al. 2008). *Blumea lacera* has been used in traditional medicine for relieving pain and swelling which could be substantiated by its analgesic and anti-inflammatory properties as reported by (Khair et al. 2014). *Carthamus tinctorius,* used by all tribes of the state for treating jaundice was reported to have hepatoprotective activity (Dehariya & Dixit 2015). *Chromolaena odorata* was commonly used by different tribes for a wide variety of disorders as a diuretic, pain reliever, headache, fever, blood coagulation and gonorrhoea. It was reported to have several pharmacological properties such as analgesic (Jena & Chakraborty 2010), antipyretic (Taiwo et al. 2000), antigonorrhoeal (Cáceres et al. 1995) diuretic (Rejitha Gopinath. 2009) and blood coagulating activity (Triratana et al. 1991), proving its wide application and effectiveness in traditional medicine. *Chrysanthemum indicum* used for relieving swelling has been reported to possess anti-inflammatory property (Wu et al. 2013). *Dichrocephala bicolor*, used for various types of digestive problems possess antihelminthic (Wabo et al. 2013), antimicrobial and antidiarrheal (Kamgang et al. 2015) activities. *Eclipta prostrata* used in traditional medicine as antiseptic in fresh cuts and wounds and for treating jaundice was reported to have antimicrobial properties (Panghal et al. 2011) and hepatoprotective activities (Chandra et al. 1987). *Laggera pterodonta* reportedly showed anti-inflammatory activity (Wu et al. 2006) which supported its traditional use in treating inflammation and swelling. *Mikania micrantha,*used traditionally for treatment of headache, skin diseases, itching and a variety of eye problems exhibited anti-inflammatory and analgesic properties antibacterial and analgesic (Zhuang et al. 2010, Li et al. 2013) and anti-inflammatory (Jyothilakshmi et al. 2016). *Solidago virgaurea* used for treatment of swelling possesses anti-inflammatory properties (El-Ghazaly 1992). *Spilanthes acmella* and *S. paniculata* were widely used in traditional medicine for relieving toothache, among other uses such as digestive problems, tongue ulcers, cough, fever, cold, bodyache, killing intestinal worms, skin infections and malaria. The pharmacological properties associated with these uses include antimicrobial (Morshed et al. 2011), antipyretic and local anaesthetic (Chakraborty et al. 2010), anti-inflammatory and analgesic (Peiris et al. 2001, Chakraborty et al. 2004) and antinociceptive properties (Hossain et al. 2012, Ahmed et al. 2004). *Taraxacum officinale*, traditionally used as liver tonic and treatment of jaundice, was reported to have hepatoprotective activity (Mahesh et al. 2010). The traditional use of *Vernonia cinerea* for headache could be linked to its reported analgesic and antipyretic activity (Iwalewa et al. 2003).

5. Conclusion and Future Prospects

The present study revealed the important role played by plants of Asteraceae family in traditional medicinal practices of the indigenous communities of Arunachal Pradesh. This family, being one of the most dominant and widely distributed family globally, provides an abundant and easily accessible resource base for use by the traditional healthcare practitioners as well as common people in the remote villages where there is unavailability of medical facilities.These species have been reported to possess a variety of pharmacological properties. Therefore, there is ample scope for further research to identify the bioactive compounds present in these plants which are responsible for the reported pharmacological properties. Among the species recorded in the present study, *Ageratum conyzoides, Bidens pilosa, Blumea lacera, Chromolaena odorata, Crassocephalum crepidioides, Eclipta prostrata* and *Mikania micrantha* have been identified as invasive alien species (Reddy et al. 2008) which are a potential threat to native ecosystem. Therefore, exploitation of these plant resources for herbal medicine industry may serve a double purpose of weed control and development of potential herbal drugs from these species. Moreover, a number of species recorded in this study possess valuable pharmacological properties such as anti-cancer, anti-diabetic, anti-malarial, etc., which offers a promising scope for selection and identification of suitable species that may be used as alternatives to the highly exploited, endangered and threatened medicinal plants thus promoting their conservation.

Abbreviation Used

TMPS: Traditional Medical Practitioners

Conflict of Interest

Authors declare no conflict of Interest

Acknowledgement

The authors express gratitude to the Director, G.B. Pant National Institute of Himalayan Environment & Sustainable Development for providing the necessary support required for carrying out this review work.

References

Abdelhamed S, Yokoyama S, Hafiyani L, Kalauni SK, Hayakawa Y, Awale S & Saiki I (2013). Identification of plant extracts sensitizing breast cancer cells to TRAIL. *Oncology Reports;* DOI: 10.3892/or.2013.2293.

Adebayo AH, Tan NH, Akindahunsi AA, Zeng GZ, Zhang YM (2010). Anticancer and antiradical scavenging activity of *Ageratum conyzoides* L. (Asteraceae). *Pharmacognosy Magazine* 6: 62-66.

Aftab-Ullah, Mahmood Ahmad, Taseer Ahmad & Faiza Naseer (2015). Hepatoprotective activity of *Sonchus asper* in paracetamol-induced hepatic damage in rabbits. *Bangladesh Journal of Pharmacology* 10: 115.

Ahmed AA, Saad MH, Haidar AAGM & Sami AAR (2016). Phytochemistry, antimicrobial, antigiardial and antiamoebic activities of selected plants from Albaha area, Saudi Arabia. *British Journal of Medicine and Medical Research* 18(11): 1-8.

Ahmed F, Selim MST, Das AK & Choudhuri MSK (2004). Anti-inûammatory and antinociceptive activities of *Lippia nodiûora* Linn.*Pharmazie* 59(4): 329–333.

Ahuja J, Suresh J, Paramakrishnan N, Mruthunjaya K & Naganandhini MN (2011). An Ethnomedical, Phytochemical and Pharmacological Profile of *Artemisia parviflora* Roxb. *Journal of Essential Oil Bearing Plants* 14 (6): 647-657.

Ali N & Ghosh B (2006) Ethnomedicinal plants in Arunachal Pradesh some tact prospects. EVNIS Bulletin: Himalayan Ecology 14(2).

Ali SA, Mahanand S & Khan SW (2012). Hepatoprotective and antioxidant activity of *Spilanthes paniculata* flower extracts on liver damage induced by paracetamol in rats. *African Journal of Pharmacy and Pharmacology* 6(42): 2905-2911.

Aparna N & Duraiswamy B (2009). Evaluation of anti-asthmatic activity of aerial parts of *Artemisia nilagirica* and seeds of *Sesamum indicum*. Proceedings of the 2nd Indian Pharmaceutical Association Students Congress, Bangalore, Indian Pharmaceutical Association.

Arumugam S, Vadivel V & Senthilkumar GP (2010). In vitro Antimicrobial Activity of *Vernonia cinerea* (L) Less. *Pharmacologyonline* 2: 957-960.

Bahar E, Akter K-M, Lee G-H, Lee H-Y, Rashid H-O, Choi M-K, Bhattarai KR, Hossain MMM, Ara J, Mazumder K, Raihan O, Chae H-J & Yoon H (2017). â-Cell protection and antidiabetic activities of *Crassocephalum crepidioides* (Asteraceae) Benth. S. Moore extract against alloxan-induced oxidative stress via regulation of apoptosis and reactive oxygen species (ROS). *BMC Complementary and Alternative Medicine* 17:179; DOI 10.1186/s12906-017-1697-0.

Bam J, Rai S, Bhattacharya D, Maiti S, Islam S, Pathak P, Bera A & Deb S (2015). Indigenous curative and prophylactic traditional practices used against haematophagous leeches in Arunachal Pradesh and Sikkim. *Indian Journal of Traditional Knowledge* 14: 493-497.

Banerjee S, Chanda A, Adhikari A, Das AK & Biswas S (2014). Evaluation of Phytochemical Screening and Anti Inûammatory Actvity of Leaves and Stem of *Mikania scandens* (L.) Willd. *Annals of Medical and Health Sciences Research* 4(4): 532-536.

Baruah S, Borthakur SK, Gogoi P & Ahmed M (2013). Ethnomedicinal plants used by Adi-Minyong tribe of Arunachal Pradesh, Eastern Himalaya. *Indian Journal of Natural Products and Resources* 4(3): 278-282.

Bhinge SD, Hogade MG, Chavan C, Kumbhar M & Chature V (2010). In vitro anthelmintic activity of herb extract of *Eclipta prostrata* L. against *Pheretima posthuma. Asian Journal of Pharmaceutical and Clinical Research* 3(3): 229–230.

Bhuyan M (2015). Comparative Study of Ethnomedicine among the Tribes of North East India. *International Research Journal of Social Sciences* 4(2): 27-32.

Cáceres A, Menendez H, Mendez E, Cohobon E, Samayoa BE, Jauregui E, Peralta E & Carillo G (1995). Antigonorrhoeal activity of plants used in Guatemala for the treatmentof sexually transmitted diseases. *Journal of Ethnopharmacology* 48(2): 85-88.

Chakraborty A, Devi BRK, Rita S, Sharatchandra K & Singh TI (2004). Preliminary studies on antiinûammatory and analgesic activities of *Spilanthes acmella* in experimental animal models. *Indian Journal of Pharmacology* 36(3): 148–150.

Chakraborty A, Devi BRK, Sanjebam R, Khumbong S & Tokchom IS (2010): Preliminary studies on local anaesthetic and antipyretic activities of *Spilanthes acmella* Murr. in experimental animal models. *Indian Journal of Pharmacology* 42 (5): 277–279.

Chandra T, Sadique J & Somasundaram S (1987). Effect of *Eclipta alba*on inflammation and liver injury. *Fitoterapia* 58(1):23-32.

Chethan J, Sampath Kumara KK, Shailasree S & Prakash HS (2012). Antioxidant, Antibacterial and DNA protecting activity of selected medicinally important Asteraceae plants. *International Journal of Pharmacy and Pharmaceutical Sciences* 4(2): 257-261.

Chiang YM, Chuang DY, Wang SY, Kuo YH, Tsai PH & Shyur LF (2004). Metabolite profiling and chemopreventive bioactivity of plant extracts from *Bidens pilosa*. *Journal of Ethnopharmacology* 95: 409–419.

Chien S-C, Young PH, Hsu Y-J, Chen C-H, Tien Y-J, Shiu S-Y & Li T-H (2009). Antidiabetic properties of three common *Bidens pilosa* variants in Taiwan. *Phytochemistry* 70(10): 1246–1254.

Choi UK, Lee OH, Yim JH, Cho CW, Rhee YK, Lim SI & Kim YC (2010). Hypolipidemic and antioxidant effects of dandelion (*Taraxacum officinale*) root and leaf on cholesterol-fed rabbits. *International Journal of Molecular Sciences* 11(1): 67-78.

Chomnawang MT, Surassmo S & Nukoolkarn VS (2005). Antimicrobial effects of Thai medicinal plants against acne-inducing bacteria. *Journal of Ethnopharmacology* 101: 330–333.

Chowdhury H, Singh RD & Biswas SK (2003). In vitro antifungal activity of two essential oils on *Drechslera sorokiniana* causing spot blotch of wheat. *Annals of Plant Protection Science* 11(1): 67-69.

Chun H & Kitts D (2003). Antioxidant, pro-oxidant, and cytotoxic activities of solvent-fractionated dandelion flower extracts in vitro. *Journal of Agricultural and Food Chemistry* 51: 301-310.

Das AK & Hui Tag (2006). Ethnomedicinal studies of the Khampti tribe of Arunachal Pradesh. *Indian Journal of Traditional Knowledge* 5(3): 317-322.

Das AK (2003). Some notes on the folk medicines of the Adis of Arunachal Pradesh, pp. 41-48. In: Mibang T (Ed.) *Ethnomedicine of the tribes of Arunachal Pradesh*. Himalayan Publishers, New Delhi.

Das M, Jaishi A & Sarma HN (2013). Traditional Medicines of herbal origin practice by the Adi tribe off East Siang District of Arunachal Pradesh, India. *Global Journal of Research on Medicinal Plants and Indigenous Medicine* 2(5): 298–310.

Dasgupta D, Dash S & Chakraborty J (2014). Evaluation of anticancer activity of *Mikania micrantha* Kunth (Asteraceae) against Ehrlich Ascites Carcinoma in Swiss Albino mice. *International Journal of Pharmaceutical Research & Allied Sciences* 3(2): 9-18.

Deba F, Xuan TD, Yasuda M & Tawata S (2008). Chemical composition and antioxidant, antibacterial and antifungal activities of the essential oils from *Bidens pilosa* Linn. var. *radiata*. *Food Control* 19: 346–352.

Dehariya R & Dixit AK (2015). A Review on Potential Pharmacological Uses of *Carthamus tinctorius* L. *World Journal of Pharmaceutical Sciences* 3(8): 1741-1746.

Devmurari VP & Jivani NP (2011). Phytochemical screening and antibacterial activity of ethanolic extract of *Artemisia nilagirica*. *Annals of Biological Research* 1(1): 10-14.

Doley B, Gajurel PR, Rethy P, Singh B, Buragohain R & Potsangbam S (2010). Lesser known Ethno medicinal Plants Used by the Nyshi community of Papumpare District, Arunachal Pradesh. *Journal of Bioscience Research* 1(1): 34-36.

Egharevba HO, Abdullahi MS,Okwute SK & Okogun JI (2010). Phytochemical analysis and broad spectrum Antimicrobial activity of *Laggera pterodonta*(DC.) Sch. Bip. (Aerial Part). *Researcher* 2(10): 35-40].

Ekanem P, Wang M, Simon JE & Moreno DA (2007). Antiobesity properties of two African plants (*Afromomum meleguetta* and *Spilanthes acmella*) by pancreatic lipase inhibition. *Phytotherapy Research* 21(12): 1253–1255.

El-Ghazaly M, Khayyal MT, Okpanyi SN & Arens-Corell M (1992). Study of the anti-inflammatory activity of *Populus tremula, Solidago virgaurea* and *Fraxinus excelsior*. *Arzneimittelforschung* 42: 333–336.

Gautam S, Bhagyawant SS & Srivastava N (2014). Detailed study on therapeutic properties, uses and pharmacological applications of safflower (*Carthamustinctorius* L.). *International Journal of Ayurveda and Pharma Research* 2(3): 5-16.

Ghosh G, Ghosh DC, Melkania U & Majumdar U (2014). Traditional medicinal plants used by the Adi, Idu and Khamba tribes of Dehang-Debang Biosphere Reserve in Arunachal Pradesh. *International Journal of Agricultural and Environmental Biotechnology* 7(1): 165-171.

Gross SC, Goodarzi G, Watabe M, Bandyopadhyay S, Pai SK & Watabe K (2002). Antineoplastic activity of *Solidago virgaurea* on Prostatic Tumor Cells in an SCID Mouse Model. *Nutrition and Cancer* 43(1): 76-81; DOI: 10.1207/S15327914NC431_9.

Gupta V (2006). Plants used in folklore medicine by Bangnis of East Kameng, Arunachal Pradesh. *Natural Product Radiance* 5(1): 52-59.

Hartwell JL (1971). Plants used against cancer. A survey. *Lloydia* 30: 379-436.

Hasan SMR, Jamila M, Majumder MM, Akter R, Hossain Md.M, Mazumder Md.EH, Alam MdA, Jahangir R, Rana Md.S, Md. Arif and Rahman S (2009). Analgesic and Antioxidant Activity of the Hydromethanolic Extract of *Mikania scandens* (L.) Willd. Leaves. *American Journal of Pharmacology and Toxicology* 4 (1): 1-7.

Hossain H, Shahid-Ud-Daula AFM, Jahan IA, Nimmi I, Hasan K & Haq MM (2012). Evaluation of antinociceptive and antioxidant potential from the leaves of *Spilanthes paniculata* growing in Bangladesh. *International Journal of Pharmacy and Phytopharmacology Research* 1(4): 178–186.

Huang D, Chen Y, Chen W, Liu Y, Yao F, Xue D & Sun L (2015). Anti-inflammatory effects of the extract of *Gnaphalium affine* D. Don in vivo and in vitro. *Journal of Ethnopharmacology* 176: 356–364.

Hui Tag & Das AK (2004). Ethnobotanical notes on the Hill Miri tribe of Arunachal Pradesh. *Indian Journal of Traditional Knowledge* 3(1): 80-85.

Hui Tag, Das AK & Hari Loyi (2007). Anti-inflammatory plants used by the Khampti tribe of Lohit District in eastern Arunachal Pradesh, India. *Natural Product Radiance* 6(4): 334-340.

Iwalewa EO, Iwalewa OJ & Adeboye JO (2003). Analgesic, antipyretic, anti-inflammatory effects of methanol, chloroform and ether extracts of *Vernonia cinerea* leaf. *Journal of Ethnopharmacology* 86: 229-234; DOI: 10.1016/S0378-8741(03)00081-3.

Jaiswal N, Bhatia V, Srivastava SP, Srivastava AK & Tamrakar AK (2012). Antidiabetic effect of *Eclipta alba* associated with the inhibition of 5ØüÞ5ØüÞ-glucosidase and aldose reductase. *Natural Product Research* 26: 2363–2367.

Janbaj KH & Gilani AH (1995). Evaluation of the protective potential of *Artemisia maritima* non acetaminophen- and CCl4– induced liver damage. *Journal of Ethnopharmacology* 47: 43-47.

Jena PK & Chakraborty AK (2010). Evaluation of analgesic activity studies of various extracts of leaves of *Eupatorium odoratum* Linn. *International Journal of Pharmacy &Technology* 2(3): 612-616.

Jeon H, Kang H, Jung H, Kang Y, Lim C, Kim Y & Park E (2008). Anti-inflammatory activity of *Taraxacum officinale*. *Journal of Ethnopharmacology* 115: 82-88.

Jeong SC, Kim SM, Tae Jeong YT & Song CH (2013). Hepatoprotective effect of water extract from *Chrysanthemum indicum* L. flower. *Chinese Medicine* 8:7;http://www.cmjournal.org/content/8/1/7.

Jimoh FO, Adedapo AA & Afolayan AJ (2011). Comparison of the Nutritive Value, Antioxidant and Antibacterial Activities of *Sonchus asper* and *Sonchus oleraceus*. *Records of Natural Products* 5(1): 29-42.

Jyothilakshmi M, Jyothis M & Latha MS (2016). *Mikania micrantha* – a Natural Remedy to Skin Infections. *International Journal of Current Microbiology and Applied Science* 5(2): 742-745; DOI: http://dx.doi.org/10.20546/ijcmas.2016.502.083.

Kagyung R, Gajurel PR & Singh B (2010). Ethnomedicinal plants used for gastro-intestinal diseases by Adi tribes of Dehang-Debang Biosphere Reserve in Arunachal Pradesh. *Indian Journal of Traditional Knowledge* 9(3): 496-501.

Kala CP (2005). Ethnomedicinal botany of the Apatani in the Eastern Himalayan region of India. *Journal of Ethnobiology and Ethnomedicine* 1:11; http://www.ethnobiomed.com/content/1/1/11.

Kamgang R, Fankem Gaëtan O, Ngo Tetka J, Gonsu Kamga H & Fonkoua MC (2015). Antimicrobial and Antidiarrheal effects of four Cameroon Medicinal Plants: *Dichrocephala integrifolia, Dioscorea preusii, Melenis minutiflora*, and *Tricalysia okelensis*. *International Journal of Current Pharmaceutical Research* 7 (1): 21-24.

Kar A & Borthakur SK (2008). Medicinal plants used against dysentry, diarrhoea and cholera by the tribes of erstwhile Kameng district of Arunachal Pradesh. *Natural Product Radiance* 5(1): 52-60.

Khair MA, Ibrahim M, Ahsan Q, Kuddus MR, Rashid RB & Rashid MA (2014). Preliminary Phytochemical Screenings and Pharmacological Activities of *Blumea lacera* (Burn.f.) DC. *Dhaka Univiversity Journal of Pharmaceutical Sciences* 13(1): 69-73.

Khongsai M, Saikia SP & Kayang H (2011). Ethnomedicinal plants used by different tribes of Arunachal Pradesh. *Indian Journal of Traditional Knowledge* 10(3): 541-546.

Kohli YP (2003). Ethnomedicines in the service of the Tani tribes,pp. 13-17. In: Mibang T & Choudhuri SK (Eds.) *Ethnomedicines of the tribes of Arunachal Pradesh*. Himalayan Publishers, Itanagar, Arunachal Pradesh.

Krishnaiah D, Sarbatly R & Nithyanandam RR (2011). A review of the antioxidant potential of medicinal plant species. *Food Bioproducts Processing* 89: 217–233.

Kviecinski MR, Felipe KB & Schoenfelder T (2008). Study of the antitumor potential of *Bidens pilosa* (Asteraceae) used in Brazilian folk medicine. *Journal of Ethnopharmacology* 117: 69–75.

Kviecinski MR, Felipe KB, Correia JFG, Ferreira EA, Rossi MH, Gatti MFD, Filho DW & Pedrosa RC (2011). Brazilian *Bidens pilosa* Linne. yields fraction containing quercetin-derived flavonoid with free radical scavenger activity and hepatoprotective effects. *Libyan Journal of Medicine* 6: 5651.

Latha LY, Darah I, Sasidharan S & Jain K (2009). Antimicrobial activity of *Emilia sonchifolia*DC.,*Tridax procumbens* L. and *Vernonia cinerea* L. of Asteraceae family: potential as food preservatives. *Malaysian Journal of Nutrition* 15: 223–231.

Lee B-R, Lee J-H &An H-J (2012). Effects of *Taraxacumofficinale* on Fatigue and Immunological Parameters in Mice. *Molecules* 17: 13253-13265; DOI:10.3390/molecules171113253.

Lee C-L, Yen M-H, Hwang T-L, Yang J-C, Peng C-Y, Chen C-J, Chang W-Y & Wu Y-C (2015). Anti-inflammatory and cytotoxic components from *Dichrocephala integrifolia. Phytochemistry Letters*; DOI: 10.1016/j.phytol.2015.04.012.

Li J, Zhao GZ, Chen HH, Wang HB, Qin S, Zhu WY, Xu LH, Jiang CL & Li WJ (2008). Antitumour and antimicrobial activities of endophytic streptomycetes from pharmaceutical plants in rainforest. *Letters in Applied Microbiology* 47: 574–580.

Li Y, Shen B-b, Li J, Li Y, Wang X-x & Cao A-c (2013). Antimicrobial potential and chemical constituents of *Mikania micrantha* H.B.K. *African Journal of Microbiology Research* 7(20): 2409-2415.

Li YL, Ooi LSM, Wang H, But PPH & Ooi VEC (2004). Antiviral activities of medicinal herbs traditionally used in southern mainland China. *Phytotherapy Research* 18: 718–722.

Liu YB, Jia W, Gao WY, Zhao AH, Zhang YW, Takaishi Y & Duan HQ (2008). Two eudesmane, sesquiterpenes from *Laggera pterodonta*. *Journal of Asian Natural Products* 8: 303-307.

Ma Q, Wei R, Zhou B, Sang Z, Liu W & Cao Z (2017). Antiangiogenic phenylpropanoid glycosides from *Gynura cusimbua*. *Natural Product Research* 33(4): 1-7; DOI: 10.1080/14786419.2017.1389931

Mahesh A, Jeyachandran R, Cindrella L, Thangadurai D, Veerapur VP & Muralidhara RD (2010). Hepatocurative potential of sesquiterpene lactones of *Taraxacum officinale* on carbon tetrachloride induced liver toxicity in mice. *Acta Biologica Hungarica* 61(2): 175-190.

Maity T, Ahmad A & Pahari N (2012). Hepatoprotective activity of *Mikania scandens* (L.) Willd. against diclofenac sodium induced liver toxicity in rats.*Asian Journal of Pharmaceutical and Clinical Research* 5(2):185-189.

Ma-Ma K, Nyunt N & Tin KM (1978). The protective effect of *Eclipta alba* on carbon tetrachloride-induced acute liver damage. *Toxicology and Applied Pharmacology* 45: 723–728.

Mazumder UK, Gupta M, Manikandan L, Bhattacharya S, Haldar PK, & Roy S (2003). Evaluation of anti-inflammatory activity of *Vernonia cinerea* Less. extract in rats. *Phytomedicine* 10: 185–188.

Meneses R, Ocazionez RE, Martínez JR & Stashenko EE (2009). Inhibitory effect of essential oils obtained from plants grown in Colombia on yellow fever virus replication in vitro. *Annals of Clinical Microbiology and Antimicrobials* 8:8; DOI: 10.1186/1476-0711-8-8.

Mishra D, Sarkar DK, Nayak BS, Rout PK, Ellaiah P & Ramakrishna S (2010). Phytochemical investigation and evaluation of anthelmintic activity of extract from leaves of *Eupatorium odoratum* Linn. *Indian Journal of Pharmaceutical Education and Research* 44(4): 369-374.

Mitscher LA, Leu RP, Bathala MS, Wu W-N, Beal JL & White R (1972). Antimicrobial agents from higher plants, I. Introduction, rational and methodology. *Lloydia* 35: 157-166.

Modaresi M & Resalatpour N (2012). The effect of *Taraxacum officinale* hydroalcoholic extract on blood cells in mice. *Advances in Hematology,*Article ID 653412, 4 pages; DOI: 10.1155/2012/653412.

Morshed MA, Uddin A, Saifur R, Barua A & Haque A (2011). Evaluation of antimicrobial and cytotoxic properties of *Leucas aspera* and *Spilanthes paniculata.International Journal of Biosciences* 1(2): 7-16.

Myers N, Mittermeier RA, Mittermeier CA, da Fonseca GAB & Kent J (2000). Biodiversity hotspots for conservation priorities. *Nature* 403: 853-858.

Nagaraj DS & Venkateswarlu B (2013). Pharmacological studies of anti-diarrhoeal activity of *Vernonia cinerea* in experimental animals. *International Journal of Pharmacological Screening Methods* 3: 16–21.

Namsa ND, Mandal M, Tangjang S & Mandal SC (2011). Ethnobotany of the Monpa ethnic group at Arunachal Pradesh, India. *Journal of Ethnobiology and Ethnomedicine* 7: 31.

Newton SM, Lau C & Wright CW (2000). A review of antimycobacterial natural products. *Phytotherapy Research* 14: 303-322.

Nguyen MT, Awale S, Tezuka Y, Tran QL, Watanabe H & Kadota S (2004). Xanthine oxidase inhibitory activity of Vietnamese medicinal plants. *Biological and Pharmacological Bulletin* 27: 1414–1421.

Nimachow G, Rawat JS, Arunachalam A & Dai O (2011). Ethnomedicine of Aka Tribe, West Kameng district Arunachal Pradesh (India). *Science and Culture* 77: 3-4.

Noor RA, Khozirah S, Mohd RM, Ong BK, Rohaya C, Rosilawati M, Hamdino I, Badrul A & Zakiah I (2007). Antiplasmodial properties of some Malaysian medicinal plants. *Tropical Biomedicine* 24: 29–35.

Nour AMM, Khalid SA, Kaiser M, Brun R, Abdalla WE, Schmidt TJ (2010). The antiprotozoal activity of methylated flavonoids from *Ageratum conyzoides* L. *Journal of Ethnopharmacology* 129(1): 127-130.

Nyunaï N, Njikama N, Abdennebic EH, Mbaford JT & Lamnaouer D (2009). Hypoglycaemic and antihyperglycaemic activity of *Ageratum conyzoides* L. in rats. *African Journal of Traditional Complementary and Alternative Medicine* 6(2): 123-130.

Odeleye OP, Oluyege JO, Aregbesola OA & Odeleye PO (2014). Evaluation of preliminary phytochemical and antibacterial activity of *Ageratum conyzoides* (L) on some clinical bacterial isolates. *The International Journal of Engineering and Science* 3(6): 01-05.

Okunade AL (2002). Review *Ageratum conyzoides* L. Asteraceae. *Fitoterapia* 73: 1–16.

Önal S, Timur S, Okutucu B & Zihnioðlu F (2005). Inhibition of á-glucosidase by aqueous extracts of some potent antidiabetic medicinal herbs. *Preparative Biochemistry and Biotechnology* 35(1): 29-36.

Ongkana R (2003). Phytochemistry and anti-malarial activity of *Eupatorium odoratum* L. Ph.D. Thesis, Faculty of Graduate Studies, Mahidol University (Unpublished).

Panghal M, Kaushal V & Yadav JP (2011). In vitro antimicrobial activity of ten medicinal plants against clinical isolates of oral cancer cases. *Annals of Clinical Microbiology and Antimicrobials* 10: 21; DOI: 10.1186/1476-0711-10-21.

Payum T, Das AK, Shankar R & Lego YJ (2015). 99 Selected Folk Medicinal Plants of East Siang District of Arunachal Pradesh, India. American Journal of *PharmTech Research* 5(1); www.ajptr.com

Peiris KPP, Silva GKJ & Ratnasooriya WD (2001). Analgesic activity of water extract of *Spilanthes acmella* ûowers on rats. *Journal of Tropical Medicinal Plants* 2: 201–204.

Perme N, Choudhury SN, Choudhury R, Natung T & De B (2015). Medicinal Plants in Traditional Use at Arunachal Pradesh, India. *International Journal of Phytopharmacy Research* 5(5): 86-98.

Phan TT, Wang L, Patrick SEE, Grayer RJ, Chan SU & Lee ST (2001). Phenolic compounds of *Chromolaena odorata* protect cultured skin cells from oxidative damage: Implication for cutaneous wound healing. *Biological and Pharmacological Bulletin* 24(12): 1373-1379.

Pires JM, Mendes RM, Negri G, Duarte-Almeida JM, Carlini EA, (2009). Antinociceptive peripheral effect of *Achillea millefolium* L. and *Artemisia vulgaris* L.: both plants known popularly by brand names of analgesic drugs. *Phytotherapy Research* 23: 212–219.

Poiatã A, Tuchilu C, Ivãnescu B, Ionescu A & Lazãr MI (2009). Antibacterial activity of some *Artemisia* species extract. *Revista Medico-Chirurgicala a Societatii de Medici si Naturalisti din Iasi* 113(3): 911-914.

Poudel BK, Sah JP, Subedi SR, Amatya MP, Amatya S & Shrestha TM (2015). Pharmacological studies of methanolic extracts of *Sonchus arvensis* from Kathmandu. *Journal of Pharmacognosy and Phytotherapy* 7(11): 263-267; DOI: 10.5897/JPP2015.0359

Pradeesh S (2018). In vitro anticancer activity of *Bidens biternata* (lour.) Merr. & Sheriff – an ethno medicinal plant of Kerala. *Open Access International Journal of Science & Engineering* 3(1): 15-20.

Priydarshi R, Melkani AB, Mohan L & Pant CC (2015). Terpenoid composition and antibacterial activity of the essential oil from *Inula cappa* (Buch-Ham. ex. D. Don) DC. *Journal of Essential Oil Research*, http://dx.doi.org/10.1080/10412905.2015.1090935

Pu HL, Zhao JH, Xu SB & Hu Q (2000). Protective actions of *Blumea* flavanones on primary cultured hepatocytes against lipid peroxidation. *Chinese Traditional and Herbal Drugs* 31: 1113–1115.

Quiming N, Asis JL, Nicolas M, Versoza D & Alvarez MR (2016). In vitro á-glucosidase inhibition and antioxidant activities of partially purified *Antidesma bunius* fruit and *Gynura nepalensis* leaf extracts. *Journal of Applied Pharmaceutical Sciences* 6: 97-101.

Ragasa C, Co ALKC & Rideout JA (2005). Antifungal metabolites from *Blumea balsamifera*. *Natural Product Research* 19: 231–237.

Rani SA & Murty SU (2006). Antifungal potential of flower head extract of *Spilanthes acmella* Linn. *African Journal of Biomedical Research* 9: 67–69.

Rao MR, Reddy IB & Ramana T (2006). Antimicrobial activity of some Indian medicinal plants. *Indian Journal of Microbiology* 46: 259-262.

Ratnasooriya WD, Pieris KPP, Samaratunga U & Jayakody JRAC (2004). Diuretic activity of *Spilanthes acmella* flowers in rats. *Journal of Ethnopharmacology* 91(2-3): 317–320.

Reddy CS, Bagyanarayana G, Reddy KN & Raju VS (2008). *Invasive Alien Flora of India*. National Biological Information Infrastructure, USGS, USA.

Reddy NS, Satya BL, Ch. Shivakoti, Ch.Venu, Reddy PV (2015). Evaluation of antimicrobial activity and *Bidens biternata* Ehrenb. leaves. *Ijsrm.Human* 1 (4): 1-7.

Rejitha Gopinath (2009). Diuretic activity of *Eupatorium odoratum* Linn. *Journal of Pharmacy Research* 2(5):844-846.

Rizvi SMD, Biswas D, Arif JM & Zeeshan M (2011). In-vitro antibacterial and antioxidant potential of leaf and flower extracts of *Vernonia cinerea* and their phytochemical constituents. *International Journal of Pharmaceutical Sciences Review and Research* 9: 164–169.

Saewan N, Koysomboon S & Chantrapromma K (2011). Anti-tyrosinase and anticancer activities of flavonoids from *Blumea balsamifera* DC. *Journal of Medicinal Plants Research* 5: 1018–1025.

Sagar MK, Ashok PK, Chopra H & Upadhyaya K (2010). Phytochemical and pharmacological potential of *Artemisia indica* in experimental animal models. *Pharmacologyonline* 2: 1-4.

Sagar MK, Ashok PK, Sangameshwaren B & Upadhyaya K (2008). Studies on gastric antiulcer effect of *Artemisia nilagirica* in experimental animal models. In: *Proceedings of60th Indian Pharmaceutical Congress,* Indian Pharmaceutical Association,New Delhi.

Sah JP, Poudel BK, Subedi SR, Amatya S, Shrestha TM &, Amatya MP (2015). Hypoglycemic and Analgesic effects of Methanolic Extracts of *Sonchus arvensis* from Nepal. *International Journal of Pharmacognosy and Phytochemical Research* 7(3): 613-615

Saklani A & Jain SK (1994). *Cross cultural ethnobotany of Northeast India.* Published by Deep Publications, New Delhi.

Saleem Md, Naseer F, Ahmad S, Nazish A, Bukhari FR, Rehman AU, Khan IM, Sadiq S & Javed F (2014). Hepatoprotective Activity of Ethanol Extract of *Conyza bonariensis* against Paracetamol Induced Hepatotoxicity in Swiss Albino Mice. *American Journal of Medical and Biological Research* 2(6): 124-127.

Sangeetha T & Venkatarathinakumar T (2011). Antitumor activity of aerial parts of *Vernonia cinerea* (L.) Less. against Dalton's ascitic lymphoma. *International Journal of PharmTech Research* 3: 2075–2079.

Sareen KN, Misra N, Verma DR, Amma MKP & Gujral ML (1961). Oral contraceptives. V. Anthelminthics and antifertility agents. *Indian Journal of Physiology and Pharmacology* 5: 125-135.

Sarmah R (2010). Commonly used non-timber forest products (NTFPs) by the Lisu tribe in Changlang district of Arunachal Pradesh, India. *SIBCOLTEJO* 5: 68-77.

Sarmah R, Adhikari D, Majumder M & Arunachalam A (2008). Traditional medicobotany of Chakma community residing in the Northwestern periphery of Namdapha National Park in Arunachal Pradesh. *Indian Journal of Traditional Knowledge* 7(4): 587-593.

Shafi PM, Nambiar MKG, Clery RA & Sarma YR (2004). Composition and anti-fungal activity of the oil of *Artemisia nilagirica* (Clarke) Pamp. *Journal of Essential Oil Research* 18(4): 377-379.

Shah NZ, Muhammad N, Azeem S, Khan AZ, Samie M & Khan H (2013). Antimicrobial and phytotoxic properties of *Conyza bonariensis*. *Pharmacy and Pharmacology Research* 1(1): 8-11.

Shankar R & Rawat MS (2008). Medicinal plants used in traditional medicine in Lohit and Dibang valley districts of Arunachal Pradesh. *Indian Journal of Traditional Knowledge* 7(2): 288-295.

Shankar R, Rawat MS & Singh VK (2003). Ethnomedicinal plants of Papum Pare district, Arunachal Pradesh,pp 30-34. In: Mibang T & Choudhuri SK (Eds.) *Ethnomedicines of the tribes of Arunachal Pradesh.* Himalayan Publishers, Itanagar, Arunachal Pradesh.

Shankar R., Rawat MS, Deb S & Sharma BK (2012). Jaundice and its traditional cure in Arunachal Pradesh. *Journal of Pharmaceutical and Scientific Innovation* 1(3): 93-97.

Sharma GP, Jain NK & Garg BD (1979). Antihelmintic activity of some essential oils. *Indian Perfume* 23: 210-212.

Sigstedt SC, Hooten CJ, Callewaert MC, Jenkins AR, Romero AE, Pullin MJ, Kornienko A, Lowrey TK, Slambrouck SV & Steelant WF (2008). Evaluation of aqueous extracts of *Taraxacum officinale* on growth and invasion of breast and prostate cancer cells. *International Journal of Oncology* 32(5): 1085-1090.

Singh BR, Kumar VOR, Sinha DK, Agrawal RK, Vadhana P, Bharadwaj M & Singh SV (2016). Antimicrobial Activity of Methanolic Extract and Ether Extract of *Ageratum conyzoides*. *Pharmaceutica Analytica Acta* 7: 471; DOI:10.4172/2153-2435.1000471.

Sinha S & Raghuwanshi R (2016). Phytochemical screening and antioxidant potential of *Eclipta prostrata* (L) L. -a valuable herb. *International Journal of Pharmacy and Pharmaceutical Sciences* 8(3): 255-260.

Sreedevi A, Bharathi K & Prasad KVSRG (2011). Effect of *Vernoniacinerea* aerial parts against Cisplatin-induced nephrotoxicity in rats. *Pharmacologyonline* 2: 548-555.

Srivastava RC & Adi community (2009). Traditional Knowledge of Adi tribe of Arunachal Pradesh on plants. *Indian Journal of Traditional Knowledge* 8(2): 146-153.

Srivastava RC & Nyshi community (2010). Traditional knowledge of Nyishi (Dafla) tribe of Arunachal Pradesh. *Indian Journal of Traditional Knowledge* 9(1): 26-37.

Stevens PF (2001 onwards). *Angiosperm Phylogeny Website*. Version 13, updated: 07/07/2016 19:32:56. http://www.mobot.org/MOBOT/research/APweb/welcome.html

Taiwo OB, Olajide OA, Soyannwo OO & Makinde JM (2000). Anti-Inflammatory, Antipyretic and Antispasmodic Properties of *Chromolaena odorata*. *Pharmaceutical Biology* 38(5): 367-370.

Tangjang S, Namsa ND, Aran C & Litin A (2011). An ethnobotanical survey of medicinal plants in the Eastern Himalayan zone of Arunachal Pradesh, India. *Journal of Ethnopharmacology* 134: 18–25.

Temraz A & El-Tantawy WH (2008). Characterization of antioxidant activity of extract from *Artemisia vulgaris*. *Pakistan Journal of Pharmaceutical Sciences* 21(4): 321-326.

Teoh WY, Sim KS, Richardson JSM, Wahab NA & Hoe SZ (2013). Antioxidant Capacity, Cytotoxicity, and Acute Oral Toxicity of *Gynura bicolor*. *Evidence-Based Complementary and Alternative Medicine*, Article ID 958407, 10 pages, http://dx.doi.org/10.1155/2013/958407.

Thakur S, Koundal R, Kumar D, Maurya AK, Padwad YS, Lal B & Agnihotri VK (2019). Volatile Composition and Cytotoxic Activity of Aerial Parts of *Crassocephalum crepidioides* growing in Western Himalaya, India. *Indian Journal of Pharmaceutical Sciences* 81(1): 167-172.

Tigno XT & Gumila E (2000). In vivo microvascular actions of *Artemisia vulgaris* L. in a model of ischemia-reperfusion injury in the rat intestinal mesentery. *Clin. Hemorheology and Microcirculation* 23: 159-165.

Triratana T, Suwannuraks R & Naengchomnong W (1991). Effect of *Eupatorium odoratum* on blood coagulation. *Journal of the Medical Association of Thailand* 74(5): 283-287.

Twaij HAA & Al-Badr AA (1988). Hypoglycemic activity of *Artemisiaherba alba*. *Journal of Ethnopharmacology* 24: 123-126.

Uddin SJ, Grice ID & Tiralongo E (2011). Cytotoxic effects of Bangladeshi medicinal plant extracts. *Evidenced Based Complementary and Alternative Medicine,*Article ID 578092, 7 pages; http://dx.doi.org/10.1093/ecam/nep111.

Valecha N, Biswas S, Badoni V, Bhandari KS & Sati OP (1994). Antimalarial activity of *Artemisia japonica, Artemisia maritima* and *Artemisia nilagirica. Indian Journal of Pharmacology* 26: 144-146.

Wabo PJ, Payne V, Mbogning Tayo G, Komtangi MC, Yondo J, Ngangout AM, Mpoame M & Bilong Bilong CF (2013). In vitro anthelminthic efficacy of *Dichrocephala integrifolia* (Asteraceae) extracts on the gastro-intestinal nematode parasite of mice: *Heligmosomoides bakeri* (Nematoda, Heligmosomatidae). *Asian Pacific Journal of Tropical Biomedicine* 3(2): 100-104.

Wang J, Petrova V, Wu S-B, Zhu M, Kennelly EJ & Long C (2014): Antioxidants from *Gerbera piloselloides:* an ethnomedicinal plant from southwestern China. *Natural Product Research* 28(22): 2072-2075;DOI: 10.1080/14786419.2014.924000.

Wanjen K, Chaudhry S, Arya SC & Samal PK (2011). A preliminary investigation on ethno medicinal plants used by Wancho tribe of Arunachal Pradesh, India. *Journal of Non-Timber Forest Products* 18(2): 129-132.

Whittle BA (1964). The use of changes in capillary permeability in mice to distinguish between narcotic & non-narcotic analgesics. *British Journal of Pharmacology and Chemotherapy* 22: 246–249.

Wongsawatkul O, Prachayasittikul S, Isarankura-Na-Ayudhya C, Satayavivad J, Ruchirawat S & Prachayasittikul V (2008). Vasorelaxant and antioxidant activities of *Spilanthes acmella* Murr. *International Journal of Molecular Sciences* 9(12): 2724–2744.

Wu J & Tang C, Yao S, Zhang L, Ke C, Feng L, Lin G & Ye Y (2015). Anti-inflammatory Inositol Derivatives from the whole plant of *Inula cappa. Journal of Natural Products*78(10):2332-2338; DOI: 10.1021/acs.jnatprod.5b00135.

Wu L-C, Fan N-C, Lin M-H, Chu IR, Huang SJ, Hu CY & Han SY (2008). Anti-inflammatory ettect of spilanthol from *Spilanthes acmella* on murine macrophage by down-regulating LPS-induced inflammatory mediators. *Journal of Agricultural and Food Chemistry* 56(7):2341–2349.

Wu X-L, Li C-W, Chen H M, , Su ZQ, Zhao XN, Chen JN, Lai XP, Zhang XJ & Su ZR (2013). Anti-Inflammatory Effect of Supercritical-Carbon Dioxide Fluid Extract from Flowers and Buds of *Chrysanthemum indicum* Linnén. *Evidence-Based Complementary and Alternative Medicine,* Article ID 413237, 13 pages; https://doi.org/10.1155/2013/413237.

Wu Y, Yang L, Wang F, Wu X, Zhou C, Shi S, Moban J & Zhao Y (2007). Hepatoprotective and antioxidative effects of total phenolics from *Laggera pterodonta* on chemical-induced injury in primary cultured neonatal rat hepatocytes. *Food and Chemical Toxicology* 45(8): 1349-1355.

Wu Y, Zhou C, Li X, Song L, Wu X, Lin W, Chen H, Bai H, Zhao J, Zhang R, Sun H & Zhao Y (2006). Evaluation of Anti-inflammatory Activity of the total Flavonoids of *Laggera pterodonta* on Acute and Chronic Inflammation Models. *PhytotherapyResearch* 20: 585–590;DOI: 10.1002/ptr.1918.

Yarnell E & Abascal K (2009). Dandelion (*Taraxacum oûcinale* and *T. mongolicum*). *Integrative Medicine* 8: 35–38.

Yashphe J, Segal R, Brever A & Erdreict-Naftali G (1979). Antibacterial activity of *Artimesia herba alba. Journal of Pharmaceutical Sciences* 68: 924-931.

Yongam D (2007). Ethnomedico-botany of the Mishing tribes of East Siang District of Arunachal Pradesh, pp. 89-96. In: Rama Shankar (Ed.) *Ayurvedic and drugs for all,* 1st Edition. Himalayan Publishers, Itanagar, Arunachal Pradesh.

Yuan L, Chen F, Ling L, Dou P, Bo H, Zhong M & Xia L (2008). Protective effects of total flavonoids of *Bidens pilosa* L. (TFB) on animal liver injury and liver fibrosis. *Journal of Ethnopharmacology* 116: 539-546.

Zahara K, Bibi Y, Tabassum S, Mudrikah, Bashir T, Haider S, Araa A, Ajmal M (2015). A review on pharmacological properties of *Bidens biternata*: A potential nutraceutical. *Asian Pacific Journal of Tropical Diseases* 5(8): 595-599.

Zeng W-C, Zhang W-C, Zhang W-H, He Q & Shi B (2013). The antioxidant activity and active component of *Gnaphalium affine* extract. *Food and Chemical Toxicology* 58: 311-317; DOI: 10.1016/j.fct.2013.05.004.

Zhuang SH, Hao CQ, Feng JT & Zhang X (2010). Active antifungal components of *Mikania micrantha* H.B.K. *Journal of Zhejiang University* 36(3): 293-298.

6

Analysis of Indigenous Uses of Ethnobotanical Herbaceous Plants and their Diversity Among Nature-Dependent Communities in Atraulia of Burhanpur Tehsil, District Azamgarh, Uttar Pradesh

Pooja Shukla and Bikarma Singh

Abstract

Wild plants of medicinal value are the embodiment of knowledge perception gift from our ancestors and these plays a significant role in curing various human health disorders. This gifted knowledge of economic plants used by the local people usually has long narrations for use in their daily activities. Considering the value of indigenous knowledge associated with plants, a study was undertaken on herbaceous communities in Atraulia area (Burhanpur tehsil), district Azamgarh, Uttar Pradesh. During the investigation, a total of 71 informants (22 females and 49 males) were interviewed and data regarding ethnobotanical information on wild plant species were documented and analysed. The results indicated a total of 169 species belonging to 41 families used in indigenous herbal preparation as medicine from Atraulia area by local tribal people. The dominant growth forms were the annual herbs represented by 91 species, followed by 46 species of perennial nature and 32 species of

Pooja Shukla

Banaras Hindu University, Varansi, UP, India (Place of Work); CSIR-National Botanical Research Institute, Lucknow, Uttar Pradesh, India

Bikarma Singh (✉)

Botanical Garden, CSIR-National Botanical Research Institute, Lucknow-226001 Uttar Pradesh, India

✉*Corresponding author email: drbikarma.singh@nbri.res.in*

Plants for Novel Drug Molecules: Ethnobotany to Ethnopharmacology
Bikarma Singh & Yash Pal Sharma (eds.), (pp. 169-189)

Email: *info@nipabooks.com* Web: *www.nipabooks.com*

biennial growth. Therefore, we can conclude that this region is rich in herbaceous plants and plant used as medicine is wide spread in the study area. This communication deals with demography of the informants, botanical name of herbaceous plants, family, part used, disease cure, use report, frequency of use and recommendation for conservation of unique valueable resources.

Keywords: Traditional knowledge, Herbaceous communities, Medicinal plant, Atraulia, Azamgarh (UP), Conservation

1. Introduction

Plants are basic requirements as they provide food, medicine and shelter for all living organisms inhibiting the earth (Singh 2019). About 20,000 plant species are utilized in different traditional medicine system around the globe and are considered as potential agents for the revelation of new medications (Yadav et al. 2019). Since time immemorial, plants of medicinal value belonged to history of mankind who tried to place them in the context in which they lived (Mautone et al. 2019). Ethnobotanical studies showed that traditional plant knowledge still survive in different regions of Himalaya, particularly among the aged people residing in interior mountains and valleys (Agelet et al. 2001). Wild plants are numerous in nature and possess widespread ecology which is used by people in different, complex, and evolving ways. But the understanding of these processes is still fundamental and the ethnobotanical research goes on to find novel or unusual employments of well-known medicinal plants (Savo et al. 2011). Traditional knowledge systems reflect the cognitive experiences of human groups worldwide. Not only do they synthesize the diverse learning, concepts, and customs attributed to interactions between local farmers and their landscapes, but they also express the different ways in which people structure and organize, in categories and hierarchies, their cognitive experiences. Therefore, the ethnobotanical research has becomes a continuous developing process, influenced by environmental and cultural factors.

The biodiversity of India are rich in plants belonging to angiosperms, lichens, mosses, pteridophytes, liverworts and gymnosperms groups. The forests play essential role in harboring approximately 45,000 floral and 81,000 faunal species of which 5150 floral and 1837 faunal species are endemic (Negi and Dhyani 2012). Vascular flora comprises of 18,664 vascular taxa, making India the world's tenth richest country in terms of plant diversity (Singh 2016). As per the India State of Forest Report-2009, the total forest cover of the country is 690,899 km^2 constituting 21.02% of the total geographic area of the country. In the Indian Himalayan region (IHR), forest is major land use/land cover category, which covers

41% of geographical area in the IHR. The Hindu Kush Himalaya (HKH) comprised of approximately 39% grassland, 20% forest, 15% shrub land and 5% agricultural land. The remaining 21% is made up of other types of land use such as barren land, rock outcrops, built-up areas, snow cover, and water bodies (Chettri et al. 2008). Elevation zones across HKH extend from tropical (6000m), with a principal vertical vegetation regime comprised of tropical and subtropical rainforest, temperate broadleaved deciduous or mixed forest, and temperate coniferous forest, including high altitude cold shrub or steppe and cold desert (Pei 1995). Only Himalayan belts supports 50% of the country's total flora and more than 4,000 species recorded from India are Himalayan endemics and their distribution restricted to higher elevations. Approximately 7,000 species documented have medicinal value in Indian traditional systems of medicine such as Ayurveda, Siddha, Unani, Homeopathy and Amchi or Sowa-Rigpa system.

Uttar Pradesh ranked as the 4th largest state of India having total geographical area of 243,290 sq.km, and known for natural beauty coupled with unique culture. The state enjoys tropical monsoon type of climate, however, humid subtropical climate with dry winter and semi-arid type dominant in western U.P. These variations favour the occurrence of unique species of plants, microbes and animals. Literature reveals several species inventory research works carried out by different botanists at different times in the past for the Uttar Pradesh (Duthie 1903-1929; Bhattacharyya 1963, 1964, Kanjilal 1966, Rau 1969; Sharma and Pandey 1984, Sharma and Dhakre 1995, Uniyal et al. 1999; Chaudhary et al. 2016) and they have documented 2,932 species under 1017 genera from the state (Sainin et al. 2011). More than 500 species have been published as medicinal plants for the state which has unique ethnobotanical applications by ethnic tribes residing in interior regions of state such as Bhotia, Bhoxa, Tharu and Jaunsars. Sporodic ethnobotanical works have been documented for the state (Singh et al. 2002, Choudhary 2010, Singh et al. 2012, Mishra et al. 2012, Singh and Dubey 2012). The floristic account of Burhanpur tehsil, district Azamgarh of Uttar Pradesh, however, is still unexplored in spite of its rich diversity. Hence, there is a need for proper inventory and documentation of all plants available in the district. Hence, efforts have been made through this communication to provide a contribution to medicinal plants through studying ethnobotanical herbaceous plants and their diversity among nature-dependent communities in Atraulia of Burhanpur tehsil, district Azamgarh, Uttar Pradesh.

2. Materials and Methods

2.1. Study Area

District Azamgarh is one among seventy five districts of Uttar Pradesh located between latitude-26° 15′16.74″N and longitude-83° 01′21.83″E). This district spreads over in 4053 sq. km area with total population of 46,13,913 people as per census 2011. Climate is predominantly tropical monsoon type. Three main seasons are winter (November–February), summer (March-June), monsoon (July-October). Temperatures fluctuating from 0 to 50 °C in many parts of the district Azamgarh. Drought and flood occurs in many parts of the state due to unpredictable rains. Flood hazard of UP is due to overflowing of main rivers such as Ganga, Yamuna, Ramganga, Gomti, Sharda, Gaghra, Rapti and Ganda. Gond, Dhuria, Nayak, Pathari and Raj Gond are major population inhabiting the area of district Azamgarh. Besides, mixed population comprising culture of other parts of India is also found.

2.2. Field Tours, Collection and Identification

Field tours covering 33 days have been undertaken between 2016 and 2018 to investigate ethnobotanical plant knowledge posses by local people of district Azamgarh. Total 346 plant voucher samples along with more than 400 digital photographs were taken during the field investigation and samples were processed following herbarium techniques proposed by Jain and Rao (1977). Twenty three localities were visited to complete the questionnaires prepared for the survey and data analysis. Hakkim, Vaid, medicine men, sarpanch and herdsman were consulted for investigation. People of age group between 15 and above years (including male & female) were consulted during the survey. The ethnobotanical information was collected from the experts and well experienced herbal practitioner, local faith healers and traditional medicine men. The information collected included habit, plant part used, preparation and mode of administration. The data collected through questionnaire and interviews were analyzed for Use Value (*UV*) and Factor Informant Consensus (Fic).

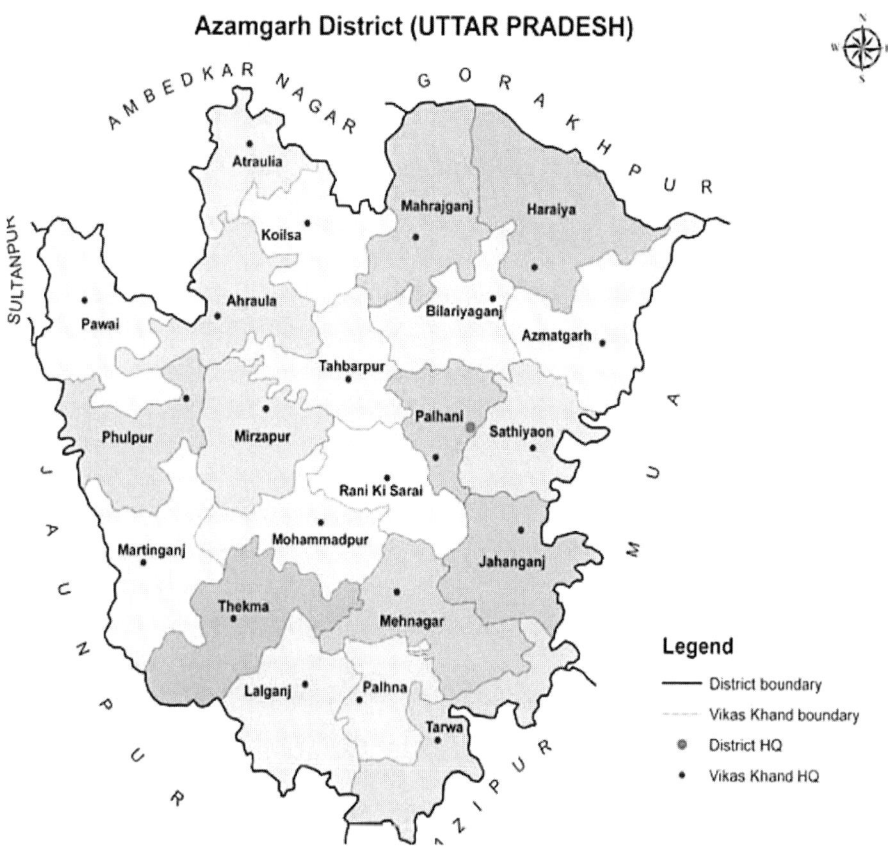

Fig. 1: Study area covered for ethnobotanical investigation

2.3. Analysis of Data

The data collected through discussions and interviews while surveying the localities were analysed by using quantitative indices such as use value, use frequency, as given below:

2.3.1. Use Value

The relative importance of the species was calculated by using the use value (Phillip et al. 1994). Use value is calculated by following formula:

$UV = \Sigma U / \eta$

U is the number of reports cited by each informant for a given plant species and n is the total number of informants interviewed. Use values are high when there are many use reports for a plant and low when there are few reports related to its use. Use value is high when there are many use-reports for a plant, indicating that the plant is important and it approaches zero (0) when there were few reports related to its use.

2.3.2. *Use Frequency*

Use Frequency is expressed in percentage and is calculated by multiplying use value by 100. This gives the overall percentage of use reports of the particular species of the study area.

2.3.3. *Factor Informant Consensus (Fic)*

This index was calculated to determine homogeneity of knowledge provided by different informants and has been calculated by the formula proposed by Heinrich et al. (1998).

Fic = Nur- Nt/ (Nur-1)

where, Fic is factor informant consensus, N_{ur} is the number of use reports for a particular category and N_t is the number of taxa used the category.

The Fic value ranges between 0 and 1, where low Fic values (near 0), indicate that plants are chosen randomly and no exchange of their use information is there among informants, and approach one (1) when there is a well defined selection criterion and exchange of information between informants.

3. Results and Discussion

3.1. Informants Demography

During the present survey, a total of 71 informants (22 females and 49 males) were interviewed and data regarding ethnobotanical information on 171 plant species were documented and analysed. The informants of the study area were divided into four age groups, viz., (1) 15-30 years, (2) 31-45 years, (3) 46-60 years and above 60 years. Among these, 8 (16.33%) males and 5 (10.20%%) females belong to group I, 13 (26.53%) males and 13 (26.53%) belong to group II, 19 (38.78%) males and 3 (6.12%) females belong to group III and 9 (18.37%) male and 1 (2.04%) female belong to group IV (Table 1).

Table 1: Demography of the informants in the study area

Informants	Number (percentage)
Female	22 (30.98%)
Male	49 (69.01%)
Total	71

Age group (years)	Male	Female	Total
15-30 Yrs	8 (16.33%)	5 (10.20%)	13 (26.53%)
31-45Yrs	13 (26.53%)	13 (26.53%)	26 (53.06%)
46-60 Yrs	19 (38.78%)	3 (6.12%)	22 (44.90%)
Above 60 Yrs	9 (18.37 %)	1 (2.04%)	10 (20.41%)

Education level	Male	Female	Total
Illiterate	19	12	31
Attended school upto 5 classes	24	9	33
Attended school for 6-10 classes	4	1	5
Intermediate(12th class) onwards	2	0	2

3.2. Floristic Analysis

Investigation results indicated a total of 169 species belonging to 41 families were recorded from Atraulia of Burhanpur tehsil from Azamgarh district. The dominant growth forms were the annual herbs represented by 91 species, followed by 46 species of perennial herbs and 32 species of biennial herbs (Fig.2). The dominant family was cyperaceae represented by 30 species, followed by poaceae 20 species and asteraceae by 10 species. The least dominant were represented by 1 species and includes Amaryllidaceae, Apocynaceae, Cannaceae, Elatinaceae, Juncaceae, Nelumbonaceae, Polygalaceae, Ranunculaceae, Hydroleaceae, Rosaceae and Sphenocleaceae (Fig.3.). In term of genera, *Cyperus* was the most frequently used group of plants in the study area. *Ageratum conyzoides* (L.) L., *Achyranthes aspera* L., *Ammannia baccifera* L., *Bacopa monnieri* (L.) Wettst., *Celosia argentea* L., *Centella asiatica* (L.) Urb., *Cyperus rotundus* L., *Eclipta prostrata* (L.) L., *Kyllinga brevifolia* Rottb., *Persicaria barbata* (L.) H.Hara, *Potentilla supina* L., *Rotala densiflora* (Roth) Koehne, *Rumex dentatus* L. and *Utricularia aurea* Lour. are the most dominant species growing in all parts of the district and are famous plants among local people for herbal preparation.

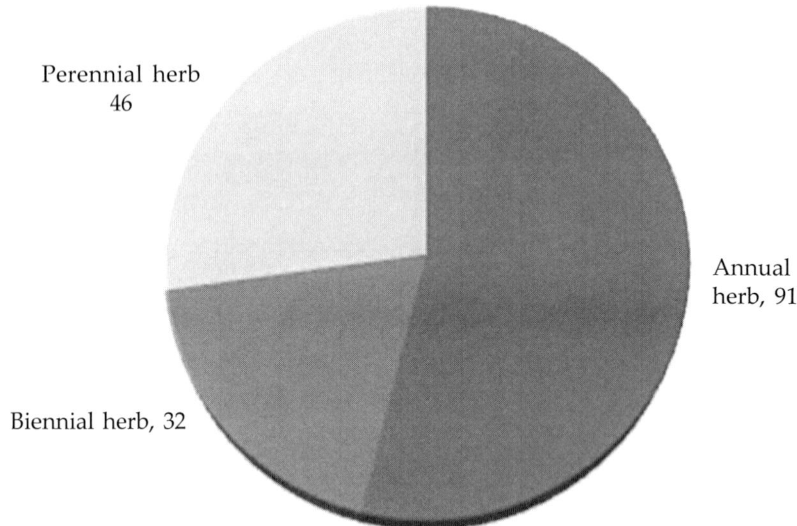

Fig. 2: Growth-form of plant species in the study area

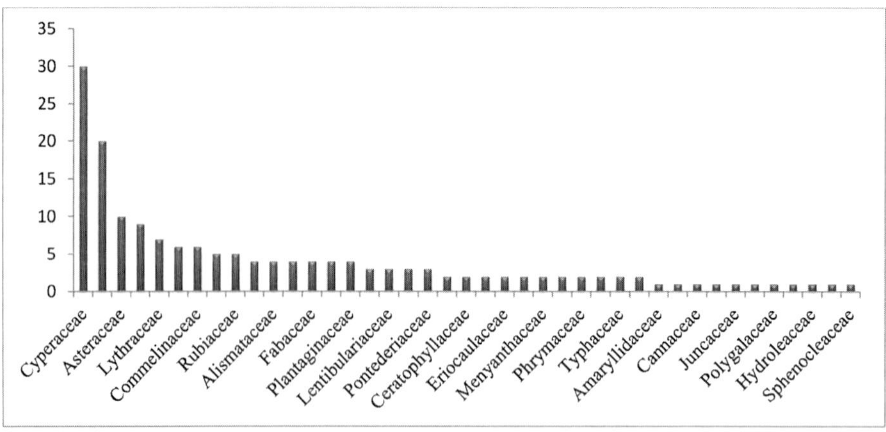

Fig. 3: Family representation in the study area

3.3. Analysis of Medicinal Plants

Out of these documented species, 67% species have medicinal and therapeutic potentials. Total 169 species documented used as ethnomedicine in treatment of local 22 types of disease (Table 1). Some of the frequent occurrence diseases are obesity, liver infection, diabetes, rheumatism, tumors, stomachache, insomnia, nerve troubles, migraine, paralysis, jaundice, skin infection, constipation, cough and asthma, cancer, swelling and inflammation. Frequently used plants for curing disease by locals on the basis of use value (UV) were *Achyranthus aspera* L., *Centella asiatica* (L.)

Table 2: Ethnobotanically important plants and their use analysis growing in Atraulia of Burhanpur tehsil, district Azamgarh, Uttar Pradesh

Botanical name/Voucher no	Family	Life form	Part Used	Mode of Use	Use Citation	Use Value	Frequency of Use (%)
Acmella paniculata (Wall. ex DC.) R.K.Jansen	Asteraceae	Annual herb	Flowers	Medicine	11	0.15	15.49
Actinoscirpus grossus (L.f.) Goetgh. & D.A.Simpson	Cyperaceae	Annual herb	Roots	Medicine	9	0.13	12.68
Aeschynomene aspera L.	Fabaceae	Perennial herb	Aerial parts	Medicine	22	0.31	30.99
Aeschynomene indica L.	Fabaceae	Perennial herb	Leaves	Medicine	17	0.24	23.94
Ageratum conyzoides (L.) L.	Asteraceae	Perennial herb	Aerial parts	Medicine	34	0.48	47.89
Alternanthera paronychioides A.St.-Hil.	Amaranthaceae	Annual herb	Whole plant	Medicine	4	0.06	5.63
Alternanthera sessilis (L.) R.Br. ex DC.	Amaranthaceae	Perennial herb	Leaves	Vegetable	19	0.27	26.76
Achyranthes aspera L.	Amaranthaceae	Biennial herb	Leaves	Medicine	12	0.17	16.90
Ammannia auriculata Willd.	Lythraceae	Biennial herb	Whole plant	Medicine	9	0.13	12.68
Ammannia baccifera L.	Lythraceae	Annual herb	Seeds	Medicine	33	0.46	46.48
Ammannia multiflora Roxb.	Lythraceae	Biennial herb	Whole plant	Medicine	14	0.20	19.72
Aponogeton abyssinicus Hochst. ex A.Rich	Aponogetonaceae	Annual herb	Whole plant	Vegetable	12	0.17	16.90
Aponogeton natans (L.) Engl. & K.Krause	Aponogetonaceae	Perennial herb	Whole plant	Medicine	13	0.18	18.31
Bacopa monnieri (L.) Wettst.	Plantaginaceae	Biennial herb	Aerial parts	Vegetable	39	0.55	54.93
Bergia ammannioides Roxb. ex Roth	Elatinaceae	Annual herb	Whole plant	Medicine	17	0.24	23.94
Bolboschoenus glaucus (Lam.) S.G.Sm.	Cyperaceae	Annual herb	Whole plant	Medicine	13	0.18	18.31
Bolboschoenus maritimus (Roth) T.Koyama	Cyperaceae	Biennial herb	Roots	Medicine	2	0.03	2.82

Contd.

Brachiaria ramosa (L.) Stapf	Poaceae	Annual herb	Whole plant	Medicine	9	0.13	12.68
Brachiaria reptans (L.) C.A.Gardner & C.E.Hubb.	Poaceae	Annual herb	Whole plant	Medicine	7	0.10	9.86
Butomopsis latifolia (D.Don) Kunth	Alismataceae	Biennial herb	Rhizomes	Medicine	14	0.20	19.72
Caesulia axillaris Roxb.	Asteraceae	Annual herb	Whole plant	Medicine	23	0.32	32.39
Caldesia parnassifolia (L.) Parl.	Alismataceae	Annual herb	Leaves	Medicine	13	0.18	18.31
Canna indica L.	Cannaceae	Perennial herb	Rhizomes	Medicine	10	0.14	14.08
Carex fedia Nees	Cyperaceae	Annual herb	Roots	Medicine	9	0.13	12.68
Celosia argentea L.	Amaranthaceae	Biennial herb	Whole plant	Medicine	11	0.15	15.49
Centella asiatica (L.) Urb.	Apiaceae	Biennial herb	Whole plant	Medicine	7	0.10	9.86
Centipeda minima (L.) A.Braun & Asch.	Asteraceae	Annual herb	Aerial parts	Medicine	19	0.27	26.76
Ceratophyllum demersum L.	Ceratophyllaceae	Annual herb	Whole plant	Medicine	16	0.23	22.54
Chenopodium ambrosioides L.	Ceratophyllaceae	Biennial herb	Aerial parts	Vegetable	24	0.34	33.80
Chrysopogon zizanioides (L.) Roberty	Poaceae	Annual herb	Whole plant	Medicine	17	0.24	23.94
Coix aquatica Roxb.	Poaceae	Perennial herb	Inflorescence	Medicine	18	0.25	25.35
Colocasia esculenta (L.) Schott	Araceae	Biennial herb	Leaves	Vegetable	33	0.46	46.48
Commelina huskarlii C.B.Clarke	Commelinaceae	Annual herb	Leaves	Medicine	12	0.17	16.90
Commelina benghalensis L.	Commelinaceae	Biennial herb	Leaves	Vegetable	17	0.24	23.94
Commelina gelatinosa Edgew.	Commelinaceae	Annual herb	Whole plant	Medicine	16	0.23	22.54
Commelina haitiensis Urb. & Ekman	Commelinaceae	Biennial herb	Leaves	Medicine	6	0.08	8.45
Corchorus aestuans L.	Malvaceae	Perennial herb	Seeds	Medicine	8	0.11	11.27
Corchorus capsularis L.	Malvaceae	Perennial herb	Leaves	Medicine	13	0.18	18.31
Crinum viviparum (Lam.) R.Ansari & V.J.Nair	Amaryllidaceae	Annual herb	Whole plant	Medicine	19	0.27	26.76
Cyanotis axillaris (L.) D.Don ex Sweet	Commelinaceae	Biennial herb	Aerial parts	Medicine	2	0.03	2.82
Cyperus compactus Retz.	Cyperaceae	Annual herb	Roots	Medicine	7	0.10	9.86
Cyperus corymbosus Rottb.	Cyperaceae	Annual herb	Whole plant	Medicine	6	0.08	8.45
Cyperus dubius Rottb.	Cyperaceae	Annual herb	Whole plant	Medicine	12	0.17	16.90

Contd.

Cyperus imbricatus Retz.	Cyperaceae	Biennial herb	Roots	Medicine	9	0.13	12.68
Cyperus iria L.	Cyperaceae	Annual herb	Whole plant	Medicine	3	0.04	4.23
Cyperus platystylis R.Br.	Cyperaceae	Annual herb	Roots	Medicine	7	0.10	9.86
Cyperus pseudokyllingiodes Kuk	Cyperaceae	Annual herb	Whole plant	Medicine	2	0.03	2.82
Cyperus rotundus L.	Cyperaceae	Perennial herb	Roots	Medicine	22	0.31	30.99
Cyperus squarrosus L.	Cyperaceae	Annual herb	Whole plant	Medicine	13	0.18	18.31
*Cyperus alulatus*J.Kern	Cyperaceae	Annual herb	Roots	Medicine	6	0.08	8.45
Cyperus difformis L.	Cyperaceae	Biennial herb	Whole plant	Medicine	11	0.15	15.49
Cyperus exaltatus Retz.	Cyperaceae	Annual herb	Roots	Medicine	13	0.18	18.31
Cyperus pulicaris Kük.	Cyperaceae	Annual herb	Leaves	Medicine	8	0.11	11.27
Cyperus pycnostachyus(Kunth) Kunth	Cyperaceae	Annual herb	Leaves	Medicine	7	0.10	9.86
Dentella repens (L.) J.R.Forst. & G.Forst.	Rubiaceae	Perennial herb	Leaves	Vegetable	9	0.13	12.68
Desmostachya bipinnata (L.) Stapf	Poaceae	Annual herb	Whole plant	Medicine	6	0.08	8.45
Digitaria ciliaris (Retz.) Koeler	Poaceae	Annual herb	Whole plant	Medicine	7	0.10	9.86
Dopatrium junceum (Roxb.) Buch.-Ham. ex Benth.	Plantaginaceae	Biennial herb	Whole plant	Medicine	4	0.06	5.63
Echinochloa colona (L.) Link	Poaceae	Perennial herb	Aerial parts	Medicine	5	0.07	7.04
Echinochloa crusgalli (L.) P.Beauv.	Poaceae	Perennial herb	Whole plant	Medicine	3	0.04	4.23
Echinochloa stagnina (Retz.) P.Beauv.	Poaceae	Annual herb	Aerial parts	Medicine	7	0.10	9.86
Eclipta prostrata (L.) L.	Asteraceae	Annual herb	Flowers	Medicine	36	0.51	50.70
Eichhornia crassipes (Mart.) Solms	Pontederiaceae	Annual herb	Stems	Vegetable	14	0.20	19.72
Eleocharis palustris (L.) Roem. & Schult.	Cyperaceae	Perennial herb	Roots	Medicine	6	0.08	8.45
Eleocharis atropurpurea (Retz.) J.Presl & C.Presl	Cyperaceae	Annual herb	Whole plant	Medicine	7	0.10	9.86
Eleocharis dulcis (Burm.f.) Trin. ex Hensch.	Cyperaceae	Annual herb	Whole plant	Medicine	16	0.23	22.54
Enydra fluctuans DC.	Asteraceae	Perennial herb	Leaves	Medicine	8	0.11	11.27

Contd.

Eragrostis truncata Hack.	Poaceae	Annual herb	Whole plant	Medicine	18	0.25	25.35
Eriocaulon cinereum R.Br.	Eriocaulaceae	Biennial herb	Whole plant	Medicine	12	0.17	16.90
Eriocaulon quinquangulare L.	Eriocaulaceae	Biennial herb	Leaves	Medicine	8	0.11	11.27
Euryale ferox Salisb.	Nymphaeaceae	Biennial herb	Stems	Vegetable	23	0.32	32.39
Fimbristylis quinquangularis (Vahl) Kunth	Cyperaceae	Annual herb	Roots	Medicine	1	0.01	1.41
Fimbristylis aestivalis Vahl	Cyperaceae	Annual herb	Whole plant	Medicine	5	0.07	7.04
Fimbristylis dichotoma (L.) Vahl	Cyperaceae	Perennial herb	Whole plant	Medicine	4	0.06	5.63
Fimbristylis ovata (Burm.f.) J.Kern	Cyperaceae	Annual herb	Leaves	Medicine	2	0.03	2.82
Fimbristylis tetragona R.Br.	Cyperaceae	Annual herb	Roots	Medicine	7	0.10	9.86
Glossostigma diandrum (L.) Kuntze	Phrymaceae	Biennial herb	Leaves	Medicine	6	0.08	8.45
Gnomophalium pulvinatum (Delile) Greuter	Asteraceae	Biennial herb	Leaves	Medicine	7	0.10	9.86
Grangea maderaspatana (L.) Poir.	Asteraceae	Annual herb	Leaves	Medicine	8	0.11	11.27
Hydrilla verticillata (L.f.) Royle	Hydrocharitaceae	Biennial herb	Leaves	Vegetable	9	0.13	12.68
Hydrolea zeylanica (L.) Vahl	Hydroleaceae	Annual herb	Aerial parts	Medicine	7	0.10	9.86
Hygrophila difformis Blume	Acanthaceae	Annual herb	Leaves	Medicine	6	0.08	8.45
Hygrophila auriculata (Schumach) Heine	Acanthaceae	Biennial herb	Whole plant	Medicine	8	0.11	11.27
Hygroryza aristata (Retz.) Nees ex Wight & Arn.	Poaceae	Biennial herb	Whole plant	Medicine	2	0.03	2.82
Ipomoea carnea Jacq.	Convolvulaceae	Annual herb	Leaves	Vegetable	7	0.10	9.86
Ipomoea aquatica Forssk.	Convolvulaceae	Annual herb	Leaves	Medicine	17	0.24	23.94
Ischaemum rugosum Salisb.	Poaceae	Annual herb	Leaves	Medicine	11	0.15	15.49
Juncus bufonius L.	Juncaceae	Annual herb	Aerial parts	Medicine	6	0.08	8.45
Justicia japonica Thunb.	Acanthaceae	Perennial herb	Stems	Medicine	11	0.15	15.49
Justicia quinqueangularis K.D.Koenig ex Roxb.	Acanthaceae	Perennial herb	Whole plant	Medicine	13	0.18	18.31
Kyllinga brevifolia Rottb.	Cyperaceae	Annual herb	Roots	Medicine	18	0.25	25.35
Leersia hexandra Sw.	Poaceae	Annual herb	Aerial parts	Medicine	2	0.03	2.82
Lemna perpusilla Torr.	Araceae	Annual herb	Roots	Vegetable	3	0.04	4.23

Contd.

Leptochloa panicea (Retz.) Ohwi	Poaceae	Annual herb	Roots	Medicine	1	0.01	1.41
Limnophila indica (L.) Druce	Plantaginaceae	Annual herb	Leaves	Medicine	8	0.11	11.27
Lindernia anagallis (Burm.f.) Pennell	Linderniaceae	Perennial herb	Aerial parts	Medicine	7	0.10	9.86
Lindernia crustacea (L.) F.Muell.	Linderniaceae	Annual herb	Whole plant	Medicine	8	0.11	11.27
Lippia javanica (Burm.f.) Spreng.	Verbenaceae	Annual herb	Aerial parts	Medicine	9	0.13	12.68
Ludwigia adscendens (L.) H.Hara	Onagraceae	Perennial herb	Seeds	Medicine	3	0.04	4.23
Ludwigia perennis L.	Onagraceae	Perennial herb	Leaves	Medicine	6	0.08	8.45
Mazus pumilus (Burm.f.) Steenis	Phrymaceae	Annual herb	Leaves	Medicine	15	0.21	21.13
Melochia corchorifolia L.	Malvaceae	Perennial herb	Seeds	Medicine	9	0.13	12.68
Mikania cordata (Burm.f.) B.L.Rob.	Asteraceae	Perennial herb	Aerial parts	Medicine	13	0.18	18.31
Monochoria hastata (L.) Solms	Pontederiaceae	Annual herb	Leaves	Vegetable	29	0.41	40.85
Monochoria vaginalis (Burm.f.) C.Presl	Pontederiaceae	Biennial herb	Rhizomes	Medicine	5	0.07	7.04
Murdannia nudiflora (L.) Brenan	Commelinaceae	Annual herb	Whole plant	Medicine	8	0.11	11.27
Najas graminea Delile	Hydrocharitaceae	Annual herb	Leaves	Medicine	9	0.13	12.68
Nelumbo nucifera Gaertn.	Nelumbonaceae	Annual herb	Stems	Vegetable	24	0.34	33.80
Neptunia oleracea Lour.	Fabaceae	Perennial herb	Seeds	Medicine	6	0.08	8.45
Nymphaea nouchali Burm.f.	Nymphaeaceae	Annual herb	Leaves	Medicine	8	0.11	11.27
Nymphaea stellata Willd.	Nymphaeaceae	Annual herb	Stems	Medicine	8	0.11	11.27
Nymphoides cristata (Roxb.) Kuntze	Menyanthaceae	Annual herb	Whole plant	Medicine	7	0.10	9.86
Nymphoides indica (L.) Kuntze	Menyanthaceae	Annual herb	Whole plant	Medicine	6	0.08	8.45
Oenanthe javanica (Blume) DC.	Apiaceae	Perennial herb	Seeds	Medicine	2	0.03	2.82
Oldenlandia ovatifolia (Cav.) DC.	Rubiaceae	Perennial herb	Leaves	Medicine	8	0.11	11.27
Oldenlandia corymbosa L.	Rubiaceae	Perennial herb	Whole plant	Medicine	9	0.13	12.68
Oldenlandia paniculata L.	Rubiaceae	Perennial herb	Leaves	Medicine	8	0.11	11.27
Oldenlandia umbellata L.	Rubiaceae	Perennial herb	Whole plant	Medicine	1	0.01	1.41
Oplismenus burmanni (Retz.) P.Beauv.	Poaceae	Annual herb	Whole plant	Medicine	9	0.13	12.68
Oryza rufipogon Griff.	Poaceae	Perennial herb	Seeds	Medicine	8	0.11	11.27

Contd.

Ottelia alismoides (L.) Pers.	Hydrocharitaceae	Annual herb	Whole plant	Medicine	13	0.18	18.31
Oxystelma secamone H. Karst.	Apocynaceae	Annual herb	Leaves	Medicine	3	0.04	4.23
Panicum paludosum Roxb.	Poaceae	Perennial herb	Whole plant	Medicine	2	0.03	2.82
Paspalidium flavidum (Retz.) A.Camus	Poaceae	Annual herb	Whole plant	Medicine	7	0.10	9.86
Paspalum scrobiculatum L.	Poaceae	Biennial herb	Whole plant	Medicine	6	0.08	8.45
Paspalum vaginatum Sw.	Poaceae	Biennial herb	Whole plant	Medicine	3	0.04	4.23
Pentapetes phoenicea L.	Malvaceae	Perennial herb	Leaves	Medicine	7	0.10	9.86
Persicaria barbata (L.) H.Hara	Polygonaceae	Annual herb	Aerial parts	Medicine	13	0.18	18.31
Persicaria glabra (Willd.) M.Gómez	Polygonaceae	Perennial herb	Whole plant	Medicine	7	0.10	9.86
Persicaria hydropiper (L.) Delarbre	Polygonaceae	Biennial herb	Aerial parts	Medicine	9	0.13	12.68
Persicaria lapathifolia (L.) Delarbre	Polygonaceae	Annual herb	Leaves	Medicine	1	0.01	1.41
Persicaria limbata (Meisn.) H.Hara	Polygonaceae	Annual herb	Whole plant	Medicine	4	0.06	5.63
Persicaria minor (Huds.) Opiz	Polygonaceae	Perennial herb	Whole plant	Medicine	5	0.07	7.04
Phragmites karka (Retz.) Trin. ex Steud.	Poaceae	Perennial herb	Leaves	Medicine	6	0.08	8.45
Phyla nodiflora (L.) Greene	Verbenaceae	Biennial herb	Aerial parts	Vegetable	7	0.10	9.86
Pistia stratiotes L.	Araceae	Annual herb	Aerial parts	Medicine	9	0.13	12.68
Polygala erioptera DC.	Polygalaceae	Annual herb	Whole plant	Medicine	3	0.04	4.23
Polygonum plebeium R.Br.	Polygonaceae	Perennial herb	Whole plant	Vegetable	2	0.03	2.82
Potamogeton crispus L.	Potamogetonaceae	Annual herb	Aerial parts	Medicine	9	0.13	12.68
Potamogeton nodosus Poir.	Potamogetonaceae	Biennial herb	Whole plant	Vegetable	7	0.10	9.86
Potentilla supina L.	Rosaceae	Perennial herb	Roots	Medicine	13	0.18	18.31
Pseudoraphis spinescens (R.Br.) Vickery	Poaceae	Annual herb	Aerial parts	Medicine	8	0.11	11.27
Pycreus sanguinolentus (Vahl) Nees	Cyperaceae	Annual herb	Roots	Medicine	6	0.08	8.45
Ranunculus sceleratus L.	Ranunculaceae	Biennial herb	Aerial parts	Medicine	17	0.24	23.94
Rotala densiflora (Roth) Koehne	Lythraceae	Perennial herb	Leaves	Medicine	19	0.27	26.76
Rotala indica (Willd.) Koehne	Lythraceae	Perennial herb	Whole plant	Medicine	6	0.08	8.45
Rotala rotundifolia (Buch.-Ham. ex Roxb.) Koehne	Lythraceae	Perennial herb	Aerial parts	Medicine	4	0.06	5.63

Contd.

Rottboellia exaltata (L.) L.f.	Polygonaceae	Perennial herb	Whole plant	Vegetable	2	0.03	2.82
Rumex dentatus L.	Polygonaceae	Perennial herb	Leaves	Vegetable	28	0.39	39.44
Sagittaria guayanensis Kunth	Alismataceae	Annual herb	Leaves	Medicine	13	0.18	18.31
Sagittaria sagittifolia L.	Alismataceae	Annual herb	Aerial parts	Vegetable	16	0.23	22.54
Schoenoplectiella articulata (L.) Lye	Cyperaceae	Perennial herb	Roots	Medicine	14	0.20	19.72
Schoenoplectiella juncoides (Roxb.) Lye	Cyperaceae	Annual herb	Whole plant	Medicine	13	0.18	18.31
Sesbania bispinosa (Jacq.) W.Wight	Fabaceae	Biennial herb	Seeds	Medicine	17	0.24	23.94
Seseli diffusum (Roxb. ex Sm.) Santapau & Wagh	Apiaceae	Annual herb	Seeds	Medicine	3	0.04	4.23
Sparganium erectum L.	Typhaceae	Annual herb	Seeds	Medicine	2	0.03	2.82
Sphaeranthus indicus L.	Asteraceae	Perennial herb	Flowers	Vegetable	11	0.15	15.49
Sphenoclea zeylanica Gaertn.	Sphenocleaceae	Annual herb	Roots	Medicine	3	0.04	4.23
Spirodela polyrrhiza (L.) Schleid.	Araceae	Annual herb	Rhizomes	Medicine	4	0.06	5.63
Trapa natans L.	Lythraceae	Perennial herb	Rhizomes	Vegetable	38	0.54	53.52
Typha domingensis Pers.	Typhaceae	Annual herb	Seeds	Medicine	37	0.52	52.11
Typhonium trilobatum (L.) Schott	Araceae	Annual herb	Rhizome	Vegetable	17	0.24	23.94
Utricularia gibbosa Hill	Lentibulariaceae	Annual herb	Whole plant	Medicine	16	0.23	22.54
Utricularia aurea Lour	Lentibulariaceae	Annual herb	Whole plant	Medicine	5	0.07	7.04
Utricularia inflexa Forssk.	Lentibulariaceae	Annual herb	Whole plant	Medicine	3	0.04	4.23
Vallisneria spiralis L.	Hydrocharitaceae	Annual herb	Leaves	Medicine	3	0.04	4.23
Veronica anagallis-aquatica L.	Plantaginaceae	Perennial herb	Aerial parts	Medicine	4	0.06	5.63
Wolffia arrhiza (L.) Horkel ex Wimm.	Araceae	Annual herb	Leaves	Vegetable	8	0.11	11.27

Urb., *Croton bonplandianum* Baill, *Cyperus rotundus* L., *Euphorbia hirta* L., *Mallotus philippinensis* Muell-Arg., *Phyla nodiflora* (L.) Greene and *Rumex dentatus* L., and several other high altitude plants. Lots of local herbal preparations are made to cure local seasonal diseases such as cold, malaria, fever, applied on cuts, burns and wounds.

Fig. 3: Medicinal plant diversity in the study area (A-*Croton bonplandianum*, B-*Euphorbia hirta*, C-*Mallotus philippinensis*, D-*Cyperus rotundus*, E-*Achyranthus aspera*, F-*Rumex dentatus*)

3.4. Plant Parts Used

It has observed that the different parts of the investigated species in the study area are used as food and medicine by the local people for their daily use. Whole plants of 60 species used as economic potential plants and is favourite among the people, followed by leaves of 40 species, aerial parts of 24 species, seeds of 11 species, roots of 9 species, rhizomes of 6 species, stems of 5 species and flowers are the least consumed or applied as medicine (4 species) in the study area. The details are given in Table 2 and Fig.4. Several species such as *Alternanthera sessilis, Aponogeton abyssinicus, Bacopa monnieri, Chenopodium ambrosioides, Colocasia esculenta, Commelina benghalensis, Eichhornia crassipes, Euryale ferox, Monochoria hastata, Nelumbo nucifera, Rumex dentatus* and several other species are used as cooked vegetable, however, 67% species of the study area is used as medicine and only 33% species used daily as vegetable.

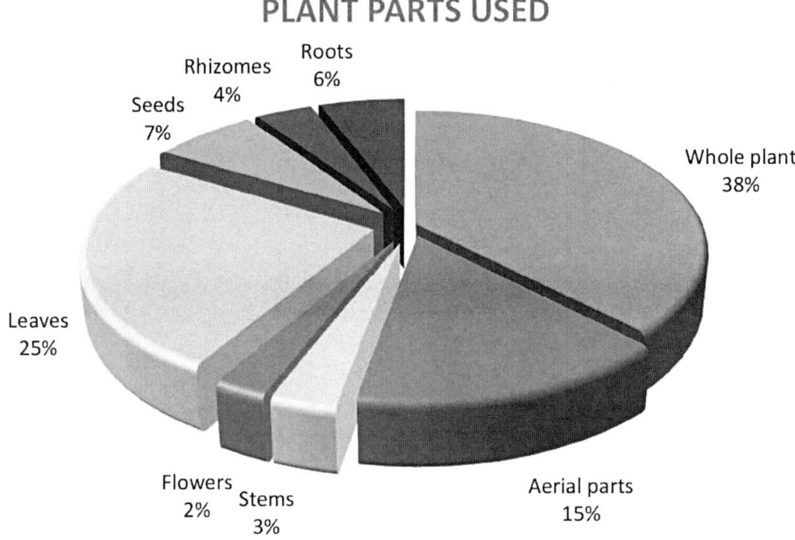

Fig. 4: Part used as medicine and vegetable in the study area

3.5. Determined Factor Informant Consensus

Analysis of data records, total 604 actual use records belonging to 17 different ailments (medicine and vegetable) have been registered through interviews conducted with 71 informants in the study area (Table 3). People of the region use maximum number of plant species (n=29) are used as vegetable and for which maximum use reports are obtained (n=51) followed by skin disease (species used= 27 and Use reports= 62) and cuts and wounds (species used= 24 and Use reports= 59). The least plant species used in case hair fall category (n=3) for which use reports analyzed indicated (n=51). The details are given in Table 3. The Fic of maximum category is more than 0.50 (towards 1.0) meaning good homogeinity of knowledge among the people of the study region regarding the use of members of family for ethnobotanical uses.

Table 3: Informant consensus factor (F_{ic}) of different plant species used in the study

Category	Number of Taxa (N_t)	Number of Use Reports (N_{ur})	Informant consensus factor (Fic)
I. Medicine			
Cancer treatment	7	19	0.667
Cattle diseases	16	47	0.674
Cuts and wounds	24	59	0.603
Eye infections	6	14	0.615
Fever and headache	12	38	0.703
Fracture	19	27	0.308
Hair fall	3	9	0.750
Joint pain	14	36	0.629
Kidney/Urine problems stone	8	26	0.720
Reptile and insects repellent	16	43	0.643
Respiraory troubles	19	54	0.660
Skin diseases	27	62	0.574
Stomach problems	17	34	0.515
Tonic/Vitamin source	8	16	0.533
II. Vegetables			
Spice and condiment	12	33	0.656
Cooked	29	51	0.440
Raw edible	13	36	0.657

4. Conclusion

This research is the first ethnobotanical study conducted in the Atraulia of Burhanpur tehsil, district Azamgarh, Uttar Pradesh, and a total of 169 species were recorded. The results show a high diversity of traditional ethnomedicinal plants. By assessing minutely the cause for endangerment status of the plants used in the traditional medicine, we recommend that folk utilization is not the main reason for the degeneration of wild resources, but may be huge collection for industrial purpose cause fast depletion of the wild plant resources. Despite the richness of medicinal plants in the study area, the inheritance of this valuable culture is facing a serious threat, mainly due to the rapid development of modern medicine. The ageing of herbalists without inheritors results in the rapid loss of valuable knowledge. In addition, the knowledge of traditional medicinal plants inherited *via* the oral mode and the accuracy of inheritance are difficult to determine. Those medicinal plants that were used were most commonly made by women to treat children's skin disorders. Unfortunately, the traditional use of plants is declining and accordingly knowledge is mainly restricted to the elderly. Moreover, the comparison of the documented species and their uses with ethnobotanical literature of other regions reveals that the traditional plant knowledge in Atraulia of Burhanpur tehsil shows strong similarities with adjacent areas. Some of the recorded species and administration processes however seem to be unique for Atraulia of Burhanpur tehsil. In current scenario, value addition and product development from locally available medicinal plants can provide an alternate source of livelihood for the local people living mostly in under-developed village areas. Since time immemorial, humans are coming using large varieties of wild fruits, vegetables, fodder and medicinal plants for meeting their day-to-day requirements. Therefore, conservation of valuable flora is the need of the hrs. There is need to put emphasis on conservation of these medicinal plants before they get depleted along with loss of traditional knowledge. Instead, it involves developing new modern methods that utilize traditional knowledge in new ways. These new methods must be community-driven. In which case, the form they take is often unpredictable. Their creation does not necessarily come from lofty planning and foresight. Rather, they develop out of practice-from people going out to the bush collecting plants and making medicine, from people telling stories of the plants they have used in the past, and from the hopes for what will come in the future.

Abbreviation Used

HKH: Hinhu Kus Himalaya; IHR: Indian Himalayan region; UP: Uttar Pradesh; UV: Use Value; Fic: Factor Informant Consensus.

Conflict of Interest

Authors declare no conflict of interest for this particular manuscript.

Acknowledgement

Authors would like to thanks people for help during field investigation and research work.

References

Agelet A and Vallès J (2001). Studies on pharmaceutical ethnobotany in the region of Pallars (Pyrenees, Catalonia, Iberian Peninsula). Part I. general results and new or very rare medicinal plants. *Journal of Ethnopharmacology* 77: 57-70.

Bhattacharyya UC (1963). A contribution to the flora of Mirzapur-I: Some new records for the district and for the Upper Gangetic Plain. *The Bulletin of the Botanical Survey of India, Dehra Dun* 5(1): 59-62.

Bhattacharyya UC (1964). A contribution to the flora of Mirzapur-II. *The Bulletin of the Botanical Survey of India, Dehra Dun* 6(2-4): 191-210.

Chaudhary LB, Kushwaha AK and Bajpai O (2016). Trees of Uttar Pradesh (Part 1). Bennett, Coleman & Co. Ltd., Lucknow, India.

Chaudhary RS (2010). Taxa of family Fabaceae: A potential of local medicinal values in Vindhya Region Uttar Pradesh, India. *International Journal of Pharma and Bio Sciences* 1(4): 46-53.

Chettri N, Shakya B, Thapa R and Sharma E (2008). Status of a protected area system in ther Hindu Kush-Himalayas: An analysis of PA coverage. *International Journal of Biodiversity Science and Management* 4: 164-178.

Duthie JF (1903-1929). Flora of upper Gangetic plain, and of the adjacent Siwalik and sub-Himalayan tracts Vol. 1-3. Superintendent of Government Printing, Calcutta, India.

Heinrich M, Ankli A, Frei B, Weimann C and Sticher O (1998). Medicinal plants in Mexico: healer's consensus and cultural importance. *Social Science and Medicine* 47: 1863-1875.

Kanjilal PC (1966). A forest flora for the plains of Uttar Pradesh-Part II & III. Superintendent Printing, Lucknow, India.

Mautone M, Martino LD and Feo VD (2019). Ethnobotanical research in Cava de' Tirreni area, Southern Italy. *Journal of Ethnobiology and Ethnomedicine* 15 (50). https://ethnobiomed.biomedcentral.com/ articles/10.1186/s13002-019-0330-3.

Mishra NK, Das R & Srivastava DK (2012). Ethno-medicinal weeds of veterinary importance from Dullapathar frontier region of district Sonebhadra (south east UP). *Indian Journal of Plant Sciences* 2(1): 109-111.

Negi GCS and Dhyani PP (2012). Glimpses of forestry research in the indian himalayan region. Bishen Singh Mahendra Pal Singh 23-A, New Connaught Place Dehra Dun - 248 001, India.

Pei S (1995). Banking on Biodiversity. Report on the regional consultations on biodiversity Assessment in the Hindu Kush-Himalaya. ICIMOD, Kathmandu, Nepal.

Phillips O, Gentry AH, Reynel C, Wilki P and Gavez-Durand CB (1994). Quantitative ethnobotany and Amazonian conservation. *Conservation Biology* 8: 225-248.

Rau MA (1969). Flora of the Upper Gangetic Plain and the adjacent Siwalik and sub-himalayan tracts- Checklist. *The Bulletin of the Botanical Survery of India, Dehra Dun* 2: 10-87.

Saini et al. 2011. Floristic Diversity of Uttar Pradesh with special reference to Eastern Uttar Pradesh. Published by Uttar Pradesh, State Biodiversity Board, Lucknow, India.

Savo V, Caneva G, Guarrera PM and Reedy D (2011) Folk phytotherapy of the Amalfi Coast (Campania, Southern Italy). *Journal of Ethnopharmacology* 35: 376-792.

Sharma AK and Dhakre JS (1995). Flora of Agra District. Botanical Survey of India, Kolkata, India.

Sharma BD and Pandey DS (1984). Exotic flora of Allahabad District. Botanical Survey of India, Kolkata, India.

Singh B (2016). Orchids of Himalaya, distribution and taxonomy. Published by Write & Print Publisher, New Delhi, India.

Singh B (2019). Plants for human survival and medicine. Published by NIPA, India and CRC Press Taylor & Francis, UK.

Singh A and Dubey NK (2012). An ethnobotanical study of medicinal plants in Sonebhadra District of Uttar, Pradesh, India with reference to their infection by foliar fungi. *Journal of Medicinal Plants Research* 6(14): 2727-2746.

Singh A, Singh GS and Singh PK (2012). Medico-ethnobotanical inventory of Renukoot forest division of district Sonbhadra, Uttar Pradesh, India. *Indian Journal of Natural Products and Resources* 3(3): 448-457.

Singh AK, Raghubanshi AS and Singh JS (2002). Medico-ethnobotany of tribal of Sonaghati of Sonbhadra district, Uttar Pradesh, India. *Journal of Ethnopharmacology* 81(1): 31-41.

Uniyal SK, Swami A and Uniyal BP (1999). Monocotyledonous plants of Uttar Pradesh: A Checklist (excluding Cyperaceae and Poaceae). Bishen Singh Mahendra Pal Singh, Dehra Dun, India.

Yadav V, Krishnan A and Vohora D (2019). A systematic review on *Piper longum* L.: Bridging traditional knowledge and pharmacological evidence for future translational research. *Journal of Ethnopharmacology* doi: https://doi.org/10.1016/j.jep.2019.112255.

7

Ethnobiology of Lichens of North Western Himalayas

Mamta Bhat and Susheel Verma

Abstract

Lichenized fungi or lichens represent the most successful symbiotic organisms on earth. They are the inconspicuous group of lower plants having numerous properties. Their different growth forms namely crustose, foliose and fruticose have been put to use in one way or the other. Since ancient times, they were used as food, fodder, dye and in perfumery. Lichens are a rich source of several important secondary metabolites because of which they are widely used as a source of medicine against several diseases and to improve the health of humans as well as animals. Most of the information related to the usage of lichens is gathered from the traditional knowledge holders. Therefore, there is a need to explore and document this traditional knowledge associated with lichens. The present paper thus makes an attempt to present and document the valued information on the ethnobiological uses of lichens from the Northwestern parts of the Himalayas, if not, the undocumented information will be lost for all the future generations. Ethnobiology of 39 species of lichens belonging to 08 families is being presented in this communication.

Keywords: Ethnobiology, Lichens, Northwestern, Himalayas

Mamta Bhat and Susheel Verma (✉)
Department of Botany, School of Biosciences and Biotechnology, BGSB University, Rajouri- 185234 Jammu and Kashmir, India

✉*Corresponding author email: eremurus@rediffmail.com*

Plants for Novel Drug Molecules: Ethnobotany to Ethnopharmacology
Bikarma Singh & Yash Pal Sharma (eds.), (pp. 191-201)

Email: *info@nipabooks.com* Web: *www.nipabooks.com*

1. Introduction

Lichens are formed by the symbiotic association between algae and fungi. However, cyanobacteria have also been reported to be associated with them. They are the most successful symbiotic organisms in nature. Considered to be the pioneers, lichens play an important role in paedogenesis and succession (Hale 1983). Ecosystem health and level of air pollution is indicated by their presence or absence (Richardson 1988). They are known to produce many secondary metabolites which provides them protection against environmental stress (Tiwari et al. 2011). Economically they are very significant, and are used in traditional medicines, in cosmetics, as dyes, as food (Llano 1948). They constitute the staple diet of the Alaskan reindeer and Himalayan musk deer (Skunke 1969; Negi 1996). Many lichens possess numerous medicinal properties such as antimicrobial (antibacterial, antifungal and antiviral), antioxidant, antitumour and immunomodulator activity (Malhotra et al. 2007). They have been widely used in folk as well as traditional medicine system in many parts of Western world, India and China (Malhotra et al. 2007). Notwithstanding their significance, the information related to their ethnobiology remains to be undocumented.

The use of lichens is not new to man but dates back to the ancient times. The local communities or tribals were extremely knowledgeable about the importance of lichens since time immemorial. Lichens were known as Shipal and their usage dates back to 1500 BC in Atharaveda (Atharavveda, 1958). Also, Sushruta Samhita (1000 BC) and Charak Samhita (300-200 BC) summarize their (Shilapushp or Shailya) use as medicines (Sharma 1998; Singh and Chunakar 1972). According to the ancient texts, these lichens were later identified as *Parmelia cirrhata* and *Parmelia perforata* (Chanda & Singh 1971; Sharma (1984a,b). Their use is also described by Egyptian and Chinese in their traditional medicine system (Dymock et al. 1890; Chopra et al. 1958; Gonzalez – Tejero et al. 1995).

Consequently, this knowledge related to the ethnobiology of lichens has been traversed downstream through people, indigenous and local communities verbally. However, with the advent of modern and technological developments the focus shifted from traditional to non-conventional approaches leading to erosion of this centuries old traditional knowledge. As ethnobiology focuses on the cultural and scientific knowledge of the ethnic communities based on plants (Martin 1995), it also let people realise the fact that they are dependent on the nature and therefore, must live in harmony with nature for their sustenance and survival. These interactions between man and nature in general and species in particular has made him to understand the origin, evolution and use of different types of plants.

Fig. 1: North West Himalaya (Map not to scale)

Many pilot studies have been carried out to document the information on higher plants of the Himalayas, leaving lower groups unexplored and undocumented (Mani 1978; Kala 2000). Since, this region is rich in endemic flora and considered as a hotspot, information availability on the ethnobiology of lower organisms including lichens is low (Sheikh et al. 2006). North western Himalayas includes Ladakh and Kashmir (J&K), Kullu valley, Lahaul and Spiti, covering Jammu and Kashmir, Himachal Pradesh and Uttarakhand in India. Beyond India, it also extends to parts of Pakistan and China and some parts of Afghanistan (Burbank et al. 1996) (Fig. 1). The typical eco-climate of this mountain range supports diverse life forms, lichens being one of them. The region falls in the lichen geographic zone which consists of mountainous to semi mountainous plains and thus houses good lichen diversity (Upreti & Chatterjee 2000). However, most of the area remains snow-clad during winters and during summers a number of large meadows appear there after melting of snow. The region is rich in flora and fauna and these elements have not been explored and documented fully. Because of the diverse kind of anthropogenic activity in the region, it is being observed that different elements of biodiversity are under stress and can disappear without even documentation. Need of the hour is therefore to have a documentation and compilation of lichens of the region.

2. Methods

Lichen based data was gathered through secondary sources of scientific literature based on free listing participatory field collection, semi structured interviews and open ended questionnaires. Large surveys were carried out by the researchers in which the questionnaires were used to gather the data from the tribal communities and local population. Interactions were made with the selected group of locals where they were requested to answer the same series of questions. The data gathered was subjected to statistical analysis.

3. Results and Discussion

This paper reports 39 lichen species belonging to 23 genera and 08 families with ethnobiological importance (Table 1). The most common used growth forms are foliose followed by fruticose and crustose (Fig 2). Depending upon the nature of usage of lichens, six main categories are observed including medicinal, edible, dye, spice & condiments and perfumes. They find use in religious & sacred ceremonies. The most common use of lichens by the people is as spice and condiments. About 22 species are used for this purpose whereas the least use has been found in perfumery, only one species is reported for the purpose. Ten species belonging to 06 families are used as medicines, 03 species of 02 families as edible, 12 species of 02 families as dyes and 05 species of 02 families are used for religious and sacred purposes (Fig 3). Twenty six species of Parmeliaceae are most commonly used by the people of the region followed by 04 lichens of Physciaceae and Ramalinaceae each. Rest of the 05 families are represented by one species each (Fig 4).

The ethnic knowledge of numerous properties lichens possess has been transcended over the centuries among communities and their generations in different parts of the world. In India, lichens are reported to be traded to different parts of the country from Uttaranchal hills for medicinal purposes (Atkinson 1882). Thus, lichens have been contributing to economy and cultural identity of local people there since long (Lal & Upreti 1995). These lichens from the wild are collected by the men, women and children who gather in groups and travel to a distance of several kilometers from their villages walking several hours. This social activity is commonly seen as a means of furnishing supplementary foods for their families for economic and social needs. However, lichens are preferred as food only during famines or period of food scarcity depending upon their abundance and availability (Ivanova and Ivanonv 2009). Similarly in India, Lepchas of Northern Sikkim and Gaddis of Kangra valley use

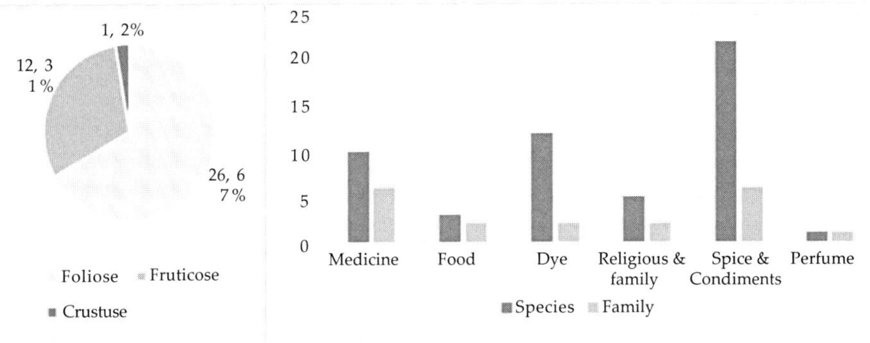

Fig. 2: Growth forms of lichens

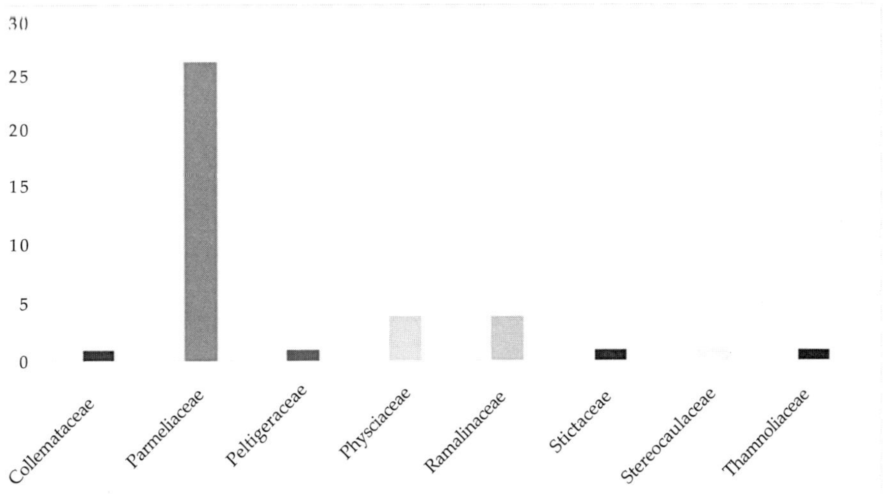

Fig. 3: Family wise use of lichens in the North West Himalayas

Table 1: Ethnobiological uses of lichens of North West Himalayas

Family	Species	Uses	References
Collemataceae	*Leptogium* sp.	SP	Upreti et al., 2005
Parmeliaceae	*Bulbothrix meizospora* (Nyl) Hale	SP	Upreti et al., 2005
Parmeliaceae	*Canomaculina subtinctoria* (Zahlbr.) Elix	SP	Upreti et al., 2005
Parmeliaceae	*Canoparmelia texana* (Tuck.) Elix& Hale	SP	Upreti et al., 2005
Parmeliaceae	*Cetrelia collate* (Nyl) Culb. & C. Culb.	RS	Upreti et al., 2005
Parmeliaceae	*Evernia mesomorpha* Nyl.	DY	Shukla et al., 2014
Parmeliaceae	*Everniastrum cirrhatum* (Fr) Hale	i.FO; ii. RS; iii. SP; iv. DY	Upreti et al., 2005; Devkotaet al., 2017; Shukla et al., 2014
Parmeliaceae	*Everniastrum nepalense* (Taylor) Hale ex Sipman	FO	et al., 2017
Parmeliaceae	*Flavoparmelia caperata* (L.) Hale	DY	Shukla et al., 2014
Parmeliaceae	*Flavopunctelia flaventior* (Stirton) Hale	SP	Upreti et al., 2005
Parmeliaceae	*Melanelia infumata* (Nyl.) Essl.	RS	Upreti et al., 2005
Parmeliaceae	*Nephromopsis nephromoides* (Nyl.) Ahti & Randl.	DY	Shukla et al., 2014
Parmeliaceae	*Parmotrema cetratum*	FO	Devkota et al., 2017
Parmeliaceae	*Parmotrema nilgharrensis* (Nyl.) Hale	i. RS; ii. DY; iii. ME; iv. SP	Upreti et al., 2005; Shukla et al., 2014; Shah, 2014
Parmeliaceae	*Parmotrema pseudonilgharrense* (Asahina) Hale	SP	Upreti et al., 2005
Parmeliaceae	*Parmotrema reticulate* (Taylor) Choisy	DY	Shukla et al., 2014
Parmeliaceae	*Parmotrema sanctae – angelii* (Lynge) Hale	ME	Upreti et al., 2005
Parmeliaceae	*Parmotrema tinctorum* (Despr. ex. Nyl.) Hale	i. SP; ii. DY	Upreti et al., 2005; Shukla et al., 2014
Parmeliaceae	*Punctelia reducta* (Ach.) Krog	DY	Shukla et al., 2014
Parmeliaceae	*Rimelia reticulata* (Taylor) Hale & Fletcher	SP	Upreti et al., 2005
Parmeliaceae	*Usnea austroindica* G. Awasthi	SP	Upreti et al., 2005
Parmeliaceae	*Usnea baileyi* (Strilon) Zahlbr.	i. ME; ii. SP	Shah, 2014
Parmeliaceae	*Usnea longissima* (L.) Ach	i. ME; ii. SP; iii. DY; iv. RS	Ji et al., 2004; Upreti et al., 2005; Bhat & Sharma, 2019; Shukla et al., 2014; Devkota et al., 2017

Contd.

Parmeliaceae	*Usnea orientates* Mot.	SP	Upreti et al., 2005
Parmeliaceae	*Usnea stigmatoides* G. Awasthi	DY	Shukla et al., 2014
Parmeliaceae	*Usnea thomsonii Stirt*	SP	Upreti et al., 2005
Parmeliaceae	*Xanthoparmelia somoloensis* (Ach.) Ahti & Hawksw.	DY	Shukla et al., 2014
Peltigeraceae	*Peltigera polydactyla* (Neck.) Hoffm.	ME	Upreti et al., 2005
Physciaceae	*Buellia subsororiodes* S. Singh and Awasthi	DY	Upreti et al., 2005
Physciaceae	*Heterodermia diademata* (Taylor) Awasthi	i. ME; ii. PE	Upreti et al., 2005; Devkota et al., 2017; Singh et al., 2015
Physciaceae	*Heterodermia leucomala* (L.) Poelt	SP	Upreti et al., 2005
Physciaceae	*Heterodermia tremularis* (Müll. Arg.) W. Culb.	SP	Upreti et al., 2005
Ramalinaceae	*Ramalina* sp.	i. FO; ii. ME	Ghorbani et al., 2012; Devkota et al., 2017
Ramalinaceae	*Ramalina conduplicans* Vainio	SP	Upreti et al., 2005
Ramalinaceae	*Ramalina sinensis* Jatta	SP	Upreti et al., 2005
Ramalinaceae	*Ramalina subcomplanata* (Nyl) Kashiw	SP	Upreti et al., 2005
Stereocaulaceae	*Stereocaulon himalayense* Awasthi & Lamb	ME	Upreti et al., 2005
Stictaceae	*Lobaria retigera* (Ach.) Treris	i. ME; ii. SP	Ji et al., 2004; Upreti et al., 2005
Thamnoliaceae	*Thamnolia vermicularis* (Sw.) Ach.	i. ME; ii. RS; iii. SP	Ji et al., 2004; Devkota et al., 2017; Upreti et al.,2005

ME – Medicinal Uses; FO – Food; DY – Dye; RS – Religious and sacred importance; SP – Spices and condiments; PE – Perfumery

Everniastrum cirrhatum as vegetable. *Cetrelia collata, Melanelia infumata, Parmotrema nilgharrensis* are used widely during religious and sacred ceremonies by the Gaddis. Garhwal herdsmen colour their body parts using *Bulbothrix meizospora* (Upreti et al. 2005; Devkota et al. 2017).

In some parts of China lying near North west Himalayas, some lichens from the wild are gathered by the locals and tribals of different ethnic groups from the thick forests for many important purposes (Ji*et* al. 2004).

The whole thallus of *Lobaria retigera* (Ach.) Treris is used as oral poultice against indigestion, nephritis, dropsy, skin diseases, scald by people of Lisu Yagong, China whereas the whole thallus of *Thamnolia vermicularia* (Sw.) Ach. is used to prepare decoction as oral poultice against falling sickness, phthisis, asthama, neurasthenic in Lisu Gualamu (Jiet al. 2004). The whole thallus of *Usnea longissima* of Lisu Jianmingyan is used as

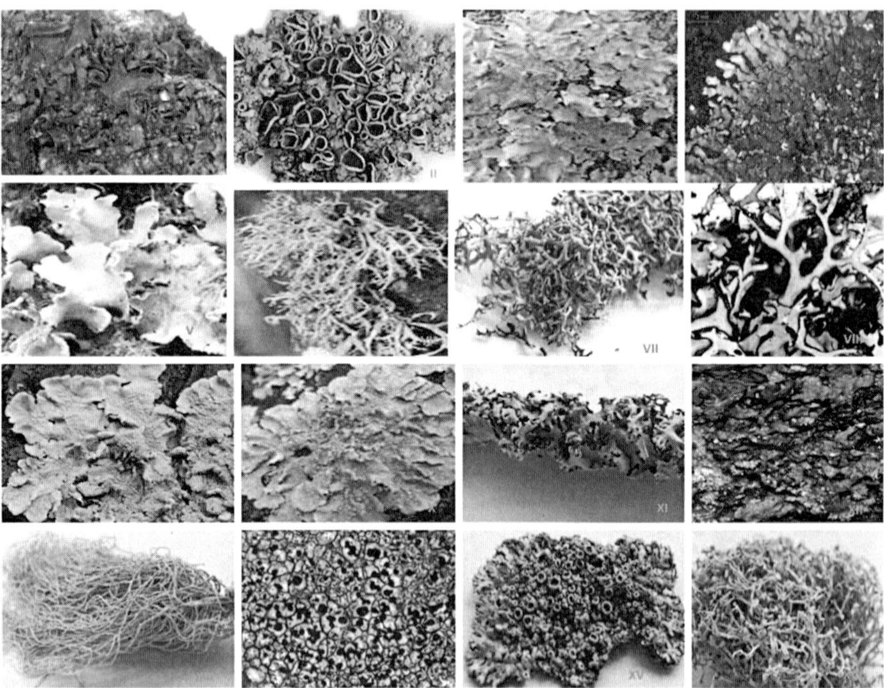

Fig. 5: Representative samples of different lichens present in North-west Himalayas (**i.** *Leptogium* sp., **ii.** *Bulbothrix meizospora* (Nyl) Hale, **iii.** *Canomaculina subtinctoria* (Zahlbr.) Elix, **iv.** *Canoparmelia texana* (Tuck.) Elix & Hale, **v.** *Cetrelia collata* (Nyl) Culb. & C.Culb, **vi.** *Evernia mesomorpha* Nyl., **vii.** *Everniastrum cirrhatum* (Fr) Hale, **viii.** *Everniastrum nepalense* (Taylor) Hale ex Sipman, **ix.** *Flavoparmelia caperata* (L.) Hale, **x.** *Flavopunctelia flaventior* (Stirton) Hale, **xi.** *Parmotrema tinctorum* (Despr. ex. Nyl.) Hale, **xii.** *Melanelia infumata* (Nyl.) Essl., **xiii.** *Usnea bailyi* (Strilon) Zahlbr., **xiv.** *Buellia subsororiodes* S. Singh and Awasthi, **xv.** *Heterodermia diademata* (Taylor) Awasthi, xvi. *Ramalina* sp.

decoction crushed as oral poultice in lymphadenitis, rheumatism, injury from falls and furumculosis (Ji et al., 2004). *Ramalina* sp. is consumed as wild food plant by the ethnic groups namely Lahu, Hani, Dai and Mountain Ham in Naban river watershed in Yunnan, China (Ghorbani et al. 2012). *Lobaria isidiophora, Lobaria kurokawae, Ramalina conduplicans, Lobaria yoshimurae* and *Ramalina sinensis* are consumed as food in Yunnan province. However, *Lethariella cashmeriana, Lethariella sernanderi, Lethariella sinensis, Thamnolia vermicularis* and *Thamnolia subuliformis* are used as tea for good health (Wang et al. 2001).

4. Conclusions

The paper reveals that the use of lichens in North western Himalayas is deep rooted and prevalent in various traditional and ethnic groups since past. Besides other uses, lichens have been predominantly used as remedies for the treatment of many ailments. The incessant overexploitation of lichens and loss of substrate are the two prominent causes for their depletion. The continued overexploitation and destruction of forests is leading to the loss of many lichen species even before their documentation and as such there will be no progression of this knowledge to test their properties. Before we lose this traditional knowledge and the species, the need is to explore the active principles so that the effective plant based drugs are developed from them.

Abbreviation

ME – Medicinal Uses; FO – Food; DY – Dye; RS – Religious and sacred importance; SP – Spices and condiments; PE – Perfumery Abbreviation Used

Competing Interest

The authors declare no competing interest.

Acknowledgements

The authors are thankful to Head, Department of Botany, BGSB University for providing necessary support. The work did not receive any specific funding.

References

Atharaveda (1958). Commentrator Shri Pad Damodar Stvalekar. In Hindi, printed at Mehra Offset Press, New Delhi.

Atkinson ET (1882). The Himalayan districts of North West Provinces of India. Cosmo Publication. New Delhi.

Bhat M & Sharma B (2019). Ethnobiology, phytochemistry and pharmacology of *Usnea longissima*: a review. *International Journal of Scientific Research in Biological Sciences* 6(1): 263 – 269.

Burbank DW, Leland J, Fielding E, Anderson RS, Brozovic N, Reid MR & Duncan C (1996). Bedrock incision, rock uplift and threshold hillspoles in the Northwestern Himalayas. *Nature* 379: 505 – 510.

Chanda S & Singh A (1971). A crude lichen drug (Charrila) from India. *Journal of Research in Indian Medicine*. 6: 209 – 215.

Chopra RN, Chopra IC, Handa KL & Kapur LD (1958). Indigenous Drugs of India. Pub. U.N. Dhur & Sons, Calcutta.

Devkota S, Choudhary RP, Werth S & Scheidigger C (2017). Indigenous knowledge and use of lichens by the lichenophillic communities of Nepal Himalayas. *Journal of Ethnobiology and Ethnomedicine* 13: 15.

Dymock W, Warden CJH & Hooper D (1890). Pharmacographia Indica. Thacker, Spink & Co., Calcutta.

Ghorbani A, Langenberger G & Sauerborn J (2012). A comparison of the wild food plant use knowledge of ethnic minorities in Naban river watershed National Nature reserve, Yunnan. *Journal of Ethnobiology and Ethnomedicine* 8:17.

Gonzàlez – Tejero MR, Martìnez-Lirola MJ, Casares-Porcel M & Molero-Mesa J (1995). Three lichens used in popular medicine in Eastern Andalucia (Spain). *Economic Botany* 49: 96 – 98.

Hale Jr EM (1983). The Biology of lichens third Ed. Edward Arnold Ltd.

Ivanova D & Ivanonv D (2009). Ethnobotanical uses of lichens: lichens for food review. *Scripta Scientifica Medica* 41(1): 11 – 16.

Ji H, Shengji P, & Chunlin L (2004). An ethnobotanical study of medicinal plants used by the Lisu people in Nujiang, Northwest Yunnan, China. *Economic Botany* (58) Supplement, S253 – S264.

Kala CP (2000). Status and conservation of rare and endangered medicinal plants in Indian trans – Himalaya. *Biological Conservation* 93(3): 371 – 379.

Lal B & Upreti DK (1995). Ethnobotanical notes on three Indian lichens. *Lichenologist* 27: 77 – 79.

Llano GA (1948). Economic uses of lichens. *Economic Botany* 2: 15 – 45.

Malhotra S, Subban R & Singh A (2007). Lichens- Role in Traditional Medicine and Drug Discovery.*The Internet Journal of Alternative Medicine* 5(2): 1 – 5.

Mani MS (1978). Ecology and Phytogeography of high altitude plants of North west Himalayas. Chapman & Hall.

Martin JG (1995). Ethnobotany A method manual . Chapman & Hall, London.

Negi HR (1996). *Usnea longissima* – the winter staple food of Musk deer ; a case study from Musk deer breeding center, Kanchula Kharak in Garhwal Himalaya. *Tiger Paper* 23: 30 – 32.

Richardson DHS (1988). Understanding the pollution sensitivity of lichens. *Botanical Journal of the Linnean Society* 96: 31 – 43.

Singh S, Upreti DK, Lehri A & Paliwal AK(2015). Quantification of lichens commercially used in traditional perfumery industries of Uttar Pradesh, India. *Indian Journal of Plant Sciences* 4(1): 29-33.

Sharma PV (1984 a). Dravyaguna Vigyan. Chaukhmbha Bharti Academy, Varanasi, India IV: 27-28.

Sharma PV (1984 b). Dravyaguna Vigyan. Chaukhmbha Bharti Academy, Varanasi, India IV:151, 198.

Sharma PV (1998). Dravyaguna Vigyan. Chaukhmbha Bharti Academy, Varanasi, India VII: 724-726.

Shah NC (2014). Lichens of commercial importance in India. *The Scitech Journal* 1(2): 32 – 36.

Sheikh MA,Upreti DK & Raina AK (2006). Lichen diversity in Jammu and Kashmir. India. *Geophytology* 36(2): 69 – 85.

Shukla P, Upreti DK, Nayaka S & Tiwari P (2014). Natural dyes from Himalayan lichens. *Indian Journal of Traditional Knowledge* 13(1): 95 – 201.

Singh TB & Chunekar KC (1972). Glossary of vegetable drugs in Brhattrayi. The Chaukhambha Sanskrit Series, Varanasi.

Skunke F (1969) Reindeer ecology and management in Sweden; Univ. Alaska. *Biological Pap*ers 8: 1 – 82.

Tiwari P, Rai H, Upreti DK, Trivedi S & Shukla P (2011). Assessment of Antifungal Activity of Some Himalayan Foliose Lichens against Plant Pathogenic Fungi. *American Journal of Plant Sciences* 2: 841-846.

Upreti, DK & Chatterjee S (2000). Distribution of lichens on Quercus and Pinustree in Almora district, Kumaon Himalaya, India. *Geophytology* 28: 41 – 49.

Upreti DK, Divakar PK & Nayaka S (2005). Commercial and ethnic use of lichens in India. *Economic Botany* 59(3): 269 – 273.

Wang L, Narui T, Harada H, Culberson CF & Culbreson WL (2001). Ethnic uses of lichens in Yunnan, China. *The Bryologist* 104(3): 345 - 349.

8

Managing Leucorrhea: A Traditional Perspective Emanating from the Ethnomedicinal Folklore in Jammu and Kashmir, India

Harpreet Bhatia and Yash Pal Sharma

Abstract

Plants are an integral component of life in many ethnic communities. Plant-based medicaments have been used since time immemorial as curative and protective agents for prolonging life of humans and animals by combating a plethora of ailments. One such ailment affecting women of reproductive age-group is leucorrhea. The study aimed at systematically documenting the precious traditional knowledge regarding the plants traditionally used for curing leucorrhea in Jammu and Kashmir. A total of 55 infomants (39 females and 16 males) between the age group 25-94 years were selected and interviewed for extracting ethnomedicinal information and 29 plant species belonging to 28 genera and 23 families were observed to be of ethnomedicinal significance in the treatment of leucorrhea in Jammu and Kashmir. Fabaceae was observed to be the frequently encountered family. Bulk of the taxa were observed to grow in wild habitat and majority of them were herbs. Seeds were observed to be of highest therapeutic significance and ethnomedicinal formulations were prepared in the form of powder, decoction, paste, etc and all of these were administered orally. The present study may serve as a base line data for further research and will go a long way in the treatment of leucorrhea thereby promoting well being of women folk. Advanced investigations are required to validate the time-tested information associated with medicinal

Harpreet Bhatia and Yash Pal Sharma (✉)

Department of Botany, University of Jammu, Jammu–180006, Jammu & Kashmir, India
✉Corresponding email: yashdbm3@yahoo.co.in

Plants for Novel Drug Molecules: Ethnobotany to Ethnopharmacology
Bikarma Singh & Yash Pal Sharma (eds.), (pp. 203-219)

Email: *info@nipabooks.com* Web: *www.nipabooks.com*

flora to determine the identity of phytochemicals linked with leucorrhea, so that compounds of pharmacological application having prospective relevance in the alleviation of human maladies may be discovered.

Keywords: Leucorrhea, Ethnomedicinal, Folklore, Jammu and Kashmir

1. Introduction

Plants are an intrinsic component of existence in many ethnic communities. Worldover, the civilizations are dependent on flora in all aspects of daily life for food, fodder, shelter, medicine, fuel, rituals, etc. Plants and plant-based medicaments have been employed since antiquity as curative and protective agents for prolonging life of humans and animals by combating a plethora of ailments. Being regarded as safe, cost-effective and easily affordable, traditional herbal medicine is still the cornerstone of more than three quarters of the world's population in developing countries for primary health care (Kamboj 2000; WHO 2002). Generally, women especially those of reproductive age group follow more precautionary health measures and resort to medical care usually more often (Chambliss 2001; Romm 2010) because in majority of the cases, the world's female population after attaining puberty, is afflicted with certain disorders which adversely affect the quality of their life (Kumar and Choyal 2012). One such prevalent disorder is leucorrhea referred to as 'shwait pradar' in sanskrit (locally known as 'safed pani'), an abnormally excessive vaginal discharge, white, yellow or sometimes greenish in colour, thick, having foul odour, non hemorrhagic in nature and often associated with irritation and pruritis (Sabaratnum et al. 1999). Leucorrhea can be physiologic or pathologic. Whereas, physiologic leucorrhea is caused by congestion of the vaginal mucosal membranes due to hormonal stimulation and may occur during ovulation and pregnancy, pathologic leucorrhea is usually due to infection of the female genital tract by *Chlamydia trachomatis, Neisseria gonorrhoeae* and *Trichomonas vaginalis* (Elkabbakh et al.1995).

India abounds in plant wealth and represents one of the richest centres of medicinally important flora. The country is being inhabited by over 54 million tribal people residing in about 5000 forest dominated villages comprising 15% of the total geographical area. Due to constant association with forests, ethnic people have colossal plantlore and folklore which they inherit and pass on from generation to generation just through oral conservation (Nath and Khatri 2010). Jammu and Kashmir State of India has a rich heritage of biodiversity and is a vast repository of medicinal plants. As far as the traditional knowledge regarding usage of plants of Jammu and Kashmir is concerned, they have been explored from various angles viz., ethnobotanical, ethnomedicinal, ethnoveterinary, ethnoreligious and find mention in literature (Gupta et al. 1980; Dar et al. 1984; Vir Jee et al. 1984; Kaul et al. 1985; Rashid et al. 2007; Ayub 2009; Gupta et al. 2013; Bhatia et al. 2014). But, no attempt has been made in the past to study the traditional perspective emanating from ethnomedicinal folklore in the management of leucorrhea. Furthermore, major chunk of this wisdom garnered over millennia by tribals and ethnic communities exists only as a verbal folklore and is passed on from one generation to another through word of mouth. As this skill is usually confined to local healers and elderly community members, it is in danger of being lost forever with the death of these knowledgeable people. Hence an initiative has been taken up to investigate, compile, analyse and document the traditionally used plants of this area effective in the treatment of leucorrhea.

2. Materials and Methods

2.1. Study Area

Jammu and Kashmir (J and K) is the northernmost state of India sharing its borders with Himachal Pradesh, Punjab and neighbouring countries of Pakistan, China and Afghanistan. Lying between 32° 17′ to 37° 06′ N latitude and 73° 26′ to 80° 30′ E longitude and spread over an area of 2,22,236 sq. km., the state consits of three divisions viz., Jammu, Kashmir and Ladakh comprising of 22 districts. The altitudinal variation ranges from 247 masl to 7742 masl and climate varies from alpine in the north-east to sub-tropical in the south-west. Whereas in the alpine region, average annual precipitation is about 75 mm, the sub-tropical zone records a precipitation of 1143 mm annually. Agriculture and livestock rearing are the mainstay of rural populace. Besides these, horticulture including produce of apples, apricots, walnuts, cherries, peaches, plums and almonds plays pivotal role in income generation. Small scale and cottage industries like carpet weaving, silk worm rearing, basketry, pottery and paper machie provide employment to large number of local inhabitants thereby helping in the development of the state's economy. The natural beauty and picturesque locations have made Jammu and Kashmir a favoured destination for tourists across the world.

2.2. Data Collection

Repeated field trips were carried out in different seasons of the year in various villages of Jammu and Kashmir to assemble information on traditionally used medicinal plant species effective in curing leucorrhea.

A total of 55 infomants (39 females and 16 males) between the age group 25-94 years were selected. Verbal approval was obtained from the participants and objectives of the study were made understandable to them. Information was obtained by organizing interviews (through questionnaire) and group discussions with the informants (local healers, nomads, community members) in their native dialect. Information of medicinal plant species collected included local name of plant species, part used, mode of use, route of drug administration and traditional method of treatment. The medicinal attribute of each plant species was substantiated, if at least five separate informants had a similar perspective. The collected plant specimens were dried and mounted on the herbarium sheets using standard protocols. For plant identification, flora of Jammu and flora of Udhampur were consulted. For confirmation of identification, herbaria of IIIM, Jammu and Department of Botany, University of Jammu were consulted. Valid botanical names with author citations were verified from The Plant List, Version 1.1 (TPL 2013). The herbarium sheets of all the collected plant specimens have been deposited in the herbarium of Department of Botany, University of Jammu. The gathered field information was analyzed and compiled to draw a clear picture of the ethnomedicinal plants effective in the treatment of leucorrhea in Jammu and Kashmir (Table 1).

2.3. Data Analysis

The data collected through interview of the informants was analyzed using a quantitative index *viz.*, use value (UV). It is a quantitative measure for the relative importance of a species known locally (Phillips et al. 1994) and was calculated using the formula: $UV = \Sigma\ U/n$ where, U is the number of use-reports cited by each informant for a given species and n refers to the total number of informants. Use values are high when there are many use-reports for a plant, implying that the plant is important, and approach zero (0) when there are few reports related to its use. The use value, however, does not distinguish whether a plant is used for single or multiple purposes (Musa et al. 2011).

3. Ethnomedicianl Plants Effective in the Treatment of Leucorrhea

During the course of investigation in the study area, 55 infomants (39 females and 16 males) between the age group 25-94 years were interviewed for extracting ethnomedicinal information regarding plants useful for the treatment of leucorrhea. Of the total informants, 70.90 % were females and all of these were mid-wives locally referred to as 'Daai' who were by and large illiterate. The male informants (29.10 %) were mainly hakims and vaidyas and were somewhat more literate than females. The patients generally favoured mid-wives over hakims or vaids due to coyness and societal norms. Illiteracy amongst female participants is an exceedingly cause of concern which might lead to depletion of the crucial knowledge they possess. Furthermore, females engaged in midwifery did not desire their offsprings to land in the same profession because of meagre income and their poor social standing in the society. Besides this, owing to escalating literacy and awareness in the society, the general local population is now consulting gynecologists for leucorrhea and other female disorders.

A total of 29 plant species belonging to 28 genera and 23 families were observed to be of ethnomedicinal significance in the treatment of leucorrhea in Jammu and Kashmir (Table1). Frequently encountered families include Fabaceae (3 species), Apiaceae, Moraceae, Poaceae and Lythraceae (2 species each). There seems to be a predisposition for a few plant families to stand out in any pharmacopoeia. Legumes produce a high diversity of secondary metabolites which not only serve as defence compounds against herbivores and microbes, but also as signal compounds to attract pollinating and fruit dispersing animals (Wink 2013). The wide utilization of species from Fabaceae might relate to the presence of effective bioactive constituents against leucorrhea (Gazzaneo et al. 2005).

With respect to diverse sources of ethnomedicinal plants in the study area, bulk of the taxa (44.83%) were growing in their natural wild habitat whereas 31.03% species were solely cultivated, 17.24% species existed in both wild and cultivated forms and 6.90% species were purchased. Yinegar et al. (2007) reported the herbalists in Ethiopia to gather healing flora from the wild pointing to the reality that the respondents have yet not started the cultivation of plant species they are using as medicine. As far as various habits of these species are concerned, 51.72% were herbaceous followed by trees (34.48%), shrubs and climbers (6.90% each). The prominence of herbs in a medicinal flora could be due to their abundance and easy accessibility in close proximity as compared to trees and shrubs that are often found to grow in woods (Simbo 2010). Amongst the ethnomedicinal plants enlisted, seeds were observed to be of highest therapeutic significance accounting for 36.37% followed by leaves and fruits (18.18% each), roots, bark and stem (6.06 % each) and proproots, flowers and rhizome (3.03% each). Utilization of roots, rhizomes, bark which contributed to harvesting of medicinal plants through destructive means could be a severe threat for survival of the often endangered and slow reproducing species (Chinsembu and Hedimbi 2010).

The ethnomedicinal formulations were prepared in the form of powder (54.55%), decoction (21.21%), paste (12.12%), infusion (6.06%), juice and grinded form (3.03% each) and all of these were administered orally exerting their beneficial effects by causing different reactions and found to be effective against leucorrhea. The foremost use of medicinal plant powder might be associated with their verified efficacy over several years of assessment and indigenous wisdom accumulated on the usefulness of such preparations as has been found by Tsobou et al. (2016). The observation that all the medicinal formulations in the study area were administered orally has also been reported by Bhattarai et al. (2010) who suggested the option of oral administration over probable alternatives due to the use of some solvents like water, butter, etc that are usually thought to aid the uptake of active principles in the herbal dose. Some ingredients viz., milk, water, honey, jaggery, sugar candy were also added in the preparation of ethnomedicinal formulations. Etana (2010) reported the usage of additives to solublize the active ingredients and improve the efficacy of the medicine.

In order to find out the relative importance of various ethnomedicinal plant species effective in the treatment of leucorrhea, a quantitative index i.e. use-value (UV) was employed. The results of the study have been presented in table 1 which indicated that the use value ranged from 0.13-1.0. On the basis of UV, the most important ethnomedicinal species of the present study area effective against leucorrhea included: *Triticum aestivum* (UV=1.0), *Oryza sativa* (UV=0.89), *Dalbergia sissoo* (UV=0.87), *Asparagus adscendens, Piper nigrum, Punica granatum* (UV=0.8 each), *Ziziphus jujuba* (UV=0.71), *Withania somnifera* (UV=0.62), *Ficus benghalensis* (UV=0.65), *F. religiosa* (UV=0.58), *Zingiber officinale* (UV=0.56), *Cicer arietinum, Vitex negundo* (UV=0.53 each) and *Trapa natans* (UV=0.51). Minimum UV of 0.13 was obtained for *Geranium wallichianum*.

Leucorrhea occurs either due to inflammation or congestion of vaginal mucosa or estrogen stimulation (Kumar and Choyal 2012). Recommendation of plants like *Triticum aestivum, Cicer arietinum* and *Oryza sativa* by the informants to treat leucorrhea was mainly due to rich nutritive value which improved the immunity of body and helped in recovering from hormonal imbalance. At times, *Talkhira*, a special preparation from *T. aestivum* given to female folk complaining of leucorrhea, was considered to be highly nutritious and is also available in the local markets. In addition to nutritive value, *Oryza sativa, Momordica charantia, Piper nigrum* and *Cuminum cyminum* possessed anti-inflammatory properties (Kirtikar and Basu 1999; Raman and Lau 1996; Panda and Kar 2003; Duke 1983).

Table 1: Ethnomedicinal plants from Jammu and Kashmir effective in the treatment of leucorrhea

Botanical name/ Family	Local name	Life-form / Source	Part used	Mode of use	Route of drug adminis-tration	Traditional method of treatment (No. of citations)	UV
Aesculus indica (Wall. ex Cambess.) Hook./ Hippocastanaceae	Goon	Tree, W	Fruit	Oral	Decoction	One teaspoon of honey is added to 200 ml of fruit decoction and taken daily on an empty stomach for a month (24)	0.44
			Fruit	Oral	Powder	Fully ripe fruits are dried, finely powdered and mixed with jaggery. One teaspoon is taken daily for two weeks (11)	0.2
Asparagus adscendens Roxb./ Asparagaceae	Sahns Pour	Herb, W	Root	Oral	Powder	One teaspoon of dried and powdered root is administered with milk for two weeks (44))	0.8
Azadirachta indica A. Juss./ Meliaceae	Nimm	Tree, W/C	Bark	Oral	Powder	Bark is shade dried, finely powdered and mixed with sugar candy. Two teaspoons of this preparation is taken daily for a fortnight (21)	0.38
Boerhavia diffusa L./ Nyctaginaceae	Ittsitt	Herb, W	Leaves	Oral	Powder	Leaves are dried and powdered. A teaspoon is administered daily with a glass of milk on an empty stomach (15)	0.27

Cardiospermum halicacabum L./Sapindaceae	Kanpatti	Climber,W	Stem	Oral	Paste	Stem is ground to a fine paste and prepared in the form of tablets. One tablet each is consumed twice a day for 3 weeks (9)	0.16
Cicer arietinum Linn./ Fabaceae	Chana	Herb,C	Seeds	Oral	Powder	Seeds are roasted and finely powdered. One teaspoon each of this powder is administered with milk twice a day for two weeks (29)	0.53
Cuminum cyminum L./ Apiaceae	Jira	Herb, C	Seeds	Oral	Powder	A teaspoon of roasted and finely powdered seeds is administered with lukewarm milk for two weeks (15)	0.27
Dalbergia sissoo DC./ Fabaceae	Tali	Tree, W	Leaves	Oral	Paste	Newly sprouted leaves are ground to a fine paste with jaggery. This preparation is made into small tablets. One tablet each is consumed with milk twice a day for 21 days (48)	0.87
Ficus benghalensis L./ Moraceae	Borh	Tree,W	Proproot	Oral	Powder	Equal quantities of dried and powdered prop roots and sugar candy are mixed well and one teaspoon is administered with milk twice a day for two weeks (36)	0.65

F. religicsa L. / Moraceae	Pipal, Barh	Tree,W	Fruits	Oral	Paste	Fully ripe fruits are finely ground with sugar candy and a teaspoon of this paste is administered with lukewarm milk for a month (32)	0.58
Geranium wallichianum D. Don ex Sweet / Geraniaceae	Rattan Jot, Lal Jari	Herb,W	Roots	Oral	Decoction	Few drops of honey are added to two teaspoons of root decoction and administered orally for a month (10)	0.18
					Powder	Roots are shade dried, finely powdered and mixed with jaggery. Five grams of this preparation is taken daily for 21 days (7)	0.13
Lepidium sativum Linn./ Brassicaceae	Halian	Herb, C/W	Seeds	Oral	Infusion	Seeds are ground and kept in water overnight. Few drops of honey are added to this infusion and 100 ml is administered daily for a month (15)	0.27
Momordica charantia Linn./ Cucurbitaceae	Karela	Climber, C	Seeds	Oral	Powder	Seeds are dried, finely powdered and mixed well with powdered jaggery. Two teaspoons of this preparation are taken on an empty stomach for three weeks (12)	0.22
Musa paradisiaca Linn./ Musaceae	Kela	Herb, C	Leaves	Oral	Decoction	Newly sprouted leaves are boiled in milk to form	0.470.45

						decoction, 150 ml of which is taken empty stomach for 21 days (26)	
			Fruits	Oral	Paste	Fully ripe fruits and honey are made into a fine paste. One teaspoon of this preparation is taken at bedtime after dinner for one month (25)	0.45
Oryza sativa L./ Poaceae	Dhan,Munji	Herb, C	Seeds	Oral	Decoction	Rice is boiled in water and seived. A cup of this left out water is taken orally for a fortnight (49)	0.89
Phyllanthus emblica L./ Euphorbiaceae	Amla	Tree, W/C	Fruits	Oral	Juice	Two teaspoonsof fresh fruit juice are administered twice a day for 21 days (21)	0.38
			Seeds	Oral	Powder	Dried seeds are powdered and mixed with sugar candy. One teaspoon of this preparation is taken after dinner for a month (25)	0.45
Piper nigrum L./ Piperaceae	Kalemaarch	Herb,P	Seeds	Oral	Powder	Fully ripe seeds are dried, powdered and mixed with equal quantity of dried and powdered leaves of *Punica granatum*. One teaspoon of this mixture is mixed with one teaspoon of honey and taken orally for two weeks (44)	0.8
Punica granatum L./ Lythraceae	Darooni	Tree, W/C	Leaves	Oral	Powder	Dried and powdered leaves are mixed with equal quantity	0.8

						of dried and powdered fully ripe seeds of *Piper nigrum*. One teaspoon of this mixture is mixed with one teaspoon of honey and taken orally for two weeks (44)	
Robinia pseudoacacia L./ Fabaceae	Kikkar	Tree,W	Seeds	Oral	Powder	A teaspoon of dried and powdered seeds is taken with a glass of lukewarm milk for a fortnight (15)	0.27
Sida cordifolia L./ Malvaceae	Tammani	Herb, W	Seeds, Leaves	Oral	Powder	Seeds and/or leaves are dried and finely powdered. A teaspoon of this powder is consumed with milk for one week (24)	0.44
Syzygium cumini (L.) Skeels/Myrtaceae	Tallay, jaamnoo	Tree,W	Bark	Oral	Decoction	A cup of barkdecoction is administered daily for two weeks (9)	0.16
Trachyspermum ammi Sprague/Apiaceae	Ajwain	Herb,C	Seeds	Oral	Infusion	Seeds are kept in water overnight. Next day they are seived and the water left out is taken orally (250 ml) daily for one month (25)	0.45
Trapa natans Roxb./ Trapaceae	Singhara	Herb,C/P	Seeds	Oral	Powder	Seeds are finely ground to obtain flour. Chapatis are made from it and consumed in breakfast for a fortnight (28)	0.51
Triticum aestivum Linn./ Poaceae	Kanak	Herb, C	Seeds	Oral	Ground	Seeds are soaked in water for five days by	1.0

						replacing water every day. These are then finely ground, sun-dried resulting in the preparation of a semi-solid sediment locally called *Talkhira*, which is cut into small pieces. One piece is consumed daily with water on an empty stomach for one month (55)	
Vitex negundo Linn./ Verbenaceae	Bana	Shrub, W	Leaves	Oral	Decoction	Leaf decoction is prepared in milk. 100 ml of this preparation is administered twice a day for 2 weeks (29)	0.53
Withania somnifera (Linn.) Dunal/ Solanaceae	Asgandh	Herb, C/W	Stem	Oral	Powder	Stem is dried and finely powdered. One teaspoon of this powder is taken orally with water on an empty stomach for a fortnight (34)	0.62
Woodfordia fruticosa (Linn.) Kurz/ Lythraceae	Dhaeen	Shrub, W	Flower	Oral	Powder	Flowers are dried and powdered. Few drops of honey are added to this preparation. One teaspoon is consumed daily on an empty stomach for a fortnight (16)	0.29
Zingiber officinale Roscoe/ Zingiberaceae	Adrak	Herb, C	Rhizome	Oral	Decoction	Decoction of fresh rhizome is prepared with seeds of *Trachyspermum ammi* and 150 ml is taken orally at bed	0.56

						time for a fortnight (31)	
Ziziphus jujuba Mill./ Rhamnaceae	Bair	Tree, W/C	Fruits	Oral	Powder	Five gram powder of shade dried fruits is mixed with 50 gram fresh banana paste and 20 gram powdered sugar candy. Two teaspoons of this preparation is taken daily with milk for a month (39)	0.71

4. Scientific Validation of Drugs Used for Curing Leucorrhea

Although there have been a number of ethnomedicinal claims regarding the usage of crude plant-based drugs for the treatment of various human/animal maladies, yet evidence based scientific validation of ethnopharmacological assertions on traditional medicine is the need of the hour for their safety assessment, global acceptance and reinforcement. A polyherbal formulation, "Sufoofe sailan", considered to be a safe and cost-effective drug used in the management of leucorrhoea also finds mention in Indian system of traditional medicine. It is prepared by mixing dried flowers of *Woodfordia fruticosa* and *Areca catechu* (12.5% each) with gum of *Bombax malabaricum* and *Mimusops elengi* (12.5 % each) and sugar crystals (50%). The effectiveness of this light colourred, odourless and sweet formulation in the management of leucorrhea has been tested as a single blind, randomized placebo controlled trial carried out on patients of leucorrhea (Rani et al. 2015). *Woodfordia fruticosa*, one of the constituents of this drug and an important medicinal plant in Ayurveda, contains flavonoids, essential oils, tannins and phenolic compounds which impart antimicrobial properties to the plant (Bajracharya et al. 2008). Flavonoids harboured in this species have been found in-vitro to be effective against *Chlamydia trachomatis* which may be due to their ability to form complex with bacterial cell wall resulting in its dissolution. Furthermore, tannins have been reported to show inhibitory activity towards DNA topoisomerase of *C. trachomatis* (Scalbert 1991).

Another plant-based formulation in Ayurveda, "Panchavalkal", a combination of barks of *Ficus benghalensis, F. religiosa, F. glomerata, F. lacor* and *Thespesia populnea*, shows immuno-modulatory, antiseptic, anti-inflammatory, anti-microbial properties and is found to be effective against *Trichomonas vaginalis*, the causal agent of vulvovaginitis. Panchavalkal choorna and ointment help in increasing local cell immunity and prevent recurrence of symptoms of leucorrhea, thereby providing relief to the patients (Gajmal et al. 2014; Dhiman 2014). Two Ayurvedic products based on this drug viz., V-gel (Himalaya Drug Company) and Pentaphyte-5 cream (Dr Paleps Research Foundation) are also commonly sought after medicines among leucorrhea patients. Combination of these five taxa was found to show superior activity as compared to individual plant extracts. The phytochemical analysis of the barks of Panchavalkal plants revealed the presence of gylcosides, tannins, triterpenes, sterols, quercetin-3-galactoside, leucoanthocyanin, aminoacids, lupeol acetate imparting antibacterial and antifungal properties thereby justifying the biological plausibility of the therapeutic use of this formulation in the treatment of leucorrhea (Joshi and Upadhye 2008; Patil and Patil 2010).

5. Conclusion

Since time immemorial, the human society has flourished amidst, and in close propinquity with plant life. Folk knowledge in a particular area is based on skill and experience which is tested time and again over centuries of use and gets tailored and adapted to local traditions and milieu. The present study pointed to the fact that the rural populace of the area has wealth of indigenous wisdom concerning medicinal plants for curing leucorrhea. Inhabitants of rural and far-flung areas still prefer traditional medicine over modern drugs because of their accessibility, safety, acceptability in society and convenience of use. This knowledge base is garnered by them over a long interval of time by experimentation and improvement. The tradition of using local medicinal plants by the inhabitants of the study area promises the incessant surge of native knowledge linked with the species. The present study may serve as a base line data for further research and will go a long way in the treatment of leucorrhea thereby promoting well being of women folk. Advanced investigations are requisite to validate the time-honored information associated with medicinal flora by executing apposite experiments and to determine the identity of phytochemicals linked with leucorrhea, so that compounds of pharmacological application having prospective relevance in the alleviation of human maladies may be discovered.

Conflict of Interests

The authors declare that they have no competing interests.

Acknowledgements

The authors would like to extend their sincere thanks to the Head, UGC SAP-DRS II Department of Botany, University of Jammu, Jammu for providing necessary laboratory facilities. The first author wishes to acknowledge the financial support through UGC-BSR fellowship.

References

Ayub H. 2009. Taxonomic studies of the Sacred Plants of district Kishtwar and their ethnoreligious significance. M. Phil. Dissertation, University of Jammu, Jammu.

Bajracharya AM, Yami KD, Prasai T, Basnyat SR, Lekhak B. 2008. Screening of some medicinal plants used in Nepalese traditional medicine against enteric bacteria. *Science World* 6: 107–10.

Bhatia H, Sharma YP, Manhas RK, Kumar K. 2014. Ethnomedicinal plants used by the villagers of district Udhampur, J and K, India. *Journal of Ethnopharmacology* 151: 1005-1018.

Bhattarai S, Chaudhary RP, Quave CL, Taylor RSL. 2010. The use of medicinal plants in the trans-Himalayan arid zone of Mustang district, Nepal. *Journal of Ethnobiology and Ethnomedicine* 6: 14.

Chambliss L. 2001. Alternative and complementary medicine: an overview. *Clinical Obstetrics Gynecology* 44: 640–652.

Chinsembu KC, Hedimbi M. 2010. An ethnobotanical survey of plants used to manage HIV/AIDS opportunistic infections in Katima Mulilo, Caprivi region, Namibia. *Journal of Ethnobiology and Ethnomedicine* 6:25.

Dar G H, VirJee, Kachroo P, Butt G M. 1984. Ethnobotany of Kashmir-I. Sind Valley. *Journal of Economic and Taxonomic Botany* 5: 668-675.

Dhiman K. 2014. Leucorrhea in Ayurvedic literatute: A review. *International Journal of Ayurveda and allied sciences* 3: 73-78.

Duke JA. 1983. Medicinal plants of the Bible. Trado-Medic Books, Buffalo, New York, pp. 233.

Elkabbakh GT, Elkabbakh GD, Broekhuizen F, Griner BT. 1995. Value of wet mount and cervical cultures at the time of cervical cytology in asymptomatic women. *Obstetrics Gynecology* 85: 449-503.

Etana B. 2010. Ethnobotanical study of traditional medicinal plants of Goma Woreda, Jimma zone of Oromia region, Ethiopia. M.Sc. thesis, School of Graduate Studies, Biology Department, Addism Ababa University, Addis Ababa, Ethiopia.

Gajmal AA, Shende MB, Chothe DS. 2014. A clinical evaluation of Panchavalkal-A review. *Unique journal of Ayurvedic and herbal medicines* 2: 6-9.

Gazzaneo LRS, Lucena RFP, Albuquerque UP. 2005. Knowledge and use of medicinal plants by local specialists in a region of Atlantic forest in the state of Pernambuco (North eastern Brazil). *Journal of Ethnobiology and Ethnomedicine*1: 1.

Gupta OP, Srivastava TN, Gupta SC, Badola DP. 1980. An ethnobotanical and phytochemical screening of high altitude plants of Ladakh-I. *Bulletin of Medico-Ethno botanical Research* 1: 301-317.

Gupta SK, Sharma OP, Raina NS, Sehgal S. 2013. Ethno-botanical study of medicinal plants of Paddar valley of Jammu and Kashmir, India. *African Journal of Taditional, Complemantary and Alternative medicine* 10: 59-65.

Joshi U, Upadhye M. 2008. Evaluation of Antioxidant Activity of Aqueous Extract Bark of *Ficus glomerata*. *Research Journal of Pharmaceutical Technology* 1:538-539.

Kamboj VP. 2000. Herbal Medicine. *Current Science*78: 35–39.

Kaul MK, Bhatia AK, Atal CK. 1985. Ethnobotanical studies in North-Western and Trans-Himalaya- Contribution to the wild Food Plants of Ladakh. *Journal of Economic and Taxonomic Botany* 6: 523-527.

Kirtikar KR, Basu BD. 1999. Indian medicinal plants. International Book Distributors, Dehradun, India, pp. 204.

Kumar N, Choyal R. 2012. Ethnobotanical notes on some plants used for the treatment of leucorrhoea and other gynecological problems in Himachal Pradesh. *Indian Journal of Fundamental and Applied Life Sciences* 2: 126-133.

Musa MS, Abdelrasool FE, Elsheikh EA, Ahmed LAMN, Mahmoud ALE, Yagi SM. 2011. Ethnobotanical study of medicinal plants in the Blue Nile State, south-eastern Sudan. *Journal of Medicinal Plants Research* 5: 4287-4297.

Nath V, Khatri PV. 2010. Traditional knowledge on ethnomedicinal uses prevailing in tribal pockets of Chhindwara and Betul districts, Madhya Pradesh, India. *African Journal of Pharmacy and Pharmacology* 4: 662-670.

Panda S, Kar A. 2003. Piperine lowers the serum concentration of thyroid hormones, glucose and hepatic 5D activity in adult male mice. *Hormone and Metabolic Research* 35: 523-526.

Patil VV, Patil VR. 2010. *Ficus benghalensis* Linn.-an overview. *International Journal of Pharma and Bio Sciences* 1: 1-11.

Phillips O, Gentry AH, Reynel C, Wilki P, Gavez-Durand CB. 1994. Quantitative ethnobotany and Amazonian conservation. *Conservation Biology* 8: 225-248.

Raman A, Lau C. 1996. Anti-diabetic properties and phytochemistry of *Momordica charantia* L. (Cucurbitaceae). *Phytomedicine* 2: 349-362.

Rani S, Rahman K, Sultana A, Younis PM, Basar SN. 2015. Physico-Chemical and Microbiological Standardization of Sufoofe sailan: A Unani Polyherbal Formulation. *World Journal of Pharmacy and Pharmaceutical Sciences* 4: 1554-1563.

Rashid A, Anand V K, Shah A H. 2007. Plant Resource Utilization in the ethnoveterinary practices by the Gujjar and Bakerwal Tribes of Jammu and Kashmir State. *Journal of Phytological Research* 20: 293-298.

Romm A. 2010. Botanical medicine for women's health. – Livingstone.

Sabaratnum, Drukumaran, Sivanesa VR, Alokananda C. 1999. Text book of Adolescent gynecology, sexually active adolescent 46(2):733.

Scalbert A. 1991. Antimicrobial properties of tannins. *Phytochemistry* 30: 3875-3883.

Sharma PK, Singh V. 1989. Ethnobotanical studies on North-Western and Trans Himalaya: Ethnoveterinary, Medcinal plants used in Jammu and Kashmir State. *Journal of Ethnopharmacology* 27: 63-70.

Simbo DJ 2010. An ethnobotanical survey of medicinal plants in Babungo, northwest region, Cameroon. *Journal of Ethnobiology and Ethnomedicine* 6: 8.

TPL. 2013. The Plant List- a working list of all known plant species, version 1.1. http://www.theplantlist.org/

Tsobou R, Mapongmetsem PM and Damme PV. 2016. Medicinal plants used for treating reproductive health care problems in Cameroon, central Africa. *Economic Botany* 70: 145-159.

VirJee, Dar GH, Kachroo P, Bhat GM. 1984. Taxo-ethnobotanical studies of the rural areas in district Rajouri. *Journal of Economic and Taxonomic Botany* 5: 831-838.

W.H.O. 2002. Traditional medicinal strategy 2002-2005. World Health Organization, Geneva, pp. 61.

Wink M. 2013. Evolution of secondary metabolites in legumes (Fabaceae). *South African Journal of Botany* 89: 164-175.

Yineger H, Kelbessa E, Bekele T, Lulekal E. 2007. Ethnoveterinary medicinal plants at Bale Mountains national park, Ethiopia. *Journal of Ethnopharmacology* 112: 55-70.

9

Traditional Usages and Chemical Constituents of High Value Medicinal Plants Recorded from Chithara Village Panchayat (Gautam Buddha Nagar) of Uttar Pradesh, India

Amit K Tripathi, Jyoti K Sharma and Mohd Ahmad

Abstract

In developing countries, about 80 percent of the world population depends on traditional medicines for their primary healthcare as estimated by World Health Organization. In a recent floristic survey undertaken in Chithara Village Panchayat recorded a very rich diversity of wild medicinal plants. Based on extensive literature search, of the total 272 plants, 96% species were known to have marginal to outstanding medicinal properties and uses. Among the plants, ten species namely, Bacopa monnieri (L.) Wettst., Cannabis sativa L., Cissampelos pareira L., Coccinia grandis (L.) Voigt, Eclipta prostrata (L.) L., Euphorbia hirta L., Hyptis suaveolens (L.) Poit., Justicia adhatoda L., Solanum americanum Mill. and Withania somnifera (L.) Dunal were found to be very important as they possess medicinal properties for curing a large number of diseases in traditional system of medicines in India and other parts of the world. The whole plants including root, stem and leaves are utilized traditionally for curing various ailments such as abdominal pain, cold, cough, diarrhoea, dysentery, inflammation, fever, rheumatism, sexual disorders, skin diseases, respiratory diseases etc. These plants are

Amit K Tripathi (✉), Jyoti K Sharma and Mohd Ahmad

Center for Environmental Sciences & Engineering, School of Natural Sciences, Shiv Nadar University, Greater Noida-201314, Uttar Pradesh, India

✉*Corresponding author email: amitnehu@gmail.com*

Plants for Novel Drug Molecules: Ethnobotany to Ethnopharmacology
Bikarma Singh & Yash Pal Sharma (eds.), (pp. 221-273)

Email: *info@nipabooks.com* Web: *www.nipabooks.com*

also reported to contain numerous important phytochemicals such as alkanes, amino acids, flavonoids, organic acids, polyphenolic compounds, polyphenols, steroids, tannins, terpenes etc. which qualifies their various medicinal properties. Plants are reported to yield potent drug such as vasicine and vasicinone (by A. vasica), bacosoids (B. monnieri) and withanolides (W. somnifera). The paper provides the details of traditional medicinal usages and chemical constituents of ten high value medicinal plants recorded from the study area.

Keywords: Medicinal plants, Traditional usages, Chemical constituents, Diseases, Chithara

1. Introduction

Plants have been rich source of medicinal agents since time immemorial. They have remained main components of various traditional systems of medicine, namely, Ayurveda, Unani, Siddha, Chinese, and so forth. World Health Organization estimated that there are about 80 percent of the world population in developing countries depends on traditional medicines for their primary healthcare. Globally, there are large number of medicinal plants have been used traditionally as folklore medicines even today. India has rich biodiversity, traditional knowledge and heritage of herbal medicines. There are about 17,000 higher plant species in India, out of which nearly 44% (7,500 plants) are known for their medicinal uses (Shiva 1996). Some of these medicinal plants are growing as weeds in various parts of the country and elsewhere. Weeds growing in wasteland, roadsides and fellow lands are known to possess numerous medicinal properties and are used worldwide in traditional system of medicines (Sharma et al. 2019). Traditional medicines are easily available and they do not possess much side effects and play significant role in the health care of rural people all over the world. Ever since the beginning of human civilization medicinal plants have been playing a vital role in their primary healthcare. Globally, traditional medicine mostly herbal medicine is known as a major healthcare provider in rural as well as remote areas and large number of people in developing and under developing countries are depends on such medicine for their healths (Sen & Chakraborty 2017).

A number of phytochemicals isolated from weeds are responsible for medicinal activity of plants. Traditional medicinal plants are known for numerous important chemicals constituents such as alkaloids, saponins, steroids, flavonoids, tannins etc. and they are the basic source of the pharmaceutical industries and these chemical constituents play a vital role in the identification of crude drugs. The various aspects of medicinal plants have been studied extensively in India and other parts of the world by different researchers (Adhikari et al. 2010, Kala et al. 2006, Larsen & Olsen 2007, Schippmann et al. 2002, Teklehaymanot & Giday 2007, Uprety et al. 2012). Inspite of tremendous development in modern world, plants still remain one of the major sources of drugs and traditional system of medicines throughout the world. The origin and evolution of many traditional herbal therapies is based on the medicinal properties of plants (Kala 2006). The paper provides the details of ten selected high value medicinal plants growing as weeds in Chithara along with their chemical constituents and traditional medicinal uses. It is hoped that the knowledge of these plants will be useful in drugs development and the cure of some serious diseases.

2. Materials and Methods

2.1. Geography of Study Area

The study area, Chithara Village Panchayat comes under Gautam Buddha Nagar District in western Uttar Pradesh, is situated in eastern part of Delhi NCR (Fig.1). Geographically, it lies between latitude 28°31' to 28°33′N and longitude 77°33' to 77°35′E, is one of the 56 Panchayats under Dadri Block Panchayat/ Tehsil and covers an area of 770.78 ha. The campus of Shiv Nadar University (SNU) is part of Chithara (Fig. 1). The climate of study area is typically monsoonic with three distinct seasons, namely, summer (March to mid-June), rainy (mid-June to September), and winter (October to February). The soil of the study area is fertile alluvial soil characteristic of soils represented in Wheat-Rice agriculture system in southwestern plain of Upper Gangetic Plain. The soil pH ranges 7.64 to 9.38 with reasonably good maximum water holding capacity. Nutrient wise the soil contained OC (0.299-0.400%) which is within the desirable range of soils of Upper Gangetic Plain. Delhi is approximately 40 Km and Noida 25 Km from Chithara Village.

2.2. Field Surveys and Data Analysis

The extensive floristic surveys were carried out in Chithara Village Panchayat including the campus of SNU during 2015-2018, covering different types of habitat such as agricultural fields, canal bunds, roadsides, village streets, wetlands, wastelands, grasslands, etc. The aquatic and terrestrial vegetation of area were surveyed frequently throughout year in all the three seasons. The documented plants were identified following regional floras (Duthie 1903-1929, Maheshwari 1966, Sharma & Dhakre 1995, Vardhana 2007) and herbaria of Botanical Survey of India (BSI), Dehradun. The accepted name of the species and family were verified from 'The Plant List, Version 1.1' (The Plant List 2013).

The traditional medicinal uses, medicinal properties and chemical constituents of each plants were gathered from the extensive literature search. Of 272 plant species recorded in the study area, the present study reports the traditional medicinal usages and chemical constituents of ten high value medicinal plants of Chithara as they are reported to cure a large number of diseases/health conditions in traditional system of medicines.

Fig. 1: Location of the study area

3. Results and Discussion

The floristic surveys of Chithara Village Panchayat (Gautam Buddha Nagar District), Uttar Pradesh revealed a total of 272 vascular plant species belonging to 203 genera and 69 families. The majority of species were Angiosperms (268 plant species: dicots- 215 spp.; monocots- 53 spp.), whereas 4 species belonged to ferns and fern allies (Table 1). Habit wise analysis of flora showed that herbs were most dominant with 147 species followed by sedges and grasses (38 sp.), climber and creepers (24), trees (22), undershrubs (20), shrubs (17) and Pteridophytic herbs (4 spp.) (Table 1). Besides twelve species (*Commelina forsskalii, Cyperus alopecuroides, Cyperus compressus, Digitaria ciliaris, Eriochloa procera, Leptochloa panicea, Mecardonia procumbens, Persicaria lanigera, Poa annua, Potamogeton crispus, Trifolium tomentosum, Utricularia stellaris*), others 260 plants (approximately 96%) possessed medicinal properties against one or more number of diseases. This is certainly a very high percentage of occurrences of medicinal plants in a small geographic area of Chithara Village Panchayat indicating that the weeds possess high medicinal properties (Stepp 2004) which have potential to drug development.

Of the 260 medicinal plant species, the present study reports the traditional medicinal uses and chemical constituents of ten selected high value medicinal plants of Chithara. Of the ten important medicinal plants (Fig.2), *Euphorbia hirta* is reported to treat highest number of diseases/health conditions (71) followed by *Solanum americanum* (69), *Cissampelos pareira* (55), *Cannabis sativa* (53), *Hyptis suaveolens* (50), *Justicia adhatoda* (49), *Eclipta prostrata* (48), *Coccinia grandis* (46), *Bacopa monnieri* (44) and *Withania somnifera* (40) in traditional systems of medicines (Fig.3). If we consider the efficacy of these ten plants, it is found that they are known to cure for a wide variety of diseases/ailments such as cuts, wounds, heart ailments, blood pressure, rheumatism, epilepsy, cancer, cold, cough, diarrhoea, dysentery, inflammation, fever, rheumatism, sexual disorders, skin diseases and respiratory diseases (Table 2). All the ten plants possessed anti-inflammatory and antipyretic properties used for various types of inflammation and treating fevers respectively. In case of snake bites, scorpion stings and dog bites some of these species are used. Of the ten, seven species are used in dysentery.

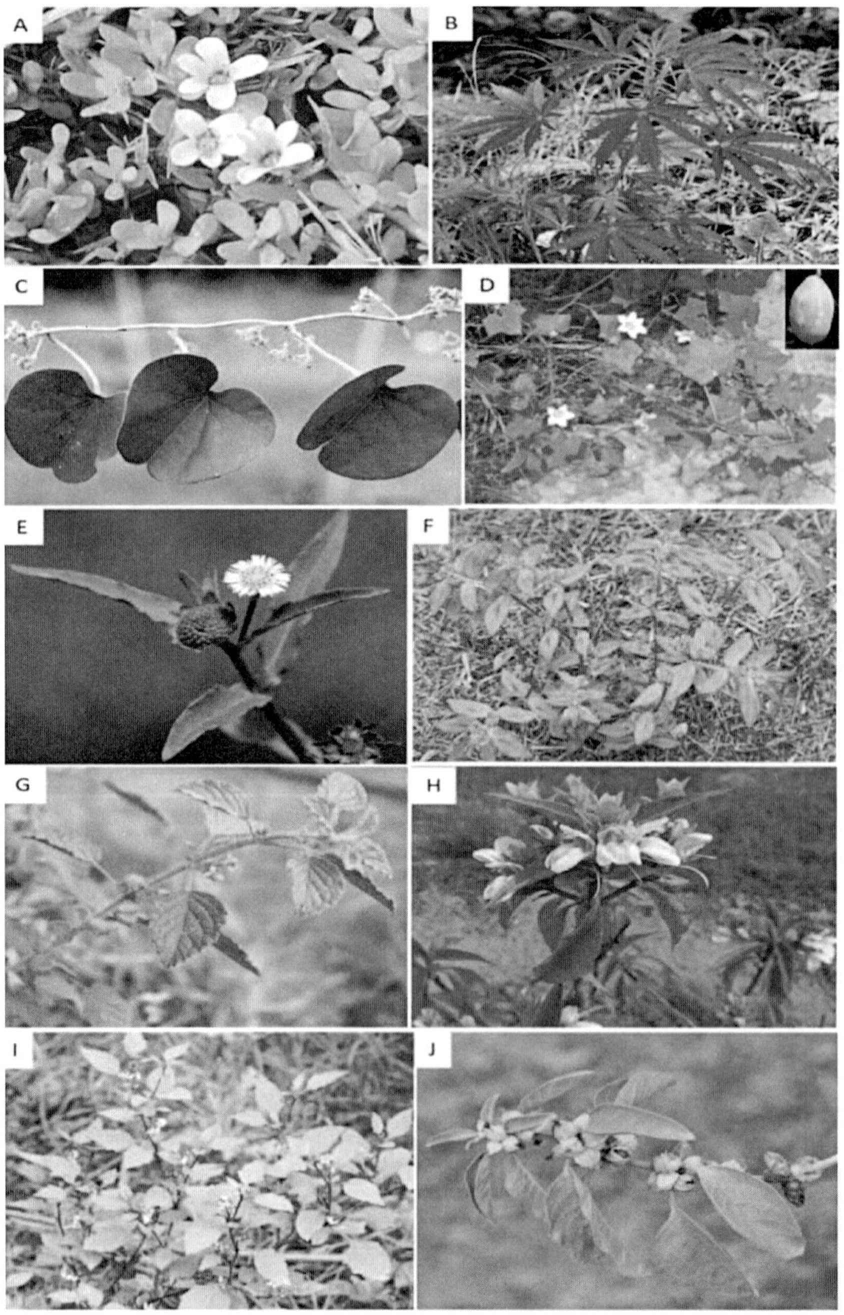

Fig. 2: High usages medicinal plants of Chithara, **A)** *Bacopa monnieri* (Brahmi), **B)** *Cannabis sativa* (Bhang), **C)** *Cissampelos pareira* (Padha), **D)** *Coccinia grandis* (Kundru), **E)** *Eclipta prostrata* (Bhringraj), **F)** *Euphorbia hirta* (Dudhi, Asthma plant), **G)** *Hyptis suaveolens* (Vilaiti Tulsi), **H)** *Justicia adhatoda* (Arus), **I)** *Solanum americanum* (Makoy), **J)** *Withania somnifera* (Ashwagandha)

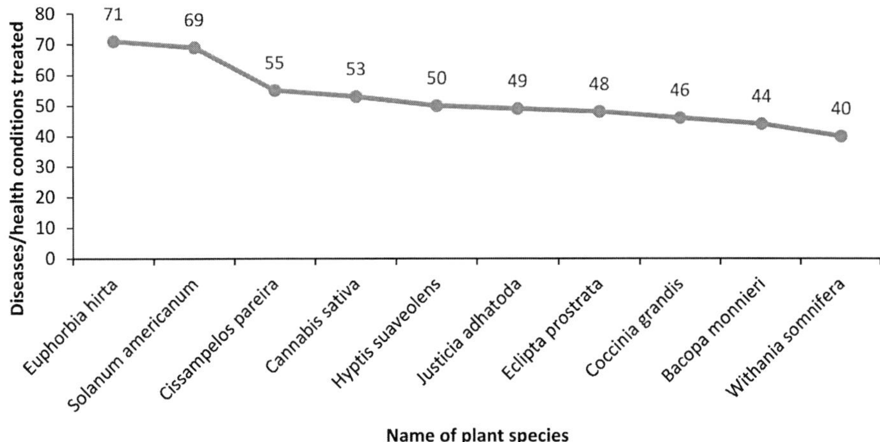

Fig. 3: Number of diseases (health conditions) treated by medicinal plants

Except *B. monnieri,* others are known to cure Asthma. *S. americanum* and *E. prostrata* is reported to be effective in hepatitis. Eight plant species had diuretic properties. Traditionally, all the plants are used to cure coughs globally. A number of skin diseases like itching, skin allergy, lesions and rashes, skin sores, skin ulcers, blisters, skin tumour, pruritus etc. were treated by these plants as reported in various published literatures. *S. americanum, C. sativa* and *H. suaveolens* is useful in case of cancer disease. In case of diabetes, *B. monnieri, C. grandis, J. adhatoda* and *S. americanum* is reported to be useful. The plant parts of *C. pareira* is used to cure menstrual/ premenstrual problems and other associated diseases in women. *E. hirta, S. americanum, C. sativa, J. adhatoda, E. prostrata* and *C. grandis* are reported to be effective in jaundice (Table 2).

Besides, these plants also have been reported to contain numerous phytochemicals such as alkanes, amino acids, flavonoids, organic acids, polyphenolic compounds, polyphenols, steroids, tannins, terpenes etc. Indian traditional medicinal plants are reported to yield new drugs (Sen & Chakraborty 2017). In present study, some of the plants reported to yield potent drug such as, vasicine and vasicinone (by *A. vasica*), bacosoids (*B. monnieri*) and withanolides (*W. somnifera*). Most of these plants are growing as weed along roadside, railway lines, wastelands, fallow fields, agricultural lands, marshy areas etc. The details of traditional medicinal usages along with medicinal properties and chemical constituents of top ten selected high value medicinal plants are given below:

3.1. Traditional Usages and Chemical Constituents of High Value Plants

3.1.1. *Bacopa monnieri* (L.) Wettst.

3.1.1.1. *Traditional Uses*

B. monnieri (Brahmi) is known to have long history of use in the Ayurveda for increasing concentration, memory capacity, anti-ageing, memory-enhancing and for reducing stress induced anxiety. The plant is mentioned in different Vedic and ancient scripters as invigorating and life sustaining in Charaka Samhita and for strengthening of body and improving seminal weakness in Atharva Veda and for increasing memory and lifespan in Sushruta Samhita. It has been mentioned in religious, social and medical treatises of India since the time of Atharva Veda (800 BC); the first clear reference to its CNS effect is found in Charak Samhita, written in the first century AD. It is mentioned in the authentic Ayurvedic treatise, Susrutu Samhita, which describes Brahmi as efficacious in the loss of intellect and memory. Brahmi has been used for centuries to benefit in epilepsy, memory capacity, increase concentration, and reduce stress-induced anxiety. Brahmi is used for the treatment of numerous complaints such as ascites, anaemia, amentia, invigorates sex; biliousness, bronchitis, constipation and indigestion, dyspepsia, dysmenorrhoea, dermatitis, diabetes, elephantiasis/filaria, flatulence, fever and general debility, erysipelas, leprosy, fever, hoarseness, insanity, epilepsy, enlargement of spleen, inflammations, leucoderma, leprosy, splenomegaly, skin diseases, syphilis, strangury, sterility, tumours, ulcers, nervous breakdown and general debility, etc. Brahmi is promising blood purifier and useful in diarrhoea and fevers; invigorating and effective against cold and cough. Plant-juice (along with ginger, sugar and bark extract of *Moringa oleifera*) is given to children in stomach disorder and leaves (fried in ghee) as a brain tonic, in nervous weakness, hysteria, epilepsy, insanity, anxiety neurosis, and to sharpen dull memory (memory enhancing). As per Unani system of medicine, Bhahmi is effective in scabies, leucoderma, syphilis, etc. (References consulted: Fern 2014, Kaul & Dwivedi 2010, Shukla et al. 2010; http://www.bsienvis.nic.in, http://medplants.blogspot.in).

3.1.1.2. *Medicinal Properties*

Brahmi, used as nervine tonic/memory enhancer, is known to be bitter, sweet, cooling, anodyne, antispasmodic, anticancer, asthenia, astringent, anti-inflammatory, anticonvulsant, blood purifier, bronchodilator, carminative, cardiotonic, diuretic, digestive, depurative, emetic, emmenagogue, febrifuge, intellect promoting, laxative, antipyretic, musculature relaxant, sudorific and tonic. It also has antioxidant properties. (References consulted: Alsnafi 2013, Fern 2014, Warrier et al. 2010).

3.1.1.3. *Chemical Constituents*

Brahmi is reported to contain triterpenoid saponins, bacosides A and B, bacogenins, betulic acid, betulinic acid, d-mannitol, stigmasterol, β-sitosterol, saponin, stigmastanol, hersaponin, monnierin, nicotine, luteoline and its glucosides. Leaves contain alkaloid brahmine, which raises blood pressure when administered in therapeutic dosage, and is hypotensive when given in very strong dosage (References consulted: Alsnafi 2013; http://www.bsienvis.nic.in).

3.1.2. *Cannabis sativa* L.

3.1.2.1. *Traditional Uses*

In traditional medicines, *C. sativa* has been used in the treatment of alcohol and morphine withdrawal, anthrax, asthma, blood poisoning, bronchitis, burns, catarrh, convulsions, coughs, cystitis, delirium, depression, diarrhoea, dysentery, dysmenorrhoea, epilepsy, fever, gonorrhoea, gout, inflammation, insomnia, jaundice, lockjaw, malaria, mania, menorrhagia, migraine, neuralgia, palsy, rheumatism, scalds, snakebite, toothache, uterine prolapse, and whooping cough. The different parts of *C. sativa* are also effective in the treatment of abdominal disorders, amenorrhea, otalgia, and stomatalgia, loss of appetite, penile dysfunction, premature ejaculation, hypertension, indigestion, piles, neuritis, headache, insanity, hiccough and cholera.

Leaves boiled with butter are given to prevent vomiting while drops leaf juice are poured in the ear to cure earache. Leaves and seeds of this plant is reported to cure old cancer and tumours. The seed paste is known as a folk remedy for tumours and cancerous ulcers whereas root decoction is considered to be a good remedy for hard tumours and knots in the joints. *C. sativa* is also effective in corns.

Contra indication: The excessive use of *C. sativa* causes cough, delirium, headache, impotence, melancholy, dyspepsia, dropsy, sleeplessness, hyperpyrexia and insanity. (References consulted: Fern 2014, Rahul 2013, Singh et al. 2010; https://www.hort.purdue.edu/newcrop/duke_energy/Cannabis_sativa.html).

3.1.2.2. Medicinal Properties

The plant has diuretic, antiemetic, antiepileptic, antiinflammatory, painkilling and antipyretic properties. Plants are tonic, intoxicant, stomachic, antispasmodic, analgesic, narcotic, sedative and anodyne. Leaves are astringent, hallucinogenic, hypnotic, sedative, analgesic, anti-inflammatory, tranquilizer, aphrodisiac, tonic, stomachic and abortifacient (References consulted: Bhutya 2011; Fern 2014, https://www.hort.purdue.edu/newcrop/duke_energy/cannabis_sativa.html).

3.1.2.3. Chemical Constituents

The phytochemical analysis shows that Cannabis contains 421 chemicals of various classes, such as cannabinoids, cannabispirans and alkaloids. More than 60 cannabinoids have been isolated; the most important one is δ-9-tetrahydrocanabinol. Most varieties contain cannabinol and cannabinin; resin contains crystalline compound cannin. The seeds are reported to have trigonelline. Cannabis also contains choline, eugenol, guaiacol, nicotine, and piperidine, all listed as toxins by the National Institute of Occupational Safety and Health. A β-resercyclic acid derivative has antibiotic and sedative properties. (References consulted: Bhutya 2011; https://www.hort.purdue.edu/newcrop/duke_energy/cannabis_sativa.html).

3.1.3. *Cissampelos pareira* L.

3.1.3.1. Traditional Uses

C. pareira (Patha) is mainly used in women aliments such as menstrual problems, painful menstruation, and hormonal imbalance, ease in childbirth, postpartum pain, to prevent threatened miscarriage, and control uterine haemorrhages, hormonal acne and premenstrual syndrome. It is very effective herb in vaginal discharge and good for overall health of women reproductive system. Leaves of the plant is used as antifertility agent because it alters the secretion of gonadotropins and estradiol. Leaves of *C. pareira* is also used for heart problems, kidney stones, kidney infections and pains, asthma, dysuria, arthritis, muscle cramps, stomach pains, non-healing ulcers, fistula, leucorrhoea, gonorrhoea, hypogalactia (agalactia or agalactorrhea), sexually transmitted diseases (STDs), snake bites and conjunctivitis. In eye trouble, skin ailments and burning sensation, wounds, fever and cold, leaves of the plant are used. Leaves are used to prepare poultice to treat, scabies, abscess, itching, acne and wounds. The plant is also used for the treatment of malaria and dengue.

Traditionally, roots are used in asthma, bronchitis, dyspepsia, indigestion, flatulence, abdominal pains, diarrhoea, dysentery, blood disorders, cardiac disorders, colic, edema, leprosy, cough, coryza, cystitis and dysuria, lactation disorders, non-healing ulcers, skin disorders, scabies, leprosy, migraine, leucorrhoea, gonorrhea, scabies and eruptions. The therapeutic uses of Patha according to the Ayurvedic Pharmacopoeia of India are in the treatment of abdominal pain, vomiting, cardiac pains, burning feeling, pruritus, poisons, breathing difficulties and worms; also effective in treating diarrhoea, skin diseases, and fever. Patha is also very useful in kushtharoga (leprosy). Patha is very effective appetizer and used to cure anorexia, indigestion and abdominal pain. The plant is a powerful expectorant and used to treat cough and dyspnoea. It is useful in blood related disorders and used as blood purifier. Paste of roots and leaves are used to reduce inflammation. It is used to purify breast milk secretion and various other disorders associated with breast. The plant has diuretic properties and used in dysuria and haematuria. Being anthelminthic, it is very effective in worm infestation. Patha is also used to cure dog bites (References consulted: Arora et al. 2012, Jhuma & Bhattacharaya 2011).

3.1.3.2. Medicinal Properties

C. pareira is bitter and retain diuretic, purgative and antiperiodic and antimalarial properties. The roots are bitter, astringent, antispasmodic, analgesic, anti-inflammatory, antilithic, expectorant, emmenagogue, rubefacient, anthelmintic, carminative, stomachic, depurative, digestive, diuretic, febrifuge and galactagogue. Leaves are good as an antiseptic against inflammation and applied on wounds in order to heal sores (References consulted: Arora et al. 2012, Bhutya 2011).

3.1.3.3. Chemical Constituents

The plant contains alkaloids *viz.*, hayatine (+ curine), hayatinine, hayatidine and other bisbenzylisoquinoline alkaloids, including some non-nitrogenous. Hayatine (dl-berberine) is the principal alkaloid of root. Its derivatives, methiodide and methochloride are potent neuro-muscular blocking agent. Other chemicals present are hayatinine, hayatinine, hayatidine, quercitol and sterol (References consulted: Arora et al. 2012, Bhutya 2011, Jhuma & Bhattacharya 2011).

3.1.4. Coccinia grandis (L.) Voigt

3.1.4.1. Traditional Uses

The plant is effective in treating anaemia, asthma, biliousness, bronchosis, catarrh, constipation, convulsion, cramp, dermatosis, diabetes, dysgeusia, earache, enterosis, fever, gas, glycosuria, gonorrhoea, gout, halitosis, inflammation, jaundice, leprosy, lowering blood cholesterol, menorrhagia, mycosis, ophthalmia, parturition, psoriasis, ringworm, sinusosis, smallpox, snakebite, sore, sore throat, stomatosis, syndrome-X, syphilis, tuberculosis, and vomiting.

The leaves are beneficial in vitiated condition of Kapha and Pitta. The leaves and fruits are used in skin diseases, fever, asthma, cough and jaundice. The fruits of this plant are used in asthma, burning sensation, intermittent fever, leprosy, skin diseases, agalactia, cough, bronchitis, consumption and jaundice. The roots are used in vomiting, burning sensation and uterine discharges. The leaves of *C. grandis* are used to treat scabies, roots are used to treat joint pain and fruits are used to treat infertility of females in Bangladesh.

Fruits are used to treat leprosy, fever, asthma, bronchitis and jaundice in traditional system of medicine. This plant helps to regulate blood sugar levels (antidiabetic). (References consulted: Ashwini et al. 2012, Selvam et al. 2014, Warrier et al. 2010, Zakaria et al. 2011; http://medplants.blogspot.in).

3.1.4.2. Medicinal Properties

The Leaves of *C. grandis* are known to have antidiabetic, antibacterial and hepatoprotective properties while fruits are cooling, sweet, astringent, depurative, antipyretic, galactagogue and expectorant. Fruit extract shows antioxidant activity while stem extracts are antispasmodic. Extract of fresh leave is reported to be anti-inflammatory and antiulcer activity. The roots are reported to have cooling and aphrodisiac properties. Due to its various properties such as bitter, hypoglycaemic, alexiteric, amebicide, antidiabetic, antidyslipidemic, antiemetic, antiprotozoal, antipyretic, antiseptic, antispasmodic, anti-hyperuricaemia and antitussive, aphrodisiac, astringent, adenopathy, depurative, diaphoretic, emetic, expectorant and hepatoprotective, the plant is used widely to cure various diseases/ailments in Ayurveda (References consulted: Selvam et al. 2014, Warrier et al. 2010; http://medplants.blogspot.in).

3.1.4.3. Chemical Constituents

The phytochemical analysis of the *C. grandis* shows the presence of alkaloids, phenols, carbohydrates, glycosides, flavonoids, tannins and saponins (Reference consulted: Selvam et al. 2014).

3.1.5 Eclipta prostrata **(L.) L.**

3.1.5.1. Traditional Uses

Eclipta is one of the ten promising herbs that establish the group dasapuspam which is considered to destroy the causative factors of all unhealthy and unpleasant features and give good health and wealth. The members of this group cure wounds and ulcers as well as fever caused by the imbalance of the tridosas - Vata, Pitta and Kapha. *E. prostrata* has been mentioned in ancient scriptures where it describes its uses against various disorders. In Atharvaveda: affects intelligence and memory, treats bile/pitta ailments and stops graying as well as falling of hairs; In Bhavaprakasha: cures problems caused by phlegm and wind, beneficial for hair, skin, teeth and eyes, removes worms, and also effective in jaundice and oedema; In Rajanighantu: beneficial for hairs, eyes, oedema and phlegm; In Kaiyadevanighantu: useful for hair, teeth, skin, cough, jaundice and oedema; In Nighanturatnakaram: plant invigorates sex; In Vaidyamanorama: drinking juice of the plant strengthens the body and secures the foetus in womb.

The plant is effective in hepatitis, enlargements of spleen, tetanus and elephantiasis, sexual weakness and various chronic skin diseases. For expulsion of worms, the juice of the plant with honey is given to children. Plant (powder) has curative effect on infective hepatitis, jaundice and viral hepatitis. Aqueous extract/juice of the leaf is known to cure shoulder pain. *E. prostrata* is is useful in blackening and strengthening of the hair. It is also used to prevent habitual abortion and miscarriage and post-delivery uterine pain. Leaf decoction of the plant is used in uterine haemorrhage. Fumigation with *E. prostrata* is considered beneficial in case of piles. In Ayurveda, it is mainly used in hair oil, while in Unani system, the juice of the plant is effective for several types of pains in the body. Other uses in Ayurveda are: cures headache, migraine; leaves remove lice and intestinal worms, cure pyorrhoea, chronic dysentery, oedema, nervous weakness, jaundice, anorexia and gum troubles.

A number of traditional uses of *E. prostrata* by different ethnic communities of India include: Leaf: in gastric troubles, hepatic disorders (by Garo tribes of Meghalaya); Stem-decoction in liver enlargement while leaf-extract in fever and cough (Manipuri tribes); Plant: in itching, Leaf: in conjunctivitis and other eye problems, in promoting hair growth (Ethnic Communities of Orissa); Leaf: in malaria and other fevers while Root: as antidote to snake bite (Ethnic Communities of Bihar); Leaf: in conjunctivitis, eye troubles (Araku Valley, Madhya Pradesh); Leaf: in sores, ulcers, wounds, spleen disorders (Tribal Societies of Eastern Rajasthan); Leaf: as antiseptic (Tribes of Kurukshetra, Haryana); Leaf: in malaria (Kol); Leaf: for promoting hair growth (Tribes of Chhindwara, Madhya Pradesh); Flower bud: in fever, headache (Tribal Societies of Anaikatty Hills, Tamil Nadu); Whole Plant: in asthma, bronchitis, leukoderma (Tribal Societies of Saurashtra, Gujarat); Whole plant: in jaundice, spleen disorders, Leaf: in leucoderma, skin diseases (Garhwali); Whole plant: in liver complaints (Tribes of Bastar, Madhya Pradesh); Plant: in toothache, headache, gland swelling, elephantiasis (Tribes of Sagar, Madhya Pradesh). The plant is also applied to insect bites and leaves are used as an antidote in scorpion stings (References consulted: Warrier et al. 2010, Sivarajan & Balachandran 1994, Singh et al. 2010, Mithun et al. 2011, Manvar et al. 2012, Neeraja & Margaret 2012, Sharma et al. 2012, Jahan et al. 2014, Ahirwar & Kujur 2015, Paliwal 2017; http://www.bsienvis.nic.in).

3.1.5.2 Medicinal Properties

The plant is bitter, acrid, thermogenic, anticatarrhal, antidontalgic, anti-inflammatory, antihepatic, anthelmintic, nematicidal, ovicidal, spasmolytic, diuretic, aphrodisiac, depurative, febrifuge, absorbent, and antipyretic. The alcoholic extract of entire plant has been reported to show antiviral activity against Ranikhet virus disease (Jahan et al. 2014, Mithun et al. 2011, Neeraja & Margaret 2012, Paliwal 2017, Sharma et al. 2012, Warrier et al. 2010; http://www.bsienvis.nic.in).

3.1.5.3. Chemical Constituents

E. prostrata is known to have a phytosterol, b-amyrin in the n-hexane extract and luteolin-7-glucoside, b-glucoside of phytosterol, a glucoside of a triterpenic acid and wedelolactone in polar solvent extract. Stigmasterol, a-terthienylmethanol, wedelolactone, dismethyl-wedelolactone and dismethylwedelolactone-7-glucoside are found in the leaves while roots contain hentriacontanol and heptacosanol, and polyacetylene substituted thiophenes. Besides nicotine and nicotinic acid, the polypeptides isolated from Bhringraj yield cystine, glutamic acid, phenyl alanine, tyrosine and methionine on hydrolysis. (References consulted: Jahan et al. 2014, Mithun et al. 2011, Neeraja & Margaret 2012, Paliwal 2017, Sharma et al. 2012; http://www.bsienvis.nic.in).

3.1.6. Euphorbia hirta L.

3.1.6.1. Traditional Uses

Traditionally, the asthma weed, *E. hirta* is useful to treat bronchitis, bronchitic and laryngeal spasm asthma, hay fever and conjunctivitis. It is very effective in the treatment of gastrointestinal disorders like diarrhoea, dysentery, intestinal parasitosis, amoebic dysentery, etc. In ayurvedic system of medicine, an extract of the Asthma weed is given to bronchitis patients; other symptoms like colds, asthma and coughs are also relieved. The extract of this herb is effective in controlling venereal diseases like gonorrhoea and syphilis. *E. hirta* is useful in relieving sexual complaints in men and also valuable to control impotency and premature ejaculation. A decoction of the root of the plant increases lactation in women nursing their babies. *E. hirta* is also known for relieving toothache, severe headache, rheumatism and colic. To relieve thrush and mouth ulcers, a decoction made from the root of this herb is used. This herb is also effective in treating urogenital diseases like kidney stones, menstrual problems and venereal diseases. The species is also effective to kills malarial parasite, Plasmodium and several types of pathogenic bacteria. Plants cure eye sores when applied on lower eyelids. Leaf tea of this plant is used in dengue fever. The decoction of whole plant is used for athlete's foot, dysentery, enteritis, fever, excrescences, influenza, gas, itch, dyspnoea eruptions, fractures and skin disorders. The plant is boiled and the solution is taken by patient who pass blood in the urine in central Province of Papua New Guinea. The extract of this plant is used as an antidote against snakebites and scorpion stings. The plant is valuable to treat fever, dysentery and several skin diseases in China whereas in Philippines and Indonesia, it is used to cure bowel problems. The flowers of *E. hirta* are used to treat infertility (barrenness) while the leaves are chewed to facilitate abortion.The latex of the plant is used to treat diseased eyes, cure wounds and soothe bruises in Malaysia. A decoction of *E. hirta* is drunk to promote urination, stop dysentery, remove blood from urine, assuage urethral pain and treat asthma. For treating dermatitis, eczema and irritated skin the herb is applied externally. A paste of the plant is used to soothe sores, and heal boils. The leaves of *E. hirta* are mixed with *Datura metel* to make cigarettes which are smoked to cure asthma in the Philippines. It is used to check bleeding, and stimulate sweating. In Vietnam, *E. hirta* is used to stop dysentery. For the treatment of cough as well as asthma in Britain, the dried entire plant was used in the form of liquid extract. The latex of the plant is very useful and used to cure diseases of urino-genitory tract, dysmenorrhoea, amenorrhea, helminthiasis, ring worm, wounds, piles, treating warts, abscesses, respiratory ailments (cough, coryza, bronchitis, and asthma), and worm infestations in children, dysentery, jaundice, pimples, gonorrhoea, digestive problems, and tumours alopecia, chronic cough, cardiac debility, acne vulgaris, laryngitis, chronic nasal and bronchial catarrh, diarrhoea, intestinal parasitosis, postnatal complaints, failure of lactation, to remove thorns from the skin. The latex of the plant is used to treat ringworm infection and heal wounds in Indonesia. Since large doses cause gastrointestinal irritation, nausea and vomiting, it has been warned that the plant should not be used without expert guidance. (References consulted: Fern 2014; Kumar et al. 2010; http:// medplants.blogspot.in).

3.1.6.2. Medicinal Properties

E. hirta has a wide spectrum of medicinal properties which include aldose-reductase-inhibitor, amebicide, analgesic, aphrodisiac, anthelmintic, antiaggregant, antiallergic, antiasthmatic, antibacterial, antifungal, anticancer, anticonvulsant, antidiabetic, antidiarrhoeal, antiedemic, antiemetic, antifertility, antihistaminic, anti-inflammatory, antileukaemic, antimalarial, antiplasmodial, antiplatelet, antipyretic, antiprostaglandin, antiseptic, antispasmodic, antiviral, anxiolytic, aphrodisiac, astringent, bronchodilator, bronchorelaxant, cardiodepressant, cicatrizant, curare, cytotoxic, diuretic, emmenagogue, expectorant, febrifuge, gram-icide, hemostat, hydragogue, hypoglycaemic, hypotensive, immunosuppressive, insecticide, lactagogue, laxative, litholytic, mastogenic, molluscicide, myorelaxant, narcotic, oxytocic, pectoral, purgative, respirostimulant, sedative, stimulant and vulnerary. Latex is vermifuge. (References consulted: Bhutya 2011, Fern 2014, Kumar et al. 2010; http://medplants.blogspot.in).

3.1.6.3. Chemical Constituents

The plant is known to have terpenes, anthocyanins, alcohols and steroids, shikimic acid, choline, L-inositol and free sugars and quercetin, dimeric-hydrolysabletannins and euphorbians. The other phytoconstituents reported are flavonoids- euphorbianin, leucocyanidol, camphol, quercitrin and quercitol; polyphenols- gallic acid, myricitrin, 3,4-di-O-galloylquinic acid, 2,4,6-tri-Ogalloyl-D-glucose, 1,2,3,4,6-penta-O-galloyl-β- D-glucose; tannins- euphorbins A, B, C, D, E; triterpenes and phytosterols-β-Amyrin, 24-methylenecycloartenol, and â-Sitosterol; and alkanes: heptacosane, n-nonacosane. (References consulted: Bhutya 2011; http://medplants.blogspot.in).

3.1.7. Hyptis suaveolens **(L.) Poit.**

3.1.7.1. Traditional Uses

In traditional system of medicine, *H. suaveolens* is used for the cure of various ailments such as headache, convulsions, migraine, uterine aûections, parasitical cutaneous diseases, catarrh and dysuria; essential oil is a remedy for toothache and dry/flaky skin . A decoction of the root is said to be emmenagogic, and a stimulant if utilized in rheumatism. Roots are useful in flatulence and other stomach problems and fevers associated with colds. Root extract of *H. suaveolens* is given for hematuria and bark for diarrhoea and dysentery. Leaf extract or juice of the plant is applied on measles for early relief and infusions for colds, flu, fever, malaria, constipation. Decoction of leaves and or stems with leaves is used for coughs, asthma and respiratory infections, skin diseases, colds, stomachache; leaves crushed and applied for sprain and swellings. Seeds soaked in water are applied on wounds allowing pus to flow out; blackish seeds soaked in a glass of milk and taken for spermatorrhea. Furthermore, the plant/leaf extracts are used to cure swellings, abscesses haemorrhoids and also as memory aid. Applied externally, it is used as a wash or poultice on skin disorders such as dermatitis and eczema, boils, headaches, etc. A poultice of the pounded fresh material is applied as a poultice on snake bites. In case of athlete's foot, the juice of leaves is applied daily between the toes.

H. suaveolens purify the blood, and is also used as a remedy for the "diseases" of women. In South America, a leaf poultice is applied to cancers and tumours. It is used in the treatment of headaches, stomachache and colic, fever and as a general beverage in Africa. The plant is popularly used in respiratory and gastrointestinal infections, indigestion, colds, pain, fever, cramps and skin ailments. The leaves are used as an anticancer and antifertility (females sex disorder) agent. In Malaysia, plant decoction is used in fever, headache and to promote digestion; also the plant is used externally to soothe skin rashes, eczema and to cure swellings. In Indonesia, the plant infusion is used to treat catarrhal (inflammation of mucous membranes, especially of the nose and throat) conditions. The leaves promote menses as well as to invigorate health in Philippines. In Taiwan, the plants are used to assuage pain and skin discomfort, while in Vietnam, the plant is used to increase milk secretion (lactagogue). The plant is also used as a veterinary medicine. The strong aromatic mint/thyme-like smell of *H. suaveolens* (leaf juice) used as an insectifuge, especially against mosquitoes. Leaves are very effective against sucking insects infesting the livestock. Leaves are used to ward off bed bugs (References consulted: Fern 2014, Prince et al. 2013, Singh et al. 2010, Wiart 2006, http://medplants.blogspot.in).

3.1.7.2. Medicinal Properties

H. suaveolens has stimulant, carminative, sudorific, galactagogue, anticatarrhal, antimalarial, anthelmintic, antitumour, anticancer, antifertility, antifungal, anti-inflammatory, antiplasmodial, antispasmodic, antinociceptive, emmenagogic, antisoporific, antirheumatic, anticephalalgic, antipyretic (fever, flu), febrifuge, stomachic, dyspepsia, carminative and insect repellant properties. The principle constituent of the essential oil is menthol (References consulted: Fern 2014, Prince et al. 2013, Wiart 2006; http://medplants.blogspot.in).

3.1.7.3. Chemical Constituents

The plant is an important source of essential oils, alkaloids, flavonoids, phenols, saponins, terpenes, tannins and sterols, for example diterpenes: suaveolic acid, suaveolol, methyl suaveolate, two steroids: β-sitosterol, β- sitosterol glycoside, two phenolic constituents: rosamarinic acid and methyl rosmarinate along with some other important constituents; also present are oleanolic acid or oleanic acid, ursolic acid, 3 β-hydroxy-lupoic acid, 3β-ursenoloic acid, 1,19 dihydroxy-ursdienoic acid and 3 β- hydroxyl lupenoic acid. Eucaliptol was the most abundant component in the oil, followed for gama-ellemene, beta-pynene, (+) 3-carene, trans-beta-cariophyllene and germacrene (Reference consulted: Prince et al. 2013).

3.1.8. Justicia adhatoda **L.**

3.1.8.1. Traditional Uses

The plant, *J. adhatoda* is a famous plant drug in Ayurvedic as well as Unani medicines for centuries, also it is mentioned in Sanskrit scriptures. It has been widely used for the treatment of respiratory tract ailments. It is a primary herb to cure cough, bronchitis, asthma and symptoms of common cold in Ayurvedic systems of Medicine. A number of herbal formulations of the plant is used for the treatment of various kinds of respiratory ailments. The source of the drug 'Vasaka' is well known in the indigenous system of medicine for its useful effects, particularly in bronchitis. In Indian traditional medicine, the various parts of the plant are used in the treatment of asthma, joint pain, lumbar pain, sprains, cold, cough, eczema, malaria, swelling, venereal diseases and rheumatism. The plant is used as an ingredient of many popular formulations including cough syrup used in combination with Ginger (*Zingiber officinale*) and Tulsi (*Ocimum sanctum*) where it employs its action as an expectorant and antispasmodic. Root of this plant is more efficient as an expectorant. In Sri Lanka, *J. adhatoda* is used in excessive phlegm and menorrhagia. The plant is also effective in case of bleeding piles and impotence.

Leaves along with roots are used in the treatment of irritable cough, bleeding in diarrhoea, bronchial and allergic asthma, fevers, vomiting, urinary disorders, tuberculosis, intestinal worms, skin diseases, cardiac tonic, and bronchitis. The leaves are used for checking postpartum haemorrhage and urinary trouble. The pregnant women in the Gora village of Lucknow (Uttar Pradesh) use the leaves to induce abortion. The leaf decoction is effective in case of throat irritation, and acts as an expectorant to loosen phlegm in the respiratory passages. The leaf powder boiled in sesame oil is used to stop bleeding, earaches as well as pus from ears and jaundice. The leaf decoction and ash of leaves are beneficial in bronchial complaints like asthma, tuberculosis, and to relieve fever and acidity. It was also used for stomach catarrh with constipation, gout, urinary stone. The warmed leaves of *J. adhatoda* used externally to cure dislocation of joint and rheumatic pains. The plant is used in Southeast Asia for curing bleeding, haemorrahge, skin diseases, wounds, headache and leprosy. The bruised fresh leaves are used for snake-bites in India and Sri Lanka. In Manipur, the plant is used as an herbal remedy for treating cold, cough, whooping cough and chronic bronchitis and asthma. In South-East Asia, the fresh flowers are used for opthalmia, cold, phthisis, asthma, bronchitis, cough, antispasmodic, fever and gonorrhoea. The fruits of this plant are used for the treatment of cold, chronic bronchitis, jaundice, diarrhoea, dysentery and fever.

The rural population used the extract of roots in diabetes, cough and liver disorders. The paste, powder and decoction of root is used for curing tuberculosis, diphtheria, malarial fever, leucorrhoea and eye diseases in Southeast Asia. In case of acute nightfall, the root paste mixed with sugar is used in Sitapur District of Uttar Pradesh. The macerated roots are applied on the pubic region and vagina to help parturition as it facilitates the expulsion of foetus. The plant is useful in intermittent, typhus fever and diphtheria. The leaves of *J. adhatoda* are used as an expectorant and spasmolytic agent in Germany (References consulted: Dhankhar et al. 2011, Fern 2014, Kaul & Dwivedi 2010, Singh & Huidrom 2013; http://medplants.blogspot.in).

3.1.8.2. Medicinal Properties

J. adhatoda, is a source of Vitamin C and has antispasmodic, fever reducer, anti-inflammatory, antibleeding, bronchodilator, antidiabetic, disinfectant, antijaundice and oxytocic properties. It is antiperiodic, astringent, diuretic and purgative. Plant and leaves are refrigerant, expectorant, antispasmodic, febrifuge, depurative, tonic, styptic, respiratory stimulant, hypotensive, antifungal, sudorific, cardiac depressant, utero-tonic and abortifacient. The primary alkaloids, vasicine and vasicinone present in the leaves possess respiratory stimulant activity and are well established as therapeutical respiratory agents. Whereas vasicine at low concentrations, induce bronchodilation and relaxation of the tracheal muscles. The leaves, flowers and roots of this plant used in herbal drugs against tubercular activities, cancer and possessed anti-helmintic properties. (References consulted: Bhutya 2011, Chanu & Sarangthem 2014, Dhankhar et al. 2011, Singh & Huidrom 2013).

3.1.8.3. Chemical Constituents

The plants contain phytochemicals such as alkaloids, tannins, saponins, phenolics and flavonoids. The leaves of *J. adhatoda* have a most important alkaloid vasicine, a quinazoline alkaloid and N-oxides of vasicine, vasicinone, deoxyvasicine and maiontone. The roots contain vasinolone, vasicol, peganine and 2-glucosyloxychalcone. The flowers contain β - sitosterol-D-glucoside, kaempferol and its glucoside and quercetin (References consulted: Bhutya 2011, Chanu & Sarangthem 2014).

3.1.9. Solanum americanum **Mill.**

3.1.9.1. Traditional Uses

In traditional and folk medicines, *S. americanum* is used widely in cardiopathy, cardiac oedema, leprosy, skin diseases, splenomegaly, rheumatalgia, swellings, cough, nasal catarrh, hiccough, asthma, bronchitis, wounds, ulcers, flatulence, dyspepsia, hepatomegaly, otalgia, ophthalmopathy, vomiting, haemorrhoids, hoarseness, nephropathy, dropsy, general debility, chronic fevers, gout and rheumatoid arthritis; also used in piles, leprosy, dysentery and liver troubles. The extract of fresh plant is effective in sore throats, coughs, digestive problems, diarrhoea and dysentery, dermatitis, heavy female discharge, inflammatory swellings, liver and spleen enlargement and in cirrhosis of liver. The whole plants are used as a decoction for wounds, tumours, cancerous growths, sores, abdominal pain, bladder inflammation, eye ailments, itch, relief of cramps, rheumatism, neuralgia, expulsion of excess fluids and skin diseases like scabies, eczema and psoriasis. The various parts of the plant such as root, stem and leaves are decocted for abscesses, cancer of the cervix, leucorrhoea and open sores. A decoction of the plant has influence on regulation of blood pressure and depresses the central nervous system and reflexes of the spinal cord. In case of rheumatic and gouty joints and skin diseases, the leaves are used as poultice. The bitter unripe berries/ fruits are useful to aching teeth and squeezed on baby's gums to ease teething pain. The fruits/berries are used as a tonic, appetite stimulant and also for curing asthma, jaundice, heart disease, liver problem, stomachache. In the treatment of blindness, conjunctivitis, glaucoma, trachoma, cataract, diarrhoea, fever and eye ailments, the diluted infusion of berries are effective. Branches and fruits applied as a vulnerary. A decoction of flowers and fruits is beneficial in cough, erysipelas, rat bite, bronchitis, pulmonary tuberculosis, fever, diarrhoea, ophthalmopathy and hydrophobia. The leaves and berries are useful in gastrohelcosis. The seeds are valuable in giddiness, dipsia, inflammations and various skin diseases while the root bark is effective in ophthalmopathy, otopathy, rhinopathy and hepatitis.

'Gewaisaag' an ointment prepared using the leaves and stem are used as the pain relieving agent by the local and tribal people of the Uttarakhand. Warmed up leaf of *S. americanum* is used to treat inflammation of testicles by Garhwalis in Uttarakhand and ethnic communities of Madhya Pradesh. They also use leaves for stomachache, female ailments and liver disorders; leaf poultice is applied to rheumatic joints, gout, rheumatic swellings and skin diseases. Leaf decoction used for body swellings; leaves and fruits extract used for tonsillitis; cooked/fried leaves eaten to cure jaundice and cough. The fresh leaves (125 gram) of the plant are boiled in about 1,000 ml of water daily for 10–15 min and when nearly 250 ml of water remains, then it is filtered and the filtrate is used to treat swelling, skin diseases, inflamed and painful body parts and to clean wounds and mouth sores. The fruits are used for hydrophobia, fever, diarrhoea and eye diseases in West Bengal. Europeans in Africa used the plant to treat convulsions. In Africa, it is used for curing headache and ulcers, also roots are boiled in milk and given to children as tonic. Young shoots are used to treat dysmenorrhoea in females. The leaves of plants are used to treat mouth ulcers. About 50-60 ml freshly prepared leaf decoction, mixed with rock salt and cumin seeds is used in spelenomegaly. In the Philippines, the berries of *S. americanum* is used to treat diabetes. A paste of the green berries is used by the Zulus for ringworm. The leaves are dried in shadow and green tea prepared using dry leaves and taken 3-4 cups in a day against asthma. The leaf and stem extracts are used on piles, gonorrhoea, dropsy and enlargements of the spleen and liver. The plant is used to cure malaria, black-water fever and dysentery in Zimbabwe (Rhodesia). In Mauritius, a poultice of the plant is applied for relief of abdominal pain and inflammation of the bladder while in Mexico the fruit is a popular remedy for erysipelas (References consulted: Dangwal & Sharma 2011, Kaul & Dwivedi 2010, Padalia 2015, Shukla et al. 2010; http://medplants.blogspot.in).

3.1.9.2. *Medicinal Properties*

S. americanum is bitter, acrid, emollient, mildly thermogenic, antiseptic, anti-inflammatory, antispasmodic, expectorant, anodyne, vulnerary, digestive, laxative, diuretic, cardio tonic, depurative, diaphoretic and febrifuge, hydragogue, rejuvenating, sedative, alterant and tonic. Berries/fruits are reported to be laxative, aphrodisiac, diuretic, antidiarrhoeal, and antipyretic. Berries and flowers are strong sudorific, analgesic and sedative with powerful narcotic properties. Plant extracts are cathartic, diuretic, and alterative (References consulted: Warrier et al. 2010; http://medplants.blogspot.in).

3.1.9.3. Chemical Constituents

A number of phytochemicals have been reported in *S. americanum* such as alkaloids, flavonoids, hydrocyanic acid, phenols, phytic acid, tannins, glycoalkaloids, glycoproteins, polysaccharides, polyphenolic compounds like gallic acid, catechin, protocatechuic acid, caffeic acid, epicatechin and rutin. The organic acids recorded are acetic acid, tartaric acid, malic acid and citric acid. The leaves and berries, especially when unripe, contain the alkaloid solanine (References consulted: Atanu et al. 2011, Chauhan et al. 2012).

3.1.10. *Withania somnifera* (L.) Dunal

3.1.10.1. Traditional Uses

W. somnifera (Ashwagandha) also known as Indian ginseng, and as Indian Winter Cherry is an important ancient herb in indigenous medical system for over 3,000 years. The roots, seeds and leaves are used in Indian traditional systems of medicines, Ayurveda and Unani. It has been used in rheumatism, consumption and in debility. In elders it provides energy, relieves inflammations, pains and aches of the back, hand and feet, and in the generative system, nervous debility and diseases due to Vata. Ashwagandha acts mainly on the reproductive and nervous systems, having a rejuvenative effect on the body, and is used to improve vitality. Internally, it is used to tone the uterus after a miscarriage and also in treating postpartum problems. The plant is also used to cure nervous exhaustion, debility, insomnia, wasting diseases, failure to thrive in children, impotence, infertility etc. It is known as a Rasayana, for strength, vigour and for rejuvenation. It is used for the treatment of various arthropathies, certain forms of hypertension, etc. It imparts resistance to infection and reduces stress, has anti-inflammatory effect, boosts sexual energy in men. It stimulates sexual impulses and increases sperm counts.

In gynaecological practice it helps in sterility, leucorrhoea, inflammation of the vagina and also helpful in breast development. Externally, it has been applied as a poultice to boils, swellings and other painful parts. The juice of leaf is useful in conjunctivitis whereas decoction of the bark is taken for asthma and applied locally to bed sores. The root powder in Ayurveda is used for semen improvement, chemopreventive efficacy against stomach and skin carcinogenesis and in many other diseases. Ashwagandha root drug also finds an important place in treatment of rheumatic pain, inflammation of joints, nervous disorders and epilepsy. The dried roots of *W. somnifera* are effective in hiccup, cold, cough, female disorders, ulcers, and in senile debility. It helps in providing progressive, long lasting results for various health concerns like aging, anaemia and slow growth, arthritis, fatigue, waning memory and stress disorders.

In Sind, Pakistan, it is used to cause abortion. The root is ground into a paste is beneficial when applied on ulcers, carbuncles and painful swellings. The infusion of the root is used for gangrenous rectitis and the whole plant for curing syphilis by Zulus. The green berries are rubbed on ringworm with beneficial effect. The leaf decoction is used externally and internally in the treatment of haemorrhoids. A paste of the leaf is applied on syphilitic sores. In southern and eastern Africa the root is used in diarrhoea and proctitis and the leaf for nausea and rheumatism. In West African local medicine, both roots and leaves are used internally, and the freshly pounded leaves also externally, against fever, chills, rheumatism, colics, etc. Some caution is advised in the use of this plant since it is toxic. (References consulted: Al-shaimaa 2013, Datta et al. 2011, Fern 2014, Gaurav et al. 2015, Gupta & Rana 2007, Kaul & Dwivedi 2010, Mirjalili et al. 2009, Qamaruddin et al. 2012, Shaheen et al. 2014, Supe et al. 2011, Umadevi et al. 2012; http://medplants.blogspot.in).

3.1.10.2. Medicinal Properties

Ashwagandha is considered to be one of the best rejuvenating agents in Ayurveda. It possesses a number of therapeutic actions which include anti-inflammatory, sedative, hypnotic, narcotic, general tonic, diuretic, aphrodisiac, alterative, deobstruent, uterine tonic and increase in production of semen. The roots are alterative, aphrodisiac, deobstruent, diuretic, narcotic, sedative and restorative in nature. Ashwagandha has anti-inflammatory, antitumour, antistress, antioxidant, mindboosting, immune-enhancing, and rejuvenating properties. Withaferin A (an active compound of *Withania*) possesses potent analgesic and antipyretic properties. The root is a tonic, alterative, aphrodisiac and used in consumption.

Table 1: List of wild plants recorded in Chithara, Gautam Buddha Nagar, Uttar Pradesh

S.N.	Botanical name	Family	Common name	Types	Growth form
1	*Abrus precatorius* L. (Syn. *Abrus abrus* (L.) Wright; *Glycine abrus* L.)	Leguminosae	Licorice, Coral Bead Vine, Crab's Eye (E); Ratti, Gunchi (H)	Dicot	CL/CR
2	*Abutilon indicum* (L.) Sweet (Syn. *Sida indica* L.)	Malvaceae	Indian Abutilon, Indian Mallow (E); Kanghi, Kakahi (H); Atibala (S)	Dicot	SH
3	*Acacia nilotica* (L.) Delile (Syn. *Acacia arabica* (Lam.) Willd.)	Leguminosae	Gum Arabic Tree, Babool (E); Babool, Kikar (H); Babbula (S)	Dicot	T
4	*Achyranthes aspera* L. (Syn. *Achyranthes acuminata* E. Mey. ex Cooke & Wright)	Amaranthaceae	Prickly Chaff Flower, Devil's Horsewhip (E); Chirchita, Latjira (H); Apamarga (S)	Dicot	H
5	*Acmella paniculata* (Wall. ex DC.) R.K.Jansen (Syn. *Spilanthes paniculata* Wall. ex DC.)	Compositae	Toothache Plant, Panicled Spot Flower (E); Pipulka (H)	Dicot	H
6	*Ageratum conyzoides* (L.) L. (Syn. *Eupatorium conyzoides* (L.) E. H. Krause)	Compositae	Billygoat-Weed, Goat Weed (E); Gha Buti, Bhakumbar (H)	Dicot	H
7	*Ageratum houstonianum* Mill. (Syn. *Ageratum mexicanum* Sims)	Compositae	Floss Flower, Blue Billygoat Weed (E); Raktarodhi (H); Nilima (S)	Dicot	H
8	*Albizia lebbeck* (L.) Benth. (Syn. *Acacia lebbeck* (L.) Willd.; *Mimosa lebbeck* L.)	Leguminosae	Lebbeck Tree, Woman's Tongues Tree (E); Siris (H); Sirisah (S)	Dicot	T
9	*Alhagi maurorum* Medik. (Syn. *Alhagi pseudalhagi* (M.Bieb.) Fisch.)	Leguminosae	Camelthorn-Bush, Persian Manna Plant (E); Bharbharra (H); Bhatuashak (S)	Dicot	US
10	*Alternanthera paronychioides* A.St.-Hil. (Syn. *Alternanthera leucantha* Moq.)	Amaranthaceae	Smooth Chaff Flower (E)	Dicot	H
11	*Alternanthera philoxeroides* (Mart.) Griseb. (Syn. *Achyranthes philoxeroides* (Mart.) Standl.)	Amaranthaceae	Alligator Weed, Pig Weed, Smooth Chaff Flower (E)	Dicot	H

Contd.

12	*Alternanthera pungens* Kunth (Syn. *Alternanthera echinata* Sm.)	Amaranthaceae	Chaff-Flower (E); Khaki-Weed, Kante Wali Santhi (H)	Dicot	H
13	*Alternanthera sessilis* (L.) R.Br. ex DC. (Syn. *Gomphrena sessilis* L.)	Amaranthaceae	Sessile Joyweed, Dwarf Copperleaf (E); Garundi, Guroo (H)	Dicot	H
14	*Alysicarpus monilifer* (L.) DC. (Syn. *Hedysarum moniliferum* L.)	Leguminosae	Necklace-Pod Alyce Clover (E); Jhuhi Ghas, Chatta Ki Ghas (H)	Dicot	H
15	*Alysicarpus vaginalis* (L.) DC. (Syn. *Hedysarum vaginale* L.)	Leguminosae	Buffalo Clover, Buffalo-Bur, One-Leaf Clover (E)	Dicot	H
16	*Amaranthus viridis* L. (Syn. *Amaranthus gracilis* Desf.)	Amaranthaceae	Slender Amaranth, Green Amaranth (E); Jangli Chaulai (H)	Dicot	H
17	*Ammannia baccifera* L. (Syn. *Ammannia vescicatoria* Roxb.)	Lythraceae	Blistering Ammannia, Acrid Weed (E); Agnibuti, Dadmari (H); Agnigarbha, Brahmasoma (S)	Dicot	H
18	*Ampelopteris prolifera* (Retz.) Copel. (Syn. *Dryopteris prolifera* (Retz.) C. Chr.)	Thelypteridaceae	Walking Fern (E); Bhaisag (H)	PH	P
19	*Anagallis arvensis* L. (Syn. *Anagallis arabica* Duby)	Primulaceae	Blue Pimpernel (E); Neel, Dharti Dhak, Krishnaneel (H)	Dicot	H
20	*Anisomeles indica* (L.) Kuntze (Syn. *Nepeta indica* L.)	Lamiaceae	Indian Catmint, Malabar Catmint (E); Kala Bhangra (H)	Dicot	US
21	*Arabidopsis thaliana* (L.) Heynh. (Syn. *Sisymbrium thalianum* (L.) J.Gay)	Brassicaceae	Thale Cress, Mouse-Ear Cress (E)	Dicot	H
22	*Arenaria serpyllifolia* L. (Syn. *Stellaria serpyllifolia* (L.) Scop.)	Caryophyllaceae	Thyme-Leaved Sandwort (E)	Dicot	H
23	*Argemone mexicana* L. (Syn. *Papaver mexicanum* (L.) E.H.L.Krause)	Papaveraceae	Mexican Prickly Poppy (E); Satyanashi, Kataila (H)	Dicot	H
24	*Artemisia scoparia* Waldst. & Kitam. (Syn. *Artemisia capillaris* Miq.)	Compositae	Redstem Wormwood (E); Seeta-Bani (H)	Dicot	US
25	*Avena sterilis* L. (Syn. *Avena fatua* var. *sterilis* (L.) Fiori & Paol.)	Poaceae	Sterile Oat, Winter Wild Oat (E); Jangali Jai (H)	Monocot	GR

Contd.

26	*Azadirachta indica* A.Juss. (Syn. *Melia azadirachta* L.)	Meliaceae	Indian Lilac, Margosa, (E); Neem (H); Arishtha (S)	Dicot	T
27	*Bacopa monnieri* (L.) Wettst. (Syn. *Lysimachia monnieri* L.; *Monniera cuneifolia* Michx.)	Plantaginaceae	Water Hyssop, Indian Pennywort (E); Brahmi (H); Brahmi, Tiktalonika (S)	Dicot	H
28	*Barleria prionitis* L. (Syn. *Barleria coriacea* Oberm.; *Barleria spicata* Roxb.)	Acanthaceae	Porcupine Flower (E); Vajradanti (H); Artagalah, Dasi Kurantakah (S)	Dicot	US
29	*Bidens pilosa* L. (Syn. *Bidens orientalis* Velen. ex Bornm.)	Compositae	Black-Jack, Beggar Tick, Cobbler's Pegs (E); Kumra, Kumur (H)	Dicot	H
30	*Blainvillea acmella* (L.) Philipson (Syn. *Blainvillea alba* Edgew)	Compositae	Para Cress Flower, Paniculated Spot Flowers (E); Kanghi (H)	Dicot	H
31	*Blumea laciniata* (Wall. ex Roxb.) DC. (Syn. *Blumea acutata* DC.)	Compositae	Cutleaf Blumea, Cutleaf False Oxtongue (E)	Dicot	H
32	*Blumea obliqua* (L.) Druce (Syn. *Erigeron obliquus* L.; *Blumea pubiflora* DC.)	Compositae	Common Floss Flower (E); Sakarkandi, Sankooli (H)	Dicot	H
33	*Blumea viscosa* (Mill.) V.M.Badillo (Syn. *Laggera aurita* (L.f.) Benth.ex C.B.Clarke)	Compositae	Sticky Blumea, Clammy False Oxtongue (E)	Dicot	H
34	*Boerhavia diffusa* L. (Syn. *Boerhavia adscendens* Willd.)	Nyctaginaceae	Red Spiderling, Spreading Hog Weed (E); Punarnava, Gadahpurna (H); Raktakanda (S)	Dicot	H
35	*Brachiaria reptans* (L.) C.A.Gardner & C.E.Hubb. (Syn. *Brachiaria prostrata* (Lam.) Griseb.)	Poaceae	Running Grass, Creeping Panic Grass (E); Para Ghas (H)	Monocot	GR
36	*Brachiaria ramosa* (L.) Stapf (Syn. *Panicum ramosum* L.)	Poaceae	Browntop Millet (E); Pedda-Sama, Popti (H)	Monocot	GR
37	*Butea monosperma* (Lam.) Taub. (Syn. *Butea frondosa* Willd.; *Erythrina monosperma* Lam.)	Leguminosae	Flame of the Forest, Bastard Teak (E); Palash, Dhak, Tesu (H); Palasha (S)	Dicot	T
38	*Caesulia axillaris* Roxb. (Syn. None known)	Compositae	Pink Node Flower (E); Gathila (H)	Dicot	H

Contd.

39	*Cajanus scarabaeoides* (L.) Thouars (Syn. *Atylosia scarabaeoides* (L.) Benth.)	Leguminosae	Showy Pigeonpea, Peanut Grass (E); Jangal Tor, Ban Kulthi (H)	Dicot	CL/CR
40	*Calotropis gigantea* (L.) Dryand. (Syn. *Asclepias gigantea* L.)	Apocynaceae	Giant Indian Milkweed, Crown Flower (E); Safed Aak (H); Sadapushpa, Arki, Alarka (S)	Dicot	SH
41	*Calotropis procera* (Aiton) Dryand. (Syn. *Asclepias procera* Aiton)	Apocynaceae	Rubber Bush, Sodom Apple (E); Aak, Madar (H); Arka, Mandara, Sadapushpa, Alarka (S)	Dicot	US
42	*Cannabis sativa* L. (Syn. *Cannabis indica* Lam.)	Cannabaceae	Marijuana, Hemp (E); Bhang, Ganja, Charas (H); Bahuvadini, Bhanga (S)	Dicot	H
43	*Capparis sepiaria* L. (Syn. *Capparis incanescens* DC.)	Capparaceae	Wild Caper Bush (E); Kanthari (H)	Dicot	SH
44	*Capsella bursa-pastoris* (L.) Medik. (Syn. *Thlaspi bursa-pastoris* L.)	Brassicaceae	Shepherd's Purse, Blind weed (E); Mumiri (H)	Dicot	H
45	*Cardamine hirsuta* L. (Syn. *Cardamine multicaulis* Hoppe ex Schur)	Brassicaceae	Hairy Bitter Cress, Flick Weed (E)	Dicot	H
46	*Carthamus oxyacantha* M.Bieb. (Syn. *Carthamus polyacantha* M.Bieb.)	Compositae	Wild Safflower, Jewelled Distaff Thistle (E); Peeli Kateri, Kandiari (H)	Dicot	H
47	*Cayratia trifolia* (L.) Domin (Syn. *Cayratia carnosa* (Lam.) Gagnep.; *Vitis trifolia* L.)	Vitaceae	Three-Leaved Wild Vine, Threeleaf Cayratia (E); Amalbel (H), Amlavetasah (S)	Dicot	CL/CR
48	*Celosia argentea* L. (Syn. *Celosia cristata* L.)	Amaranthaceae	Silver Cock's Comb, Quail Grass (E); Safed Murga (H)	Dicot	H
49	*Cenchrus ciliaris* L. (Syn. *Pennisetum ciliare* (L.) Link)	Poaceae	Buffel Grass, African Foxtail Grass (E); Anjan, Dhaman, Kusa (H)	Monocot	GR
50	*Centaurium pulchellum* (Sw.) Druce (Syn. *Gentiana pulchella* Sw.)	Gentianaceae	Pink Centaury, Field Centaury (E); Barik Chirayata, Khet Chirayata (H)	Dicot	H
51	*Centella asiatica* (L.) Urb. (Syn. *Hydrocotyle asiatica* L.)	Apiaceae	Indian Pennywort, Gotu kola (E); Brahmi, Brahma Manduki (H); Bhandi, Bhandiri, Mandukaparni (S)	Dicot	H
52	*Ceratophyllum demersum* L. (Syn. *Ceratophyllum verticillatum* Roxb.)	Ceratophyllaceae	Hornwort, Coon's Tail (E); Brihatpushpi (H)	Dicot	H

Contd.

53	*Chenopodium album* L. (Syn. *Atriplex alba* (L.) Crantz)	Amaranthaceae	Goosefoot, Wild Spinach, Lamb's-Quarters (E); Bathua (H)	Dicot	H
54	*Chenopodium murale* L. (Syn. *Atriplex muralis* (L.) Crantz)	Amaranthaceae	Nettle-Leaved Goosefoot, Australian Spinach (E); Jangli Bathua, Khartua (H)	Dicot	H
55	*Chloris barbata* Sw. (Syn. *Chloris inflata* Link)	Poaceae	Finger Grass, Feather Finger Grass, Airport Grass (E)	Monocot	GR
56	*Chrysopogon zizanioides* (L.) Roberty (Syn. *Andropogon zizanioides* (L.) Urb.; *Phalaris zizanioides* L.; *Vetiveria zizanioides* (L.) Nash)	Poaceae	Vetiver, Khas-Khas Grass (E); Khas-Khas Ghas (H)	Monocot	GR
57	*Cirsium arvense* (L.) Scop. (Syn. *Cnicus arvensis* (L.) Hoffm.)	Compositae	Creeping Thistle, Californian Thistle (E); Kandai (H)	Dicot	H
58	*Cissampelos pareira* L. (Syn. *Cissampelos argentea* Kunth; *Cissampelos hirsuta* Buch.-Ham. ex DC.)	Menispermaceae	Velvet Leaf, False Pareira Brava (E); Bhatvel, Patha (H); Laghu Patha (S)	Dicot	CL/CR
59	*Citrullus colocynthis* (L.) Schrad. (Syn. *Colocynthis vulgaris* Schrad.; *Cucumis colocynthis* L.)	Cucurbitaceae	Colocynth, Desert Gourd (E); Indrayan (H); Gavakshi, Indravaruni (S)	Dicot	CL/CR
60	*Cleome viscosa* L. (Syn. *Polanisia viscosa* (L.) Blume)	Cleomaceae	Asian Spider Flower, Dog Mustard (E); Bagra, Pili Hurhur (H)	Dicot	US
61	*Clerodendrum phlomidis* L.f. (Syn. *Volkameria multiflora* Burm.f.)	Verbenaceae	Arni (E); Arni (H); Agnimantha (S)	Dicot	SH
62	*Coccinia grandis* (L.) Voigt (Syn. *Bryonia grandis* L.; *Coccinia indica* Wight & Arn.)	Cucurbitaceae	Ivy Gourd, Scarlet Gourd (E); Kundru (H)	Dicot	CL/CR
63	*Commelina benghalensis* L. (Syn. *Commelina canescens* Vahl; *Commelina hirsuta* R.Br.)	Commelinaceae	Bengal Dayflower, Tropical Spiderwort (E); Kankawwa, Buchna (H)	Monocot	H
64	*Commelina forsskalii* Vahl (Syn. *Commelina falcata* Hassk.)	Commelinaceae	Kanpet, Day flower (E); Kankawwa (H)	Monocot	H
65	*Convolvulus arvensis* L.	Convolvulaceae	Field Bindweed (E); Hiranpug (H)	Dicot	CL/CR

Contd.

	(Syn. *Convolvulus chinensis* Ker Gwl.)				
66	*Convolvulus prostratus* Forssk. (Syn. *Convolvulus pluricaulis* Choisy)	Convolvulaceae	Prostrate Bindweed (E); Shankhpushpi (H)	Dicot	H
67	*Corchorus aestuans* L. (Syn. *Corchorus acutangulus* Lam.)	Malvaceae	East Indian Mallow, Wild Jute (E); Jangli Jute, Chonch (H); Chunchuh (S)	Dicot	H
68	*Corchorus olitorius* L. (Syn. *Corchorus catharticus* Blanco)	Malvaceae	Nalta Jute, Tossa Jute (E); Pat-Sag, Mitha-Pat (H); Mahachanchu (S)	Dicot	H
69	*Corchorus tridens* L. (Syn. *Corchorus burmanni* DC.)	Malvaceae	Wild Jute, Jew's Mallow, African Jute (E); Kadvapat (H)	Dicot	H
70	*Cordia dichotoma* G.Forst. (Syn. *Cordia obliqua* Willd.; *Cordia wallichii* G.Don)	Boraginaceae	Indian Cherry, Glue Berry (E); Lasoda, Tenti, Dela (H); Bahuvarah, Bahuvaraka (S)	Dicot	T
71	*Crateva adansonii* subsp. *odora* (Buch.-Ham.) Jacobs (Syn. *Crateva roxburghii* R.Br.)	Capparaceae	Garlic Pear Tree, Caper Tree, Three-Leaf Caper (E); Barna, Barni (H)	Dicot	T
72	*Crotalaria medicaginea* Lam. (Syn. *Crotalaria foliosa* Willd.)	Leguminosae	Trefoil Rattlepod (E); Gulali, Adbaumethi (H)	Dicot	H
73	*Crotalaria mysorensis* Roth (Syn. *Crotalaria hirsuta* Roxb.)	Leguminosae	Rattlebox, Mysore Rattlepod (E)	Dicot	H
74	*Croton bonplandianus* Baill. (Syn. *Croton sparsiflorus* Morong)	Euphorbiaceae	Croton, Rushfoil (E); Ban Tulsi, Kala Bhangra (H)	Dicot	H
75	*Cucumis melo* L. var. *agrestis* Naudin (Syn. *Cucumis melo* subsp. *agrestis* (Naudin) Pangalo; *Cucumis pubescens* Willd.)	Cucurbitaceae	Small Gourd, Wild Musk Melon (E); Kachari, Kachariya (H); Karkati (S)	Dicot	CL/CR
76	*Cuscuta chinensis* Lam. (Syn. *Cuscuta carinata* R. Br.)	Convolvulaceae	Common Dodder (E); Amar Bel (H)	Dicot	CL/CR
77	*Cuscuta reflexa* Roxb. (Syn. *Cuscuta verrucosa* Sweet)	Convolvulaceae	Giant Dodder (E); Amar bel, Akashbel (H); Amaravela, Akashballi (S)	Dicot	CL/CR

Contd.

78	*Cyanotis axillaris* (L.) D.Don ex Sweet (Syn. *Commelina axillaris* L.)	Commelinaceae	Creeping Cradle Plant (E); Kana (H)	Monocot	H
79	*Cyanthillium cinereum* (L.) H.Rob. (Syn. *Vernonia cinerea* (L.) Less.)	Compositae	Vernonia, Little Ironweed, Purple Feabane (E); Sahadevi, Ankari (H)	Dicot	H
80	*Cynodon dactylon* (L.) Pers. (Syn. *Panicum dactylon* L.)	Poaceae	Bermuda Grass, Bahama Grass (E); Doob (H); Durva (S)	Monocot	GR
81	*Cyperus alopecuroides* Rottb. (Syn. *Juncellus alopecuroides* (Rottb.) C.B.Clarke)	Cyperaceae	Nut Grass, Yellow Nut Sedge (E)	Monocot	SE
82	*Cyperus compressus* L. (Syn. *Cyperus brachiatus* Poir.)	Cyperaceae	Poorland Flat Sedge (E); Mothi (H)	Monocot	SE
83	*Cyperus difformis* L. (Syn. *Cyperus Oryzetorun* Stued.)	Cyperaceae	Small-Flowered Nut Sedge, Small-Flower Umbrella-Sedge (E); Dila, Motha (H)	Monocot	SE
84	*Cyperus rotundus* L. (Syn. *Cyperus tuberosus* Rottb.; *Pycreus rotundus* (L.) Hayek)	Cyperaceae	Nut Sedge, Common Nut Sedge, Nut Grass (E); Bara Nagar Motha, Motha (H); Mustaka, Varida, Chakranksha (S)	Monocot	SE
85	*Cyperus iria* L. (Syn. *Chlorocyperus iria* (L.) Rikli)	Cyperaceae	Rice Flat Sedge (E); Morphula (H)	Monocot	SE
86	*Dactyloctenium aegyptium* (L.) Willd. (Syn. *Cynosurus aegyptius* L.)	Poaceae	Crowfoot Grass, Finger Comb Grass (E); Makra Ghas (H)	Monocot	GR
87	*Dalbergia sissoo* DC. (Syn. *Amerimnon sissoo* (Roxb.) Kuntze)	Leguminosae	Indian Rosewood, Sisu (E); Shisham, Sisu (H)	Dicot	T
88	*Datura innoxia* Mill. (Syn. *Datura meteloides* DC. ex Dunal)	Solanaceae	Pricklyburr, Thorn-Apple (E); Safed Dhatura (H)	Dicot	H
89	*Desmodium gangeticum* (L.) DC. (Syn. *Hedysarum gangeticum* L.)	Leguminosae	Sal Leaved Desmodium (E); Dirghamuli, Shalaparni (H); Vidarigandha, Shalaparni (S)	Dicot	US
90	*Desmodium triflorum* (L.) DC. (Syn. *Hedysarum triflorum* L.)	Leguminosae	Creeping Tick Trefoil (E); Tipatiya (H); Hamsapadi, Tripadi (S)	Dicot	H

Contd.

91	*Desmostachya bipinnata* (L.) Stapf (Syn. *Briza bipinnata* L.; *Eragrostis cynosuroides* (Retz.) P.Beauv.)	Poaceae	Daabh, Salt Reed-Grass (E); Dabh, Davoli (H)	Monocot	GR
92	*Dichanthium annulatum* (Forssk.) Stapf (Syn. *Andropogon annulatus* Forssk.)	Poaceae	Marvel Grass, Ringed Dichanthium (E); Karad, Delhi Grass (H)	Monocot	GR
93	*Digera muricata* (L.) Mart. (Syn. *Achyranthes muricata* L.; *Digera arvensis* Forssk.)	Amaranthaceae	False Amaranth (E); Lahsuva, Latmahuria (H); Aranyavastuka, Kuranjara (S)	Dicot	H
94	*Digitaria ciliaris* (Retz.) Koeler (Syn. *Digitaria adscendens* (Kunth) Henrard; *Panicum ciliare* Retz.)	Poaceae	Wild Crabgrass, Summer Grass, Tropical Finger Grass (E)	Monocot	GR
95	*Diospyros montana* Roxb. (Syn. *Diospyros cordifolia* Roxb.)	Ebenaceae	Mountain Persimmon, Bombay Ebony (E); Bistendu (H); Tumala (S)	Dicot	T
96	*Dysphania ambrosioides* (L.) Mosyakin & Clemants (Syn. *Chenopodium ambrosioides* L.)	Amaranthaceae	Mexican Tea (E); Sugandha Vastooka (H); Sugandha Vastooka (S)	Dicot	US
97	*Echinochloa colona* (L.) Link (Syn. *Panicum colonum* L.)	Poaceae	Jungle Rice, Shama Millet (E); Jangli Jhangora, Jharwa (H)	Monocot	GR
98	*Echinochloa crus-galli* (L.) P.Beauv. (Syn. *Panicum crus-galli* L.)	Poaceae	Barnyard Grass, Japanese Millet, Jungle Rice, Prickly Grass (E); Sanwak, Kayada (H); Varuka (S)	Monocot	GR
99	*Eclipta prostrata* (L.) L. (Syn. *Eclipta alba* (L.) Hassk.)	Compositae	False Daisy, Trailing Eclipta (E); Bhringaraj, Kesharaj (H); Bhringarajah, Tekarajah (S)	Dicot	H
100	*Eichhornia crassipes* (Mart.) Solms (Syn. *Pontederia crassipes* Mart.)	Pontederiaceae	Water Hyacinth (E); Jal Kumbhi (H); Jalakumbhi, Variparni (S)	Monocot	H
101	*Eleusine indica* (L.) Gaertn. (Syn. *Cynosurus indicus* L.)	Poaceae	Goosegrass, Indian Goosegrass, Crowfoot Grass (E); Jangali Marua (H); Nandimukhi (S)	Monocot	GR

Contd.

102	*Emex australis* Steinh. (Syn. *Emex centropodium* Meisn.)	Polygonaceae	Three Corner Jack, Devil's Thorn (E)	Dicot	H
103	*Equisetum ramosissimum* (Desf.) (Syn. *Hippochaete ramosissima* Desf.) Börner)	Equisetaceae	Horsetail, Branched Horsetail (E); Putod, Sumbok (H)	PH	P
104	*Equisetum ramosissimum* subsp. *debile* (Roxb. ex Vaucher) Hauke (Syn. *Hippochaete ramosissima* subsp. *debilis* (Roxb. ex Vaucher) Á. Löve & D. Löve)	Equisetaceae	Horsetail (E)	PH	P
105	*Eragrostis amabilis* (L.) Wight & Arn. (Syn. *Eragrostis tenella* (L.) P.Beauv. ex Roem. & Schult.)	Poaceae	Japanese Love grass (E); Bharbhusi (H)	Monocot	GR
106	*Eragrostis japonica* (Thunb.) Trin. (Syn. *Poa interrupta* Lam.; *Poa japonica* Thunb.)	Poaceae	Pond Lovegrass (E); Panghas (H)	Monocot	GR
107	*Erigeron bonariensis* L. (Syn. *Conyza bonariensis* (L.) Cronquist)	Compositae	Flax-Leaf Fleabane, Wavy-Leaf Fleabane (E); Makshikaa-Visha (H)	Dicot	H
108	*Eriochloa procera* (Retz.) C.E.Hubb. (Syn. *Agrostis procera* Retz.)	Poaceae	Tropical Cupgrass, Cup Grass, Spring Grass (E)	Monocot	GR
109	*Euphorbia heterophylla* L. (Syn. *Poinsettia heterophylla* (L.) Klotzsch & Garcke)	Euphorbiaceae	Wild Poinsettia, Lesser Green Poinsettia (E)	Dicot	H
110	*Euphorbia hirta* L. (Syn. *Chamaesyce hirta* (L.) Millsp.; *Euphorbia capitata* Lam.)	Euphorbiaceae	Asthma Weed, Asthma Plant (E); Dudhi, Bara Dudhi (H); Dugadhika (S)	Dicot	H
111	*Euphorbia hypericifolia* L. (Syn. *Chamaesyce hypericifolia* (L.) Millsp.)	Euphorbiaceae	Graceful Spurge, Large Spotted Spurge (E); Dudh Mogra (H)	Dicot	H
112	*Euphorbia prostrata* Aiton (Syn. *Chamaesyce prostrata* (Aiton) Small)	Euphorbiaceae	Prostrate Sandmat, Prostrate Spurge (E); Choti Dudhi (H)	Dicot	H

Contd.

113	*Euphorbia serpens* Kunth (Syn. *Chamaesyce serpens* (Kunth) Small; *Euphorbia orbiculata* var. *jawaharii* Rajagopal & Panigrahi)	Euphorbiaceae	Matted Sandmat, Roundleaf Spurge (E); Dudhi (H)	Dicot	H
114	*Evolvulus nummularius* (L.) L. (Syn. *Convolvulus nummularius* L.)	Convolvulaceae	Roundleaf Bindweed, Agracejo Rastrero (E); Musakarni (H); Vishnukrantha (S)	Dicot	H
115	*Ficus palmata* Forssk. (Syn. *Ficus pseudocarica* Miq.)	Moraceae	Indian Fig (E); Jangli Anjir (H)	Dicot	SH
116	*Ficus religiosa* L. (Syn. *Ficus caudata* Stokes; *Ficus peepul* Griff.)	Moraceae	Peepal, Holy Fig Tree, Sacred Fig Tree (E); Peepal (H); Plaksha, Bodhivriksha (S)	Dicot	T
117	*Fimbristylis dichotoma* (L.) Vahl (Syn. *Fimbristylis diphylla* (Retz.) Vahl)	Cyperaceae	Common Fringe Sedge, Tall Fringe Rush (E)	Monocot	SE
118	*Fimbristylis ferruginea* (L.) Vahl (Syn. *Scirpus ferrugineus* L.)	Cyperaceae	Rusty Sedge, West Indian Fimbry (E)	Monocot	SE
119	*Fumaria indica* (Hausskn.) Pugsley (Syn. *Fumaria parviflora* var. *indica* (Hausskn.) Parsa)	Papaveraceae	Indian Fumitory (E); Papara, Pittpapra (H); Parpata (S)	Dicot	H
120	*Gnaphalium purpureum* L. (Syn. *Gamochaeta purpurea* (L.) Cabrera)	Compositae	American Cudweed, Purple Cudweed (E)	Dicot	H
121	*Gomphrena celosioides* Mart. (Syn. *Gomphrena alba* Peter)	Amaranthaceae	Prostrate Globe-Amaranth (E)	Dicot	H
122	*Gonostegia pentandra* (Roxb.) Miq. (Syn. *Pouzolzia pentandra* (Roxb.) Benn.)	Urticaceae	Melastome Pouzolz's Bush (E)	Dicot	H
123	*Grangea maderaspatana* (L.) Poir. (Syn. *Artemisia maderaspatana* L.)	Compositae	Madras Carpet (E); Mustaru, Bhediachim (H)	Dicot	H
124	*Heliotropium ellipticum* Ledeb. (Syn. *Heliotropium strictum* Ledeb.)	Boraginaceae	Heliotrope (E); Hathisunda (S)	Dicot	H
125	*Hemarthria compressa* (L.f.) R.Br. (Syn. *Rottboellia compressa* L.f.)	Poaceae	Jove Grass, Whip Grass (E); Biksa, Pansheru (H)	Monocot	GR

Contd.

126	*Holoptelea integrifolia* Planch. (Syn. *Ulmus integrifolia* Roxb.)	Ulmaceae	Indian Elm, Indian Beech Tree (E); Chilbil, Papri (H); Chirivilva (S)	Dicot	T
127	*Hydrilla verticillata* (L.f.) Royle (Syn. *Serpicula verticillata* L.f.; *Vallisneria verticillata* (L.f.) Roxb.)	Hydrocharitaceae	Waterthyme, Hydrilla (E); Sevar, Jhangi (H)	Monocot	H
128	*Hydrocotyle sibthorpioides* Lam. (Syn. *Hydrocotyle rotundifolia* Roxb. ex DC.)	Araliaceae	Lawn Pennywort, Lawn Marsh pennywort (E)	Dicot	H
129	*Hyptis suaveolens* (L.) Poit. (Syn. *Ballota suaveolens* L.)	Lamiaceae	American Mint, Bush Mint (E); Vilaiti Tulsi (H); Bhustrna (S)	Dicot	US
130	*Imperata cylindrica* (L.) Raeusch. (Syn. *Imperata arundinacea* Cirillo; *Lagurus cylindricus* L.)	Poaceae	Blady Grass, Cogon Grass, Japanese Blood Grass (E); Dabh(H); Darbha (S)	Monocot	GR
131	*Indigofera astragalina* DC.	Leguminosae	Silky Indigo (E); Dagadia (H)	Dicot	US
132	*Indigofera linifolia* (L.f.) Retz. (Syn. *Hedysarum linifolium* L.f.)	Leguminosae	Narrowleaf Indigo (E); Ratnamala, Pandarphali (H)	Dicot	H
133	*Indigofera linnaei* Ali (Syn. *Hedysarum prostratum* L.; *Indigofera enneaphylla* L.)	Leguminosae	Birdsville Indigo, Bird's-foot Indigo (E); Leel (H); Vasuka (S)	Dicot	H
134	*Indigofera tinctoria* L. (Syn. *Indigofera indica* Lam.)	Leguminosae	True Indigo, Common Indigo (E); Neelini, Neel (H); Neelika, Nilapushpa (S)	Dicot	US
135	*Ipomoea aquatica* Forssk. (Syn. *Ipomoea reptans* Poir.)	Convolvulaceae	Water Morning Glory, Swamp Cabbage, Chinese Water Spinach (E); Kalmi Sag, Nali (H)	Dicot	CL/CR
136	*Ipomoea carnea* Jacq. (Syn. *Ipomoea carnea* subsp. *carnea*)	Convolvulaceae	Bush Morning Glory (E); Behaya, Besharam (H)	Dicot	SH
137	*Ipomoea coptica* (L.) Roth ex Roem. & Schult. (Syn. *Convolvulus copticus* L.)	Convolvulaceae	Egyptian Morning Glory (E)	Dicot	CL/CR
138	*Ipomoea eriocarpa* R. Br. (Syn. *Convolvulus eriocarpus* (R. Br.) Spreng.)	Convolvulaceae	Tiny Morning Glory, Woolly Fruited Morning Glory (E); Buta (H)	Dicot	CL/CR

Contd.

139	*Ipomoea obscura* (L.) Ker Gawl. (Syn. *Convolvulus obscurus* L.)	Convolvulaceae	Obscure Morning Glory (E); Pan Bel (H); Vachagandha, Lakshmana (S)	Dicot	CL/CR
140	*Ipomoea pes-tigridis* L. (Syn. *Convolvulus pes-tigridis* (L.) Spreng.)	Convolvulaceae	Tiger Foot Morning Glory (E); Panchpatia (H)	Dicot	CL/CR
141	*Ipomoea triloba* L. (Syn. *Convolvulus trilobus* (L.) Desr.)	Convolvulaceae	Little Bell Morning Glory, Three-Lobe Morning Glory (E)	Dicot	CL/CR
142	*Jatropha gossypiifolia* L. (Syn. *Manihot gossypiifolia* (L.) Crantz)	Euphorbiaceae	Bellyache Bush, Cotton-Leaf Physic Nut (E); Ratanjoti (H); Rakta-Vyaaghrairanda (S)	Dicot	SH
143	*Justicia adhatoda* L. (Syn. *Adhatoda vasica* Nees; *Adhatoda zeylanica* Medik.)	Acanthaceae	Malabar Nut, Adhatoda, Adulsa (E); Arus, Arusa (H); Vasaka, Atarusa, Pra-Madya (S)	Dicot	SH
144	*Justicia japonica* Thunb. (Syn. *Justicia simplex* D. Don)	Acanthaceae	Common Small Justicia (E)	Dicot	H
145	*Lantana camara* L. (Syn. *Camara vulgaris* Benth.)	Verbenaceae	Lantana, Common Lantana (E); Raimuniya (H)	Dicot	SH
146	*Lathyrus aphaca* L. (Syn. *Orobus aphaca* (L.) Doll)	Leguminosae	Yellow Pea, Yellow Vetchling (E); Jangli Mattar (H)	Dicot	H
147	*Launaea procumbens* (Roxb.) Ramayya & Rajagopal (Syn. *Launaea fallax* (Jaub. & Spach) Kuntze)	Compositae	Creeping Launaea (E); Peeli Duddhi, Jangali Gobhi (H)	Dicot	H
148	*Lemna perpusilla* Torr. (Syn. *Hydrophace perpusilla* (Torr.) Lunell)	Araceae	Common Duckweed, Duck Meat (E)	Monocot	H
149	*Lepidium didymum* L. (Syn. *Coronopus didymus* (L.) Sm.)	Brassicaceae	Bitter Cress, Lesser Swine-cress (E); Jangli Hala (H)	Dicot	H
150	*Leptochloa panicea* (Retz.) Ohwi (Syn. *Cynodon filiformis* (Roxb.) Voigt; *Poa panicea* Retz.)	Poaceae	Mucronate Sprangletop, Sprangle Top, Thread Sprangletop (E)	Monocot	GR
151	*Leucaena leucocephala* (Lam.) de Wit (Syn. *Acacia frondosa* Willd. *Leucaena*	Leguminosae	Leucaena, Ipil-Ipil (E); Subabul (H)	Dicot	T

Contd.

	glauca Benth., *Mimosa leucocephala* Lam.)				
152	*Leucas cephalotes* (Roth) Spreng. (Syn. *Leucas capitata* Desf.)	Lamiaceae	Head Leucas (E); Guma, Goma Madhupati (H); Katumba (S)	Dicot	H
153	*Lindernia ciliata* (Colsm.) Pennell (Syn. *Bonnaya brachiata* Link & Otto)	Linderniaceae	Fringed False Pimperne (E)	Dicot	H
154	*Lindernia crustacea* (L.) F.Muell. (Syn. *Vandellia crustacea* (L.) Benth.)	Linderniaceae	Brittle False Pimpernel, Malaysian FalsePimpernel (E)	Dicot	H
155	*Ludwigia hyssopifolia* (G.Don) Exell (Syn. *Jussiaea hyssopifolia* G.Don; *Jussiaea linifolia* Vahl)	Onagraceae	Linear Leaf Water Primrose, Swamp Primrose (E); Ban Long, Bishkatali (H); Bhu Lavangah (S)	Dicot	H
156	*Ludwigia octovalvis* (Jacq.) P.H.Raven (Syn. *Jussiaea suffruticosa* L.)	Onagraceae	Mexican Primrose-Willow, Water Primrose (E); Ban Long (H); Bhu Lavangah (S)	Dicot	H
157	*Malva parviflora* L. (Syn. *Althaea parviflora* (L.) Alef.)	Malvaceae	Cheeseweed Mallow, Small-Flowered Mallow (H); Panirak (H)	Dicot	H
158	*Malvastrum coromandelianum* (L.) Garcke (Syn. *Malva coromandeliana* L.)	Malvaceae	False Mallow, Broom Weed, Prickly Malvastrum (E); Kharenti (H)	Dicot	H
159	*Marsilea quadrifolia* L.	Marsileaceae	Four Leaf Clover, Water Shamrock (E); Susni, Chaupatiya (H); Chatuh Patri, Chatuspatri (S)	PH	P
160	*Mazus pumilus* (Burm.f.) Steenis (Syn. *Lobelia pumila* Burm.f.; *Mazus japonicus* (Thunb.) Kuntze; *Mazus rugosus* Lour.)	Phrymaceae	Japanese Mazus, Asian Mazus (E)	Dicot	H
161	*Mecardonia procumbens* (Mill.) Small (Syn. *Bacopa procumbens* (Mill.) Greenm.)	Plantaginaceae	Baby Jump Up (E)	Dicot	H
162	*Medicago monantha* (C.A.Mey.) Trautv. (Syn. *Trigonella incisa* Benth.)	Leguminosae	Medick, Single-Flowered Medick (E)	Dicot	H
163	*Medicago polymorpha* L. (Syn. *Medicago denticulata* Willd.; *Medicago polymorpha* var. *polymorpha*)	Leguminosae	Bur Clover, Toothed Medic (E)	Dicot	H

Contd.

164	*Medicago lupulina* L. (Syn. *Medicago cupaniana* Guss.)	Leguminosae	Black Medic, Hop Clover (E); Lusan Ghas (H); Ashvabala (S)	Dicot	H
165	*Melia azedarach* L. (Syn. *Azedarach sempervirens* Kuntze)	Meliaceae	Chinaberry Tree, Persian Lilac, Bead Tree (E); Bakain, Mahaneem (H); Mahanimbah (S)	Dicot	T
166	*Melilotus indicus* (L.) All. (Syn. *Melilotus indica* (L.) All.; *Melilotus parviflorus* Desf.; *Trifolium indicum* L.)	Leguminosae	Indian Sweet Clover, Yellow Sweet Clover (E); Ban Methi (H)	Dicot	H
167	*Melilotus albus* Medik. (Syn. *Melilotus leucanthus* DC.)	Leguminosae	White Sweet Clover, White Melilot (E); Safed Ban Methi (H)	Dicot	H
168	*Melochia corchorifolia* L. (Syn. *Sida cuneifolia* Roxb.)	Malvaceae	Chocolate Weed (E); Chitrabeej (H)	Dicot	H
169	*Merremia hederacea* (Burm. f.) Hallier f. (Syn. *Ipomoea chryseides* Ker Gawl.)	Convolvulaceae	Ivy Woodrose (E)	Dicot	CL/CR
170	*Mimosa rubicaulis* Lam. (Syn. *Mimosa octandra* Roxb.)	Leguminosae	Himalayan Mimosa (E); Aila, Shiahkanta, Arlu (H); Rala- Arlu (S)	Dicot	SH
171	*Mitragyna parvifolia* (Roxb.) Korth. (Syn. *Stephegyne parvifolia* (Roxb.) Korth.)	Rubiaceae	Kaim, True Kadamb (E); Kadamb (H); Kadamba, Vitanah (S)	Dicot	T
172	*Mollugo nudicaulis* Lam. (Syn. *Lampetia nudicaulis* (Lam.) Raf.)	Molluginaceae	Naked-Stem Carpetweed, Daisy-leaved Chickweed (E)	Dicot	H
173	*Morus alba* L. (Syn. *Morus intermedia* Perr.)	Moraceae	White Mulberry, Silkworm Mulberry (E); Shahtoot (H)	Dicot	T
174	*Mukia maderaspatana* (L.) M.Roem. (Syn. *Cucumis maderaspatanus* L.; *Melothria maderaspatana*(L.) Cogn.)	Cucurbitaceae	Madras Pea Pumpkin (E); Aganaki, Agumaki, Bilari (H)	Dicot	CL/CR
175	*Murdannia nudiflora* (L.) Brenan (Syn. *Commelina nudiflora* L.)	Commelinaceae	Naked-Stem Dewflower, Doveweed (E); Kansura (H); Koshapushpi (S)	Monocot	H
176	*Nepeta hindostana* (B.Heyne ex Roth) Haines (Syn. *Glechoma hindostana*	Lamiaceae	North Indian Catmint, Catmint (E); Billilotan (H); Badranj Boya (S)	Dicot	H

Contd.

	B.Heyne ex Roth; *Nepeta ruderalis* Buch.-Ham. ex Benth.)				
177	*Nicotiana plumbaginifolia* Viv. (Syn. *Nicotiana tenella* Cav.)	Solanaceae	Tex-Mex Tobacco (E); Jangli Tambakoo (H)	Dicot	H
178	*Nymphaea pubescens* Willd. (Syn. *Nymphaea lotus* var. *pubescens* (Willd.) Hook. f. & Thomson)	Nymphaeaceae	White Water Lily, White Lotus (E); Ko Kaa (H); Kumud (S)	Dicot	H
179	*Nymphoides cristata* (Roxb.) Kuntze (Syn. *Limnanthemum cristatum* (Roxb.) Griseb.)	Menyanthaceae	Crested Floating Heart (E); Kumudini (H)	Dicot	H
180	*Oldenlandia corymbosa* L. (Syn. *Hedyotis corymbosa* (L.) Lam.)	Rubiaceae	Diamond Flower, Corymbose Hedyotis (E); Pitpapra (H); Parpatah (S)	Dicot	H
181	*Operculina turpethum* (L.) Silva Manso (Syn. *Convolvulus turpethum* L.; *Ipomoea turpethum* (L.) R. Br.)	Convolvulaceae	Indian Jalap, White day glory (E); Nisoth, Pithori (H); Triputa, Nishotra (S)	Dicot	CL/CR
182	*Oxalis corniculata* L. (Syn. *Oxalis repens* Thunb.)	Oxalidaceae	Indian Sorrel,Yellow Wood Sorrel (E); Amrul, Khati-buti(H); Asmanthaka, Kushali (S)	Dicot	H
183	*Oxalis debilis* var. *corymbosa* (DC.) Lourteig (Syn. *Oxalis corymbosa* DC.; *Oxalis martiana* Zucc.)	Oxalidaceae	Large-Flowered Pink Sorrel, Lilac Oxalis, Pink Wood Sorrel (E); Khatti-mithi (H)	Dicot	H
184	*Oxystelma esculentum* (L. f.) Sm. (Syn. *Oxystelma wallichii* Wight)	Apocynaceae	Rosy Milkweed Vine (E); Dudhialata (H), Dudhi Bel (H)	Dicot	CL/CR
185	*Panicum antidotale* Retz. (Syn. *Panicum miliare* Lam.)	Poaceae	Blue Panic Grass, Giant Panic (E); Kutki, Hadjodi (H)	Monocot	GR
186	*Parthenium hysterophorus* L. (Syn. *Argyrochaeta bipinnatifida* Cav.)	Compositae	Carrot Grass, Congress Grass, Wild Carrot Weed (E); Gajar Ghas (H)	Dicot	H
187	*Paspalum distichum* L. (Syn. *Digitaria paspalodes* Michx.; *Paspalum paspalodes* (Michx.) Scribn.)	Poaceae	Knot Grass, Ginger Grass, Seashore Paspalum, Salt Jointgrass (E); Besak (H)	Monocot	GR

Contd.

188	*Paspalum scrobiculatum* L. (Syn. *Paspalum orbiculare* G.Forst.)	Poaceae	Kodo Millet, Scrobic Paspalum (E); Kodo (H)	Monocot	GR
189	*Peristrophe bicalyculata* (Retz.) Nees (Syn. *Dianthera bicalyculata* Retz.)	Acanthaceae	Panicled Foldwing, Kali Anghedi (E); Atrilal (H); Nadikanta, Kakajangha (S)	Dicot	H
190	*Persicaria barbata* (L.) H. Hara (Syn. *Polygonum barbatum* L.)	Polygonaceae	Bearded Knotweed (E)	Dicot	H
191	*Persicaria lanigera* (R.Br.) Soják (Syn. *Polygonum lanigerum* R.Br.)	Polygonaceae	African Persicaria, Smartweed, Knotweed (E)	Dicot	H
192	*Phalaris minor* Retz.	Poaceae	Dwarf Canary Grass, Little-Seed Canary Grass (E); Mandusi, Kanaki (H)	Monocot	GR
193	*Phoenix sylvestris* (L.) Roxb. (Syn. *Elate sylvestris* L.)	Arecaceae	Wild Date Palm, Silver Date Palm (E); Khajoor (H)	Monocot	T
194	*Phragmites karka* (Retz.) Trin. ex Steud. (Syn. *Phragmites roxburghii* (Kunth) Steud.)	Poaceae	Tall Reed (E); Narkul, Nal (H)	Monocot	GR
195	*Phyla nodiflora* (L.) Greene (Syn. *Lippia nodiflora* (L.) Michx.; *Verbena nodiflora* L.)	Verbenaceae	Frog Fruit, Creeping Lip Plant (E); Jal Buti (H)	Dicot	H
196	*Phyllanthus amarus* Schumach. & Thonn. (Syn. *Phyllanthus nanus* Hook.f.; *Phyllanthus niruri* var. *amarus* (Schumach. & Thonn.) Leandri)	Phyllanthaceae	Black Catnip, Stone Breaker (E); Bhuiaonla, Hajarmani (H); Bahupatra, Bhumyaamalaki (S)	Dicot	H
197	*Phyllanthus maderaspatensis* L. (Syn. *Diasperus maderaspatensis* (L.) Kuntze)	Phyllanthaceae	Madras Leaf-Flower (E); Hajarmani (H); Bhumyaamalaki (S)	Dicot	H
198	*Phyllanthus reticulatus* Poir. (Syn. *Kirganelia reticulata* (Poir.) Baill.)	Phyllanthaceae	Black Honey Shrub, Black Berried Featherfoil (E); Kale Madhu Ka Per (H); Krishna-Kamboji (S)	Dicot	SH
199	*Phyllanthus tenellus* Roxb. (Syn. *Diasperus tenellus* (Roxb.) Kuntze; *Phyllanthus minor*	Phyllanthaceae	Mascarene Island leaf-flower, Long Stalked Phyllanthus (E)	Dicot	H

Contd.

	Fawc. & Rendle)				
200	*Physalis angulata* L. (Syn. *Physalis lanceifolia* Nees)	Solanaceae	Angular Winter Cherry, Cutleaf Groundcherry (E); Rasbhari (H)	Dicot	H
201	*Physalis peruviana* L. (Syn. *Physalis latifolia* Lam.; *Physalis tomentosa* Medik.)	Solanaceae	Cape Gooseberry, Peruvian Ground Cherry (E); Rasbhari (H); Kuntali, Tankari (S)	Dicot	H
202	*Pistia stratiotes* L. (Syn. *Pistia aegyptiaca* Schleid.)	Araceae	Water Cabbage, Water Lettuce (E); Jalkumbhi (H)	Monocot	H
203	*Pluchea lanceolata* (DC.) C.B.Clarke (Syn. *Berthelotia lanceolata* DC.)	Compositae	Rasna (E); Phaar (H); Sugandha, Surabhi, Surasa (S)	Dicot	US
204	*Poa annua* L. (Syn. *Ochlopoa annua* (L.) H.Scholz)	Poaceae	Annual Bluegrass, Annual Meadow Grass (E)	Monocot	GR
205	*Polygonum plebeium* R.Br. (Syn. *Polygonum roxburghii* Meisn.)	Polygonaceae	Small Knotweed (E); Machechi (H)	Dicot	H
206	*Portulaca oleracea* L. (Syn. *Portulaca consanguinea* Schltdl.)	Portulacaceae	Purslane (E); Lunia, Noni (H); Brihalloni, Lonamala (S)	Dicot	H
207	*Portulaca quadrifida* L. (Syn. *Portulaca meridiana* L.f.)	Portulacaceae	Chicken Weed, Wild Purslane (E); Paviri (H)	Dicot	H
208	*Potamogeton crispus* L. (Syn. *Potamogeton crenulatus* D.Don)	Potamogetonaceae	Curly-Leaf Pondweed (E)	Monocot	H
209	*Prosopis cineraria* (L.) Druce (Syn. *Mimosa cineraria* L.; *Prosopis spicigera* L.)	Leguminosae	Khejri Tree (E); Khejri, Sami (H)	Dicot	T
210	*Prosopis juliflora* (Sw.) DC. (Syn. *Mimosa juliflora* Sw.)	Leguminosae	Algaroba, Mesquite (E); Junglee Kikar, Vilayati Babul (H)	Dicot	T
211	*Ranunculus sceleratus* L. (Syn. *Ranunculus indicus* Roxb.)	Ranunculaceae	Cursed Buttercup, Poisonous Buttercup (E); Jaldhaniya (H); Kandakatuka, Kandira (S)	Dicot	H
212	*Rhynchosia capitata* (Roth) DC. (Syn. *Glycine capitata* Roth)	Leguminosae	Papra (E); Kulata, Kulthi (H)	Dicot	CL/CR
213	*Rhynchosia minima* (L.) DC.	Leguminosae	Burn-Mouth-Vine, Rhynchosia (E);	Dicot	CL/CR

Contd.

	(Syn. *Dolicholus minimus* (L.) Medik.; *Rhynchosia medicaginea* (Lam.) DC.)		Kulata, Kulthi		
214	*Ricinus communis* L. (Syn. *Cataputia major* Ludw.)	Euphorbiaceae	Castor Bean, Castor Oil Plant (E); Arandi, Chiyan (H); Gandharva Hasta (S)	Dicot	SH
215	*Rorippa palustris* (L.) Besser (Syn. *Cardamine palustris* Bubani; *Nasturtium palustre* (L.) DC.)	Brassicaceae	Yellow Marsh Cress, Bog Yellow-cress (E); Chamsuru (H)	Dicot	H
216	*Rumex dentatus* L. (Syn. *Rumex nipponicus* Franch. & Sav.)	Polygonaceae	Toothed Dock, Aegean Dock (E); Jangli Palak (H)	Dicot	H
217	*Rungia pectinata* (L.) Nees (Syn. *Rungia parviflora* Nees; *Justicia pectinata* L.)	Acanthaceae	Comb Rungia (E)	Dicot	H
218	*Saccharum bengalense* Retz. (Syn. *Erianthus munja* (Roxb.) Jeswiet; *Saccharum munja* Roxb.)	Poaceae	Munj (E); Moonj/Munj, Sarkanda, Sarpat (H)	Monocot	GR
219	*Saccharum spontaneum* L. (Syn. *Imperata spontanea* (L.) P. Beauv.)	Poaceae	Kans Grass, Thatch Grass, Tiger Grass (E); Kaans (H)	Monocot	GR
220	*Sagittaria guayanensis* Kunth (Syn. *Lophiocarpus guayanensis* (Kunth) Micheli)	Alismataceae	Guyanese Arrowhead (E)	Monocot	H
221	*Salix tetrasperma* Roxb. (Syn. *Pleiarina tetrasperma* (Roxb.) N. Chao & G.T. Gong)	Salicaceae	Indian Willow (E); Laila, Bains (H); Jalavetasa (S)	Dicot	T
222	*Scoparia dulcis* L. (Syn. *Scoparia ternata* Forssk.)	Plantaginaceae	Sweet Broom Weed (E); Mithi Patti, Ghoda Tulsi (H)	Dicot	US
223	*Senna occidentalis* (L.) Link (Syn. *Cassia occidentalis* L.)	Leguminosae	Coffee Senna (E); Kasunda, Bari Kasondi (H); Kasamarda, Vimarda (S)	Dicot	US
224	*Senna obtusifolia* (L.) H.S.Irwin &	Leguminosae	Sicklepod, Chinese Senna (E);	Dicot	H

Contd.

	Barneby (Syn. *Cassia obtusifolia* L.)		Chakwar (H);Cakramarda, Prapunnata (S)		
225	*Sesamum indicum* L. (Syn. *Sesamum orientale* L.)	Pedaliaceae	Sesame (E); Safed Til (H)	Dicot	H
226	*Sesbania bispinosa* (Jacq.) W.Wight (Syn. *Aeschynomene bispinosa* Jacq.; *Sesbania aculeata* (Willd.) Pers.)	Leguminosae	Prickly Sesban (E); Dhaincha, Daden (H)	Dicot	SH
227	*Sesbania sesban* (L.) Merr. (Syn. *Aeschynomene sesban* L.; *Sesban aegyptiaca* Poir.)	Leguminosae	Common Sesban, Egyptian Riverhemp (E); Jayanti, Rawasan (H); Jayantika (S)	Dicot	T
228	*Setaria pumila* (Poir.) Roem. & Schult. (Syn. *Setaria pallidifusca* (Schumach.) Stapf & C.E.Hubb.)	Poaceae	Yellow Fox-Tail Grass, Yellow Bristle Grass (E); Bandra, Ban Kauni (H)	Monocot	GR
229	*Setaria verticillata* (L.) P.Beauv. (Syn. *Panicum verticillatum* L.)	Poaceae	Bristly Foxtail, Bur Bristle Grass (E); Laptuna, Latkaunya (H)	Monocot	GR
230	*Sida acuta* Burm.f. (Syn. *Sida carpinifolia* L.f.)	Malvaceae	Common Wireweed, Common Fanpetals (E); Baraira (H); Bala (S)	Dicot	US
231	*Sida cordata* (Burm.f.) Borss.Waalk. (Syn. *Melochia cordata* Burm.f.; *Sida veronicifolia* Lam.)	Malvaceae	Long-Stalk Sida (E); Bhuinii, Kharenti (H); Bhumibala, Nagabala (S)	Dicot	H
232	*Sida cordifolia* L. (Syn. *Sida herbacea* Cav.; *Sida rotundifolia* Lam.)	Malvaceae	Heart-Leaf Sida, Bala, Country Mallow (E); Kharenti (H); Balaa, Bhadrabalaa (S)	Dicot	US
233	*Silene conoidea* L. (Syn. *Conosilene conoidea* Fourr.)	Caryophyllaceae	Cone Catchfly, Weed Campion (E)	Dicot	H
234	*Sisymbrium irio* L. (Syn. *Sisymbrium irioides* Boiss.)	Brassicaceae	London Rocket (E); Khubkalan (H), Khakasi, Khubakalan (S)	Dicot	H
235	*Solanum americanum* Mill. (Syn. *Solanum nigrum* L.)	Solanaceae	Black Nightshade (E); Makoy (H); Kakamachi (S)	Dicot	H
236	*Solanum surattense* Burm. f. (Syn. *Solanum jacquini* Willd.)	Solanaceae	Thorny Nightshade, Yellow Berried Nightshade (E); Kateli, Oonth	Dicot	H

Contd.

			Kateli (H); Kantakari (S)		
237	*Solanum torvum* Sw. (Syn. *Solanum ficifolium* Ortega)	Solanaceae	Turkey Berry, Susumber (E); Bhurat, Bhankatiya (H); Brihati (S)	Dicot	SH
238	*Sonchus asper* (L.) Hill (Syn. *Sonchus spinosus* Lam.)	Compositae	Prickly Sow-Thistle, Rough Sow-Thistle (E); Dudhi (H).	Dicot	H
239	*Sonchus brachyotus* DC. (Syn. *Sonchus arvensis* var. *glaber* Haines)	Compositae	Sadhi (E); Dudhali (H)	Dicot	H
240	*Sorghum halepense* (L.) Pers. (Syn. *Andropogon halepensis* (L.) Brot.)	Poaceae	Johnson Grass, Aleppo Grass (E); Jangli Jowar (H)	Monocot	GR
241	*Spergula arvensis* L. (Syn. *Spergularia arvensis* (L.) Cambess.)	Caryophyllaceae	Corn Spurry (E); Ban Dhania (H)	Dicot	H
242	*Spermacoce articularis* L.f. (Syn. *Borreria articularis* (L.f.) F.N.Williams)	Rubiaceae	Shaggy Buttonweed, False Buttonweed (E); Guthari (H); Madanaghanti (S)	Dicot	H
243	*Sphenoclea zeylanica* Gaertn. (Syn. *Gaertnera pangati* Retz.; *Pongatium indicum* Lam.)	Sphenocleaceae	Goose Weed, Chickenspike, Wedgewort (E); Phulanghas (H)	Dicot	H
244	*Spirodela polyrrhiza* (L.) Schleid. (Syn. *Lemna polyrrhiza* L.)	Araceae	Common Duck Meat, Duckweed (E)	Monocot	H
245	*Stellaria media* (L.) Vill. (Syn. *Alsine media* L.)	Caryophyllaceae	Chickweed, Starweed, Passerina (E); Buch-Bucha (H)	Dicot	H
246	*Streblus asper* Lour.	Moraceae	Sand Paper Tree, Toothbrush Tree (E); Dahia, Sihora (H); Shakotakah (S)	Dicot	T
247	*Stuckenia pectinata* (L.) Börner (Syn. *Potamogeton pectinatus* L.)	Potamogetonaceae	Sago Pondweed, Fennel Pondweed (E)	Monocot	H
248	*Syzygium cumini* (L.) Skeels (Syn. *Eugenia jambolana* Lam.)	Myrtaceae	Java Plum, Black Plum (E); Jamun (H); Jambula (S)	Dicot	T
249	*Tamarix indica* Willd. (Syn. *Tamarix gallica* var. *indica* (Willd.) Ehrenb.)	Tamaricaceae	Indian Tamarisk (E); Jhau (H); Aphalah, Jhavukah (S)	Dicot	SH
250	*Tephrosia pumila* (Lam.) Pers. (Syn. *Tephrosia procumbens*	Leguminosae	Indigo Sauvage (E)	Dicot	H

Contd.

	(Buch.-Ham.) Gamble)				
251	*Tephrosia purpurea* (L.) Pers. (Syn. *Cracca purpurea* L.)	Leguminosae	Wild Indigo, Fish Poison (E); Sarphonk, Sharpunkha (H)	Dicot	US
252	*Tephrosia villosa* (L.) Pers. (Syn. *Cracca villosa* L.; *Tephrosia hirta* (Buch.-Ham.) Benth.)	Leguminosae	Shaggy Wild Indigo, Hoary Tephrosia (E)	Dicot	US
253	*Teramnus labialis* (L.f.) Spreng. (Syn. *Glycine labialis* L.f.)	Leguminosae	Rabbit Vine, Horse Vine (E); Mashani, Van Udad (H); Mashaparni, Kalyani (S)	Dicot	CL/CR
254	*Trianthema portulacastrum* L. (Syn. *Trianthema monogyna* L.)	Aizoaceae	Giant Pigweed, Horse-Purslane (E); Vishakhapara, Sabuni (H); Chiratika, Dhanapatra (S)	Dicot	H
255	*Tribulus terrestris* L. (Syn. *Tribulus lanuginosus* L.)	Zygophyllaceae	Puncture Vine, Caltrop (E); Gokhru (H); Gokshura (S)	Dicot	H
256	*Trichosanthes cucumerina* L. (Syn. *Trichosanthes cucumerina* var. *cucumerina*)	Cucurbitaceae	Wild Snake Gourd (E); Jangali Chachinda, Jangali Chichonda (H)	Dicot	CL/CR
257	*Tridax procumbens* (L.) L. (Syn. *Balbisia divaricata* Cass.)	Compositae	Tridax Daisy, Mexican Daisy (E); Khal-muriya, Tal-muriya, Ghamra (H); Jayanti Veda (S)	Dicot	H
258	*Trifolium tomentosum* L. (Syn. *Trifolium curvisepalum* Tackh.)	Leguminosae	Woolly Clover, Cottonball Clover (E); Tipatiya Ghaas (H)	Dicot	H
259	*Triumfetta rhomboidea* Jacq. (Syn. *Bartramia indica* L.; *Triumfetta indica* Lam.)	Malvaceae	Diamond Burbark, Chinese Bur (E); Chiki (H); Jhinjharita, Gippit (S)	Dicot	US
260	*Typha domingensis* Pers. (Syn. *Typha angustata* Bory & Chaub.)	Typhaceae	Elephant Grass, Reed, Narrow leaf cattail (E); Patera (H)	Monocot	H
261	*Urena lobata* L. (Syn. *Urena trilobata* Vell.)	Malvaceae	Caesar weed (E); Bachita, Lapetua (H)	Dicot	SH
262	*Utricularia stellaris* L.f. (Syn. *Utricularia inflexa* var. *stellaris* (L.f.) P.Taylor)	Lentibulariaceae	Bladderwort (E)	Dicot	H

Contd.

263	*Vallisneria spiralis* L. (Syn. *Vallisneria aethiopica* Fenzl)	Hydrocharitaceae	Tape Grass, Straight Vallisneria (E); Sival, Shaival (H)	Monocot	H
264	*Verbascum chinense* (L.) Santapau (Syn. *Verbascum coromandelianum* (Vahl) Kuntze)	Scrophulariaceae	Chinese Mullein (E); Kokhima, Gadar tambaku, Kalhar (H); Bhutakeshi (S)	Dicot	H
265	*Verbascum thapsus* L. (Syn. *Verbascum simplex* Hoffmanns. & Link)	Scrophulariaceae	Great Mullein, Common Mullein, Adam's Flannel (E); Jangli Tambaku (H); Ekula Veer (S)	Dicot	H
266	*Veronica agrestis* L. (Syn. *Pocilla agrestis* (L.) Fourr.)	Plantaginaceae	Field Speedwell, Green Field Speedwell (E)	Dicot	H
267	*Veronica anagallis-aquatica* L. (Syn. *Veronica anagallidiformis* Boreau)	Plantaginaceae	Water Speedwell, Brook-Pimpernel (E); Sada, Sadevi (H)	Dicot	H
268	*Vicia sativa* L. (Syn. *Vicia alba* Moench)	Leguminosae	Common Vetch, Black-Pod Vetch (E); Chatri, Matri (H)	Dicot	H
269	*Withania somnifera* (L.) Dunal (Syn. *Physalis somnifera* L.)	Solanaceae	Winter Cherry, Indian Ginseng (E); Aswagandha, Asgandh (H); Ashwagandhi (S)	Dicot	US
270	*Xanthium strumarium* L. (Syn. *Xanthium indicum* Roxb.)	Compositae	Common Cocklebur, Clotbur (E); Chota Dhatura, Sankhahuli (H); Arishta Medhya, Sarpakshi (S)	Dicot	H
271	*Ziziphus jujuba* Mill. (Syn. *Rhamnus jujuba* L.; *Ziziphus mauritiana* Lam.)	Rhamnaceae	Indian Jujube, Indian Plum (E); Ber (H)	Dicot	T
272	*Ziziphus nummularia* (Burm.f.) Wight & Arn. (Syn. *Rhamnus nummularia* Burm.f.; *Ziziphus rotundifolia* Lam.)	Rhamnaceae	Jhar Beri, Chani Bor (E); Jhar Ber (H); Sukshmaphala, Bhukartaka (S)	Dicot	SH

(H-Herb; SH-Shrub; US-Undershrub; T-Tree; CL/CR-Climber/Creeper; GR-Grass, SE-Sedge, PH-Pteridophytic herb)
(Common name: E-English; H-Hindi; S-Sanskrit)

Table 2: Number of diseases/health conditions could be treated by high value medicinal plants of Chithara

Botanical Name	No. of diseases/health conditions could be treated	Total
Euphorbia hirta L.	Asthma (bronchitic/laryngeal spasm asthma), bronchitis, hay fever, conjunctivitis, diarrhoea, dysentery, intestinal parasitosis, colds, coughs, gonorrhea, syphilis, sexual disorders, impotency, premature ejaculation, lactation problem in women, thrush, mouth ulcers, toothache, headache, rheumatism, colic, kidney stones, amenorrhea, dysmenorrhea, venereal diseases, snakebites, scorpion stings, urino-genitory tract disorders, helminthiasis, wounds, piles, coryza, worm infestations in children, jaundice, pimples, digestive problems, alopecia, cardiac debility, acne vulgaris, laryngitis, nasal catarrh, postnatal complaints, malaria, bacterial diseases, dengue fever, athlete's foot, enteritis, fever, gas problems, itch, dyspnoea, excrescences, influenza, fractures, thorn in skin, blood in urine, bowel problems, warts, abscesses, ringworm, facilitate abortion, infertility (barrenness), eye diseases, sweating (stimulate), bruises, urethral pain, dermatitis, eczema, sores, boils, bleeding.	71
Solanum americanum Mill.	Swellings, cough, asthma, bronchitis, wounds, flatulence, dyspepsia, hepatomegaly (enlarged liver), otalgia, hiccough, nasal catarrh, vomiting, cardiopathy, leprosy, haemorrhoids, hoarseness, nephropathic problems, dropsy/oedema, general debility, gout, piles, dysentery, sore throats, digestive problems, diarrhoea, dermatitis, enlargement of spleen, cirrhosis, tumours, abdominal pain/stomachache, psoriasis, scabies, eczema, cramps, rheumatism, nerve diseases, abscesses, cancer, leucorrhoea, blood pressure, toothache, baby's teething pain, appetite loss, jaundice, blindness, conjunctivitis, eye diseases (glaucoma, trachoma, cataract), fever, erysipelas, rat bite, tuberculosis, hydrophobia, gastrohelcosis, giddiness, dipsia, inflammations, rhinopathy, hepatitis, tonsillitis, mouth sores/ulcer, ringworm, convulsions, headache, dysmenorrhea, diabetes, gonorrhoea, malaria	69
Cissampelos pareira L.	Menstrual/premenstrual problems, hormonal imbalance, postpartum pain, miscarriage, uterine haemorrhages, women reproductive system, heart problems, kidney stones, dysuria, arthritis, muscle cramps, ulcers, fistula, leucorrhoea, gonorrhoea, hypogalactia (agalactia or agalactorrhea), sexually	55

Contd.

	transmitted diseases, snake bites, conjunctivitis, dyspepsia, indigestion, flatulence, diarrhoea, blood disorders, cardiac disorders/pain, edema, cough, coryza, asthma, bronchitis, cystitis, scabies, skin disorders (eruption, pruritus), migraine, eye trouble, burning sensation, wounds, cold, malaria, dengue, fever, vomiting. abdominal/stomach/colic pains, poisons, kushtharoga (leprosy), anorexia, dysentery, dyspnea, inflammation, breast disorder (milk secretion), haematuria, worm infestation, abscess, itching, acne, dog bites	
Cannabis sativa L.	Alcohol withdrawal, morphine withdrawal, anthrax, asthma, blood poisoning, bronchitis, burns/scalds, catarrh, convulsions, cystitis, delirium, depression, diarrhoea, dysentery, dysmenorrhea/menorrhagia, fever, gonorrhoea, gout, inflammation, insomnia, jaundice, lockjaw, malaria, mania, migraine, neuralgia, palsy, rheumatism, snakebite, toothache, uterine prolapse, cough/whooping cough, hiccough, abdominal disorders, otalgia, stomatalgia, loss of appetite, penile dysfunction, premature ejaculation, hypertension, indigestion, piles, epilepsy, amenorrhea, neuritis, headache, insanity, cholera, vomiting, cancer, tumours, ulcers, corns	53
Hyptis suaveolens (L.) Poit.	Headache, convulsions, migraine, uterine affections, parasitical cutaneous diseases, catarrh, toothache, rheumatism, flatulence, hematuria, diarrhoea, dysentery, measles, cold, flu, fever, malaria, constipation, cough, asthma, sprain, boils, wounds, spermatorrhea, swellings, abscesses, haemorrhoids, memory aid, infections of the uterus, colic, dermatitis, eczema, snake bites, athlete's foot, body weakness, blood disorder (purify the blood), cancers, tumours, female sex disorder (antifertility), dysuria, gastrointestinal infections, indigestion, pain, cramps, skin rashes/dry/flaky skin, respiratory infections diseases, promote menses, lactagogue, veterinary diseases	50
Justicia adhatoda L.	Bronchitis, asthma (bronchial/allergic asthma), common cold, joint pain, lumbar pain, sprains, eczema, malaria, rheumatism, swelling, venereal diseases, phlegm, menorrhagia, bleeding piles, impotence, vomiting, tuberculosis, intestinal worms, skin diseases, postpartum haemorrhage, urinary trouble, induce abortion, ophthalmia, throat irritation, earaches, jaundice, acidity, stomach catarrh, constipation, gout, urinary disorders/urinary stone, wounds, headache, leprosy, snake-bites, cough/ whooping cough, diarrhoea, dysentery, phthisis, gonorrhoea, diabetes, liver disorders, malarial	49

Contd.

	fever, leucorrhoea, eye diseases, acute nightfall, parturition, typhus fever, diphtheria	
Eclipta prostrata (L.) L.	Wounds, ulcers, fever, hepatitis, spleen enlargements, skin diseases, sores, tetanus, elephantiasis, bile (pitta) disorders, falling of hairs, teeth problems, worms, jaundice, oedema, cough, invigorates sex, weakness of the body, secures the foetus in womb, myocardial depressant, shoulder pain, graying of the hair, haemorrhages, gum troubles, sexual weakness, vision deterioration, prevent habitual abortion, miscarriage, post-delivery uterine pain, headache, migraine, lice problem, pyorrhoea, dysentery, nervous weakness, anorexia, gastric troubles, scorpion sting, itching, malaria, sores, snake bite, asthma, bronchitis, leukoderma, hepatic problems, gland swelling, insect bites	48
Coccinia grandis (L.) Voigt	Anaemia, asthma, biliousness, catarrh, constipation, convulsion, cramp, dermatosis, dysgeusia, earache, enterosis, fever, gas, glycosuria, gonorrhoea, gout, halitosis, inflammation, leprosy, lowering blood cholesterol, menorrhagia, mycosis, ophthalmia, parturition, psoriasis, ringworm, sinusosis, smallpox, snakebite, sore throat, stomatosis, syndrome-X, syphilis, tuberculosis, vomiting, cough, jaundice, burning sensation, agalactia, bronchitis, consumption, uterine discharges, scabies, infertility of females, joint pain, diabetes	46
Bacopa monnieri (L.) Wettst.	Mind concentration, stress, body weakness, seminal weakness,epilepsy, ascites, anaemia, amentia, sexual dystunction biliousness, bronchitis, constipation, indigestion, dyspepsia, dysmenorrhoea, dermatitis, diabetes, elephantiasis /filaria, flatulence, fever, general debility, erysipelas, leprosy, insanity, inflammations, leucoderma, splenomegaly, skin diseases, syphilis, strangury, sterility, tumours, ulcers, sterility, stomach disorder, nervous weakness, hysteria, neurosis, dull memory, scabies, blood impurity, diarrhoea, cold, cough	44
Withania somnifera (L.) Dunal	Rheumatism, general debility, inflammations, body pains, nervous debility, improve vitality, miscarriage, postpartum difficulties, insomnia, wasting diseases, impotence, vigour, hypertension, failure to thrive in children, leucorrhoea, breast development, boils, carbuncles, conjunctivitis, asthma, epilepsy, hiccup, cold, cough, female disorders, ulcers, senile debility, anaemia, arthritis, fatigue, waning memory, stress disorders, abortion, ringworm, haemorrhoids, syphilitic sores, diarrhoea, nausea, fever, colic	40

It has also narcotic, diuretic and deobstruent properties. Ashwagandha root has also been noted to have sex-enhancing properties. The whole plant, but especially the leaves and the root bark, are abortifacient, adaptogen, antibiotic, aphrodisiac, deobstruent, diuretic, narcotic, strongly sedative and tonic (References to consulted: Bano et al. 2015, Datta et al. 2011, Fern 2014, Gupta & Rana 2007, Mirjalili et al. 2009, Mishra et al. 2000, Qamaruddin et al. 2012, Rahul 2013, Sabina et al. 2009, Supe et al. 2011, Umadevi et al. 2012; http://medplants. blogspot.in).

3.1.10.3. Chemical Constituents

The chemistry of *W. somnifera* has been extensively studied. Indian ginseng's pharmacological activity has been attributed to two main withanolides, withaferin A and withanolide D. The leaves of *W. somnifera* contain 12 withanolides, alkaloids, glycosides, glucose and free amino acids. The phytochemicals present in *W. somnifera* include alkaloids, such as choline, tropanol, pseudotopanol, cuscokygrene, 3- tigioyloxytropana, isopelletierine and several other steroidal compounds. The pharmacological activity of the root is attributed to the alkaloids and steroidals lactones. From the roots, 12 alkaloids, 35 withanolides and several sitoindosides have been isolated. In addition to alkaloids, the roots are reported to contain starch, reducing sugars, glycosides, dulcitol, withancil, an acid and a neutral compound. Berries contain a milk coagulating enzyme, two esterases, free amino acids, fatty oil, essential oil and alkaloids. The amino acids such as aspartic acid, glycine, tyrosine, proline, tryptophan, alanine, glutamic acid and cysteine have been reported from the root. (References consulted: Bano et al. 2015, Datta et al. 2011, Gupta & Rana 2007, Mirjalili et al. 2009, Supe et al. 2011, Umadevi et al. 2012).

4. Conclusions

India has great pool of diverse medicinal plant resources, a long and well characterized traditional medicinal system which opens enormous opportunities for new drug discoveries that can bring India on the forefront of drug development. In present study ten species viz., *B. monnieri, C. sativa, C. pareira, C. grandis, E. prostrata, E. hirta, H. suaveolens, J. adhatoda, S. americanum* and *W. somnifera* possessed outstanding medicinal values in traditional medicines as they are being used to cure a large number of diseases. Some of these plants can be grown easily in fellow and waste lands. Marginal farmers can harvest these plants on sustainable basis and supply to Ayurvedic herb dealers to generate additional income. Uncontrolled anthropogenic factors and unsustainable harvest regimes lead to habitat destruction and loss of important medicinal plant resources day by day. Hence, a scientific approach for exploiting the medicinal weeds in a sustainable manner and their large scale cultivation and propagation is needed to meet the market demand and drug development.

Acknowledgements

This work forms part of the research project "Biodiversity documentation of Chithara Gram Panchayat, Dadri Block Panchayat, Tehsil Dadri, District Gautam Buddha Nagar, Uttar Pradesh and preparation of a Biodiversity Register" funded by Shiv Nadar University, India. The authors are grateful to Prof. Rupamanjari Ghosh, Founding Director, School of Natural Sciences, and Vice-Chancellor of Shiv Nadar University for her unstinted support during the course of this investigation.

Conflict of Interest

The authors declare that they have no conflict of interest to publish this manuscript.

References

Adhikari BS, Babu MM, Saklani PL & Rawat GS (2010). Medicinal plants diversity and their conservation status in Wildlife Institute of India (WII) Campus, Dehradun. *Ethnobotanical Leaflets* 14: 46-83.

Ahirwar RK & Kujur M (2015). Ethnomedicinal uses of some plants species by the tribes of Amarkantak district Anuppur, Madhya Pradesh, India. *International Journal of Science and Research* 4(8): 1648-1651.

Al-shaimaa HA, Ali M, Hassan S, Aboubaker D, Djama M, Asfaw Z & Kelbessa E (2013). Medicinal plants and their uses by the people in the region of Randa, Djibouti. *Journal of Ethnopharmacology* 148: 701-713.

Al-Snafi AE (2013). The pharmacology of *Bacopa monniera*. a review. *International Journal of Pharma Sciences and Research* 4: 154-159.

Arora M, Sharma T, Devi A, Bainsal N & Siddiqui AA (2012). An inside review of *Cissampelos pareira* Linn: A potential medicinal plant of India. *International Research Journal of Pharmacy* 3(12): 38-41.

Ashwini M, Lather N, Bole S, Vedamutrhy AB & Balu S (2012). In vitro antioxidant and anti-inflammatory activity of *Coccinia grandis*. *International Journal of Pharmacy and Pharmaceutical Sciences* 4(3): 239-242.

Atanu FO, Ebiloma UG & Ajayi EI (2011). A review of the pharmacological aspects of *Solanum nigrum* Linn. *Biotechnology and Molecular Biology Review* 6(1): 1-7.

Bano A, Sharma N, Dhaliwal HS & Sharma V (2015). A systematic and comprehensive review on *Withania somnifera* (L.) Dunal- An Indian Ginseng. *British Journal of Pharmaceutical Research* 7(2): 63-75.

Bhutya RK (2011). Ayurvedic medicinal plants of India, Vol. 1-2. Published by Scientific Publishers, 649 pages.

Chanu WS & Sarangthem K (2014). Phytochemical constituents of *Justicia adhatoda* Linn. found in Manipur. *Indian Journal of Plant Sciences* 3(2): 19-21.

Chauhan R, Ruby, Shori A & Dwivedi A (2012). *Solanum nigrum* with dynamic therapeutic role: A review. *International Journal of Pharmaceutical Sciences Review and Research* 15(1): 65-71.

Dangwal LR & Sharma A (2011). Indigenous traditional knowledge recorded on some medicinal plants in Narendra Nagar block (TehriGarhwal), Uttarakhand. *Indian Journal of Natural Products and Resources* 2(1): 110-115.

Datta AK, Das A, Bhattacharya A, Mukherjee S & Ghosh BK (2011). An Overview on *Withania somnifera* (L.) Dunal-The Indian ginseng. *Medicinal and Aromatic Plant Science and Biotechnology* 5(1): 1-15.

Dhankhar S, Kaur R, Ruhil S, Balhara M, Dhankhar S & Chhillar AK (2011). A review on *Justicia adhatoda:* A potential source of natural medicine. *African Journal of Plant Science* 5(11): 620-627.

Duthie JF (1903-29). Flora of the upper gangetic plain and of the adjacent Siwalik and Sub-Himalayan tracts. Calcutta, India.

Fern K (2014). Useful tropical plants database. Available at http://tropical.theferns.info.

Gaurav N, Kumar A, Tyagi M, Kumar D, Chauhan UK & Singh AP (2015). Morphology of *Withania somnifera* (distribution, morphology, phytosociology of *Withania somnifera* L. Dunal). *International Journal of Current Science Research* 1(7): 164-173.

Gupta GL & Rana AC (2007). *Withania somnifera* (Ashwagandha): A review. *Pharmacognosy Reviews* 1: 129-136.

http://medplants.blogspot.in. Medicinal plants. Medicinal plants with usage, patents and their publications.

http://www.bsienvis.nic.in.ENVIS centre on floral diversity

https://www.hort.purdue.edu/newcrop/duke_energy/Cannabis_sativa.html

Jahan R, Al-Nahain A, Majumder S & Rahmatullah M (2014). Ethnopharmacological significance of *Ecliptaalba* (L.) Hassk. (Asteraceae). *International Scholarly Research Notices* 1-22.

Jhuma S & Bhattacharya S (2011). *Cissampelos pareira*: A promising antifertility agent. *International Journal of Research in Ayurveda and Pharmacy* 2(2): 439-442.

Kala CP, Dhyani PP & Sajwan BS (2006). Developing the medicinal plants sector in northern India: challenges and opportunities. *Journal of Ethnobiology and Ethnomedicine* 2:32, doi.10.1186/1746-4269-2-32.

Kaul S & Dwivedi S (2010). Indigenous ayurvedic knowledge of some species in the treatment of human diseases and disorders. *International Journal of Pharmacy & Life Sciences* 1(1): 44-49

Kumar S, Malhotra R & Kumar D (2010). *Euphorbia hirta*: Its chemistry, traditional and medicinal uses, and pharmacological activities. *Pharmacognosy Reviews* 4(7): 58-61.

Larsen HO & Olsen CS (2007). Unsustainable collection and unfair trade? uncovering and assessing assumptions regarding Central Himalayan medicinal plant conservation. *Biodiversity and Conservation* 16: 1679-1697.

Maheshwari JK (1966). Illustrations of the flora of Delhi. Published by CSIR, New Delhi, India, 282 pages.

Manvar D, Mishra M, Suriender Kumar S & Pandey VN (2012). Identification and evaluation of anti-Hepatitis C Virus phytochemicals from *Eclipta alba. Journal of Ethnopharmacology* 144(3): 545–554.

Mirjalili MH, Moyano E, Bonfill M, Cusido RM & Palazón J (2009). Steroidal lactones from *Withania somnifera,* an ancient plant for novel medicine. *Molecules* 14: 2373-2393.

Mishra LC, Singh BB & Dagenais S (2000). Scientific basis for the therapeutic use of *Withania somnifera* (Ashwagandha): A review. *Alternative Medicine Review* 5(4): 334-346.

Mithun NM, Shashidhara S & Kumar RV (2011). *Eclipta alba* (L.) A review on its phytochemical and pharmacological profile. *Pharmacologyonline* 1: 345-357.

Neeraja PV & Margaret E (2012). *Eclipta alba* (L.) Hassk: A valuable medicinal herb. *International Journal of Current Pharmaceutical Review and Research* 2(4): 188-197.

Padalia K (2015). Gewaisaag: A folk medicine used by the tribal people of Central Himalayan region. *Indian Journal of Traditional Knowledge* 14(1): 144-146.

Paliwal DNK (2017). *Eclipta alba* (Linn.) Hassk.– A review. *World Journal of Pharmaceutical and Life Sciences* 3(1): 713-721.

Prince S, Roy RK, Anurag, Gupta D & Sharma VK (2013). *Hyptis suaveolens* (L.) Poit: A phyto-pharmacological review. *International Journal of Chemical and Pharmaceutical Sciences* 4(1): 1-11.

Qamaruddin, Samiulla L, Singh VK & Jamil SS (2012). Phytochemical and pharmacological profile of *Withania somnifera* Dunal: A review. *Journal of Applied Pharmaceutical Science* 2(1): 170-175.

Rahul J (2013). An ethnobotanical study of medicinal plants in Taindol Village, district Jhansi, region of Bundelkhand, Uttar Pradesh, India. *Journal of Medicinal Plants Studies* 1(4): 59-71.

Sabina EP, Chandel S & Rasool MK (2009). Evaluation of analgesic, antipyretic and ulcerogenic effect of Withaferin A. *International Journal of Integrative Biology* 6(2): 52-56.

Schippmann U, Leaman DJ & Cunningham AB (2002). Impact of cultivation and gathering of medicinal plants on biodiversity: global trends and issues. In: Biodiversity and the ecosystem approach in Agriculture, forestry and fisheries. Food and Agriculture Organization (FAO), Rome.

Selvam V, Sumathy R & Kumuthakalavalli R (2014). Phytochemical constituents and in vitro anti - inflammatory activity of an Ivy gourd, *Coccinia grandis*. *World Journal of Pharmacy and Pharmaceutical Sciences* 3(7): 1173-1179.

Sen S & Chakraborty R (2017). Revival, modernization and integration of Indian traditional herbal medicine in clinical practice: Importance, challenges and future. *Journal of Traditional and Complementary Medicine* 7(2): 234-244.

Shaheen H, Qureshi R, Akram A & Gulfraz M (2014). Inventory of medicinal flora from Thal Desert, Punjab, Pakistan. *African Journal of Traditional, Complementary and Alternative Medicine* 11(3): 282-290.

Sharma AK & Dhakre JS (1995). Flora of Agra District. Published by Botanical Survey of India, Calcutta, India, 356 pages.

Sharma JK, Tripathi AK & Ahmad M (2019). Illustrated flora: Part of Western Uttar Pradesh and Delhi NCR, India. Published by Astral International Pvt. Ltd. and Shiv Nadar University, 606 pages.

Sharma M, Yusuf M, Hussain S & Hussain A (2012). Phytochemical constituents and pharmcological activities of *Eclipta alba* Linn. (Asteraceae): A review. *International Journal of Pharmaceutics* 3(12): 5-53.

Shiva MP (1996). Inventory of forestry resources for sustainable management and biodiversity conservation. New Delhi: Indus Publishing Company.

Shukla AN, Srivastava S & Rawat AKS (2010). An ethnobotanical study of medicinal plants of Rewa district, Madhya Pradesh. *Indian Journal of Traditional Knowledge* 9(1): 191-202.

Singh KJ & Huidrom D (2013). Ethnobotanical uses of medicinal plant, *Justicia adhatoda* L. by Meitei community of Manipur, India. *Journal of Coastal Life Medicine* 1(4): 322-325.

Singh PK, Kumar V, Singh RH, Tiwari RK, Sharma A, Rao CV & Singh RH (2010). Medico-ethnobotany of 'Chatara' block of district Sonebhadra, Uttar Pradesh, India. *Advances in Biological Research* 4(1): 65-80.

Sivarajan VV & Balachandran I (1994). Ayurvedic drugs and their plant sources. Published by Oxford and IBH Publishing Co. Pvt. Ltd., New Delhi, 570 pages.

Supe U, Dhote F & Roymon MG (2011). A review on micro propagation of *Withania somnifera* – A medicinal plant. *Journal of Agricultural Technology* 7(6): 1475-1483.

Teklehaymanot T & Giday M (2007). Ethnobotanical study of medicinal plants used by people in Zegie Peninsula, Northwestern Ethiopia. *Journal of Ethnobiology and Ethnomedicine* 3: 12-23.

The Plant List (2013). Version 1.1. Published on the internet. http://www.theplantlist.org.

Umadevi M, Rajeswari R, Rahale CS, Selvavenkadesh S, Pushpa R, Kumar KPS & Bhowmik D (2012). Traditional and medicinal uses of *Withania somnifera. The Pharma Innovation* 1(9): 102-110.

Uprety Y, Asselin H, Dhakal A & Julien N (2012). Traditional use of medicinal plants in the boreal forest of Canada: review and perspectives. *Journal of Ethnobiology and Ethnomedicine* 8: 1-14.

Vardhana R (2007). Flora of Ghaziabad District. Published by Shree Publishers and Distributors, New Delhi, 639 pages.

Warrier PK, Nambiar VPK & Ramankutty C (2010). Indian medicinal plants: A compendium of 500 species (Arya Vaidya Sala), Vol. 1-5. Universities Press (Private Limited), Himayatnagar, Hyderabad, India.

Wiart C (2006). Medicinal plants of the Asia-Pacific: Drugs for the future? World Scientific, 756 pages.

Zakaria DM, Islam M, Anisuzzaman SMD, Kundu SK, Khan MS & Begum AA (2011). Ethnomedicinal survey of medicinal plants used by folk medical practitioners in four different villages of Gazipur district, Bangladesh. *Advances in Natural and Applied Sciences* 5(4): 392-399.

10

Ethnobotanical Studies on Members of Family Apiaceae from Jammu Division (Jammu and Kashmir) Western Himalaya India

Sajan Thakur, Anjina Devi, Nawang Tashi and Harish Chander Dutt

Abstract

Family Apiaceae (or Umbelliferae) is one of the largest plant group represented by 418 genera and 3257 species (TPL 2013). The plants members are mostly aromatic, used by local populace as traditional medicines as well as spice and condiments in their routine food preparations. They have characteristic aroma due to the presence of some essential oils or oleoresin. Members of this family possess large number of compounds having antibacterial, hepato-protective, vaso-relaxant, inducing apoptosis and antitumor activities. Although the family is distributed globally but Indian subcontinent is a major center of its diversification, where more than 200 species are distributed specifically in the Himalayan eco-terrains. Jammu province (Jammu and Kashmir, India) in North Western Himalaya having unique climatic conditions supports a large variety of medicinal and aromatic plants, and amongst these Apiaceae are well known. Local people uses these plants as vegetables, for culinary

Sajan Thakur(✉), Anjina Devi, Nawang Tashi and Harish Chander Dutt (✉)

[1]*Ecological Engineering Laboratory, Department of Botany, University of Jammu, Jammu-180006 Jammu and Kashmir, India*

Anjina Devi

Plant Cytology and Genetics Laboratory, Department of Botany, University of Jammu Jammu-180006, Jammu and Kashmir, India

✉*Corresponding author email: hcdutt@rediffmail.com; sajan0007thakur@gmail.com*

Plants for Novel Drug Molecules: Ethnobotany to Ethnopharmacology
Bikarma Singh & Yash Pal Sharma (eds.), (pp. 275-286)

Email: *info@nipabooks.com* Web: *www.nipabooks.com*

purposes and valued as traditional medicine like Bunium persicum (black cumin), Carum carvi (caraway), Coriandrum sativum (coriander), Cuminum cyminum (cumin), Foeniculum vulgare (fennel), Narthex asafoetida (Heeng), Trachyspermum ammi (ajwain) and Centella asiatica *(Brahmi). People in Jammu province have their unique traditional way of using plants, knowledge of which is being transferred from generation to generation through oral dialect. As the data on the numbers of Apiaceae is scattered in the literature and difficult to be procured by the researchers. Therefore the current attempt is made so that the data along with some quantitative statistics can be provided at one platform only.*

Keywords: Ethnobotany, Apiaceae, Jammu division, Western Himalaya.

1. Introduction

The relationship between plants and humans is not only limited to the use of plants for food, clothing and shelter but also includes their use for religious ceremonies, ornamentation and health care. With the advent of modern drugs and allopathic medicine their use and practice experienced a decline but at the same time, their harmful effects and toxicity brought traditional medicinal plants again to the fore front of healthcare system (Hassan et al. 2013). The use of herbal medicine for curing various ailments is the old age culture of people residing in Jammu province. Most of the villages in this region are cut off from the routine visits to towns due to long distance and poor transport facilities. The rural populace of the area therefore depends mostly on wild plants for food, medicine as well for various cultural and religious practices. Because of the easy and free availability as well as the vast knowledge of the people in the use of these plants as folk medicine inherited from their forefathers, they are still used in various practices. In Jammu province, the presence of diverse climatic zones and ecological variability supports a rich and spectacular biodiversity of great scientific interest and promising economic benefits. This variability in turn supports various medicinally important taxa which are used by the locals for various purposes. Due to geographical remoteness of most of the areas of this province and socioeconomic conditions, local populace relied mostly on folk medicine for various practices. Since ages, through trial and error, people in this region have learned the practice of using plants growing in their close vicinity for various medicinal purposes. Amongst the various plant species used by the local inhabitants for various medicinal and culinary purposes, the members of family Apiaceae are valued the most particularly *Bunium persicum, Carum carvi, Coriandrum sativum, Cuminum cyminum, Funiculum vulgare, Narthex asafoetida, Trachyspermum ammi* and *Centella asiatica.* Both wild as well as cultivated species of this particular family have their role to be used as medicine and as a source of spice and condiment. This precious ancient wisdom, disseminated through the word-of-mouth, requires to be documented urgently. During the last half a century, only few studies have been carried out to document the ethnomedicinal uses of the plant species growing in the region (Kumar et al. 2009, Malik et al. 2011, Bhellum and Singh 2012, Bhatia et al. 2014, Kumar et al. 2015, Dutt et al. 2015, Bhatia et al. 2018). In this backdrop, the main objective of the present study is to fill knowledge gaps of the plants of this particular family in this region and to insist the farmers to cultivate them on commercial scale. When supplemented and validated with the latest scientific insights, this precious indigenous knowledge, can offer new holistic models of sustainable development that are economically viable, environmentally benign and socially acceptable (Shinwari and Gilani 2003).

2. Material and Methods

2.1. Study Area

The present study was conducted in Jammu province of Jammu and Kashmir, India. According to the previous literature, different regions of Jammu province were again visited during April 2017- October 2018 to carry out systematic and extensive ethno-botanical survey of plants belonging to family Apiaceae. Jammu province includes 10 districts whose climatic conditions ranges from sub-tropical to alpine at higher peaks and altitude from 300-6000 masl (Jammu, Kathua, Samba, Rajouri, Poonch, Reasi, Udhampur, Ramban, Doda and Kishtwar) (Figure 1). Due to this variability, Jammu province supports different members of Apiaceae some of which (*C. carvi, C. sativum, C. cyminum, F. vulgare*) have adapted themselves to wide range of habitat diversity whereas others are restricted in their distribution either at low (*C. asiatica*) or higher (*B. persicum, C. carvi, N. asafoetida, T. ammi*) altitudes. In all these districts the people have more or less similar way of using these plants for various ethnobotanical purposes.

Fig. 1: Study Area

2.2. Data Collection

The information was collected from different villages of Jammu province regarding the various ethnobotanical uses of wild as well as cultivated members of the family in the study area. During the field surveys, informants were selected randomly and information was gathered through interviews and group discussions regarding the indigenous knowledge of these plants. A total of 67 informants (42 females and 25 males) between the age group 22-70 years were interviewed through a semi-structured questionnaire. The ethnobotanical information was collected from the experts and well experienced people, local faith healers and traditional medicine men. The information collected included their local names, habit, plant part used, preparation and mode of administration. The data collected through questionnaire and interviews was then analyzed for Use Value (*UV*) and Factor Informant Consensus (*Fic*).

2.3. Data analysis

The data collected through discussions and interviews was analysed using quantitative indices as given below.

2.3.1. *Use Value*

It is used to calculate the relative importance of species among the informants according to Phillips *et al.*, 1994, using formula given below.

$UV = \Sigma U/n$

where U= No. of reports cited by each informant for a given species and n = Total no. of informants.

Use value is high when there are many use-reports for a plant, indicating that the plant is important and it approaches zero (0) when there are few reports related to its use.

2.3.2. *Factor informant consensus (Fic)*

This index was calculated to determine homogeneity of knowledge provided by different informants and has been calculated by the formula proposed by Heinrich et al. (1998).

$Fic = Nur- Nt/ (Nur-1)$

[where, *Fic* is factor informant consensus, *nur* is the number of use reports for a particular category and *nt* is the number of taxa used the category.

The *Fic* values ranges between 0 and 1, where low *Fic* values (near 0), indicates that plants are chosen randomly and no exchange of their use information is there among informants, and approach one (1) when there is a well defined selection criterion and exchange of information between informants.

3. Results and Discussion

3.1. Demography of Informants and Analysis

During the present survey, a total of 67 informants (42 females and 25 males) were interviewed and data regarding ethnobotanical information on 18 different members of Apiaceae were documented (Fig.2). The informants included in the study were divided into four age groups: Group I, 20-35 years; Group II, 36-45 years; Group III, 46-55 years and Group IV, above 55 years. Among these, 7 (28.0%) males and 12 (28.57%) females belong to group I, 11 (44.0%) males and 19 (45.23%) belong to group II, 5 (20.0%) males and 11 (26.19%) females belong to group III and 2 (8.0%) male and no female belong to group IV (Table 1).

Fig. 2: Some ethnobotanically used plants of family Apiaceae from Jammu province- **(A)** *Ferula jaeschkeana* Vatke. **(B)** *Anethum sowa* Roxb. ex Fleming **(C)** *Bunium persicum* (Boiss.) B.Fedtsch. **(D)** *Hydrocotyle javanica* Thunb. (E) *Chaerophyllum reflexum* var. *acuminatum* (Lindl.) Hedge & Lamond **(F)** *Centella asiatica* (L.) Urb.

Table 1: Demography of the informants

Informants	**Number**		
Female	42 (62.68%)		
Male	25(37.31%)		
Total	**67**		
Age group (years)	**Male**	**Female**	**Total**
Group I (20-35)	7 (28.0%)	12 (28.57%)	19 (28.35%)
Group II (36-50)	11 (44.0%)	19 (45.23%)	30 (44.77%)
Group III (51-65)	5 (20.0%)	11(26.19%)	16 (22.88%)
Group IV (above 65)	2 (8.0 %)	-	2 (2.98%)
Education level	**Male**	**Female**	
Never attended school	8	21	
Attended school for 1-5 classes	11	13	
Attended school for 6-10 classes	4	4	
Intermediate (12th class)	2	4	

Botanical name, vernacular name, and method of usage were recorded for 18 members of family Apiaceae (Table 2). The Use Value (*UV*) and *Fic* were also calculated in order to document the importance of these species among the local inhabitants of that area.

As evident from table 1, among 67 informants, the uses of different plant parts for medicinal purposes vary. Different plant parts are used for curing various diseases and among these seeds (44.4%) followed by roots and leaves (27.77% each) are the most commonly used plant part (Figure 3). The great demand of seeds is may be due to the presence of essential oils in them.

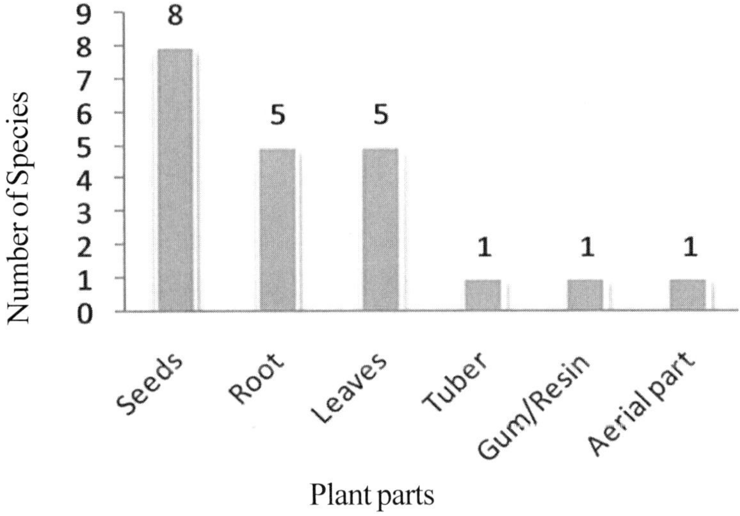

Fig. 3. Bar diagram showing plant parts of different species of family Apiaceae used in the treatment of different ailments

3.2. Use Value

Use Value (*UV*) of 18 taxa (Table 1) indicates that *T. ammi* with use value of 1.73 came out to be medicinally important species followed by *D. carota* (*UV*=0.88) *and C. asiatica* (*UV*=0.76) whereas *C. maculatum* and *C. reflexum* var. *acuminatum* with Use Value 0.15 and 0.22 can be considered as least important species (Table 2). Therefore on the basis of calculated Use Value we can conclude that *T. ammi* and *D. carota* are comparatively most used species in the area for curing various ailments and also cultivated by the locals for their use in day to day life. The maximum use of these species is also due to their occurrence in almost all the parts of the region.

Table 2: Ethnobotanical uses of some members of family Apiaceae along with their Use value from Jammu division

S.No	Species Name	Common names	Plant part used	Ethnobotanical Uses(Use report)	Total use reports	Use value
1.	*Anethum sowa* Roxb. ex Fleming	Soa	Seeds	Fruit extract is used to remove kidney stone (5). Fresh seeds are fed to cattle during pregnancy (4). A mixture of its seeds along with black pepper, ajwain, mint, banafsha is bolied in water to make decoction (arak/ Kada) (13). Also used in making pickles and kheer (10).	32	0.47
2.	*Angelica glauca* Edgew.	Chorai	Root	Root powder mixed with sugar is eaten against stomachache (7). Root is also burnt to remove snakes from houses (17).	24	0.35
3.	*Bunium persicum* (Boiss.) B.Fedtsch.	Kala Zeera	Seeds and Tubers	Used as spice and also against indigestion and stomachache (18). Starchy tubers are eaten as vegetable (10).	28	0.41
4.	*Bupleurum falcatum* L.	Jard Ziri, Pipllu	Root and leaves	Decoction is used against indigestion (8), diarrhea (5) and constipation (9).	22	0.32
5.	*Carum carvi* L.	Shia Zeera	Seeds	Used as spice and condiment (12).	12	0.18
6.	*Centella asiatica* (L.) Urb.	Brahmi	Leaves	Used as brain tonic (22), against skin diseases (13) and hair fall (16).	51	0.76
7.	*Chaerophyllum reflexum* var. *acuminatum* (Lindl.) Hedge & Lamond	Nyochha	Seeds	Used as condiment (9) and also in stomach pain (6).	15	0.22
8.	*Conium maculatum* L.	-	Root	Used in skin allergies (3) and to remove unwanted body hairs (7).	10	0.15
9.	*Coriandrum sativum* L.	Dhaniya	Seeds and Leaves	Used as carminative (8), liver tonic (5), against indigestion (7) and also in flavoring food (24).	44	0.65

Contd.

10.	*Cuminum cyminum* L.	Zeera	Seeds	Used to treat cold and fever (17). Also used as spice (25).	42	0.62
11.	*Daucus carota* L.	Gaajar	Root	Used as salad (29) and is useful against painful eyes (12) and to improve eyesight (18).	59	0.88
12.	*Ferula jaeschkeana* Vatke.	Heeng	Root	Roots are used as aphrodisiac for cattle (6) and for sterility (7).	13	0.19
13.	*Foeniculum vulgare* Mill.	Saunf	Seeds	Used as condiment (24) and against constipation (18).	42	0.62
14.	*Hydrocotyle javanica* Thunb.	Bankehri	Leaves	Infusion is used against indigestion (12) and dysentery (16).	28	0.42
15.	*Narthex asafoetida* Falc. ex Lindl.	Heeng	Gum resin	Used to make perfumes (9), flavoring food (21) and against indigestion (13).	43	0.64
16.	*Psammogeton canescens* Vatke	Gargera	Aerial part	Used as disinfectant (8) and flavoring agent (10).	18	0.27
17.	*Scandix pecten-veneris* L.	Bhuss	Whole Plant	Eaten as wild vegetable (9).	9	0.13
18.	*Trachyspermum ammi* (L.) Sprague	Ajwain	Seeds and Leaves	Used against cough (18), cold (18), fever (18), asthma (12), bronchitis (8), stomachache (22), and indigestion (20).	116	1.73

3.3. Factor Informant Consensus

A total of 614 actual use records belonging to 18 different ailments have been registered through interviews conducted with 67 informants in the study area (Table 3). People of the region use maximum number of plant species (n=8) are used as spice and condiment and for which maximum use reports are also obtained (n=135) followed by Indigestion (Plant species used= 6 and Use reports= 72) and Abdominal/ Stomach pain (Plant species used= 5 and Use reports= 61) (Table 3). The Fic of all the Use category is more than 0.87 (towards 1.0) meaning good homogeneity of knowledge among the people of Jammu division regarding the use of members of family Apiaceae for ethnobotanical uses.

Table 3: Informant consensus factor *(F_{ic})*

Category	Number of Taxa (N_t)	Number of Use Reports (N_{ur})	F_{ic}
Abdominal/stomach pain	5	61	0.93
As vegetable	3	48	0.96
Brain tonic	1	22	1.00
Cattle diseases	2	17	0.94
Constipation	3	45	0.95
Cough and cold	1	12	1.00
Diarrhea and dysentery	2	21	0.95
Disinfectant	1	8	0.87
Eye problems	1	30	1.00
Fever	3	48	0.96
Indigestion	6	72	0.93
Kidney stone	1	5	1.00
Liver tonic	1	5	1.00
Perfumes	1	9	1.00
Respiratory diseases	1	20	1.00
Skin and hair problems	2	39	0.97
Snake repellent	1	17	1.00
Spice and condiment	8	135	0.95

Current research investigations on useful plants of family Apiaceae in different disorders indicate Jammu division is a rich as per floristic diversity of medicinal plants. Seeds, roots and leaves of the investigated plants are most used parts in terms of medicine. *Trachyspermum ammi* (*UV*=1.73), *Daucus carota* (*UV*= 0.88) and *Centella asiatica* (*UV*= 0.76) are evaluated as relatively important members of family Apiaceae used as ethnimedicine in the study area. The homogeneity of traditional knowledge among informants regarding use of particular plant species in treatment of a particular ailment has been calculated and it is revealed that traditional knowledge is homogeneous for the treatment of different ailments as *Fic* value for each category is more than 0.87 indicating that the homogeneity of knowledge is maintained among the informants.

4. Conclusion

A total of 18 members of family Apiaceae are being used ethno-botanically by the people of Jammu province. *T. ammi* is one of the important species as per traditional medicinal practices are concerned. Data analysis revealed that there is good homogeneity of knowledge among the people of the region. People share and pass on their traditional knowledge through oral dialect from generation to generation. Members of family Apiaceae generally belong to aromatic plants group and contain a number of photochemical and essential oils. Further investigation and extraction of essential compounds from them will help in future drug discovery.

Conflict of Interest

Authors declare no conflict of interest

Acknowledgements

Authors are thankful to Head Department of Botany University of Jammu and Co-ordinator UGC-SAP(Des-II) for providing proper research and field facilities along with herbarium access.

References

Bhatia H, Sharma YP, Manhas RK & Kumar K (2014). Ethnomedicinal plants used by the villagers of district Udhampur, J&K, India. *Journal of Ethnopharmacology* 151: 1005–1018.

Bhatia H, Sharma YP, Manhas RK & Kumar K (2018). Traditionally used wild edible plants of district Udhampur, J&K, India. *Journal of Ethnobiology and Ethnomedicine* 14 (73): 1-13.

Bhellum BL & Singh S. (2012) Ethnomedicinal plants of district Samba of Jammu and Kashmir state (List-II). *International Journal of Scientific and Research Publications* 2(9): 1-8.

Dutt HC, Bhagat N & Pandita S. (2015). Oral traditional knowledge on medicinal plants in jeopardy among Gaddi shepherds in hills of northwestern Himalaya, J&K, India. *Journal of Ethnopharmacology* 168: (2015), 337–348.

Hassan GA, Ahmad TB & Mohi-ud-din A. (2013). An ethnobotanical study in Budgam district of Kashmir valley: An attempt to explore and document the traditional knowledge of the area. *International Research Journal of pharmacy* 4 (1): 201-204.

Heinrich M, Ankli A, Frei B, Weimann C & Sticher O. (1998). Medicinal plants in Mexico: healer's consensus and cultural importance. *Social Science and Medicine* 47: 1863-1875.

Kumar K, Sharma YP, Manhas RK, & Bhatia H. (2015) Ethnomedicinal plants of Shankaracharya Hill, Srinagar, J&K, India. *Journal of Ethnopharmacology* 170: 255–74.

Kumar M, Paul Y & Anand VK (2009) An ethnobotanical study of medicinal plants used by the locals in Kishtwar, Jammu and Kashmir, India. *Ethnobotanical leaflets* 13:1240-1256.

Malik AH, Khuroo AA, Dar GH & Khan ZS. (2011) Ethnomedicinal uses of some plants in the Kashmir Himalaya. *Indian Journal of Traditional Knowledge* 10: 362-366.

Phillips O, Gentry AH, Reynel C, Wilki P & Gavez-Durand CB. 1994. Quantitative ethnobotany and Amazonian conservation. *Conservation Biology* 8: 225-248.

Shinwari SK & Gilani SS (2003) Sustainable harvest of medicinal plants at Bulashbar Nullah, Astore (Pakistan), *Journal of Ethnopharmacology* 84 (3): 289-298.

11

Investigation of Traditional Knowledge of Economically Important Plants in Proper Neelum Valley, District Bandipora, Jammu and Kashmir, North-Western Himalaya, India

Shiekh Marifatul Haq, Eduardo Soares Calixto and Bikarma Singh

Abstract

Traditional knowledge on plants and their uses by tribal indigenous culture is helpful in the conservation of biodiversity. During 2017-2018, ethnobotanical investigations of the plants growing in Neelum Valley in District Bandipora (a part of Kashmir Himalaya) were carried out to collect information regarding different usages of the plants species growing in the region through questionnaire and interviews. Floristically, a total of 57 species belonging to 54 genera and 38 families were investigated to be used as medicine,

Shiekh Marifatul Haq

Department of Botany, University of Kashmir, Hazratbal, Srinagar-190006, Jammu & Kashmir India

Eduardo Soares Calixto

Postgraduate Program in Entomology, Faculdade de Filosofia, Ciências e Letras. Universidade de São Paulo. 3900 Bandeirantes av., Zipcode 14040-901, RibeirãoPreto, SP, Brazil

Bikarma Singh (✉)

Botanical Garden, CSIR- National Botanical Research Institute, Lucknow-226001 Uttar Pradesh, India

✉*Corresponding author email: drbikarma.singh@nbri.res.in*

Plants for Novel Drug Molecules: Ethnobotany to Ethnopharmacology
Bikarma Singh & Yash Pal Sharma (eds.), (pp. 287-302)

food, fodder, timber, fuel wood, vegetable/edible fruit, resin yield and herbal tea purposes. On the basis of the floristic analysis of the study area, the leading family recorded was Asteraceae with (11%) species co-dominated by Rosaceae (9%). In terms of habit forms, herbaceous (70%) growth form is the most dominant. The most frequently used plant parts are leaves (26%) followed by whole plant (23%) and roots (19%). The plant documented cured different categories of diseases, the most common diseases such as diarrhea, fever, digestive problems, urinary troubles, rheumatism, eye diseases, skin diseases and wounds. Based on investigated knowledge, the data are discussed under sub-head as single, double and multi-diseases curing plants. The present study can provide a baseline data for future researchers, policy makers, common public, land managers, and the other stakeholders to develop scientifically-informed strategies for conservation of natural resources and sustainable use of plant resources in hotspot regions like Himalayas and other similar biodiversity rich sites.

Keywords: Bioresources, Ethnobotanical usage, Forests, Neelum Valley, Kashmir Himalaya.

1. Introduction

Ethnobotanical investigation is an alternate to documents and conserve imperative traditional knowledge before it gets vanished (Kayan et al. 2015). Ethnobotanical analyses have developed key association between the traditional use of plants and tribal communities (Turner and Tjorve 2005, Pieroni et al. 2002, Verpoorte et al. 2005). The field surveys played an important role in conservation of plant resources, particularly biodiversity of economically important plants (Leonti 2011). Besides, the surveys are crucial in enlightening significant role of indigenous plant species, primarily for finding new crude drugs (Heinrich and Gibbons 2001, Leonti et al. 2002). The ethnic tribal communities have adopted local plants as the traditional means of healing systems and are broadly used by all sections of community, whether directly as folk remedies or the medicaments of the different indigenous systems as well as in modern medicine system (Singh and Lal 2008, Mahmood et al. 2011).

The traditional information occupies vital position at the global level right from the primitive past to present (Ghorbani 2005). Local resident committees have been utilizing plants sources for a variety of purposes over many generations (Kayani et al. 2015). Ethnobotanical research explore how these plants resources are used for things such as medicine, fuel wood, food, shelter, agriculture, timber, furniture, fodder and religious ceremonies (Amsbara et al. 2003).Native plants as a bioresource are accountable for the socio-economic improving of the region and its resident people (Awan et al. 2011). Ethnobotanical knowledge does not only entail listing the usage of plants but also helps taxonomists, ecologists, pharmacologists and wildlife managers in the conservation of biodiversity and preservation of traditional cultures (Ajaib et al. 2010).

Western Himalayan arc is a hub of unique floristic diversity (Haq et al. 2018), however, this area is neglected for ethnobotanical information by ecologists and ethnobotanists till few decades ago. Kashmir Himalaya is located along the north-western boundary of the Western Himalayan biodiversity hotspot and is known for its rich biodiversity. Keeping in view, the role of the region in the conservation of biodiversity and its varied potential of ecosystem services, the present study was conducted with the specific objectives of quantifying the floristic diversity and ethnobotanical usage of plants. Hopefully, the present study can provide a baseline data for the future researchers, policy makers, common public, land managers and the other stakeholders to develop scientifically-informed strategies for conservation of natural resources and sustainable use of plant diversity in this Himalayan region.

2. Material and Methods

2.1. Study Area

The region is the remotest tehsil of District Bandipora of Jammu and Kashmir (J&K). The valley is located in the high altitude Kashmir Himalaya, about 86 Kms from Bandipora and 146 kms from summer capital (Srinagar). The valley is mostly inhabited by Pahari and Gujjar tribes. With total human inhabitants of 30,000, the study area comprises of fifteen Panchayats (village administrative units). Dawar is the central township in the area. Due to heavy snowfall and closure of Razdan Pass in winter, the valley remains cut off for six months in a year. The valley is surrounded by snow-capped mountains drained by Kishanganga river. The climate is dry and mild, excellent of growth of diverse vegetation.

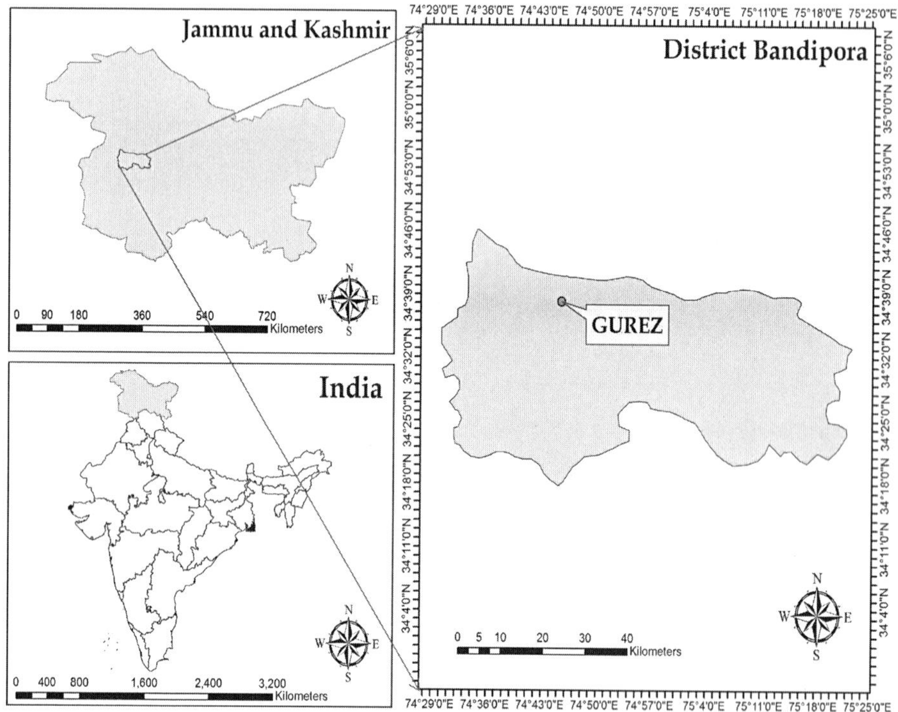

Fig. 1: Map of the study area

2.2. Methodology: Survey, Collection and Data Investigation

The reconnaissance field surveys were carried out to get an understanding about the nature of terrain, species composition, accessibility and distribution of various forests of the study area. Forest working plan was consulted for authenticating of geographical location, administrative jurisdiction and forest vegetation types. The representative forest sites were visited in three trips from year 2017 to 2018. During field studies the detailed relevant data about the plant specimens were recorded. Specimens were collected from field and were dried, preserved, labeled and mounted on herbarium sheets following standard herbarium techniques (Bridson and Forman 1999). Specimen was identified using relevant taxonomic literature (Stewart 1972; efloras). The specialized taxonomic database of The Plant List (www.theplantlist.org) was used for the updated nomenclature of taxa.

To document the ethnobotanical information of the plant resources of the study area, questionnaires and interviews method was used for collecting indigenous knowledge about plant species. The study area was visited on regularly basis and wild plants were collected. Ethnobotanical knowledge regarding plants were collected from diverse ethnic groups of the area, i.e. Gujars, herbal practitioners, hakims, shopkeepers, farmers, labours and wood sellers etc. by interviewing and filing a questionnaire. Informants were asked about their common uses of plant species, e.g. as medicinal, food, fodder, timber, fuel wood, vegetable/edible fruit, resin yield and herbal tea plants. Field based personal findings also added more information to the research work. The respondent were further asked about the part of the plant preferred if they utilized a species for usage – stem, bark, leaves, flowers, fruits or whole plant in cases of herbs.

The contribution of different growth form and plant part usage was developed using *ggplot2* package and chord diagrams were designed using the package *circlize* (Gu et al. 2014)in R software 3.6.1(R Core Team 2018).

3. Results

The and present study recorded a total of 57 species belonging to 54 genera and 38 families (Table 1). The species distribution pattern was unequal across the families, with 8 families contributes half of the species, and while remaining were represented by 30 families. Total 29 families are monotypic (Fig. 2). On the basis of the floristic analysis of the study area the dominant family reported was Asteraceae with 6 (11%) species, co-dominated by Rosaceae with 5 (9%) species, Pinaceae with 4 (7%) species, and Geraniaceae with 3(5%) species. Brassicaceae, Fabaceae, Plantaginaceae, Poaceae and Ranunculaceae were represented by 2 species each. Remaining of species was formed by other families, in which each represented by single species (Table 1; Fig. 2).

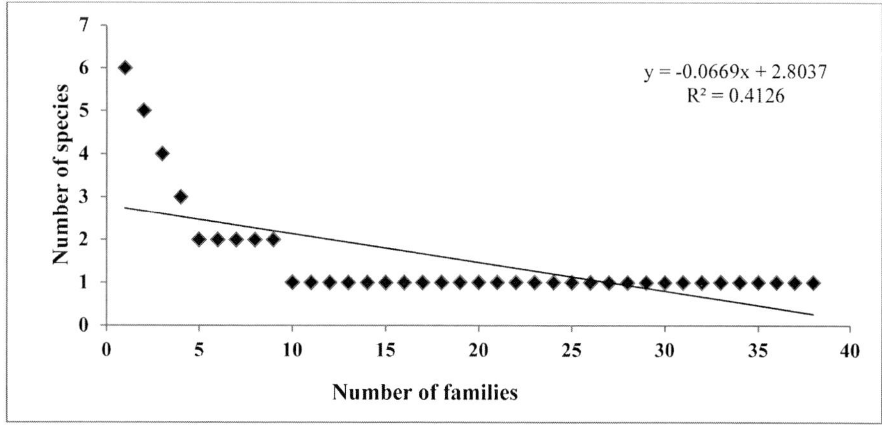

Fig. 2: Species-family relationship of flora in the study area

Based upon plant habits, the flora of the region can be classified into herbs with 40(70%) species, followed by trees with 8 (23%), shrubs with 7(12%), sub shrub and climbers with 1(2%) each, respectively (Fig.3). The principal coniferous and broad leaved tree species are *Abies pindrow*, *Cedrus deodara*, *Pinus wallichiana*, *Acer caesium*, *Ailanthus altissima* etc. Shrubs include *Rosa webbiana*, *Berberis lycium*, *Indigofera heterantha*, *Viburnum grandiflorum*; Sub-shrub *Astragalus grahamianus*, *Thymus linearis* while the herbs consists on *Aster falconeri*, *Taraxacum officinale*, *Geranium nepalense*, *Capsella bursa-pastoris*, *Cannabis sativa*, *Cichorium intybus*, *Rumex dentatus* and *Viola biflora* etc. The important medicinal plants growing in the valley are *Aconitum chasmanthum*, *Ajuga parviflora*, *Bergenia ciliata*, *Corydalis govaniana*, *Podophyllum hexandrum* and *Heracleum candicans*.

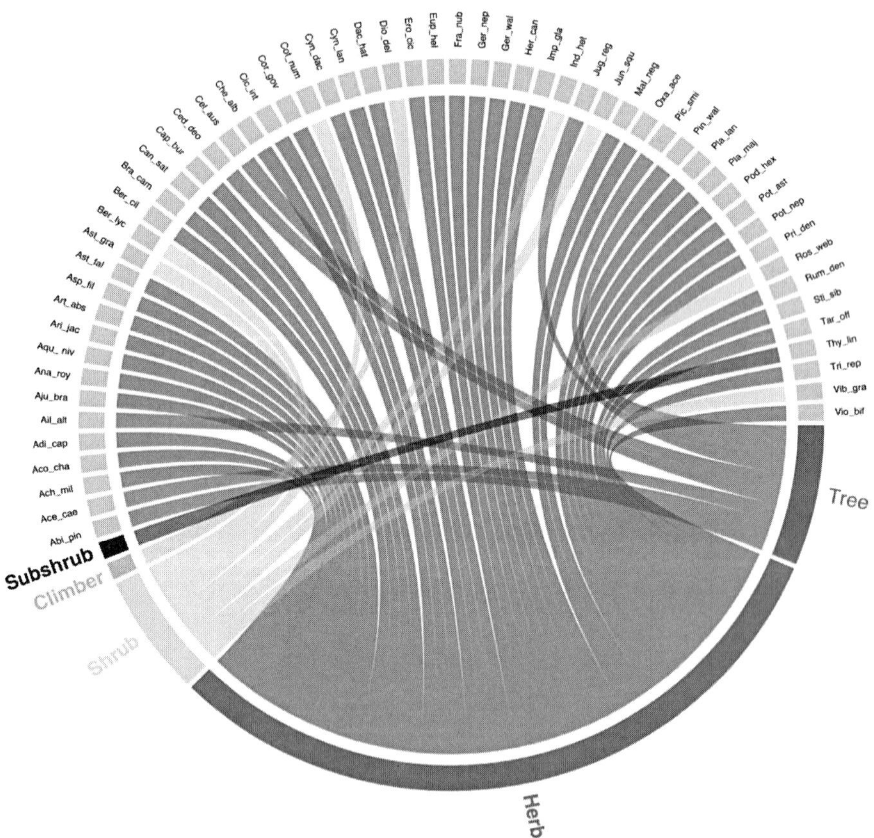

Fig. 3: Plant species distribution (57 species) according to different growth form in Neelum Valley, Himalaya, India

Table 1: Floristic and ethno-botanical usage of flora in Neelum Valley in District Bandipora, Kashmir Himalaya, India

Name of taxa	Family	Growth form	Part used	Ethno botanical importance
Abies pindrow Royle	Pinaceae	Tree	Stem, bark	Stem used for construction purposes and bark for tea.
Acer caesium Wall. ex Brandis	Aceraceae	Tree	Stem, leaves	Wood used for making agricultural implements and also as fuel and fodder.
Achillea millefolium L.	Asteraceae	Herb	Leaves	Gastrointestinal disorders, stomach ache, headache, wounds, fever.
Aconitum chasmanthum Stapfex Holmes	Ranunculaceae	Herb	Rhizome	Body tonic.
Adiantum capilus-veneris L.	Adiantaceae	Herb	Whole plant	Leaves are employed as febrifuge with pepper and in catarrhal affections along with honey.
Ailanthus altissima Swingle	Simaroubaceae	Tree	Whole plant	It is cultivated on degraded slopes for prevention of soil erosion and also as fuel and fodder.
Ajuga parviflora Wall. ex Benth.	Lamiaceae	Herb	Whole plant	Fever.
Anaphalis royleana DC.	Asteraceae	Herb	Whole plant	Diarrhoea, pulmonary infections, burns, ulcers, headaches.
Aquilegia nivalis Falc. ex Jackson.	Ranunculaceae	Herb	Flowers, rhizome	Tonic, fever, chest diseases.
Arisaema jacquemontii Blume	Araceae	Herb	Fruits, rhizome	Muscular strength, skin problems, body pain.
Artemisia absinthium L.	Asteraceae	Herb	Leaves	It is used for inflammation in liver and menstrual disorders.
Asparagus filicinus (Ham.) D.Don.	Asparagaceae	Herb	Root and stem	Fever, rheumatism and as sexual tonic.
Aster falconeri Hutch.	Asteraceae	Herb	Leaves	Anaemia.
Astragalus grahamianus Royle ex Benth.	Fabaceae	Shrub	Young twigs, leaves	Throat diseases.

Contd.

Berberis lycium Royle	Berberidaceae	Shrub	Roots	Extract used as cooling agent or tonic. It is also used as an eye lotion.
Bergenia ciliata (Haw.) Sternb.	Saxifragaceae	Herb	Whole plant	Fever, cough, cold, asthma, wound, dysentery, urinary troubles.
Brassica campestris L.	Brassicaceae	Herb	Whole plant	Leaves and young stem are cooked as spinach (Sag). Whole plant is given as fodder to increase milk production.
Cannabis sativa L.	Cannabaceae	Herb	Leaves, flowers,	Strong narcotic (Charas) is obtained by rubbing of the leaves, young twigs and flowers.
Capsella bursa-pastoris Medic.	Brassicaceae	Herb	Whole plant	The plant used as a cooked vegetable and fodder.
Cedrus (Roxb. ex D.Don)	Pinaceae	Tree	Stem	It is used for construction and furniture purposes.
Celtis australis L.	Ulmaceae	Tree	Leaves, stem	Leaves are used as fodder and stem is used for making agricultural implements.
Chenopodium album L.	Chenopodiaceae	Herb	Leaves	Used as spinach (Sag).
Cichorium intybus L.	Asteraceae	Herb	Roots	Root is used for stomach diseases and as a diuretic.
Corydalis govaniana Wall.	Fumeraceae	Herb	Roots, leaves	Eyes irritation, mythological
Cotoneaster nummularia Fisch. & Mey.	Rosaceae	Shrub	Whole plant	Lungs disorder, stomachic. Also used to control soil erosion.
Cynodon dactylon (L.) Pers.	Poaceae	Herb	Whole plant	Whole plant is grazed by cattle.
Cynoglossum lanceolatum Forssk.	Boraginaceae	Herb	Whole plant	Eye disease.
Dactylorhiza hatagirea (D.Don) Soo	Orchidaceae	Herb	Root	Wounds and cuts purposes.
Dioscorea deltoidea Wall. ex Kunth.	Dioscoreaceae	Climber	Leaves, roots	Rheumatism, tonic, appetizer and cooling agents.
Erodium cicutarium L'Herit. ex Ait.	Geraniaceae	Herb	Whole plant	Dysentery, internal illness.
Euphorbia helioscopia L.	Euphorbiaceae	Herb	Whole plant	Skin diseases and fresh wounds.

Contd.

Fragaria nubicola Lindel ex. Lacaita.	Rosaceae	Herb	Whole plant	Whole plant is grazed by cattle. Fruit is edible.
Geranium nepalense Sweet	Geraniaceae	Herb	Root	Diarrhoea, cholera treatment.
Geranium wallichianum D.Don.	Geraniaceae	Herb	Leaves, roots	Toothache, joint pains, diarrhoea, cholera, body vigour.
Heracleum candicans Wall. ex DC.	Apiaceae	Herb	Root, Leaves	Snakebites, skin diseases.
Impatiens glandulifera Royle	Balsaminaceae	Herb	Root, Leaves	Depressions, Snakebite, cooling effect, sleep-enhancing
Indigoferaheterantha Wall. ex Brandis	Papilionaceae	Shrub	Roots	Skin allergy, dysentery, warts, leprosy, anti-cancerous.
Juglans regia L.	Juglundaceae	Tree	Whole plant	The wood is excellent for furniture making, carving. Bark used for tooth cleaning.
Juniperus squamata L.	Cupressaceae	Shrub	Young twigs, berries	Urinary problem, stomach cramps, asthma, kidney diseases.
Malva neglecta Wall.	Malvaceae	Herb	Leaves, roots	Constipation, skin rashes, coughs, reduce inflammation in respiratory track.
Oxalis acetosella L.	Oxalidaceae	Herb	Whole plant	Diuretic, refrigerant, urinary problems,
Picea smithiana (Wall.) Boiss.	Pinaceae	Tree	Stem	Joints pain, stomach pain.
Pinus wallichiana A.B. Jackson	Pinaceae	Tree	Resin, bark, leaves	Cough, scorpion and snake bite, fever, asthma.
Plantago lanceolata L.	Plantaginaceae	Herb	Whole plant	Seeds are used as purgative and haemostatic. Leaves are applied to wounds.
Plantago major L.	Plantaginaceae	Herb	Leaves, seeds	Leaves are cooling, alternative and diuretic. The seeds are used to cure the stomach disorder.
Podophyllum hexandrum Royle	Podophyllaceae	Herb	Root	Antiseptic (wounds)
Potentilla astroguinea Lodd.	Rosaceae	Herb	Roots, leaves	Wounds, pain killer, toothache.
Potentilla nepalensis Hook.f.	Rosaceae	Herb	Roots	Burns, looseness of bowel.

Contd.

Primula denticulata Sm.	Primulaceae	Herb	Flowers, rhizome, leaves	Coughs, sleep enhancing, bronchitis, eye diseases.
Rosa webbiana Wall. ex Royle	Rosaceae	Shrub	Flower, fruits, seeds	Digestive problems, heart diseases
Rumex dentatus L.	Polygonaceae	Herb	Whole plant	Used as fodder and for treatment of wounds.
Stipa sibirica (L.) Lam.	Poaceae	Herb	Whole plant	Fodder, fuelwood.
Taraxacum officinale Weber.	Asteraceae	Herb	Leaves, roots	Jaundice, kidneycomplaints, Liver disorders, leprosy
Thymus linearis Benth.	Scrophulariaceae	Subshrub	Whole plant	Extensively used to cleaning cooking utensils.
Trifolium repens L.	Fabaceae	Herb	Whole plant	Extensively cultivated as an important fodder plant.
Viburnum grandiflorum Wall. ex DC	Caprifolialeae	Shrub	Fruit,stem,leaves	The fruit is sweetish and edible. Its leaves and branches may be used as fodder for livestock.
Viola biflora L.	Violaceae	Herb	Whole plant	Fever, wounds, fractured bone, cold, cough.

In the present study, different parts of plants were documented for ethnobotanical usage (Fig.4). The most frequently plant parts used are leaves (26%) followed by whole plant (23%), roots (19%), stem (9%), flower and rhizomes (5%) each, fruits (4%), bark (3%), seeds and young twigs (2%) each; and less used plant parts such as resin and berries contributes 1% each (Fig.4). The plant documented cured different categories of diseases, the most common diseases such as diarrhea, fever, digestive problems, urinary troubles, rheumatism, eye diseases, skin diseases and wounds. Based on investigated knowledge, the data are discussed under sub-head as single, double and multi- diseases curing plants. Plants such as *Ajuga parviflora, Aster falconeri, Cynoglossum lanceolatum, Astragalus grahamianus* and *Dactylorhiza hatagirea* are the single disease usage plant. The plants used for more than one purpose for example *Arisaemajacquemontii, Cichorium intybus, Erodium cicutarium* and *Euphorbia*

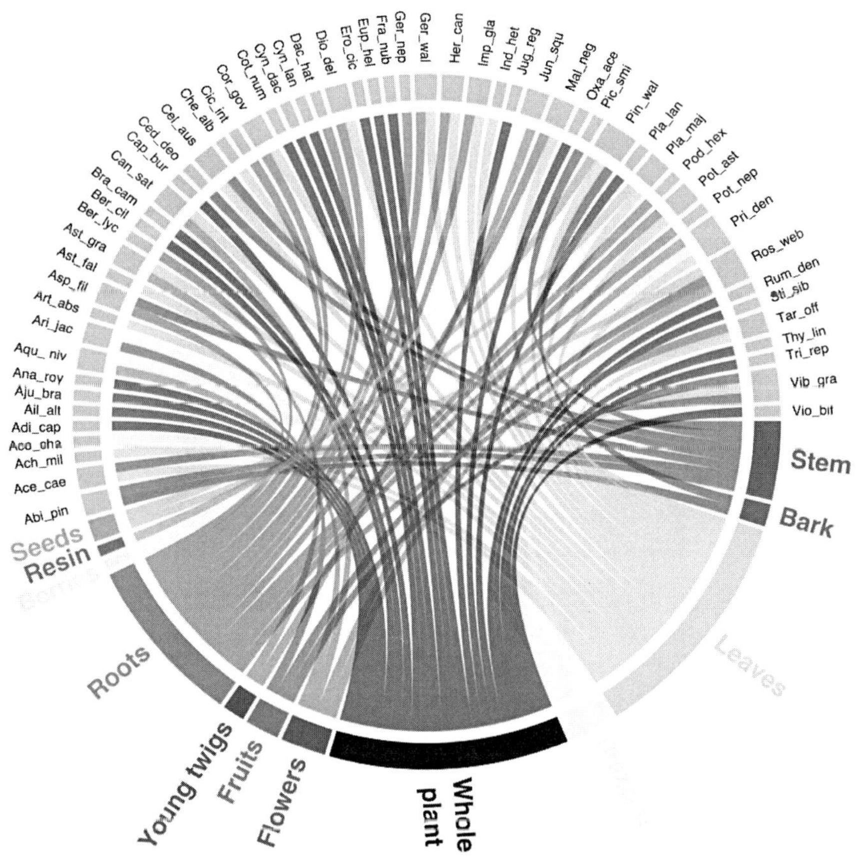

Fig. 4: Plant parts usage in Neelum Valley, Himalayan, India

helioscopia etc. Maximum of the plant species reported in the study area had multi-usage used cure more than two diseases. The examples of such plants are *Anaphalis royleana, Bergenia ciliata, Dioscorea deltoidea, Geranium wallichianum, Heracleum candicans* and *Primula denticulata*. In addition the plant species used for the fodder purpose are *Acer caesium, Brassica campestris, Cynodon dactylon, Fragaria nubicola* and *Viburnum grandiflorum*.

4. Discussion

Floristic studies of a particular area refer to the collection and complication of all plants of an area/region and when combined with ethnoecological knowledge, form an obligatory measure and a tool for conversation and sustainable utilization of plant diversity (Jayanthi and Rajendran 2013). Thus, documenting the distribution of floristic composition is the fundamental and most effective step for the future conservation of plants (Myers et al. 2000). The present study of floristic composition reported 57 plant species belonging to 54 genera and 38 families from the study area. The species richness recorded is similar to those reported by several ethnobotanical workers from different region of Himalaya (Ajaib et al. 2010; Khan et al. 2012; Begum et al. 2014; Ajaib et al. 2015). However, the plant number of this study is higher than Mahmood et al. (2011) who reported traditional use 38 plants from district Bhimber Azad Kashmir, Pakistan. Shaheen et al. (2014) reported the usage of 36 species from tribal communities of Kashmir Himalayas. Jabeen et al. (2015) surveyed a total of 49 enthnobotanically important plants of District Ghizer, Gilgit-Baltistan. For instance, the ethnobotanical information of plant resources in our studies are less than Khan et al. (2013) who reported ethnoecological knowledge of 101 species from Naran Valley, Western Himalaya, Pakistan.Awan et al. (2011) reported traditional knowledge of 102 economically important plants from Mansehra District, Pakistan. Khan et al. (2012) studied medicinal usage of 68 plants from Poonch valley Azad Kashmir, Pakistan. Haider and Qaiser (2009) reported that 83 ethno botanical uses 83 plant species from Chitral Valley, Pakistan.

However, in terms of floristic distribution patterns, these research observations can be compared with those from mountain regions in Himalaya, where floristic groups such as Asteraceae, Rosaceae and Pinaceae were the most dominant representative families. Ajaib et al. (2016) in Jatlan, Azad Kashmir, Pakistan; Gairola et al. (2010) in Uttarakhand, Garhwal Himalaya, India, reported Asteraceae as leading family. Farooq et al. (2019) in Upper Tanawal, Pakistan Himalaya reported Pinaceae as leading gymnosperm family. The floristic analyses revealed the unequal distribution of species across families and with about 29 families were monotypic. These values are similar to the previous observed values from different region of Himalaya (Gairola et al. 2010; Haq et al. 2018; Rahman et al. 2018).Maximum of the plants species recorded were herbs (70%) followed by trees (23%), and shrubs (12%). Our results are further supported by Begum et al. (2014) who carried out an ethnomedicinal study from temperate Himalayan of Pakistan. In the present study, the trees were in higher proportion (23%) than the shrubs (12%) that clearly depicts smooth functioning of a forest ecosystem. However, this finding is in support to the results of Khan et al. (2015) from Abbottabad, Pakistan.

A maximum population of the valley is living in the remote areas still relies on medicinal plants for curing different aliments. Present study includes some of important medicinal plants, which were commonly used in the valley for treatment of multiple of diseases. Some key medicinal plants were collected and used in the study area were *Anaphalis royleana, Bergenia ciliata, Dioscorea deltoidea, Geranium wallichianum, Heracleum candicans* and *Primula denticulata.* These plants are reported as potential medicinal plant in other region of Himalaya (Khan et al. 2012; Khan et al. 2013; Shaheen et al. 2014; Begum et al. 2014; Ajaib et al. 2015). The most extensively used plant parts in ethnobotanical usage were the leaves (26%). Similar results were reported by Begum et al. (2014), thus supporting our findings. The local Pahari and Gujjar tribes of Neelum Valley have been using plant resources for their various livelihoods such as medicinal, food, fodder, timber, fuel wood, vegetable/edible fruit, resin yield and herbal tea plants. Various previous ethnobotanists reported the traditional usage of bio source in various parts Himalaya for example, Haider and Qaiser (2009); Qasim et al. (2010) assessed the different purposes of plant resources in tribal communities. Out of the total reported plant species majority were used for double and multiple purposes. Our results are in agreement with Ajaib et al. (2016) who carried their study in Puna Hills, Azad Kashmir, Pakistan. The 'hakims' collects selective plants, which are profitable and can be sold easily in the local markets. The poor villagers collect plants during their work in the fields or in the surrounding forests. The local inhabitants are well versed with the valuable knowledge of plants and their economical usage through experience and ancestral long utility. The residents of the valley particularly the older generation prefer to use wild native plants because of the existing provisioning ecosystem benefits and their traditional ethnoecological knowledge.

5. Conclusion

This study provides new contributions on floristic composition and plant use by locals in the Himalayan mountainous regions. These regions represent a hotspot and therefore need to be preserved due to their high potential for biodiversity and degradation. Studying the floristic composition and use of these plants by the local human populations is fundamental to understand how the plant communities in these regions are being affected by human actions. We have shown that plants are used for different purposes, including for treating of diseases. Considering that the low social conditions of these populations and the difficulty of reaching centers with better health conditions, the use these vegetables as local medicine becomes increasingly essential for the populations of these regions. On the other hand, improper and negligent use causes the plant community to be drastically affected, changing its composition over the years. Thus, we highlight the importance of our study as a tool that will help in understanding the structuring of plant communities in these regions, and how mitigation and conservation measures can be taken for the recovery and preservation of the Himalayan plant communities.

Conflict of Interest

Authors declare no conflict of interest.

Acknowledgements

Authors would like to thanks local people for sharing their indigenous knowledge for documentation of this research.

References

Ajaib M, Abid A & Ishtiaq M (2016). Ethnobotanical studies of wild plant resources of Puna Hills, District Bhimber, AJK. *FUUAST Journal of Biology* 6(2):257-264.

Ajaib M, Khan Z, Khan N & Wahab M (2010). Ethnobotanical studies on useful shrubs of district Kotli, Azad Jammu & Kashmir, Pakistan. *Pakistan Journal of Botany* 42(3):1407-1415.

Ambara K, Gilani SS, Hussain F, Durrani MJ (2003). Ethnobotany of Gokand Valley, District Buner, Pakistan. *Pakistan Journal of Biological Sciences* 6(4): 363- 369.

Awan MR, Iqbal Z, Shah SM, Jamal Z, Jan G, Afzal M, Majid A & Gul A (2011). Studies on traditional knowledge of economically important plants of Kaghan Valley, Mansehra District, Pakistan. *Journal of Medicinal Plants Research,* 5(16): 3958-3967.

Begum S, AbdEIslam NM, Adnan M, Tariq A, Yasmin A & Hameed R (2014). Ethnomedicines of highly utilized plants in the temperate Himalayan region. *African Journal of Traditional, Complementary and Alternative Medicines* 11:132-142

Bridson D & Forman L (1999). The Herbarium Handbook. Royal Botanic Gardens, Kew. 3rd Edn (Edi.), pp: 4.

Core Team (2018) R: a language and environment for statistical computing. R Foundation for Statistical Computing, Vienna, Austria.http://www.R-project.org.

efloras. Available from: www.efloras.org (Accessed on 06-10-2018).

Farooq M, Anjum W, Hussain M, Saqib Z, Khan KR, Shah AH, Gul S & Jabeen S (2019) Forest situation analysis and future forecasting of famous Upper Tanawal forests ecosystems on western banks of lesser Himalaya. *Acta Ecologica Sinica* 39: 9-13.

Gairola S, Sharma CM, Rana CS, Ghildiyal SK & Suyal S (2010). Phytodiversity (Angiosperms and Gymnosperms) in Mandal-Choptaforest of Garhwal Himalaya, Uttarakhand, India. *Journal Nature and Science* 8: 1-17

Ghorbani A (2005). Studies on pharmaceutical ethnobotany in the region of Turkmen Sahra, north of Iran (Part 1): General results. *Journal of Ethnopharmacology* 102: 58-68.

Gu Z, Gu L, Eils R, Schlesner M & Brors B (2014) circlize implements and enhances circular visualization in R. *Bioinformatics* 30:2811–2812

Haider A & Qaiser M (2009). The enthnobotany of Chitral Valley,Pakistan with particular references to medicinal Plants. *Pakistan Journal of Botany* 41(4): 2009-2041.

Haq SM, Malik AH, Khuroo AA & Rashid I (2018). Floristic composition and biological spectrum of Keran-a remote valley of northwestern Himalaya. *Acta Ecologica Sinica* in press.

Heinrich M & Gibbons S (2001). Ethnopharmacology in drug discovery: an analysis of its role and potential contribution. *Journal of Pharmacy and Pharmacology* 53 (4): 425–432.

Jabeen N, Ajaib M, Siddiqui MF, Ulfat M & Khan B (2015). A survey of enthnobotanically important plants of District Ghizer, Gilgit- Baltistan. *FUUAST Journal of Biology* 5(1): 153-160

Jayanthi P & Rajendran A (2013). Life-forms of Madukkarai Hills of Southern Western Ghats, Tamil Nadu India. *Life Sciences Leaflets* 9:57-61.

Kayani S, Ahmad M, Sultana S, Shinwari ZK, Zafar M, Yaseen G, Hussain M & Bibi T (2015). Ethnobotany of medicinal plants among the communities of Alpine and Sub-alpine regions of Pakistan. *Journal of ethnopharmacology* 164: 186-202.

Khan MA, Khan MA & Hussain M (2012). Medicinal plants used in folk recipes by the inhabitants of Himalayan Region Poonch Valley Azad Kashmir (Pakistan). *Journal of Basic & Applied Sciences* 8: 35-45

Khan SM, Page S, Ahmad H, Shaheen H, Ullah Z, Ahmad M & Harper DM (2013). Medicinal flora and ethnoecological knowledge in the Naran Valley, Western Himalaya, Pakistan. *Journal of Ethnobiology and Ethnomedicine* 9(1): 4.

Khan W, Khan SM & Ahmad H (2015). Altitudinal variation in plant species richness and diversity at Thandiani sub forests division, Abbottabad, Pakistan. *Journal of Biodiversity and Environmental Sciences* 7: 46–53.

Leonti M (2011). The future is written: impact of scripts on the cognition, selection, knowledge and transmission of medicinal plant use and its implications for ethnobotany and ethnopharmacology. *Journal of Ethnopharmacology* 134(3): 542–555.

Leonti M, Sticher O &Heinrich M (2002). Medicinal plants of the Popoluca, México: organoleptic properties as indigenous selection criteria. *Journal of Ethnopharmacology* 81(3): 307–315.

Mahmood A, Mahmood A, Shaheen H, Qureshi RA, Sangi Y & Gilani SA (2011). Ethno medicinal survey of plants from district Bhimber Azad Jammu and Kashmir, Pakistan. *Journal of Medicinal Plants Research* 5(11): 2348-2360.

Myers N, Mittermeier RA, Mittermeier CG, da Fonseca GAB & Kent J (2000). Biodiversity hotspots for conservation priorities. *Nature* 403: 853–858.

Pieroni A, Nebel S, Quave C, Munz H & Heinrich M (2002). Ethnopharmacology of liakra: traditional weedy vegetables of the Arbereshe of the Vulture are a in southern Italy. *Journal of Ethnopharmacology* 81: 165–185.

Qasim M, Gulzar S, Shinwari Z, Aziz I & Khan MA (2010). Traditional Ethnobotanical Uses of Halophytes from Hub, Balochistan. *Pakistan Journal of Botany* 42(3): 1543-1551.

Rahman IU, Afzal A, Iqbal Z, Ijaz F, Ali N, Asif M, Alam J, Majid A, Hart R & Bussmann RW (2018). First insights into the floristic diversity, biological spectra and phenology of Manoor valley, Pakistan. *Pakistan Journal of Botany* 50: 1113-1124

Shaheen H, Islam M, & Ullah Z (2014). Indigenous ethnobotanical remedies practiced to cure feminine diseases in tribal communities of Kashmir Himalayas. *International Journal of Phytomedicine* 6(1): 103.

Singh KN & Lal B (2008). Ethnomedicines used against four common ailments by the tribal communities of Lahaul-Spiti in western Himalaya. *Journal of Ethnopharmacology* 115(1):147-159.

Stewart RR (1972) An annotated catalogue of the vascular plants of West Pakistan and Kashmir. Fakhri Printing Press, Karachi.

Turner WR & Tjørve E (2005) Scale dependence in species are a relationships. *Ecography* 28(6): 721–730.

Verpoorte R, Choi YH & Kim HK (2005).Ethnopharmacology and systems biology: a perfect holistic match. *Journal of Ethnopharmacology* 100(1): 53–56.

12

Ethnobotany and Tribals of Ladakh-Need for Scientific Initiatives and Interventions

Veenu Kaul and Sonam Tamchos

Abstract

Ladakh, a geographically isolated region, is known for its tough terrain and harsh climatic conditions. It remains cut off from rest of the country for 6-7 months in a year during winters. With limited food, fuel and other resources the people here have developed and sustained a long unique relationship with plants. They have been using both cultivated and wild ones as medicine, food, and fodder since centuries. The local populace possesses good knowledge of traditional methods of utilization of plant resources for diverse purposes, which they have acquired from their ancestors. However, this indigenous knowledge and age old practices are fast disappearing from the tribal communities, and there is an urgent need to save this treasure from getting irretrievably lost. Therefore, the present study is an attempt to document the various wild and cultivated plants along with their ethnobotanical uses as per the ancient practices of these tribal communities. For this purpose, an ethnobotanical survey was conducted in various villages of lower Ladakh. The investigation revealed a host of plants put to use by the natives in different indigenous ways. The scientific, family and local names of 50 species belonging to 48 genera and 31 families along with their uses have been enlisted in the present communication. As per the information gathered from the local inhabitants, these plants are categorized into five different groups. More than 50% of these

Veenu Kaul (✉) and Sonam Tamchos

Department of Botany, University of Jammu, Jammu-180006, Jammu and Kashmir, India

✉*Corresponding author email: veenukaul@yahoo.co.in*

Plants for Novel Drug Molecules: Ethnobotany to Ethnopharmacology
Bikarma Singh & Yash Pal Sharma (eds.), (pp. 303-330)

Email: *info@nipabooks.com* Web: *www.nipabooks.com*

are used as medicines, 38% are edible, 24% as fuel and 20% as fodder. Approximately 38% plants are employed in construction, as ornamentals, or sources of dyes, incense, cosmetics, etc.

Keywords: Ladakh, wild, cultivated, ethnobotany, traditional knowledge, tribals

1. Introduction

Ladakh, the cold desert of India, is situated between large tracts of high mountains in northern most Trans Himalayan region. It is, therefore, also known as "Land of High Passes". Covering about 70% geographical area of the erstwhile Jammu and Kashmir state, it lies between 31°44` - 32°59`N and 76°46`- 80°41` E at an altitude of 2300 to 5000 masl (Ballabh et al., 2007, Ballabh and Chaurasia, 2009; Ali, 2013; Tamchos, 2018). The region is characterized by its own typical barren topography, dry climate, high radiation, limited forest cover, very low precipitation and relative humidity; strong winds; and extreme temperature ranges from below -30°C in winter to above +35°C in summer (Kumar et al., 2009; Tamchos, 2018; Tamchos and Kaul, 2019). The economy of the region was and is largely based on agriculture, animal husbandry of yaks, double humped camels, cows, sheep, goats and donkeys (Statistical Hand Book, 2007-2008). Despite facing several seasonal constraints Ladakh Himalaya is rich in floral diversity and houses diverse groups of many interesting wild plant species. The inhabitants here, equally rich in their tribal and cultural heritage, depend on these for food, fodder and other uses. Consequently they have a long history of relationship with plants (Dorjey et al., 2012).

Several researchers (Gohil and Qadri, 1992; Chaurasia and Singh, 1996; Kala, 2005; Klimes and Dickore, 2005; Ballabh et al, 2007; Chaurasia et al., 2008; Ballabh and Chaurasia, 2009; Kumar et al., 2009; Kumar, 2012; Dorjey et al., 2012; Singh, 2012; Srivastava et al, 2016; Bashir et al., 2018) have worked on the diversity of cold desert flora and their ethno botanical values from some parts of Ladakh. But many areas are still un- or under-explored. Therefore, the present communication endeavors to give a comprehensive account of documentation of indigenous uses of some wild and cultivated plants in Sham region of Ladakh. This part of Ladakh also called lower Ladakh covers Sham and Aryan valleys and some portions of Kargil.

2. Study Area

Different villages like Khaltsi, Skurbuchan, Leido, Achina Thnag of Sham valley, Hanoo and Beema of Aryan valley of Leh and some of Kargil districts of Ladakh were covered during the exploratory trips between April and July of 2016-18. These villages are located along the basins of Indus River and are considered relatively warmer in entire Ladakh. Information on the ethno botanical usage of different plants was gathered and documented through general discussions with the village heads, elders, traditional practitioners and other local inhabitants during the frequent visits to these areas. A total of 155 people spanning young and old generations, male and female genders spread across five age groups were taken into consideration for the said purpose. The breakup of these informants has been summarized in Table 1. Perusal of this table reveals that locals interviewed in the present investigation differed in age, gender, religion, region, tribe; marital, educational and professional status. Thus, information from all the social strata could be gathered. Local language "Ladakhi" and its other dialects were used for conversation since the people are well versed and more comfortable with their native language. The plants were collected, identified and their local names obtained from the natives. A single species is known by one or more vernaculars in a particular locale and/or a single vernacular has many dialects. For example, seabuckthorn (*Hippophae rhamnoides*) is called as Tsetalulu, Tsermang, Tschermang, etc (Abassi and Kaul, 2015). Identification was subsequently authenticated by consulting experts and making comparative analyses with different floras and other published literature.

Table 1: Demographic details of the informants interviewed from the Sham region of Lower Ladakh

Attributes	**Number of informants from**			**Total**
	Sham Valley	**Aryan Valley**	**Some parts of Kargil**	
Gender				
1. Male	50	25	15	90
2. Female	25	18	22	65
Marital status				
1. Married	43	33	25	101
2. Unmarried	32	10	12	54
Religion				
1. Buddhism	75	30	0	105
2. Muslim	0	13	37	50
Tribe/community				
1. Boto	75	0	0	75
2. Brokpa	0	43	0	43
3. Balti	0	0	37	37
Age Group				
1. 20-30	8	3	5	16

Contd.

2. 30-40	10	8	2	20
3. 40-50	25	10	14	49
4. 50-60	15	12	7	34
5. Above 60	17	10	9	36
Education qualification				
1. Illiterate	25	21	17	63
2. Primary	18	11	10	39
3. Matriculate	14	5	4	23
4. Intermediate	7	4	2	13
5. Graduate	9	2	4	15
6. Post graduate	2	0	0	2
Professional status				
1. Farmer	22	15	15	52
2. Employed	20	7	10	37
3. Un/ self employed	26	18	12	56
4. Student	7	3	0	10

3. Ethnobotanical Profile

The surveyed region is extremely rich in floral diversity. The list is too long to be accommodated. Therefore, for the sake of convenience only 50 plant species belonging to 48 genera are included here. The 50 encompass herbs, shrubs and trees. Table 2 enlists these plants in alphabetical order by their scientific names along with family, vernaculars and ethno botanical uses. All the listed species break up into two divisions; 48 angiosperms and 2 gymnosperms represented by *Ephedra* and *Juniperus.* The former belongs to Ephedraceae and the latter to Cupressaceae. For angiosperms the family-wise distribution puts the 48 taxa into 29 different groups (Fig.1). Asteraceae predominates with eight species followed by four of Rosaceae and three each of Polygonaceae and Solanaceae. Five families Apiaceae, Elaeagnaceae, Fabaceae, Orobanchaceae and Salicaceae comprise of two species each. The remaining 20 are represented by single species each (Fig. 1). Plate 1(A-D) bearing Figs 3-44 show the enlisted plants growing in native habitats of the study area. The plants enlisted in Table 2 are categorized into five different groups according to the pattern/ mode of uses that is, fodder, fuel, food (edible), medicinal and others (construction purposes, tools, ornamentals, etc). Of the total taxa studied, around 54% of plants are used against various health disorders; 38% are edible and consumed as food directly or cooked; 24% as fuel and 20% as fodder. Nineteen of the fifty (38%) plants have been clubbed into others since they are used for miscellaneous purposes like as ornamentals, as incense, making of various agricultural implements, for construction, etc. (Fig. 2). Most of these plant species are collected from wild, while some herbs like *Alcea rosea, Fagopyrum esculentum, Hordeum vulgare, Medicago sativa, Melilotus officinalis, Physalis alkekengi, Tagetes erecta* are cultivated for either food or for fodder. Similarly various tree lets and tree species like *Hippophae rhamnoides, Juglans regia, Prunus armeniaca, Populus nigra, Salix alba, Vitis vinifera,* etc are planted for their multipurpose usage.

The leaves of *Allium carolinianum* and seeds of *Carum carvi* are used as condiments in different ethnic dishes and recipes where as leaves and buds of *Dracocephalum moldavica* along with coriander, onion, tomato and salt are mixed for preparation of local pickle or Chatani. Similarly, young tender leaves of *Medicago sativa, Melilotus officinalis* and *Mentha longifolia,* widely distributed along the small water channels are occasionally used in the preparation of a local recipe "Tangthurr". This dish is also prepared from leaves of *Rhodiola* after properly rinsing and boiling them in water for some time. Those of *Capparis spinosa* are used as vegetable (Kabra-tsonma) after soaking them in water for 3-5 days. The ones of *Urtica hyperborea* are collected for preparation of a very delicious recipe called as "Za-tsot Thukpa" which helps in keeping the body warm during severe winters. The flour obtained from seeds of *Fagopyrum esculentum* (Da-phay) and *Hordeum vulgare* (Tsam-phay) are widely used in the preparation of delicious ethnic dishes like "Paba", "Pra-pu", Kholak and "Tain-tan" which are of enormous economic importance. The fruits of *Echinops cornigerus, Elaeagnus angustifolia, Ephedra gerardiana* and *Morus alba* are consumed directly by children and adults alike. Both the fresh and dried fruits of *Prunus armeniaca* highly consumed by the locals are also sold in the markets. The oil extracted from kernels commonly called as "Tsegumar" is multi-purpose in utility like body massage, preparation of Phay-mar and lighting lamps in Gompas (monasteries). Fruits, kernels and oil fetch a very good price in the markets. The seeds of *Juglans regia* are consumed as dry fruit and oil extracted from seeds is also used in Gompas sometimes.

Several plant taxa like *Echinops*, buckwheat, barley, *Medicago, Melilotus, Populus, Salix* and others are used as fodder crops (see Table 2 for details). Many like *Acantholimon lycopodioides, Artemisia* sp., *Juglans regia, Morus alba, Prunus armeniaca, Ribes orientalis* and *Salix alba* are good sources of fuel. Burning of the twigs and older branches of these plants generate enough heat to keep the people warm during extremely harsh winters. Some among these are used in the making of various agricultural tools, implements and furniture while others are employed as construction and roofing materials. *Physalis alkekengi* is cultivated in the so called Aryan valley where people decorate their hats with its fruits. Still others, on account of their attractive colors and/or pleasant odors fulfill the aesthetic and decorative requirements of these tribals. Two plant taxa namely, *Rheum webbianum* and *Arnebia guttata* are used for dyeing the fabrics. The fruits of *Vitis vinifera* are edible and consumed directly; local wine is also prepared from its fruits by Aryans. Similarly, the locally prepared wine famous as "Chang" is made from seeds of *Hordeum vulgare.*

As mentioned before more than 50% of the plants mostly herbs and shrubs are used against various health problems. Most of these plant species have different medicinal properties and are being used since decades. The ash of *Acantholimon* along with milk is effective against cardiac disorders. Various herbs like *Aconitum, Artemisia, Cirsium, Geranium, Mentha, Plantago, Pedicularis* are helpful in treating abdominal pain, vomiting, other stomach and intestinal disorders, toothache and headache, while *Arnebia* and *Geranium* relieve cough and cold. The root of *Arnebia* is particularly found to cure other lung and pulmonary ailments. *Ephedra,* the single species of Gymnosperms, is widely used by the Amchis against fever, hepatic diseases, rheumatism, and bronchial asthma. The powdered root tuber of *Dactylorhiza hatagirea* and, juice and herbal tea prepared respectively from fruits and leaves of *Hippophae rhamnoides* (Seabuckthorn) are used as energy boosters and to improve the overall health (Gupta and Kaul, 2016a, b). The extract of various plant species are applied for healing wounds (*Echinops cornigerus* and *Waldheimia tomentosa*) and UV protection (*Solanum villosum*). Many taxa also find use during various religious practices by Buddhist community (Table 1). Some of these wild and cultivated plants of lower Ladakh are represented in Figs. 3-44 (Plate 1A-D).

One of the taxa mentioned above deserves special attention for the fact that it has the potential to sustain more than one industry. This has been turned into a reality in many countries particularly China and Russia to generate several billion dollar industries. *Hippophae rhamnoides* ssp. *turkestanica* widely distributed in Ladakh is popular as the 'Ladakh Gold' or 'Wonder Plant' for its multifarious uses. All the parts of the plant are medicinally important. Fruits are edible and used for preparation of juice, jam, jelly, marmalade and so on. The leaves and fruits are considered to be store house of vitamins, antioxidants, phenolics, omega-3, 6, 9 fatty acids etc. Seed oil serves as an important ingredient of sun screen lotions, lip balms, UV protecting and anti-ageing creams. The leaves are used in preparation of herbal tea, and also constitute a good forage crop for livestock. Its branches put up as barricades are also an efficient source of fuel. At the plant level, it is very effective in reclaiming poor lands (Kaul and Ali, 2013; Abassi and Kaul, 2015; Gupta and Kaul, 2016a, b; Tamchos and Kaul, 2015, 2019).

4. Past and Present Scenario

The number of people relying on the traditional health care in the developing world has been estimated to be more than 3.5 billion by the World Health Organization (Bekalo et al., 2009 and references therein). The popularity of this system can be gauged by 70-80% Africans referring themselves for treatment to the TMPs or Traditional Medical Practitioners (Bekalo et al., 2009).

Table 2: Summary of the ethnobotanical details of some wild and cultivated plants of lower Ladakh.

Botanical Name	Family	Local name	Mode of use	Plant part utilized	Traditional/Ethno Botanical Usage
Acantholimon lycopodioides (Girard) Boiss.	Plumbaginaceae	Longzeh	Fuel	Whole plant	The plant is usually found in high altitude rocky slopes and is a good source of fuel during winters.
			Medicinal		The plant ash along with milk is used against cardiac disorders.
Aconitum heterophyllum Wall. ex Royle	Ranunculaceae	Bana Karpo	Medicinal	Roots	In the traditional medicinal system of Amchis, the dried roots are powdered and effectively used against stomachache, abdominal pain, gastric problems, vomiting, toothache, etc.
Alcea rosea L.	Malvaceae	Halo Mentok	Ornamental	Whole plant	Fresh flowers along with flowers of marigold, wild rose (*Rosa ecae*), etc are used in decorating the "Gompa" (Monastery), "Chot-khang" and "guest room".
Allium carolinianum DC.	Alliaceae	Skotse	Edible	Leaves	The leaves are picked, washed and prepared into a paste "Sko-kir" and then sun dried. Small portions of the size of a small cookie are cut, stored and used as a spice to enhance the flavor of "Thukpa" and "Thangthurr".
			Medicinal		The paste of leaves is given to people suffering from indigestion.
Arnebia guttata Bunge.	Boraginaceae	Dimok	Others	Roots	Roots are used as dyeing agents in Ladakh. Lamas use it for coloring the "Chotpa" along with Yak butter in

Contd.

					several religious ceremonies.
			Medicinal		Amchis treat various ailments like cold, cough, lung and pulmonary problems with preparations made from the roots
Artemisia sp.	Asteraceae	Khampa	Fuel	Whole plant	The entire plant is used as fuel in winter.
			Medicinal	Leaves	The paste of leaves is used to alleviate stomach and intestinal disorders.
Capparis spinosa L.	Capparaceae	Kabra	Edible	Leaves Buds Roots	Young leaves and buds are collected, soaked in running water for three to five days and then cooked and consumed as vegetable. The dry roots are used for fuel in some villages.
Carum carvi L.	Apiaceae	Kosnyot	Edible	Seeds	The locals use the seeds as condiment and for enhancing the flavor of their unique ethnic dishes.
			Medicinal		Seeds are used variously by Amchis to treat people with weak eye sight, indigestion and other stomach and intestinal problems.
Cirsium arvense (L.) Scop.	Asteraceae	Kakar	Medicinal	Whole plant	The roots are eaten raw after removing their skin. Amchis use this herb against stomach ailments.
Dactylorhiza hatagirea (D. Don.) Soo.	Orchidaceae	Ambu-lakpa	Medicinal	Tuberous root	The tuberous roots are powdered and used for preparation of health tonic by the Amchis. The concoction acts as an energy booster and helps in improving overall health of an individual.
Dracocephalum moldavica L.	Lamiaceae	Tsamik	Edible	Leaves	The fresh leaves of this plant are mixed with coriander, tomato, onion and salt for preparation of local pickle or "Chatni". Dried leaves are ground

Contd.

					along with dried chilies and coriander. These are stored and used as condiment.
Echinops cornigerus DC.	Asteraceae	Akzema	Edible Fodder	Whole plant	The entire plant is used as fodder. Seeds are edible and consumed mostly by children.
			Medicinal	Leaves	The extracts prepared from leaves of this plant by Amchis are used for healing cuts and wounds.
Elaeagnus angustifolia L.	Elaeagnaceae	Sartsings	Edible	Fruits	Fruits are edible and are known to be good source of vitamins A, C and E.
			Ornamental	Flower Leaves	The twigs containing flowers and leaves are pleasant in odor and, therefore, used as an offering "Chot-khang" in Gompas (Monastery).
Ephedra gerardiana Wallich ex C.A. Meyer	Ephedraceae	Chappat; Charai	Edible Fodder	Whole plant	Ripened fruits are sweet in taste, picked from wild and eaten by children. The adults, however, harvest these fruits and subject them to short term storage and usage.The entire plant is good source of fodder.
			Medicinal		It is administered by Amchis in treating fever, hepatic diseases, rheumatism, and bronchial asthma. It is also used as a stimulant.
Fagopyrum esculentum Moench.	Polygonaceae	Ero	Edible Fodder	Seeds	It is mainly cultivated in Sham region (Lower Ladakh). The seeds, highly proteinaceous, are ground into flour "Bra-phay" and used for preparation of delicious local dishes like "Paba" "Taintan" and "Prapu".

Contd.

				Straw	It is used as fodder in winter.
Geranium himalayense Klotzsch	Geraniaceae	Gugchuk	Medicinal	Leaves Roots	The leaves and roots are ground to make a paste which is prescribed by Amchis for treating diarrhoea, cough, cold and stomach disorders.
Hippophae rhamnoides ssp. *turkestanica* Rousi	Elaeagnaceae	Tsermang, Tsar-kar, Tsesta-lulu	Edible	Leaves Fruits Seed	Fruits are edible and used for preparation of juices, jams, jellies, smarmalades, etc. The powdered seeds are mixed with flour for preparation of small cookies. Presence of thorn on the stem and the branches makes it a choice material for fencing in agricultural fields, demarcating by lanes, private lands and gardens. Two to three leaves boiled with water is a refreshing herbal tea. Rose petals added while boiling adds unique aroma to it. Leaves are a rich source of fodder for wild and domesticated livestock alike. The hard stem is an efficient source of fuel during prolonged winters. Its roots are associated with nitrogen fixing bacteria (*Frankia*); a feature known by only few locals. On account of this property it is used to reclaim and replenish depleted lands.
			Medicinal		The leaves and fruits are rich repositories of vitamins, phenolics, omega-3 fatty acids, antioxidants, etc. Fruit pulp and seed oil are popular as

Contd.

					energy boosters, cardio- and UV protective, anti-aging and anti-cancer agents, and so on.
Hordeum vulgare L.	Poaceae	Nass	Edible Fodder	Seeds	Flour of the seed (Tsam-phay, sNam phay) is widely used in preparation of local dishes like "Paba", "Kholak" and "Thukpa" and, "Chotpa" required during various religious practices by Buddhist community. Local wine called "Chang" is prepared from the seeds. Some roasted in sand, commonly called as "Yozza", are taken as dessert with seeds of apricot (*Prunus armeniaca*) and Starga (*Juglans regia*).
				Straw	It is a good source of fodder in winters known as "Phugma".
Iris lactea Pall.	Iridaceae	Tasma	Ornamental	Leaves	Plants are usually grown as ornamentals. Its long leaves are used in the preparation of sieves called locally as "Tsaks-ma".
Juglans regia L.	Juglandaceae	Starga	Edible Others*	Seeds Wood	The seeds are consumed as dry fruit and as dessert after mixing with roasted barley seeds and apricots. The oil extracted from seeds is offered in prayers by the Buddhists. Wood is utilized for making various agricultural implements, furniture and other construction purposes.
			Medicinal	Seeds, Leaves	The seeds are effective against frequent urination, asthma, and

Contd.

					constipation. Leaf extracts act as anti inflammatory agents.
Juniperus indica Bertol.	Cupressaceae	Shukpa	Others*	Leaves	The plant is considered as sacred by the Buddhist community. Its leaves and bark are used as incense "Phog" along with flowers of rose, marigold, etc. and Palu in various religious practices by the Buddhists.
Lancea tibetica Hook.f. & Thomson	Mazaceae	Spayang sta	Medicinal	Whole plant	Whole plant extracts are strongly recommended as tonic by the Amchis.
Leontopodium nanum (Hook. f. & Thomson) Hand.-Mazz.	Asteraceae	Palu	Others*	Whole plant	The entire plants in the powdered form along with Juniper are used as incense.
Medicago sativa L.	Fabaceae	Ool	Edible Fodder	Leaves	Young leaves are collected and consumed as vegetables while mature plants are a rich source of fodder and of extensive usage.
Melilotus officinalis (L.) Pall.	Fabaceae	Ool	Edible Fodder	Leaves	The young leaves are used as vegetables and in the preparation of local recipe "Tang-thurr". The mature plants are also used as fodder like *Medicago*.
Mentha longifolia L.	Lamiaceae	Pholo ling-ling	Edible	Shoots Leaves	Leaves and young shoots are used in preparation of "Tang-thurr" which is mixed with leaves of *Allium* (Skotse) fried in hot oil and consumed with *Kholag* of barley flour.
			Medicinal		Extracts of leaves and tender shoots mixed with garlic are used against dysentery diarrhoea, stomachache, vomiting and headache by Amchis in Ladakh

Contd.

Morus alba L.	Moraceae	Osay	Edible Others*	Fruits Wood	Fruits are edible. Wood is a source of fuel in winter and also for making various tools and small implements.
Myricaria squamosa Desv.	Tamaricaceae	Umbu	Others*	Young branches	The young branches are used for making local baskets called "Tsepo" and in the construction of small huts in summer.
Oxyria digyna (L.) Hill	Polygonaceae	Lamanchu	Medicinal	Leaves	The leaf extracts are taken with boiled water to improve digestion.
Plantago depressa Willd.	Plantaginaceae	Kara-ru-tse	Edible	Leaves	The young tender plants are used as vegetables.
			Medicinal		Amchis prescribe the plant extract against stomach pain and dysentery.
Pedicularis pectinata Wall. ex Benn.	Orobanchaceae		Medicinal	Whole	The entire herb in the powdered form is administered against stomach or intestinal infections.
Pedicularis longiflora Var. *tubiformis* (Klotzsch) Tsoong.	Orobanchaceae	Lughru-ser-po	Medicinal	Whole	The flowers are used for treatment against heat disorders and inflammation of the liver and gall bladder.
Physalis alkekengi L.	Solanaceae	Brokpa Mendok	Ornamental	Fruits	Fruits of the plant are used for decoration especially by the "Aryan" people.
Physochlaina praealta (Decne.) Miers.	Solanaceae	Langtang	Fodder Medicinal	Whole plant Leaves	The entire plant is used as fodder. Amchis employ leaves in the treatment of ulcers, eye diseases, and various other health issues.
Potentilla anserina L.	Rosaceae	Toma	Medicinal	Roots	Roots are consumed by children and also used against diarrhoea.
Populus nigra L.	Salicaceae	Yulat, Yarpa	Fuel, Others*	Woods	Wood, a source of fuel is also used for making various agricultural tools,

Contd.

					furniture and construction purposes. The long and strong woods are utilized for preparation of "Dongmo" (Gur-gur), a small implement employed for making special tea; typical of and essential in every Ladakhi house hold.
			Fodder	Leaves	Young shoots and leaves are used as fodder
Prunus armeniaca L.	Rosaceae	Chulli, Phating	Edible and fuel	Fruits, seeds and wood	Kernels are consumed as dry fruit and also as dessert by mixing with roasted barley seed and Starga. The oil extracted from kernels is used in *Chotmay* by Buddhists (butter lamp in monastery) and for massage. Wood is a good source of fuel during winters and also for making various tools.
Rheum webbianum Royle.	Polygonaceae	Lachu	Others*	Roots	The roots are boiled in water and used as dye for coloring clothes mainly the "Namboo" prepared from wool.
			Medicinal	Whole plant	Amchis use this plant against indigestion and abdominal disorders. The purple colored roots are effective in treating rheumatism.
Rhodiola imbricata Edgew.	Crassulaceae	Shrolo	Edible	Shoot Leaves	Young shoots and leaves are thoroughly rinsed and then boiled in water. Later these are used in the preparation of a delicious local recipe "Tang-thurr".
Ribes orientale Desf.	Grossulariaceace	Ags-ku-ta	Others*	Whole plant	The plants are used as roofing material along with willow and poplar, and for making huts in summers. The young

Contd.

					stems are used in making "Shak", a kind of sieve.
Rosa ecae Aitch.	Rosaceae	Sa-serpo Mendok	Ornamental	Flowers	Fresh flowers along with those of marigold, hollyhock etc. are used in decorating the Gompa (Monastery) "Chotkhang", and mixed with Juniper for use in "Phog".
Rosa webbiana (Wall. ex Royle)	Rosaceae	Sia- mentok	Others*	Dried flowers	Used as an ornamental and also mixed with Juniper for incense. The young stems after removing their outer layer are eaten raw.
Salix alba L.	Salicaceae	Lchangma	Fodder, fuel and others*	Whole plant	The fresh shoots are used as fodder. The young stems are used in making "Shak" (a kind of sieve). The hard stems are used for construction of windows, doors and in flooring, making huts in summers and various agriculture tools. The plant is considered multipurpose.
Solanum villosum Mill.	Solanaceae	Tsigma	Medicinal	Seeds	Paste of seeds is applied on face for protection against UV rays.
Tanacetum gracile Hk.f.&T.	Asteraceae	Khamchu	Fodder Medicinal	Whole plant Leaves	Whole plant is use as fodder. Dried leaves and flowers are used against intestinal worms.
Tagetes erecta L.	Asteraceae	Mentok chenmo	Ornamental	Flowers	Mostly cultivated for ornamental purposes. Dried petals are mixed with Juniper for incense (Phog) making by the Buddhist community.

Contd.

Taraxacum officinale Weber ex F.H. Wigg.	Asteraceae	Dandalion, Khur-mang	Medicinal	Whole plant	Used as an anti-fungal agent, diuretic, liver tonic and for controlling fever.
Urtica hyperborea Jacq. ex Wedd.	Urticaceae	Za-tsot	Edible	Leaves	Leaves collected during summer are dried and used in preparation of delicious local recipe called "Thukpa".
			Medicinal		Amchis use it for relief against rheumatism.
Vitis vinifera L.	Vitaceae	Gyun	Edible	Fruits	Fruits are edible and are consumed as such or used in preparation of local wine "Gyun-chang" by Aryans. The fruits have a very high market value.
Waldheimia tomentosa L.	Asteraceae	Palu-karpo	Others*	Whole plant	The whole plant is used in incense making.
			Medicinal		Paste of whole plant is used in healing wounds.

Others*: Building material, Incense, Dyes, Making

And, this is the trend in several areas of many other countries like Nepal, Mexico, Guatemala, Iran, etc (Heinrich, 2000; Kufer et al., 2005; Bhattarai et al., 2006; Ghorbani et al., 2006; Uprety et al., 2010). In Ladakh (India), about 30-60% people depend on this medicinal system (Chaurasia and Singh, 1996; Kala, 2005; Kumar et al., 2011; Angmo et al., 2012).

It is in this context that wild plants of Himalayan region assume immense importance particularly from medicinal and edible perspectives. Consisting of many attractive and important taxa, tribal communities of Ladakh have their own traditional ways of cooking and consuming them. Forming a composite group of myriad tribal communities like Balti, Beda, Boto, Brokpa, Changpa, Dard, Gara, Mon, Purik, etc, their food preferences are, thus, ethnic and unique in the entire country. Guided by the availability of the plants in different times of the year, their dishes are season specific. They also have rich indigenous knowledge of converting plant parts into medicine to meet their health care requirements. This traditional medicinal system of Ladakh commonly known as "Amchi system of medicine" (Namgyal and Phuntsog, 1990) is very popular for past so many centuries. Perhaps it is among the oldest surviving, well practiced, traditional healing system related to the Tibetan and Mongolian medicinal systems. It is usually known as Sowa-Rigpa (Art of Healing), and has similarities with "Unani" and "Ayurvedic" systems of medicine (Bashir et al., 2018; Kumar, 2012). Mainly based on Jung-wa-Lna (five elements) and Nespa-gsum (three humours) theories (Gurmet, 2004), the practitioners of this system are known as Amchis.

In Ladakh every major village has a trained Amchi who serves the community in his own traditional way. And this has been a common practice since ages. Every plant part such as root, stem, leaf, flower, fruit and seed is used for specific medicinal purpose. The selection of a particular plant and/or its parts is subject to the fulfillment of certain well defined criteria. Amchis administer herbs and others in various forms under a set of strict rules and regulations. Time, dosage amount and duration are set with utmost care and tend to vary with age and socio-economic status of the patient. Strict instructions are to be followed by the patients for speedy recovery. Restrictions are particularly in the type of diet to be taken by ailing people. These are largely in terms of the quantum and nature of spices, condiments used in and the mode of cooking undertaken. All the patients are regularly and stringently monitored by the Amchis. And, this calls for patience and perseverance on the part of both the ailing person and the healer.

The knowledge, skills and practice of Amchis would be handed over to the next generation through "*rgudpa*" (lineage system) in the family, and it would take several years to train a youngster and transform him into an expert Amchi. Many young Amchis would then go to Tibet to pursue further studies (Gurmet, 2004). Although predominantly males some females also qualify as Amchis. As per the study of Ballabh and Chaurasia (2011), there are about 150 Amchis residing in 226 different villages; and a single Amchi looks after the health of 2 or more villages in far flung areas. Of the 150, 23 happened to be women (Ballabh and Chaurasia, 2011). According to some workers, 60% of health care is provided by these Amchis (Chaurasia and Singh, 1996; Kala, 2005; Kumar et al., 2011). Angmo et al (2012), however, differ on this issue. As per them only 30% of the public health is taken care of by this traditional system. Nonetheless, the indigenous knowledge and age old practices are fast disappearing from the tribal communities. And there is an urgent need to save this treasure from getting irretrievably lost. Many factors contribute towards this decline. Some prominent ones among these include lack of awareness and proper documentation of this treasure; disinterest among the young generation, gradual disappearance of lineage system in the family and reluctance to share the knowledge outside the family; availability of different food stuffs in the market, introduction of western medicinal system and so on. While sharing some of these fears, Angmo and her team (2012) report a drastic reduction in the number of these traditional healers to just 36 in Western Ladakh. They largely attribute this decline to their dwindling incomes and social status thereof.

Due to the modernization and on-going developmental activities, over exploitation of natural resources and population explosion, many important wild plants are facing high rate of extinction. The newly emerging concept of natural being better than synthetic has boosted the extraction of the plants from wild. Elvin-Lewis (2001) has called this as Neo-Western herbalism. He opines that this has not only eroded the natural resources fast but has led to the availability of several formulations guised as food or food supplements in the western market. With no regulation in place, this is likely to be life-threatening (Elvin-Lewis, 2001).

In Ladakh, according to Bashir and co-authors (2018), there are many regions where the ancestral methods of plant application and consumption still exist. However, with the passage of time these have become very scarce. Many factors have contributed to the decline in the consumption of traditional dishes and/or medicines. These include non availability or lack of the documented recipes and formulations, introduction of and easy access to western and/or allopathic medicines and flooding of different food stuffs in markets. Only 20% high altitude wild plants are now used for different medicinal purposes (Bashir et al., 2018). Therefore, proper quantification and documentation coupled with the historical approach is of utmost importance for conservation of

Fig. 3: *Acantholimon lycopodiodies*

Fig. 4: *Physochlaina praealta*

Fig. 5: *Arnebia guttata*

Fig. 6: *Artemisia* sp.

Fig. 7: *Pedicularis longiflora*

Fig. 8: *Pedicularis pectinata*

Fig. 9: *Myricaria squamosa*

Fig. 10: *Morus alba*

Fig. 11: *Lancea tibetica*

Fig. 12: *Iris lactea*

Fig. 13: *Dracocephalum moldavica*

Fig. 14: *Cirsium arvense*

Plate 1A: Plant diversity of the study area

Fig. 15: *Leontopodium nanum*
Fig. 16: *Taraxacum officinale*
Fig. 17: *Ribes orientale*
Fig. 18: *Tanacetum gracile*
Fig. 19: *Solanum villosum*
Fig. 20: *Rosa ecae*
Fig. 21: *Juniperus indica*
Fig. 22: *Rheum webbianum*
Fig. 23: *Potentilla anserina*
Fig. 24: *Plantago depressa*
Fig. 25: *Echinops cornigerus*
Fig. 26: *Ephedra gerardiana*

Plate 1B: Plant diversity of the study area

Fig. 27: *Oxyria digyna* **Fig. 28:** *Geranium himalayense* **Fig. 29:** *Urica hyperborea*

Fig. 30: *Elaeagnus angustifolia* **Fig. 31:** *Physalis alkekengi* **Fig. 32:** *Vitis vinifera*

Fig. 33: *Dactylorhiza hatagirea* **Fig. 34:** *Hippophae rhamnoides* **Fig. 35:** *Prunus armeniaca*

Fig. 36: *Fagopyrum esculentum* **Fig. 37:** *Mentha longifolia* **Fig. 38:** *Rosa webbiana*

Plate 1C: Plant diversity of the study area

Fig. 39: *Tagetes erecta*

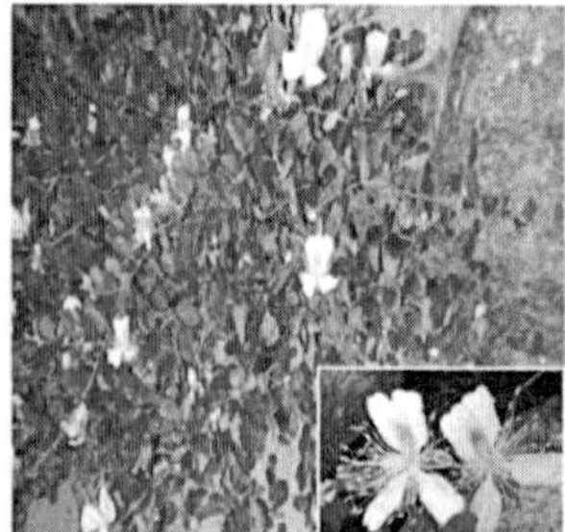

Fig. 40: *Capparis spinosa*

Fig. 41: *Melilotus officinalis*

Fig. 42: *Medicago saliva*

Fig. 43: *Populus nigra.*

Fig. 44: *Juglans regia*

Plate 1D: Plant diversity of the study area

traditional knowledge possessed by the tribal communities. In this regard the study of Soelberg and his co-workers (2016) is worth mentioning. These researchers identified 'pre-1900 medicinal plants, their historical uses and traced their status in the traditional medicine' till 2014 in St. Croix, USA. They have made attempts to assess the magnitude of loss and the extent to which traditional practices and/or knowledge has been preserved in this region for the past 114-5 years. And their findings are interesting and highly significant. An enormous percentage (63%) of medicinal usage carried out before 1900 was found to be 'forgotten or discontinued or lost'. Only 37% was in and preferred. While 20% of this loss could be attributed to dwindling biodiversity, others in all possibility seemed to have become obsolete.

This scenario is to be looked at from another perspective of social and historical importance. And, that is fragmentation of these indigenous cultures on account of increased modernization and globalization pressures. The indigenous practitioners as per Ghorbani and co-workers

(2006) seem to be more at risk than the biodiversity itself. It is essential to look for the issues plaguing these people and devise suitable remedial measures to ensure their survivability and sustainability. Therefore, the upliftment of these tribals, need to be prioritized; their enormous contribution to the scientific community highlighted, and their rights preserved. The knowledge and experience these people have gained through years of rigorous hard work should not go unacknowledged.

Nevertheless, at present ethnobotany has graduated from being a mere listing of useful plants to a multi disciplinary science linking systematic botany to botany, conservation biology, ecology, anthropology, economics, pharmacology and so on. The first ethnobotanists, therefore, are none other than the tribals or indigenous groups or natives or aborigines of their respective regions. Not only have they helped conserve the biodiversity but they have been instrumental in opening vistas for drug discovery and alleviating people from pain besides providing their counterparts with basic necessities of living. In his review, Balick (1996) has very beautifully explained the emerging trends of this science. Some of these include management and conservation of these resources, quantification and 'pharmaceutical prospecting, multiple usage and valuation studies, development of forest-based traditional medicine industry and establishment of an ethno-biomedical forest reserve' (Balick, 1996). Together with collaboration from the government and policy making bodies these would create opportunities for the local populace to get benefitted.

5. Need for Scientific Initiatives and Interventions

A number of taxa put on record here have also been reported by several researchers who have worked on the ethno botanical aspects of plants from different regions of Ladakh (Gohil and Qadri, 1992; Chaurasia and Singh, 1996; Kala, 2005; Ballabh et al, 2007; Chaurasia et al., 2008; Ballabh and Chaurasia, 2009, 2011; Dorjey et al., 2012; Srivastava et al., 2016). Many similarities emerge from the comparative analyses of the work presented here with that published before. Gastro-intestinal disorders and other illnesses associated with pain and fever are not only the most important groups of ailments but also the most common ones cutting across cultural and geographical barriers. The works of Heinrich (2000) on Mexican Indians, Kufer et al. (2005) on indigenous groups of Eastern Guatemala; Bhattarai et al. (2006) on locals of Manang and Uprety et al. (2010) on those of Rasuwa, both from Central Nepal; Bekalo et al. (2009) on Konta Special Woreda; Soelberg et al. (2016) on US Virgin Islands; Gohil and Qadri (1992), Kaul (1997), Chaurasia and Singh (1996), Kala (2005), Ballabh and Chaurasia (2007, 2009, 2011), Ballabh et al. (2007,

2008), Kumar et al. (2011), Angmo et al. (2012), Srivastava et al (2016), Bashir et al. (2018) and many more on those of Ladakh are a testimony to this. Intra- and inter- cultural comparison of the taxa administered as medicines has revealed a commonality among them with respect to their mode and purpose of usage. Taking this into cognizance, therapeutic claims of the indigenous formulations and/or decoctions after evaluation have been found to be really so. This has played a highly crucial role in the development of new drugs. Studies in this direction have identified specific inhibitors (sesquiterpene lactones such as parthenolide) of the transcription factor, NF-Kappa B (Heinrich, 2000). Similarly many more compounds are awaiting discovery.

Thus, it would not be out of place to say that impact of traditional knowledge in the quest for safe and new drugs has been enormous. According to Ghorbani et al. (2006 and references therein), many drugs like aspirin, ephedrine, atropine, digitoxin, reserpine and many more administered by the doctors today owe their origin and discovery to the leads from the medicinal folk lore. The underlying basis of this rests on the scientific validation, authentication and certification of the therapeutic claims of several medicines and/or toxins produced or used by the native healers. And this entails characterization of the phytochemical and pharmacological properties of these 'medicines' or 'drugs' or toxic substances and development of their bioactive profiles followed by their improvement and/or up gradation or refinement; then clinical and safety trials; and optimization of dosage. This, however, is possible only through multi- and inter-disciplinary approaches involving critical exploratory, ethno botanical, ethno-pharmacological, medical or pharmaceutical, anthropology based and other related investigations.

Pharmacological screening is not conducted haphazardly. It is instead conducted on plants chosen through either of the following selection strategies; random approach, phytochemical targeting, ethno directed sampling, chemotaxonomic approach and the one targeting specific plant parts (see Ghorbani et al., 2006 and references therein). However, combinations of these methods prove more fruitful in the search for novel bioactive compounds of medicinal importance.

Thus, it is of utmost concern to harness the benefits of the hard earned rich traditional knowledge of the ancestral lineage of traditional healers in general and Ladakhis in particular for the overall good of the human race before it is lost. Equally important is the national and international recognition and/or confidentiality of these tribals, safeguarding their centuries old practices, reducing their socio-economic burden and protecting their ownership rights under the aegis of Intellectual Property Rights.

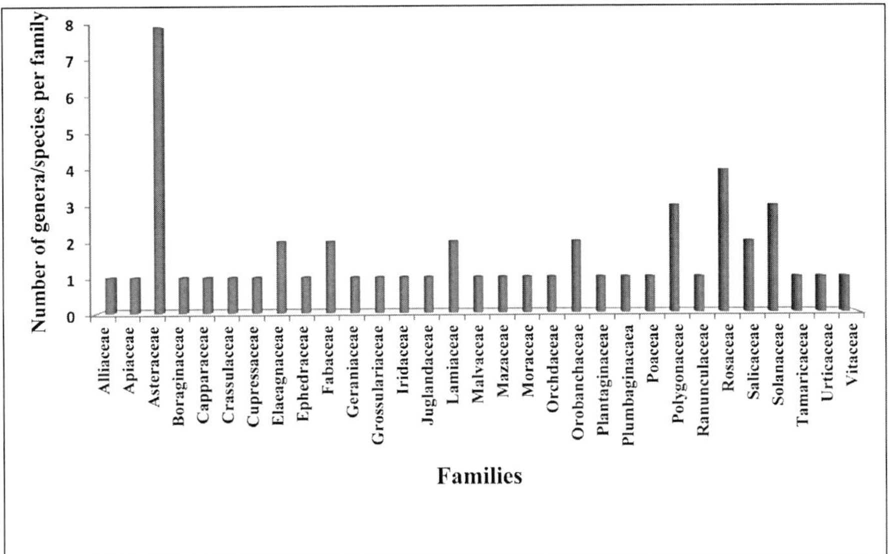

Fig. 1: Bar diagram depicting the representation of species of an individual family relative to that of others

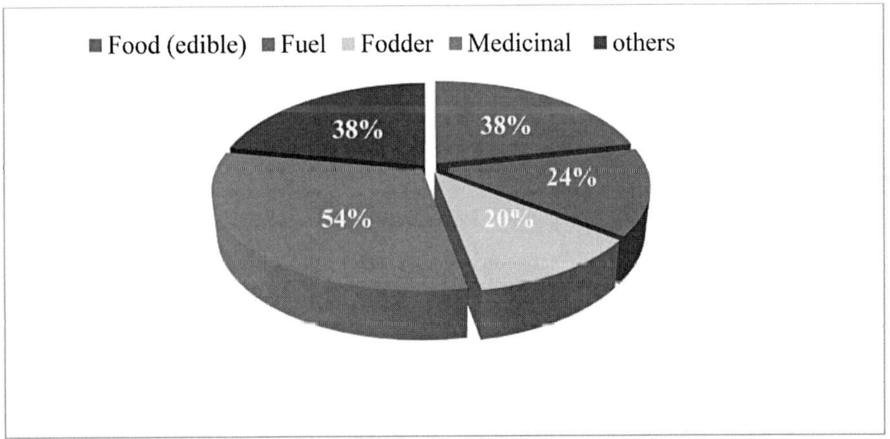

Fig. 2: Break up of plants (in percentage frequency) used for various purposes

6. Conclusions

The Sham region of Ladakh notwithstanding its harsh climatic conditions and short summers is a rich repository of various important plant species. Since decades the local people have been extremely dependent on these plants for food, fuel, fodder, as material for construction, dyeing and other purposes. Lack of proper medicinal system led Ladakhis develop a deep understanding of the native flora. This strong linkage helped them in becoming knowledgeable about the medicinal properties of the locally available plants and thence the origin of traditional therapeutic methods. Skills acquired by long term association with plants used to get transferred from generation to generation, was carefully guarded and subsequently got deep rooted within a family. However, many factors have contributed towards the decline in the number of Amchis which in turn has resulted in the loss of important information. The extensive utilization of plants without knowing their importance, urbanization, overexploitation and other anthropogenic activities are creating tremendous stress on the plant habitats these days. If such activities continue at the same rate, loss of biodiversity and traditional knowledge is inevitable. Therefore, the conservation and sustainable use of these species by taking various scientific initiatives is the need of the hour. In parallel, measures to impart education and renew interest among the young, ensure proper and optimal means of livelihood and grant recognition, should be devised and implemented for the continuity of this system of medicine, its preachers and practitioners.

Acknowledgements

The authors are thankful to the local informants for cooperating and sharing their valuable information. Department of Botany (UGC-SAP), University of Jammu is acknowledged for providing the necessary laboratory and library facilities. Dr. Harsh Chander Dutt, Assistant Professor, Department of Botany, University of Jammu and Dr. Konchok, Department of Botany, Government Degree College Leh are acknowledged for identification of some of the plants.

References

Abassi AA and Kaul V (2015). tSermang-a priceless gift to lackadaisical Ladakhis. *Current Science* 108 (2): 163-164.

Ali A (2013). Assessment of morphological and cytological variability in seabuckthorn (*Hippophae rhamnoides* L.) from some areas of Ladakh. Ph. D. Thesis University of Jammu, Jammu, J & K (Unpublished).

Angmo K, Bhupendra SA and Rawat GS (2012). Changing aspects of traditional health care system in western Ladakh. *Journal of Ethnopharmacology* 143(2): 621-630.

Balick MJ (1996). Transforming ethnobotany for the new millennium. *Annals of Missouri Botanical Garden* 83: 58-66.

Ballabh B and Chauarasia OP (2007). Traditional medicinal plants of cold desert Ladakh-used in treatment of cold, cough and fever. *Journal of Ethnopharmacology* 112(2) 341-349.

Ballabh B and Chaurasia OP (2009). Medicinal plants of cold desert Ladakh used in the treatment of stomach disorders. *Indian Journal of Traditional Knowledge* 8(2): 185–190.

Ballabh B and Chauarasia OP (2011). Herbal formulations from cold desert plants used for gynecological disorder. *Ethnobotany Research and Applications* 9: 059-066.

Ballabh B, Chauarasia OP, Ahmed Z and Singh SB (2008). Traditional medicinal plant of cold desert Ladakh- used against Kidney and urinary disorders. *Journal of Ethnopharmacology* 118(2): 331-339.

Ballabh B, Chaurasia OP, Pande PC and Ahmed Z (2007). Raw edible plants of cold desert Ladakh. *Indian Journal of Traditional Knowledge* 6(1) 182–184.

Bashir A, Singh C, Chauhan N and Rani A (2018). A review: ethnobotanical study on medicinal plants of Kargil district, Ladakh, India. *Journal of Emerging Technologies and Innovative Research* 5 (12): 181–196.

Bekalo TH, Woodmatas SD and Woldemariam ZA (2009) An ethnobotanical study of medicinal plants used by local people in the lowlands of Konta Special Woreda, southern nations, nationalities and peoples regional state, Ethiopia. *Journal of Ethnobiology and Ethnomedicine* 5:26; doi:10.1186/17466-4269-5-26

Bhattarai S, Chaudhary RP and Taylor RSL (2006) Ethnomedicinal plants used by the people of Manang district, central Nepal. *Journal of Ethnobiology and Ethnomedicine* 2:41; doi:10.1186/1746-4269-2-41

Chauarasia OP and Singh B (1996). Cold desert plants V. I-IV, FRL, DRDO, Leh India.

Chaurasia OP, Khatoon N and Singh SB (2008). Field guide Plant biodiversity of Ladakh. WWF-India, Department of Wildlife Protection, Government of Jammu & Kashmir and Defense Institute of High Altitude Research (DRDO), 196 pages.

Dorjey K, Tamchos S and Kumar S (2012). Ethnobotanical Observation in Trans-Himalayan Region Ladakh. *Journal of Plant Development Sciences* 4(4): 459–464.

Elvin-Lewis M (2001). Should we be concerned about herbal remedies? *Journal of Ethnopharmacology* 75(2-3): 141-164.

Ghorbani A, Naghibi F and Mosaddegh M (2006). Ethnobotany, ethnopharmacology and drug discovery. *Iranian Journal of Pharmaceutical Sciences* 2(2): 109-118.

Gohil RN and Qadri, M.A (1992). Ethnobotany of Kargil- Medicinal plants used by Balti, Dard, Brokpa races. *Journal of Economic and Taxonomic Botany Additional Series* 10: 301-306.

Gupta D and Kaul V (2016a) Seabuckthorn leaves- better substitute for green tea. *Current Science* 110 (4): 506-507.

Gupta D and Kaul V (2016b) Antioxidant activity vis-à-vis phenolic content in leaves of seabuckthorn from Kargil (J&K, India) –a preliminary study. *National Academy of Science Letters* 40 (1): 53-56.

Gurmet P (2004). Sowa-Rigpa; Himalayan art of healing. *Indian Journal of Traditional Knowledge* 3(2): 212–218.

Heinrich M (2000). Ethnobotany and its role in drug development. *Phytotherapy Research* 14(7): 479-88.

Kala CP (2005). Health traditions of Buddhist community and role of Amchis in Trans-Himalayan region of India. *Current Science* 80: 1331-1338.

Kaul MK 1997 Medicinal plants of Kashmir and Ladakh (Temperate and cold Himalaya) Indus publishing Company, New Delhi India.

Kaul V and Ali A (2013). Seabuckthorn- Greening Ladakh. *Science Reporter* 50(3): 44-45,CSIR Publication.

Klimes L and Dickore BA (2005). A contribution to the vascular plant flora of Lower Ladakh (Jammu & Kashmir, India). *Botanic Garden and Botanical Museum Berlin (BGBM), Willdenowia* 35(1): 125–153, https://doi.org/10.3372/wi.35.35110

Kufer J, Forther H, Poll E and Heinrich M (2005). Historical and modern medicinal plant uses- the example of the Ch'orti Maya and Ladinos in Eastern Guatemala. *The Journal of Pharmacy and Pharmacology* 57(9): 1127-52.

Kumar GP, Gupta S, Murugan P and Singh SB. (2009). Ethnobotanical Studies of Nubra Valley - A Cold Arid Zone of Himalaya. *Ethnobotanical Leaflets* 13: 752–65

Kumar GP, Kumar R and Chauarasia OP (2011). Conservation status of medicinal plants in Ladakh: Cold Arid zone of Trans-Himalayas. *Research Journal of Medicinal Plants* 5(6): 685-694.

Kumar S (2012). Traditional medicinal plants of Zanskar (Ladakh). *Annals of Pharmacy and Pharmaceutical Sciences* 3(2): 55–58.

Namgyal G and Phuntsog ST (1990). In Amchi Pharmaco-therapeutics Central Council for Research in Ayurveda and Siddha. Cambridge Printing Works, New Delhi.

Singh KN (2012) Traditional knowledge on ethnobotanical uses of plant biodiversity: a detailed study from the Indian western Himalaya. *Biodiversity Research and Conservation* 28: 63–77.

Soelberg J, Davis O and Jager AK (2016). Historical versus contemporary medicinal plant uses in the US Virgin Islands. *Journal of Ethnopharmacology* 192: 74-89.

Srivastava SK, Shabir M and Shukla, AN (2016). Some notable ethnomedicinal plants used by the tribes of Suru valley, Kargil, Ladakh, Jammu & Kashmir, India. *Ethnobotany* 28: 23-27.

Statistical Hand Book (2007-2008). District evaluation and statistical agency. Director of Economics and Statistics. District Leh and Kargil, LAHDC Leh and Kargil, Govt. of J&K, India.

Tamchos S (2018). Studies on morpho-taxonomic evaluation of Seabuckthorn (*Hippophae* L.) growing in Leh and Nubra valleys of Ladakh. Ph.D. Thesis University of Jammu, Jammu, J&K (Unpublished).

Tamchos S and Kaul V (2015). Seabuckthorn- the natural soil fertility enhancer. *Current Science* 108 (5): 763-764.

Tamchos S and Kaul V (2019). Seabuckthorn: opportunities and challenges in Ladakh. *National Academy of Science Letters* 42(2):175-178.

Uprety Y, Asselin H, Boon EK, Yadav S and Shrestha KK (2010). Indigenouu use and bio-efficacy of medicinal plants in Rasuwa District, Central Nepal. *Journal of Ethnobiology and Ethnomedicine* 6:3.

13

Ethnoveterinary Knowledge and Herbal Practices Prevalent Among Tribal Communities of District Poonch in Himalaya

Abhishek Dutta, Yash Pal Sharma and Bikarma Singh

Abstract

India harbors a wide variety of medicinal plants used traditionally in various remedies to treat livestock ailments. The therapeutic value of these plants attributes due to the concentration of several bioactive compounds present in them. Indigenous tribal communities are the treasurer of age-old traditional knowledge which inhabits Asian country since centuries. District Poonch exhibits a rich repository of traditional knowledge as major part of the population comprised of nomadic tribes such as Gujjars and Bakerwals which are still practicing the primitive traditional ways to treat diseases. In the present investigation, the traditional herbal plants and associated remedies for treatment of various veterinary disorders are presented. A total of 34 plant species belonging to 30 genera and 21 families have been reported which are being used by indigenous people for ethnoveterinary purposes. The majority of ethnoveterinary plants recorded from the district belong to families such as

Abhishek Dutta and Yash Pal Sharma (✉)

Department of Botany, University of Jammu, Jammu-180006, Jammu and Kashmir, India

Bikarma Singh (✉)

Botanical Garden, CSIR-National Botanical Research Institute, Lucknow-226001 Uttar Pradesh, India

✉*Corresponding author(s) email: drbikarma.singh@nbri.res.in; yashdbm3@yahoo.co.in*

Plants for Novel Drug Molecules: Ethnobotany to Ethnopharmacology
Bikarma Singh & Yash Pal Sharma (eds.), (pp. 331-343)

Email: *info@nipabooks.com* Web: *www.nipabooks.com*

Ranunculaceae and Araceae (4 species each), Acoraceae, Amaryllidaceae, Apiaceae, Fagaceae, Lamiaceae, Primulaceae, Rutaceae (2 species each) and rest of the families members were represented by one species each. The frequently used plant parts are roots (30.23%), followed by tubers (18.6%), leaves (16.28%), fruits and whole plant (13.95% each), stems (4.65%) and seeds (2.33%). It has been observed that numerous formulations are made from these parts which are used for the treatment of various cattle infections such as abdominal colic, allergic infection, anestrous, chest diseases, constipation, cough, digestive troubles, FMD, fracture, galactagogue, gaseous bloat, induce puberty, internal injury, lung disease, pneumonia, quick discharge, removal of Leech, sexual stimulant, skin infection, snake-bite, stomach-ache and weakness with snake bite being dominant. The present study infers that the documentation of ethnoveterinary knowledge from the district requires utmost attention and may give leads or serves for further biochemical studies for the discovery of novel plant based drugs which will certainly helps in commercialization and betterment of tribals through benefit resource sharing.

Keywords: Ethnoveterinary, Herbal practices, Poonch, Gujjar, Bakerwal, J&K.

1. Introduction

The science of ethnoveterinary knowledge has been practiced since antiquity through conventional use of plants for animal husbandry and animal healthcare facilities. Herbal remedies still play a key role in indigenous cattle care practices due to their excellent therapeutic efficacy and nominal side effects (Gurib-Fakim 2006). The traditional knowledge regarding healing properties of plants is transmitted orally through generations. The application of this knowledge has been practiced since centuries to keep cattle healthy, free from infectious agents and diseases. The documentation of ethnoveterinary knowledge gained attention in early 1980s after which several explorations and studies were conducted and presented through various conferences and workshops (Toyang et al. 1995, Phondani et al. 2010). Some of the ethnoveterinary knowledge has been saved due to these activities but it still faces threat of extinction due to death of elder members of the tribe or community as they are treasure house of it.

The use of plants for the treatment and cure of various diseases in both humans and livestock is well known since the entire course of history of mankind. There has been greater acknowledgement of ethnoveterinary knowledge in recent years, particularly in developing countries, as it is time tested with minimal side effects (Masika and Afolayan 2003, Tabuti et al. 2003). There are many reports of ethnoveterinary studies being carried out in several parts of India by different workers (Ganesan et al. 2008, Geetha et al. 2006, Kiruba et al. 2006, Phondani et al. 2010). But several factors such as rapid modernization, lack of interest of younger generation towards traditional practices, oral transmission and lack of proper documentation have resulted in fading away of this knowledge and is presently at the verge of being lost forever. Therefore greater attention is required for the documentation of traditional knowledge particular in the field of animal healthcare practices along with their pharmacological authentication which can add a new dimension in unearthing novel drugs.

The tribals of Jammu and Kashmir depend heavily on traditional phytotherapies for the treatment of infections and diseases of their livestock (Sharma et al. 2012). Their dependency attributes to their inaccessibility to allopathic healthcare facilities. There are few reports of ethnoveterinary knowledge from Jammu and Kashmir (Sharma and Singh 1989, Beigh et al. 2003, Khuroo et al. 2007, Rashid et al. 2007, Sharma et al. 2012). In district Poonch, there are only four reports of ethnoveterinary documentation viz., Khan and Kumar (2012), Shah et al. (2015), Khan and Paul (2017) and Manzoor and Ali (2017). The traditional knowledge regarding healing properties of plants is transmitted orally through generations. The application of this knowledge has been practiced since centuries to keep cattle healthy, free from infectious agents and diseases. So an attempt was made in the present study to compile all the reports from the district pertaining to ethnoveterinary knowledge.

2. Material and Methods

2.1. Study Area

Poonch, the border district of J&K situated with an altitudinal range of 800 to 4,750 m above mean sea level (AMSL). Geographically, the district lies between the longitudes of 73°58' to 74°35'E and the latitude of 33°25' to 34°01'N, and situated on the southerly foothills of Pir Panjal mountain range of Western Himalaya. The total geographical area of the district is 1,674 km^2.

2.2. Climate

The district is endowed with sub-tropical to temperate climate regime with an average temperature around 30°C and 10 to -8°C during summer and winter months, respectively. An annual precipitation ranges from 150 to 200 cm, the bulk of which is received during monsoon period (July to September).

2.3. People

Administratively the area is divided into 6 tehsils and 11 blocks comprising of 178 villages and 51 panchayats with a total population of 4,75,835 (2011 Census). Majority of the population relies on agriculture and animal husbandry for basic needs. The community of *Gujjars* and *Bakerwals* constitute the major proportion of population, where in *Gujjars* are semi-nomadic and *Bakerwals* are true nomadic in their habit. These tribes are totally dependent on plants for their daily needs as they rear cattle at high altitudes, which are inaccessible to modern facilities (Shah et al. 2015).

2.4. Forest Types and Vegetation

The total geographical area is 1143.870 sq km out of which 951 km^2 comes under forest cover in the district i.e., 56.81%. The vegetation usually comprises of coniferous forests (*Abies pindrow, A. spectabilis, Cedrus deodara, Pinus roxburghii, P. wallichiana, Picea smithiana* and *Taxus wallichiana*), broad-leaved evergreen forests (*Buxus wallichiana, Ilex dipyrena, Pyrus cerasoides and Quercus semi-carpifolia, Q. incana*), deciduous forests (*Acer caesium, Aesculus indica, Populus alba* and *Platanus orientalis*), scrub forests, interspersed with frequent grassland patches and agricultural croplands.

2.5. Methodology

The present study is a collective representation of published literature and our own field collections regarding ethnoveterinary studies conducted in district Poonch, J&K, retrieved from several online databases *viz.*, Google Scholar, PubMed, Sci Finder, Scopus, Science direct, etc. Keywords like ethnobotany, ethnoveterinary plants of Poonch were searched in the databases. A total of 04 research articles published in different journals during 2012 to 2017 (Khan and Kumar 2012, Shah et al. 2015, Khan and Paul 2017, Manzoor and Ali 2017) were reviewed. These articles include primary information on plants used for ethnoveterinary purposes by indigenous people of the district. Each plant species is provided with latest accepted name following databases like The Plant list (www.theplantlist.org) and Tropicos (www.tropicos.org).

3. Research Findings and Discussion

3.1. Ethnomedicinal Plant Diversity and Floristic Analysis

In total, 34 plant species belonging to 30 genera and 21 families have been reported from district Poonch being used by indigenous people for ethnoveterinary purposes. The majority of ethnoveterinary plants recorded from the district belong to Ranunculaceae and Araceae (4 species each), Acoraceae, Amaryllidaceae, Apiaceae, Fagaceae, Lamiaceae, Primulaceae, Rutaceae (2 species each) and rest of the families were represented by one species each. Out of the various growth forms, herbs were the most dominant (67.65%), followed by shrubs (14.71%), trees (11.76%) and climbers (5.88%) (Fig.1). The reason for herbs being the most dominant life form attributes to their accessibility, efficacy in treatment of diseases and presence of bioactive compounds as compared to other life forms (Thomas et al. 2009, Giday et al. 2010, Simbo 2010, Singh et al. 2012).

3.2. Parts Used

The frequently used plant part were roots (30.23%), followed by tuber (18.6%), leaf (16.28%), fruit and whole plant (13.95% each), stem (4.65%) and seeds (2.33%)(Fig 2). Leaves are the most exploited and preferred part in medicinal remedies due to their ease and speed of harvest (Gazzaneo et al. 2005). Also, they contain maximum proportion of majority of alkaloids, glycosides and other essential oils (Ould et al. 2003, Bhattarai et al. 2006).

3.3. Disease and Herbal Cure

A total of 22 cattle ailments *viz.*, abdominal colic, allergic infection, anestrous, chest diseases, constipation, cough, digestive troubles, FMD, fracture, galactagouge, gaseous bloat, induce puberty, internal injury, lung disease, pneumonia, quick discharge, removal of Leech, sexual stimulant, skin infection, snake-bite, stomach-ache and weakness were reported to be cured by different herbal remedies from the district. Maximum numbers of use reports were observed for the treatment of snake bite (20), followed by weakness (3), allergic infection, digestive trouble, FMD and lung disease (2 use reports each) and others with single use report.. Maximum cited plant species was *Arisaema propinquum* (3 citations), followed by *Acorus calamus, Arisaema flavum, Arisaema jacquemontii, Thymus linearis* and *Vitex negundo* (2 citations each), while rest of the species were cited only once.

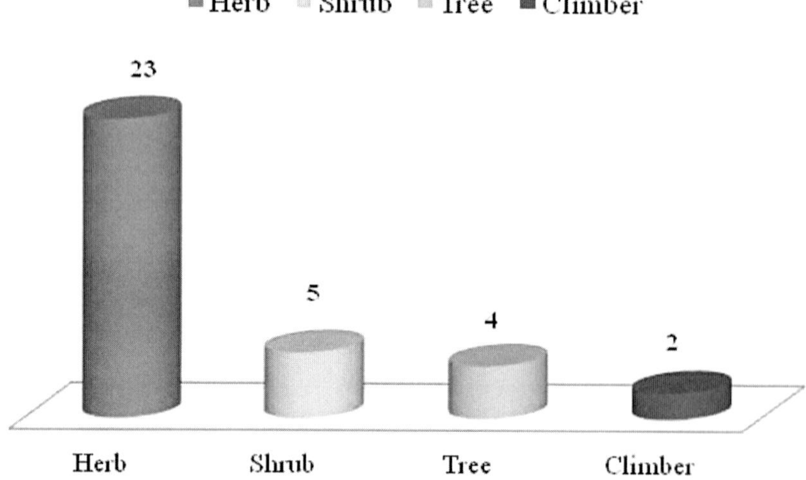

Fig.1: Analysis of habit with respect to number of species.

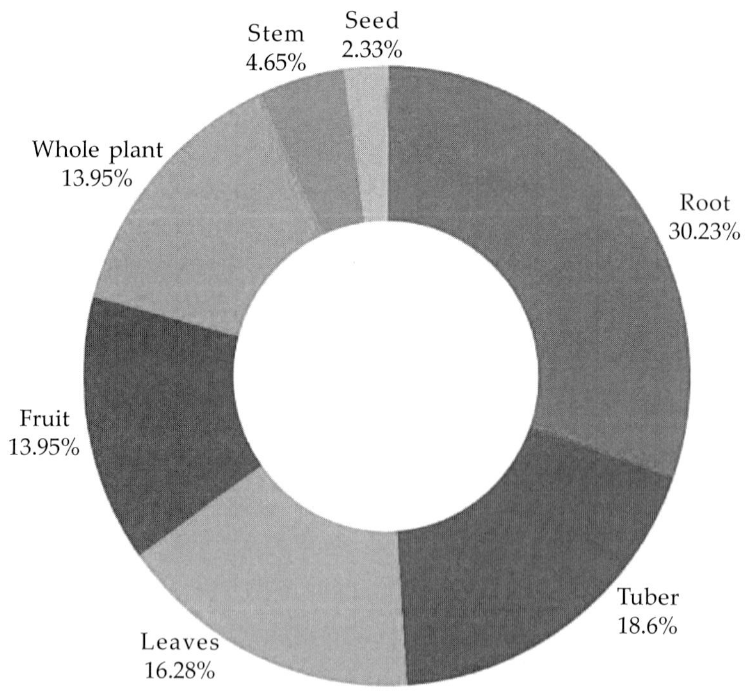

Fig.1: Plant parts used in ethnoveterinary medicine

Table 1: Plant species along with their Botanical name, family, local name, habit, parts used, ethnoveterinary use and references.

Botanical name/ Family	Local name	Habit	Part used	Ethnoveterinary use	Reference
Achillea millefolium L./ Asteraceae	Chau, Pehlkach	Herb	Root	About 50-60 gram of root crushed with stones and given orally to cows and buffaloes to treat snake bite	(Khan and Kumar 2012)
Aconitum heterophyllum Wall. ex Royle/ Ranunculaceae	Patrees	Herb	Root	About 50- 80 gram powdered or crushed root given to buffaloes, ox and horse against snake bite.	(Khan and Kumar 2012)
Aconitum violaceum Jacquem. ex Stapf./ Ranunculaceae	Mori	Herb	Root	Paste of root applied externally on cattle to treat snake bite. About 4-5 gram given internally by wrapping in wet wheat flour or butter but taking into consideration the physical activity of the victim after bite.	(Khan and Kumar 2012)
Acorus calamus L./ Acoraceae	Pyonzkartal, Bach	Herb	Rhizome	About 100 to 150 gram of crushed rhizome given to horse orally against snake-bite.	(Khan and Kumar 2012)
	Bach, Pyozkartal		Rhizome	Rhizome grinded and given orally to the animal by wrapping in wheat dahlia on allergy.	(Khan and Paul 2017)
Aesculus indica (Wall. ex Cambess.) Hook./ Sapindaceae	Bankhori	Tree	Leaves	Leaves given to cure chest diseases of horses, donkey, also act as colic.	(Manzoor and Ali 2017)
Allium cepa L./ Amaryllidaceae	Payaz	Herb	Bulb	Paste applied externally and 100-200 grams given internally followed by 10-15 grams Gol Mirch to treat snake bite in cattle.	(Khan and Kumar 2012)

Contd.

Allium sativum L./ Amaryllidaceae	Thoom	Herb	Bulb	Paste of 50-60 gram bulb given orally along with butter to treat snake bite in cattle.	(Khan and Kumar 2012)
Angelica glauca Edgew./ Apiaceae	Choura	Herb	Root	Root grinded given orally on pneumonia on abdominal colic and foot and mouth disease.	(Khan and Paul 2017)
Arisaema flavum (Forssk.) Schott/ Araceae	Hathbis	Herb	Seed, rhizome	Mixture of seed and rhizome given to cattle for increasing milk.	(Manzoor and Ali 2017)
	Sap ni mak		Tuber	About 100 to 150 gram crushed tubers given orally by wrapping in wet wheat flour or butter. 30 to 35 gram dried and powdered tuber also given by wrapping in wheat flour to treat snakebite in cattle.	(Khan and Kumar 2012)
Arisaema jacquemontii Blume/ Araceae	Sap ni mak	Herb	Tuber	Freshly collected tuber crushed with stones and about 100-150 gram given orally by wrapping in wet wheat flour. 30 to 35 gram powdered tuber also given orally by wrapping in butter to cure snake bite in cattle.	(Khan and Kumar 2012)
	Sap ki mak		Fruit	Berries wrapped in wheat dahlia and given orally on snake bite in cattle.	(Khan and Paul 2017)
Arisaema propinquum Schott/ Araceae	Surumgundo	Herb	Tuber	1 kg tuber crushed and mixed with 2 kg of milk or lassi; 2 glasses of preparation given to the cattle once in a day for 3 or 4 days alternatively. It acts as best sexual stimulant for buffaloes.	(Shah et al. 2015)
	Sap ni mak		Tuber	About 100 to 150 gram seeds given orally by crushing with stones and warping in wet wheat flour. Paste applied externally to cure snakebite in cattle.	(Khan and Kumar 2012)

Contd.

	Sarf makyoth		Fruit	Berries grinded in water and given orally by wrapping in wheat dahlia on allergic infection to cattle.	(Khan and Paul 2017)
Barleria cristata L./ Acanthaceae	Sap ni jari	Shrub	Whole plant	About 200 to 300 hundred gram whole plant made into past and given orally to treat snake bite in cattle	(Khan and Kumar 2012)
Bupleurum falcatum L./ Apiaceae	Peeley phul wali jari, nagdun	Herb	Whole plant	About 100 to 200 gram whole plant grinded with stones and given orally to cow's buffaloes and horse.	(Khan and Kumar 2012)
Capsicum annuum L./ Solanaceae	Marchi	Herb	Fruit	Fruits burnt in mustard oil and applied externally to treat pneumonia.	(Khan and Paul 2017)
Delphinium denudatum Wall. ex Hook.f. & Thomson/ Ranunculaceae	Sap ni jari	Herb	Root	10 to 20 gram powdered root given orally. Paste of root is also applied externally to cure snake bite in cattle	(Khan and Kumar 2012)
Dioscorea bulbifera L./ Dioscoreaceae	Chachla ganda, Kala ganda, Kithi ganda	Herb	Tuber	About 200 to 300 gram tuber given orally by mixing in 10 to 15 gram Gol Mirch (*Pipper nigrum*) to treat snakebite in cattle.	(Khan and Kumar 2012)
Gloriosa superba L./ Colchicaceae	Sap ki jari	Herb	Root	About 20 gram root grounded and given orally to cure snakebite in cattle.	(Khan and Kumar 2012)
Grewia optiva J.R. Drumm. ex Burret/ Malvaceae	Thaman	Tree	Leaves	Leaves given to young animals to induce puberty and to cattle for the quick discharge off after birth.	(Manzoor and Ali 2017)
Hedera nepalensis K.Koch/ Araliaceae	Batulo	Climber	Fruit	5-10 g fine powder of berries put into the nostril chambers. It helps in the removal of Leech from the nostril chambers and stops nose bleeding in cattle.	(Shah et al. 2015)
Primula denticulata Sm./ Primulaceae	Landanposh	Herb	Whole plant	150 to 200 gram plant crushed with stones and given orally to cattle to treat snake bite.	(Khan and Kumar 2012)

Contd.

Primula macrophylla D. Don/ Primulaceae	Ladanposh	Herb	Whole plant	About 200 gram whole plant grinded and given orally to cure snake bite in cattle.	(Khan and Kumar 2012)
Pueraria tuberosa (Willd.) DC./ Leguminosae	Bidh	Climber	Tuber	About 3-5 kg of crushed tuber mixed with 1 kg milk given once in a month in case of weak and anestrous animals. It makes the animals sexually receptive.	(Shah et al. 2015)
Quercus oblongata D.Don; syn. *Quercus leucotrichophora* A.Camus/ Fagaceae	Ree	Tree	Stem Bark	For cattle, a cup of bark decoction administered for twice in a day for 2 weeks to cure internal injury.	(Shah et al. 2015)
Quercus semecarpifolia Sm./ Fagaceae	Kharsu	Tree	Leaves	Young leaves fed to the cattle in case of gaseous bloat. Cattle are not allowed to drink water during feedings.	(Shah et al. 2015)
Ranunculus laetus Wall. ex Hook. f. & J.W. Thomson/ Ranunculaceae	Khand barian	Herb	Root	1/2 kg fresh root cooked for about 2-3 h with milk/ lassi. The suspension so obtained administered orally for 2-3 times in a week to cure weakness in animals.	(Shah et al. 2015)
Rheum australe D.Don/ Polygonaceae	Pamb-e-chari	Herb	Rhizome	Paste of fresh rhizome applied on fractured part of the body by inhabitants, who used to go in upper reaches along with their cattle	(Shah et al. 2015)
Sauromatum venosum (Dryand. ex Aiton) Kunth; syn. *Sauromatum pedatum* (Link & Otto) Schott/ Araceae	Surganda	Herb	Fruits	Fruits given orally by wrapping in wheat dahlia on foot and mouth diseases (FMD).	(Khan and Paul 2017)
Sinopodophyllum hexandrum (Royle) T.S.Ying/ Berberidaceae	Bankhakri wanwangun	Herb	Rhizome	Past of rhizome applied externally and about 30 to 40 gram given orally to treat snake bite in animals.	(Khan and Kumar 2012)

Contd.

Skimmia anquetilia Tayl. & Airy Shaw/ Rutaceae	Nera, patla	Shrub	Leaves	Leaves grinded and paste given orally on cough and other lungs diseases.	(Khan and Paul 2017)
Sorghum halepense (L.) Pers/ Poaceae	Baru	Herb	Root	About 100 to 150 gram root grinded and given orally to treat snakebite in cattle.	(Khan and Kumar 2012)
Thymus linearis Benth./ Lamiaceae	Chickni	Herb	Whole plant	100 to 150 gram given orally to treat snake bite	(Khan and Kumar 2012)
	Chicken, chikney		Whole plant	Paste of whole plant given orally on lungs disease.	(Khan and Paul 2017)
Viburnum grandiflorum Wall. ex DC./ Adoxaceae	Kuchh	Shrub	Leaves, Twig	Leaves given to cattle for constipation and stomach-ache; twig extract used to cure skin infection.	(Manzoor and Ali 2017)
Vitex negundo L./ Lamiaceae	Bana	Shrub	Leaves	Leaves paste of 100 to 200 gram given orally to treat snake bite in cattle.	(Khan and Kumar 2012)
	Bana		Leaves	Paste of leaves given orally on indigestion.	(Khan and Paul 2017)
Zanthoxylum armatum DC./ Rutaceae	Timer	Shrub	Fruits	Fruits grinded and given orally on digestive troubles.	(Khan and Paul 2017)

4. Conclusion and Future Perspectives

Domestic livestock plays an imperative role in the life of tribals like Gujjar and Bakerwal. They are still dependent on local herbal flora for the treatment of various ailments of livestock due to inaccessibility to modern veterinary healthcare facilities. The present study reports 34 species of plants being used by the locals of district Poonch for ethnoveterinary purposes. The study reveals that majority of the herbal veterinary remedies are used to treat snake bite in cattle. Majority of the compilations lack information regarding dosage and side effects of the remedies. Similarly, there is lack of proper validation of the knowledge through quantitative and statistical analysis of the data. Moreover, the available information that has been documented till date is not enough and there is an urgent need of holistic documentation of ethnoveterinary plant knowledge as it gets lost with course of time and death of elder members of community.

Conflict of Interest

The authors declare that they have no competing conflict of interest.

Acknowledgements

The authors wish to thank the Head, Department of Botany (UGC-SAP-DRS II), University of Jammu, Jammu and CSIR-IIIM, Jammu for providing necessary laboratory facilities. The first author sincerely acknowledges the financial support received as JRF from CSIR, New Delhi, India

References

Beigh SY, Nawchoo I and Iqbal M (2003). Traditional veterinary medicine among the tribes of Kashmir Himalaya. Journal *of Herbs, Spices and Medicinal Plants* 10: 121–127.

Bhattarai S, Chaudhary RP and Taylor RSL (2006). Ethnomedicinal plants used by the people of Manang district, central Nepal. *Journal of Ethnobiology and Ethnomedicne* 2: 41.

Ganesan S, Chandhirasekaran M and Selvaraju A (2008). Ethno-veterinary health care practices in Southern districts of Tamil Nadu. *Indian Journal of Traditional Knowledge* 7: 347-354.

Gazzaneo LRS, Lucena RFP and Albuquerque UP (2005). Knowledge and use of medicinal plants by local specialists in a region of Atlantic forest in the state of Pernambuco (North eastern Brazil). *Journal of Ethnobiology and Ethnomedicine* 1: 1-9.

Geetha S, LakshmiG and Ranjithakani P (2006). Ethnoveterinary medicinal plants of Kollihills, Tamil Nadu. *Journal of Economic and Taxonomic Botany* 12: 284-291.

Giday M, Asfaw Z and Woldu Z (2010). Ethnomedicinal study of plants used by Sheko ethnic group of Ethiopia. *Journal of Ethnopharmacology* 132: 75–85.

Gurib-Fakim A (2006). Medicinal plants: traditions of yesterday and drugs of tomorrow. *Molecular Aspects of Medicine* 27: 1–93.

Khan JA and Kumar S (2012). Ethnoveterinary values of some plants used against snake bite in Poonch district of Jammu and Kashmir (India). *Journal of Plant Development Sciences* 4: 111-114.

Khan JA and Paul R (2017). Ethnoveterinary medicinal uses of some medicinal plants used by the Gujjar and Pahari tribes of Poonch district of Jammu and Kashmir. *Internattional journal of advance research in science and engineering* 6: 377-381.

Khuroo AA, Malik AH, Dar, AR, Dar GH and Khan ZA (2007). Ethnoveterinary medicinal uses of some plant species by the Gujjar Tribe of the Kashmir Himalaya. *Asian Journal of Plant Science* 6: 148–152.

Kiruba S, Jeeva S and Dhas SSM (2006). Enumeration of ethnoveterinary plants of Cope Comorin, Tamil Nadu. *Indian Journal of Traditional Knowledge* 7: 576-578.

Manzoor J and Ali B (2017). Traditional use of medicinal plants: A report from Pahari community of subdivision Mendhar, District Poonch, Jammu & Kashmir, India. *Medicinal Plants - International Journal of Phytomedicines and Related Industries* 9: 2016-220.

Masika PJ and Afolayan AJ (2003). An ethnobotanical study of plants used for the treatment of livestock diseases in the Eastern Cape Province, South Africa. *Pharmaceutical Biology* 41: 16-21.

Ould EHM, Hadj-Mahammed M and Zabeirou H (2003). Place of spontaneous plants in the traditional medicine of the region of Ouargla (Northern Sahara). *Courrier du Savoir* 03: 47–51.

Phondani PC, Maikhuri RK and Kala CP (2010). Ethnoveterinary Uses of Medicinal Plants Among Traditional Herbal Healers in Alaknanda Catchment of Uttarakhand, India. *African Journal of Traditional, Complementary, and Alternative Medicines* 7:195–206.

Rashid A, Anand VK and Shah AH (2007). Plant resources utilization in the ethnoveterinary practices by the Gujjar and Bakarwal tribes of Jammu and Kashmir State, India. *Journal of Phytological Research* 20: 293–298.

Shah A, Bharati KA, Ahmad J and Sharma MP (2015). New ethnomedicinal claims from Gujjars and Bakerwals tribes of Rajouri and Poonch districts of Jammu and Kashmir, India. *Journal of Ethnopharmacology* 166: 119-128.

Sharma PK and Singh V (1989). Ethno-botanical studies in northwest and trans-Himalaya-V. Ethno-veterinary medicinal plants used in Jammu and Kashmir, India. *Journal of Ethnopharmacology* 27: 63–70.

Sharma R, Manhas RK and Magotra R (2012). Ethnoveterinary remedies of diseases among milk yielding animals in Kathua, Jammu and Kashmir, India. *Journal of Ethnopharmacology* 141: 265– 272.

Simbo DJ (2010). An ethnobotanical survey of medicinal plants in Babungo, northwest region, Cameroon. *Journal of Ethnobiology and Ethnomedicine* 6: 8.

Singh AG, Kumar A and Tewari DD (2012). An ethnobotanical survey of medicinal plants used in Terai forest of western Nepal. *Journal of Ethnobiology and Ethnomedicine* 8: 19.

Tabuti JRS, Dhillion SS and Lye KA (2003). Ethnoveterinary medicines for cattle (Bos indicus) in Bulamogi county, Uganda: plant species and mode of use. *Journal of Ethnopharmacology* 88: 279- 286.

Thomas E, Vandebroek I, Sanca S and Van Damme P (2009). Cultural significance of medicinal plant families and species among the Quechua farmers in apillampampa, Bolivia. *Journal of Ethnopharmacology* 122: 60-67.

Toyang NJ, Nuwanyakpa M, Ndi C, Django S and Kinyu WC (1995). Ethnoveterinary medicine practices in northwest province of Cameroon. *Indigenous Knowledge and Development Monitor. Nuffic – CIRAN, The Hague, Netherland* 1: 3-19.

14

Eco-Taxonomy, Ethnobotany and Active Chemical Constituents of Ten High Value Plants of Kathua District (J&K) of India in Southeast Asia With Special Reference to Jasrota Wildlife Sanctuary

Anand Kishor and Bishander Singh

Abstract

For thousands of years, plants have provided medicine to human and animals. People still depend on plants for food and shelter. The present surveys and investigation covers Jasrota Wildlife Sanctuary, Kathua District (J&K) situated in Southeastern parts of Asia (India). This communication mainly focused on ten high value plant species belonging to eight families of medicinal value being used to cure different types of ailment growing in Jasrota Wildlife Sanctuary and surrounding areas of Kathua district. These species are Aerva sanguinolenta (L.) Blume, Centella asiatica L. (Urb.) Fl., Cissampelos pareira L., Colebrookea oppositifolia Sm., Cryptolepis dubia M.R.Almeida, Holarrhena pubescens Wall. ex G.Don, Justicia adhatoda L., Tinospora cordifolia (Willd.) Miers ex Hook.f. & Thomson, Vitex negundo L. and Woodfordia fruticosa (L.) Kurz. It also deal with morphological enumeration, ecology and habitat, ethnobotanical uses and active chemical constituent present in studied plant species.

Keywords: Medicinal plants, Jasrota Wildlife Sanctuary, Kathua, Active chemicals, Conservation

Anand Kishor (✉) and Bishander Singh (✉)

Department of Botany, Veer Kunwar Singh University, Ara-802301, Bihar

✉*Corresponding author(s) email: kishoranand04@gmail.com; bishander85@gmail.com*

Plants for Novel Drug Molecules: Ethnobotany to Ethnopharmacology
Bikarma Singh & Yash Pal Sharma (eds.), (pp. 345-361)

Email: *info@nipabooks.com* Web: *www.nipabooks.com*

1. Introduction

Regardless of the remarkable advancement in modern medicinal science, approximately 70-80% human population still depend on traditional medicine for their primary health care needs (Mdluli 2002, Bhat et al. 2019). Currently a substantial number of drugs are developed from plants which are active against many diseases. In the developed countries, 25% of the medical drugs are based on the plant compounds and their derivatives (Pan et al. 2013) and in rural areas of many developing countries the use of medicinal plants are well known among the indigenous people (Bhat et al. 2019). The development of many important classes of drugs depend on the plants (Swain 1972). Ethnomedicine practice is an immemorial mode of curing various types of ailments (Green and Makhubu 1983). Traditional medical practices still serve as the main medium for the treatment of various types of disease especially in the mountainous and rural areas where modern medical facilities are not available. For human as well as livestock, plants are the main sources for food and healthcare, and important components of global biodiversity (Lambert et al. 2005). Mostly tribal peoples used stems and leaves as medicine. The most common way of using the medicinal plants are decoction in water.

Knowledge of ethnomedicine is the first remedies for the curing of aliment such as fever, toothache, asthma, skin disorders, bodyaches, tonic for memory improving, dropsy, leprosy, kidney problems, wound, bruises, boils, cough, dysentery, headaches, swelling, scabies and so many other types of disease. Modern medicine also developed from the ethnomdicinal knowledge. Chemical constituents of plant provide the scientific basis of the use for a disease in traditional medical practice and as a guide to bioprospecting for drugs. The plants collected from field survey extracted with suitable solvents and analyzed for bioactive molecules. Different types of bioactive molecules found in the plants species are alkaloids, essential oil, anthranoids, flavonoids, sesquiterpenes, glycosides, Methyl ester, saponins, steroids and tannins. An important source of compounds, which are useful as medicinal agents are antiviral, cardiotonic, antihelminitic, sedative, laxative, diuretic, anticancer, and so many others biological & function. They are important sources of bio-molecules, used for development of new drugs and cosmetic product (Heinrich and Gibbons 2001). The presence of these bioactive molecules in the plants helps in understanding the use of the plants in ethnomedicine practice and may serve as a lead in discovering of new pharmaceutical products of herbal origin.

This communication mainly focused on ten high value plant species belonging to eight families of medicinal value being used to cure different types of ailment growing in Jasrota Wildlife Sanctuary and surrounding areas of Kathua district. This species are *Aerva sanguinolenta* (L.) Blume, *Centella asiatica* L. (Urb.) Fl., *Cissampelos pareira* L., *Colebrookea oppositifolia* Sm., *Cryptolepis dubia* M.R.Almeida, *Holarrhena pubescens* Wall. ex G.Don, *Justicia adhatoda* L., *Tinospora cordifolia* (Willd.) Miers ex Hook.f. & Thomson, *Vitex negundo* L. and *Woodfordia fruticosa* (L.) Kurz. It also deals with morphological enumeration, ecology and habitat, ethnobotanical uses and active chemical constituents present in them.

2. Material and Methods

The frequent field surveys were carried out to get information about the nature of terrain, species composition, accessibility and distribution of various plant species of medicinal importance growing in the study area. During field studies the detailed relevant data on the plant specimen was recorded. Specimens were collected from field and were dried, preserved, labeled and mounted on herbarium sheets following standard herbarium techniques (Jain and Rao 1977). Specimen was identified using relevant taxonomic literatures (Stewart 1972; efloras). The specialized taxonomic database of The Plant List (www.theplantlist.org) was used for the updated nomenclature of taxa. To document the ethnobotanical information of the plant resources of the study area questionnaires completed and interviews method used for collecting indigenous knowledge about plant species. The study area was visited on regularly basis and wild plants were collected. Ethnobotanical knowledge regarding plants were collected from diverse ethnic groups called Duggurs and Paharis of the area categorized as herbal practitioners, hakims, shopkeepers and farmers residing in the study area by interviewing and filing a questionnaire. Informants were asked about their common uses of plant species, e.g. as medicine, food, fodder, timber, fuel wood, vegetable/edible fruit, resin yield and herbal tea plants. Field based personal findings also added more information to the research work. The respondent were further asked about the part of the plant preferred if they utilized a species for usage – stem, bark, leaves, flowers, fruits or whole plant. All data were analyzed and kept in record for studying taxonomy and active chemical constituents.

Table 1: High Value selected medicinal plants growing in Kathua district (Jasrota Wildlife Sanctuary), J&K (India) in Southeast Asia

Botanical name	Vernacular name	Family	Habit	Phenology	Global distribution	Thretened Status	
						Regional status	IUCN Red List Status
Aerva sanguinolenta (L.) Blume	Bui	Amaranthaceae	Herb	Fl.: November-April Fr.: December	Australia, Pakistan, India and Africa.	Common	Catalogue 2019
Centella asiatica L. (Urb.) Fl.	Gotu kola	Apiaceae	Herb	Fl.: April-June Fr.: June-October	India, China, Nepal, Bangladesh, Bhutan, Nepal, Pakistan, Sri Lanka and Indonesia.	Scarce	Catalogue 2019
Cissampelos pareira L.	Laghupta	Menispermaceae	Liana	Fl.: June - September Fr.: September-October	India, Pakistan, Myanmar and China.	Common	Catalogue 2019
Colebrookea oppositifolia Sm.	Pansre	Lamiaceae	Shrub	Fl.: January-May Fr.: May-August	Nepal, Assam, India, Bangladesh, Pakistan, Myanmar, Bhutan and Thailand.	Common	Catalogue 2019
Cryptolepis dubia M.R.Almeida	Karilata	Asclepiadaceae	Liana	Fl.: May to July Fr.: September to November.	Pakistan, India, Nepal, Bhutan, Myanmar, China and Sri Lanka.	Scarce	Catalogue 2019
Holarrhena pubescens	Kurchi	Apocynanceae	Tree	Fl.: May-June	Bangladesh,	Scarce	Catalogue 2019
Wall. ex G.Don				Fr.: September-October	Cambodia, China, Malaysia, Myanmar, Nepal, South Africa, Thailand and Zambia.		

Contd.

Justicia adhatoda L.	Vasa	Acanthaceae	Shrub	Fl.: February-August Fr.: April-December	India, Sri Lanka, Malaysia, Myanmar, Bangladesh, Nepal, Thailand and Pakistan.	Common	Catalogue 2019
Tinospora cordifolia (Willd.) Miers ex Hook.f. & Thomson	Guduchi	Menispermaceae	Liana	Fl.: June Fr.: November	India, Myanmar and Sri Lanka.	Common	Catalogue 2019
Vitex negundo L.	Nirgundi	Lamiaceae	Tree	Fl.: April to June Fr.: July-September	Afghanistan, Pakistan, India, Sri Lanka, Thailand, Malaysia, Eastern Africa and Madagascar, America, Europe, China and West Indies	Common	Catalogue 2019
Woodfordia fruticosa (L.) Kurz.	Dhai	Lythraceae	Tree	Fl.: January-April Fr.: May-July	Bangladesh, Bhutan, Myanmar, China, India Madagascar, Nepal, Pakistan, Sri Lanka, Tanzania and Thailand.	Common	Catalogue 2019

3. Results and Discussion

3.1. Phenology and Distribution Status

There are several plants growing in the Jastrota which has tremendous potential and application in different system of medicine. Ten high Value selected medicinal plants growing in Kathua district *Aerva sanguinolenta, Centella asiatica, Cissampelos pareira, Colebrookea oppositifolia, Cryptolepis dubia, Holarrhena pubescens, Justicia adhatoda, Tinospora cordifolia, Vitex negundo* and *Woodfordia fruticosa* is given in Table 1 along with vernicular name, habit, phenological period and distribution status.

3.2. Eco-taxonomy of High Value Medicinal Plants

3.2.1. *Aerva sanguinolenta*

Morphology: Herbaceous plant having terete stem, 20-70 cm tall, much branched, covered with stellate tomentum, woolly at the base, herbaceous above, tomentose. Leaves simple, alternate, sessile, differ in size, 1-3.5 cm long, 0.7-1.8 cm broad, linear-oblong, cuneate to attenuate at base, apex obtusely apiculate, margin- entire, densely tomentose, pale-white; petiole 0.3-0.8 cm long, hairy. Inflorescence- stalked, spikes 0.8-2.5 cm long, 0.3-0.4 cm broad, white, woolly. Flowers unisexual, dense, usually dioecious, white and sessile; bracts broadly ovate, white, apex acute and hyaline; ovary deeply forked, as long as style. Fruits utricle, orbicular-ovoid, thin; seeds minute, ovoid, lenticular with brown-black shining.

Ecology and Habitat: The plants are usually semi-perennial, commonly found on sandy or calcareous soils along the roadsides and disturbed areas.

3.2.2. *Centella asiatica*

Morphology: Herbaceous plants with slender stems and creeping, diffuse, nodes rooting. Leaves petiolate, forming rosettes along the creeping stem, 2-3 leaflets arise from each node of stems; leaflets are 0.6-1.2 cm long, 1-2.5 cm broad, orbicular-reniform, sheath at leaf base, crenate along margins, glabrous on both sides. Flowers fascicled umbels, white pink flowers. Fruits borne throughout the growing season, reniform or globose, base cordate to truncate, strongly laterally compressed, strongly thickened pericarp; seeds have pendulous embryo, laterally compressed.

Ecology and Habitat: The plants are perennial, stoloniferous, faintly aromatic herb; commonly found in shady, marshy, damp and wet places.

3.2.3. *Cissampelos pareira*

Morphology: Perennial climbing, slender tomentose stems; petioles pubescent. Leaves peltate, 1-4 cm long, 1-4.5 cm broad, triangularly broad-ovate, or orbicular, obtuse, mucronate, base cordate or truncate, tomentose on both sides. Flowers small, pedicels filiform; male flowers clustered, born in axil of a small leaf; sepals- 4 in number, obovate-oblong, hairy outside; petals-4 in number, united to form a 4-toothed cup, hairy outside; stamens- 4, column short, anthers connate, encircling the top of the column; female flowers clustered in the axils of orbicular, hoary imbricate bracts, racemes; sepal- 1, petal- 1; carpel- 1, densely hairy; style shortly 3-fid. Fruits drupaceous, subglobose, compressed, hairy-pubescent, black when dry, endocarp transversely ribbed, tuberculate; seeds horseshoe shaped.

Ecology and Habitat: Grows in forests of secondary vegetation and near rock outcrops; common in orchards, hedges, parks and gardens on moist soils.

3.2.4. *Colebrookea oppositifolia*

Morphology: Woody shrubby plants, 1.5-3 m tall, much branched; petiole 0.8-2.5 cm; stem and leaves densely tomentose. Leaves elliptic, opposite, 5 to 11.5 cm long, 2.5-6.5 cm wide, acuminate, crenulate, base acute to acuminate. Flowers ca. 2 mm. Pistillate flowers: calyx campanulate, ca. 1.5 mm, to 6 mm in fruit, tube very short, becoming conspicuously ribbed; teeth subulate, later spinescent, purple; corolla tube puberulent, lower lip slightly longer than upper lip, with middle lobe ovate; stamens inserted on apical part of tube, included; style twice as long as corolla; bisexual flowers: calyx minute, 0.5- 0.6 mm; corolla to 3 mm; upper lip ovate-oblong, ca. 0.5 mm, straight, emarginate; lower lip elongated, spreading, ca. 1.5 mm, middle lobe ovate-oblong, twice as lateral lobes; style erect, slightly longer than corolla. Fruits 4, obovoid, 1-2 mm, yellow-brown, with a small basal white scar, one seeded.

Ecology and Habitat: Plants commonly found along the forest margins.

3.2.5. *Cryptolepis dubia*

Morphology: Scandent lianas, 20-100 m; bark smooth, copper coloured, peeling off in papery rolls in old stem; branches lenticellate; petiole 1.5-2.2 cm long. Leaves opposite, 8-9 cm long, 3.2-4 cm broad, oblong-elliptic, apiculate, coriaceous, smooth and glossy above, glaucous beneath, base acute. Flowers pale yellow in a lax dichotomous cyme. Fruits follicular, 5-10 cm long, 0.8-1.5 cm broad, stout, divericate, tapering and pointed at apex; seeds compressed, oblong-ovate.

Ecology and Habitat: Perennial plants growing scramble over ground and twine into other plants for support.

3.2.6. *Holarrhena pubescens*

Morphology: Small to medium sized trees, 8-12 cm tall, with abundant white latex; bark smooth, corky, pale to dark grey; branches shortly hairy; petiole up to 1 cm long, shortly hairy, glandular at base. Leaves opposite, simple, 12-31 cm long, 5.5-12 cm broad, base rounded, apex acuminate to acute, shortly hairy to glabrous. Flowers bisexual, regular, pentamerous, fragrant; sepals elliptical to linear, free or fused at base. Fruits follicular composed of two slender 10-20 cm long follicles, pendulous, dehiscent, pale grey to dark brown, many-seeded; seeds narrowly oblong, grooved, glabrous and dense tuft hairs at apex.

Ecology and Habitat: Plant grows along forest margin in the deciduous forests and drier regions.

3.2.7. *Justicia adhatoda*

Morphology: Shrubby plants with erect and cylindrical branches, 1.5-2.5 m tall, swollen at nodes, green to pale-green. Leaves simple, exstipulate, opposite, 2-12 cm long and 1.5-5 cm broad, lanceolate to ovate, entire. Flowers sessile, pentamerous, large, white. Fruits capsular; seeds-large with hooks.

Ecology and Habitat: Plants grows in moist places of river banks to dry slopes and disturbed areas.

3.2.8. *Tinospora cordifolia*

Morphology: Plants climbers, upto 50 m in length on fences; exstipulate with petioles 6-12 cm long, roundish and pulvinate. Leaves simple, alternate, 10-20 cm long, 8-15 cm broad, ovate or ovate cordate, seven nerved and deeply cordate at base, membranous, whitish tomentose with a prominent reticulum beneath. Flowers are unisexual, small, greenish yellow on axillary and terminal racemes. Male flowers are clustered and female flowers are usually solitary; sepals and petals six in two series of three each, obovate, and membranous. Fruits clusters of one to three and ovoid smooth drupelets on thick stalks with sub terminal style scars and orange colored.

Ecology and Habitat: The plants commonly found in moist deciduous forests and scrub jungles; can be seen growing on fences.

3.2.9. *Vitex negundo*

Morphology: A bushy shrub or small tree growing up to 4-6 m tall; barks thin, grayish white; branches terete or obtusely quadrangular and slightly pubescent; petioles 1.5-3.5 cm long. Leaves palmately compound with 3 to 5 foliolate, lanceolate-elliptic or ovate-lanceolate, base cuneate or acute, margin entire or serrate, apex acuminate, chartaceous. Inflorescence terminal panicles, trichotomously branched, 10-30 cm long, peduncles slender, pubescent, about 3 to 8 cm long. Flowers are bisexual, many, fragrant, sepals campanulate, 5 toothed, teeth acute, pubescent outside, petals 5 lobed, 2-lipped, blue; stamens 4, didynamous, exserted, filaments slender, filiform. Fruits drupaceous, obovoid or subglobose, 3-5 mm in diameter, green, glabrous, black when ripe.

Ecology and Habitat: Plants commonly found in riverbanks, wetlands and moist place along roadsides as hedges.

3.2.10. *Woodfordia fruticosa*

Morphology: An evergreen shrub, growing up to 4 meters with spreading branches; stems brownish. Leaves opposite, 3.2-10.6 cm long, 0.6 to 2.9 cm broad, lanceolate, pointed, rough upper surface, vein raised gland-dotted, pale beneath; Inflorescence conspicuous when in flower with short dense axillary clusters of bright red tubular flowers often borne on lower leafless branches. Flowers are bright red, 9 to 13mm long, with 6 short triangular lobes; sepals tubular, petals small, red, scarcely as long as calyx-lobes; stamens with red filaments, much longer than calyx-tube. Fruits capsular, ovate or elliptic.

Ecology and Habitat: Plants commonly found along the forest margins and on hill slopes in subtropical forests.

3.3. Medicinal Uses and Chemical Constituents

Plant synthesizes hundreds of chemical compounds for functions including defence against insects, fungi, diseases, and for grass eating animals. Several active chemical constituents of plants have been identified and their biological activity has been established. Medicinal plants have been discovered and used in traditional medicine practices since prehistoric times. Medicinal plants are of great importance to the health of individuals and communities. The medicinal value of these plants lies in some chemical substances that produce a definite physiological action on the human body. However, since a single plant contains widely diverse phytochemicals, the effects of using a whole plant as medicine are uncertain. Further, the phytochemical content and pharmacological actions, if any, of many plants having medicinal potential remain un-inventorized by rigorous scientific research to define efficacy and safety (Ahn 2017). Alkaloids, tannins, saponins, steroid, terpenoid, flavonoids, phlobatannin and cardie glycoside are major chemical ingredients in medicinal plants (Hill 1952). Many of these indigenous medicinal plants are used as spices and food plants (Dhandapani and Sabna 2008).

Table 2: Medicinal uses and chemical constituents of selected plants

Botanical name	Parts used	Ethobotanical Use	Used in ailments/Disease	Chemical Constituents
Aerva sanguinolenta	Whole plant	Juice, decoction	Whole parts used in curing boils, wounds, diabetes worms, asthma and cough, strangury, indigestion. Decoction acts as efficacious diuretic and useful in treating catarrh of bladder. Roots show diuretic, demulcent and credited with tonic properties. The roots are used to cure headache. Flowers are used to cure headache and also for the removal of kidney stones. Besides, also cures mouth ulcer, toothaches, and chest problems.	Active constituents are tannin, β-sitosteryl palmitate, α-amyrin and β-sitosterol, campesterol, chrysin. Four flavonoid glycosides (I, II, III, IV) isolated and characterized.
Centella asiatica	Whole plants	Decoction, Juice, Wild eating	Leaves are useful in insomnia, cardiac debility, epilepsy hoarseness, asthma, bronchitis, hiccough; amentia, abdominal disorders, leprosy, strangury, fever to increase memory power. The cold poultice used as an external application in rheumatism, elephantiasis and hydrocele. Internally exacts used in leucorrhoea, kidney troubles, urethritis dropsy. Decoction of young shoots given in haemorrhoids. In paste form applied on boils and tumours. Leaves paste applied on forehead to treat severe headache, used in eczema. Leaf extract is used in the preparation	Active nutrients includes 3-glucosylquercetin, 3-glucosyl-and 7-glucosyl-kaemfrol, polyacetylenes I-V and nine other acetylenes; amino acids, asiatic, centic, centellic, centoic, pectic, madasiatic acids, carotene, centellose (oligosaccharide), hydrocotylin (alkaloid), lipid, protein, pectin, saponins, vallerine, asiaticoside (2,3,23-trihydroxy-urs-12-en-28-oic-acid-o-6-deoxy-α-L-mannopyranoyl-(1→4)-o-β-D-glucopyrono-syl-(1→6)-o-β-D-gluco-pyranosyl ester), aslaticosides A [o-á-L-rhamnopyranosyl-(1→4)-o-β-D-glucopyranosyl(1→6)]-o-β-D-

Contd.

			of medicated oil for bone fracture.	glucopyranose ester of 2α, 3β, 6β, 23β-tetra hydroxyurs-12-en-28-oic acid; asiaticoside B ([o-α-L-rhamnopyranosyl-(1→4)-o-β-D-glucopyranosyl (1→6)]-o-β-D-glucopyranose ester of 2α, 3β, 6β, 23α-tetrahydroxyolean-12-en-28-oic acid, indocentelloside, brahmoside, brahminoside, isothankuniside, medecassoside, thankcuriside, brahmic (= made cassic or made gascaric acid, 2α, 3β, 6β, 23- tetrahydroxy-urs-12-en-28-oic acid), and thankunic acids; β-caryophyllene, trans-β-farnesene, germacrene D, mesoinositol, polyphenols, α-terpinene, thymol methyl ether, betulic acid, β-sitosterol, campesterol, stigmasterol, tannins, vitamins B1, B2 and C.
Cissampelos pareira	Leave, Root		Extract, decoction	Leaves and roots used to cure dyspepsia, diarrhoea, dropsy and snake-bite. Stem yields a strong fibre. Active chemical constituent includes tubocurarine, cycleanine, hayatidin, hayatinin, hayatin, seeprine, bebeerine and cissampeline.
Colebrookea oppositifolia	Leave	Extract	Known for antimicrobial, antifungal and antioxidant properties, and traditionally used for treating sore eyes, corneal opacity or conjunctivities. Epilepsy, fever, headache, urinary problems and shows active properties of hepatoprotective, cardioprotective	Active phyto-constituents are chrysin, triacontanol, negletein, oleic acids, landenein, 5,6,7,4'-tetramethoxy flavone, acteoside, 5,6,7-trimethoxy flavone, quercetin, β-sitosterol, stearic, palmitic, n-triacontane, and b-acetyl alcohol, 32-hydroxydotri-acontyl ferulate.

Contd.

			and anti-inflammatory attributes can be cured. Essential oils can be extracted from leaves and young flowering twigs and these have fungitoxic properties. Also used in the management of ringworms, and employed in treatment of dermatitis, nose bleeds coughs and dysentery. In Ayurvedic traditional system of medicine, leaves used for treatment of wounds and fractures, where as roots helps in curing epilepsy.	
Cryptolepis dubia	Leave	Decoction	Used in treatment of muscle and tendon pains by decoction. Also taken to heal fresh wound and rickets.	Active alkaloids are buchananine, - 1, 3, 6-O-trinicotinoyl-a-D-glucopyranose; Cardenolides includes cryptosin, sarmentogenin, sarmentocymarin, cardenolide glycosides (cryptanoside A,B,C,D).
Holarrhena pubescens	Stem, leave, seed	Decoction, Paste	Used to cure anaemia, colic pain, diarrhoea, haematuria, menorrhagia, obstetric conditions, spermatorrhoea, splenomegaly. Seeds useful in amoebic dysentery, diabetes, diarrhoea, asthma, bronchopneumonia, hepatopathy, gastropathy, hepatosplenomegaly, internal haemorrhage, haemorrhoids, rheumatism, malaria, vomiting, verminosis, uropathy, skin diseases, bleeding piles colitis, bark paste applied in pruritus, rheumatism ulcers, uterine discharge. Fresh leaves used in chronic bronchitis, applied to boils	Active phytochemicals in woody stem includes conessine, conimine, isoconessimine, conessidine, conessimine, conkurchine, holadiene, holarrhonine, holarrhimine, holafebrine, N3-methyl-holarrhimine, holacine, holacimine, regholarrhenines A, B & C, conamine, conarrhimine, conkurchinine, holarrhin, holarrhessimine, kurchine, kurchicine, lettocine, norconessine, holarrhidine, holarrifine, kurchinidine, kurchinine, kurcholessine, holarrhetin, N,N,N′,N′-tetra methylholarrhimine, etc. Leaves contain kurchiphyllamine,

Contd.

			and ulcers.	kurchiphylline, holantosines A & B, N-acetylholantosine C, N-acetylholantosine D, N-acetylholarosine A, N-acetyl-holantosamine and its α and β-methyl derivatives, holarosineB, holantosines E & F, etc. Seeds contain holarricine, triacanthin, conomine, conarrhimine, irehline, lettocine, kurcine, linoceric, linolenic, linolei, oleic, palmitic and stearic acids, 9-D-hydroxy-cis-12-octadecenoic acid, etc.
Justicia adhatoda	Leave	Decoction	Leaves extensively used in the treatment of cough and other respiratory ailments such as asthma and bronchitis. Liquid extract of the plant is used in many pharmaceutical formulations as an expectorant. It was also observed to act as an antihistamine agent.	Active chemical constituents are tannins, saponins, vasicine, oxyvascicine, asicinonea and α-phellandrene
Tinospora cordifolia	Stem	Decoction	Used in treatment of diabetes, act as anti-periodic, anti-spasmodic, anti-inflammatory, anti-arthritic, anti-oxidant, anti-allergic, anti-stress, anti-leprotic, anti-malarial, hepatoprotective, immune-modulatory and anti-neoplastic activities	Most active alkaloids are aporphine, clerodane diterpenes, berberine, palmatine, tembertarine, tinosporin, magniflorine, choline, berberine, palmatine, tembetarine, magnoflorine choline, tinosporin etc. and glycoside like tinocordiside, tinocordifolioside.
Vitex negundo	Whole plant	Decoction	Used in treatment of diseases such as headache, skin affections, wounds, swelling, asthmatic pains, male and female sexual and reproductive problems.	Major phytochemical constituents are flavonoids, casticin, chryso-splenol and vitexin, Chrysophenol D, nishindine and hydrocotylene. It also contains monoterpenes agnuside, eurostoside, and aucubin. Vitex contains the

				flavonoids, casticin, chryso-splenol and vitexin. The main compounds are viridiflorol (19.55%), beta-caryophyllene (16.59%), sabinene (12.07%), 4-terpineol (9.65%), gamma-terpinene (2.21%), caryophyllene oxide (1.75%), 1- oceten-3- ol (1.59%), and globulol (1.05%). The major constituents of the leaf oil were α-pinene (35.9%), limonene (12.2%) and bicyclogermacrene (9.5%), while the fruit oil contained α-pinene (31.7%), bicyclogermacrene (14.5%) and limonene (11.5%), and the flower oil contained α-pinene (14.7%), bicyclogermacrene (8.3%) and limonene (5.8%)
Woodfordia fruticosa	Flower	Decoction	Possess health tonic stimulating properties and used to treatment of one or more systemic disorders. Flowers help in treating burning sensations in stomach. Plant is pungent, acrid, uterine sedative, cooling and useful in leprosy, toothache, leucorrhea, fever, dysentery and blood disease. Used for treating stomach disorders and with honey helpful in pediatric diarrhea.	Active compounds are phenolics, particularly hydrolysable tannins and flavonoids. Octacosanol, β-sitosterol, hecogenin, meso-inositol, triterpenoids lupeol, betulin, betulinic acid, oleanolic acid, ursolic acid, flavonoids or tannins, gallic acid, ellagic acid, bergenin, norbergenin, chrysophanol-8-O-β-dglucopyranoside, flavonoid quercetin glycosides, 3-β-l-arabinoside, 3-O-α-l-arabinopyranoside, 3-O-β-dxylopyranoside, 3-O-(6″-galloyl)-â-d-glucopyranoside, and 3- O-(6″-galloyl)-β-d-galactopyranoside, and 3-O-α-l-arabinopyranoside.

Image 1: High Value selected medicinal plants growing in Kathua district (Jasrota Wildlife Sanctuary), J&K (India) in Southeast Asia; **(A)** *Aerva sanguinolenta,* **(B)** *Centella asiatica,* **(C)** *Cissampelos pareira,* **(D)** *Colebrookea oppositifolia,* (E) *Tinospora cordifolia,* **(F)** *Holarrhena pubescens,* **(G)** *Justicia adhatoda,* **(H)** *Vitex negundo,* **(I)** *Woodfordia fruticosa,* **(J)** *Cryptolepis dubia.*

4. Conclusion

Traditional medicine is the knowledge, skills and practices based on the theories, beliefs and experiences indigenous to different cultures, used in the maintenance of health and in the prevention, diagnosis, improvement or treatment of physical and mental illness. Herbs and ethnobotanicals have been used since the early days of humankind and are still used throughout the world for health promotion and treatment of disease. Plants and natural sources form the basis of today's modern medicine and contribute largely to the commercial drug preparations manufactured today. About 25% of drugs prescribed worldwide are derived from plants. Still, herbs, rather than drugs, are often used in health care. Plants may contain toxic ingredients in the form of phytochemicals, which, if introduced into the body of any animal or human, can act by injuring the protoplasm of the cell. The harmful effects may be immediate, or slow accumulation of phytochemicals in the body can result in toxicity or even be fatal (Singh et al. 2016). Concentrations of phytochemicals in plants vary with the species and depend on the growing conditions, and in many instances, the toxins may not be present in sufficient quantities to be harmful in humans, or sometimes cooking or fermentation processes may destroy or dissipate these chemicals (Singh et al. 2016). Same plant can be used for multiple purposes. Illustrating further, stem bark of *Vitex negundo* is prescribed for lumbago and sciatica in Ayurvedic medicine, where as leaves and fruits of *V. negundo* are used as an effective medicine for rheumatoid arthritis. For some, herbal medicine is their preferred method of treatment. For others, herbs are used as adjunct therapy to conventional pharmaceuticals. However, in many developing societies, traditional medicine of which herbal medicine is a core part is the only system of health care available or affordable. Regardless of the reason, those using herbal medicines should be assured that the products they are buying are safe and contain what they are supposed to, whether this is a particular herb or a particular amount of a specific herbal component. To achieve such objectives, global harmonization of legislation is needed to guide the responsible production and marketing of herbal medicines. Further, we feel that if sufficient scientific evidence of benefit is available for any botanicals, then legislation should allow for this to be used appropriately to promote the use of that herb so that these benefits can be realized for the promotion of public health and the treatment of disease

Conflict of Interest

Authors declare no conflict of interest

References

Ahn K (2017). The worldwide trend of using botanical drugs and strategies for developing global drugs. *BMB Reports* 50 (3): 111-116.

Bhat S, Mulgund GS & Bhat P (2019) Ethnomedicinal Practices for the Treatment of Arthritis in Siddapur Region of Uttara Kannada District, Karnataka, India. *Journal of Herbs, Spices and Medicinal Plants* 25(4): 316-329.

Dhandapani R & Sabna B (2008). Phytochemical constituents of some Indian medicinal plants. *AMJ* 27(4): 1-8.

Green EC & Makhubu LP (1983). Traditional Healers in Swaziland: Towards improved cooperation between the traditional and modern health sectors. Ministry of Health, Mbabane.

Heinrich M & Gibbons S (2001). Ethnopharmacology in drug discovery: An analysis of its role and potential contributions. *Journal of Pharmacy and Pharmacology* 53:425-432.

Hill AF (1952). Economic Botany. A Textbook of useful plants and plant products, 2nd edn, McGraw – Hill Book Company Inc, New York. pp.32-35 (1952)

Jain SK & Rao RR (1977). A Handbook of Field and Herbarium Methods. Today and Tomorrow's Printer, New Delhi, India, 157 pp.

Lambert JDH, Ryden PA & Esikuri EE (2005). Capitalizing on the bioeconomic value of multi-purpose medicinal plants for rehabilitation of dry lands in sub-Saharan Africa. The international bank for reconstruction and development. The World Bank, Washington, DC, USA. 11-14.

Mdluli K (2002). Management of HIV/AIDS with traditional medicine. In: Makhubu L P, Amusan OOG, Shongwe MS (Eds). Proceedings of the workshop on the management of HIV/AIDS with traditional medicine, Kwaluseni, Swaziland. February 2002.

Pan SY, Zhou SF, Gao SH, Yu ZL, Zhang SF, Tang MK, Sun JN, Ma DL, Han YF, Fong WF & Kao KM (2013). New Perspectives on How to Discover Drugs from Herbal Medicines: CAM's Outstanding Contribution to Modern Therapeutics. *Evidence-based Complementary and Alternative Medicine* 627375.

Singh B, Sultan P, Hassan QP, Gairola S & Bedi YS (2016) Ethnobotany, Traditional Knowledge, and Diversity of Wild Edible Plants and Fungi: A Case Study in the Bandipora District of Kashmir Himalaya, India. *Journal of Herbs, Spices & Medicinal Plants* 22, 247-278, DOI: 10.1080/10496475. 2016.1193833.

Stewart RR (1972) An annotated catalogue of the vascular plants of West Pakistan and Kashmir. Fakhri Printing Press, Karachi.

Swain T (1972). Plants in the Development of Modern Medicines, Harvard University Press, Cambridge.

15

Nutraceutical Value of Wild Plants Used as Food and Medicine

Ashutosh K. Dash, Sunil Kumar and Debaraj Mukherjee

Abstract

Wild plants have a great role in the lives of tribal peoples living in and around the world by way of providing products for both food and medicine. Such herbal plants are the core of research, which can solve several health issues as well as nutritional supplements. Keeping on the view we have presented a summary of medicinal plants used as nutraceuticals collected from various regions of India. This communication describes 35 herbal plants used as a medicine against different diseases as well as their nutritional applicability.

Keywords: Nutraceutical, Wild plant, Food, Medicine

1. Introduction

Medicinal plants play an important role in health care and prevention of different types of diseases and these have very few side effects (Bhattachariya & Borah 2008). Plant extracts are therapeutically used and they play an important role in the world health care (Yadav et al.

Ashutosh K. Dash and Sunil Kumar

School of Pharmaceutical Sciences, Department of Pharmaceutical Chemistry, Shoolini University Solan Himachal Pradesh, 173229 India.

Debaraj Mukherjee (✉)

[a]*Academy of Scientific and Innovative Research (AcSIR)*

[b]*Natural Product Chemistry Division, CSIR-Indian Institute of Integrative Medicine, Jammu Jammu and Kashmir 180 001, India*

✉*Corresponding author email: : dmukherjee@iiim.ac.in*

Plants for Novel Drug Molecules: Ethnobotany to Ethnopharmacology

Bikarma Singh & Yash Pal Sharma (eds.), (pp. 363-378)

Email: *info@nipabooks.com* Web: *www.nipabooks.com*

2006). Plants have several type of phytoconstituents present. They not only provide nutrition but also have several health benefits, such as antioxidant, antimalarial, anticancer and antimutagenic properties (Bernal et al. 2011). Since time immemorial, mankind has used plant extracts from different plants to cure many diseases and thus relieve him from physical agony. The healing power of traditional herbal medicines has been realized and documented since Rigveda and Atharvaveda. India has about 45,000 plant species and more than 35,000 plant species have been claimed to possess medicinal properties and are being used in various human cultures around the world for medicinal purposes (Lewington 1993). The word "Nutraceuticals" is derived from two different words i.e. nutrient and pharmaceuticals. Nutraceuticals are also known as functional foods which are used singly combination of several dietary components utilizes for the treatment of several diseases or inborn deficiencies (Pandey et al. 2011). Nutraceuticals or functional foods can be classified based on their natural sources, pharmacological parameters or according to their chemical constitution. The most usual nutraceuticals are nutrients, herbals, dietary supplements, functional food and natural chemicals derived from different medicinal plants (Bickford et al. 2006). The chemical characteristics can be used for grouping nutraceuticals based upon their chemical nature. This approach allows nutraceuticals to be categorized under different molecular/elemental groups and subgroups. To cite few examples are (i) Amino acid-based substances: amino acids, ally-S compounds, indole, folate and choline (ii) Carbohydrates and derivatives: ascorbic acid, oligosaccharides, non-starch-polysaccharides (iii) Fatty acids and structural lipids: lecithin, fatty acids, oil (iv) Isoprenoid derivatives: carotenoids, saponins, tocopherols, simple terpenes (v) Phenolic substances: coumarins, tannins, lignin, anthocyanins, isoflavones, flavones, flavonols (vi) Microbes: probiotics, prebiotics (vii) Minerals: calcium, selenium, potassium, copper, zinc and several others in minor quantity. Several herbs, shrubs, and trees play a significant role in the cure/prevention of different diseases. Earlier report proclaims that the nutrients available in wild vegetables of South Africa, *Chenopodium album, Sonchus asper, Solanum nigrum,* and *Urtica urens* are comparable with or higher than those of commonly used vegetables such as spinach, lettuce and cabbage (Afolayan & Jimoh 2009).

2. Nutritional Values of Selected Indian Based Herbal Plants

In terms of antinutritional principles, all the vegetables had comparatively lower concentrations of phytate, alkaloids, and saponins (Misra et al. 2008). Nutritional values of some Indian based herbal plants which are also used as ayurvedic medicines have been depicted in Table 1. A detailed discussion of some medicinally important herbs used as nutraceuticals are discussed below:

2.1.*Trichosanthes tricuspidata* Lour.

Trichosanthes tricuspidata (syn.-*T. palmata* Roxb., *T. bracteata* Lam., *T. pubera* Blume) belongs to the family cucurbitaceae is a vine which is found between an elevation of 1200 to 2300 m above sea level (ASL). The distribution of this species extends from eastern Himalayas in India to southern China, southern Japan, Malaysia and tropical Australia (Bhandari et al. 2008). *T. tricuspidata* is medicinally important in several traditional systems. In ayurvedic medicines, the fruits are used in the treatment of asthma, ear ache, and ozoena (intranasal crusting, atrophy, and fetid odor). In the Unani system of medicine, the fruits are used as a carminative (an agent that relieves flatulence), a purgative, and an abortifacient, to lessen inflammation, cure migraines, and reduce heat of the brain, as a treatment for opthalmia (inflammation of the eye), leprosy (infectious disease caused by *Mycobacterium leprae*), epilepsy (episodic impairment or loss of consciousness, abnormal motor phenomenon) and rheumatism (painful local inflammation of joints and muscles) as well as other uses. The seeds are emetic and a good purgative. In the Thai traditional system of medicine, the plant is used as an anti-fever remedy, a laxative, an anthelmintic as well as in migraine treatments. (Kanchanapoom et al. 2002) The roots of the plant are used to treat lung diseases in cattle and for the treatment of diabetic carbuncles and headaches (Chopra et al. 1956, Gaur1999) has reported the use of this plant in curing bronchitis, and the application of seed paste for hoof and mouth disease in cattle. The *vaidyas*, or practitioners of ayurveda, also use the fruits in treating stomatitis. The oil extracted from the roots is used as a pain killer. In Bastar district, Chhattisgarh, India, the plant is used for curing snakebite poisoning and the juice of the plant is applied externally for skin eruptions. In Nepal, the roots are used to cure bleeding in chickens.

2.2. *Rubus ellipticus* Sm.

R. ellipticus belongs to family Rosaceae, and commonly known as yellow Himalayan raspberry. This species is one of the common wild edible fruits of Himalaya and its distribution recorded from Uttarakhand, J&K, Punjab and Uttar Pradesh (Samant & Dhar 1997). The fruits are shown to be highly nutritious, delicious, and rich in vitamins and sugars (Parmar & Koushal 1982), their antioxidant and antiproliferative potentials remain underexplored.

2.3. *Acorus calamus* L.

A. calamus is a perennial plant belonging to family Acoraceae with creeping extensively branched aromatic rhizomes. The plants grows in wild in marshy places upto 2000 m altitude in Himalayas, Manipur, Naga Hills and in some parts of South India (Balakumbahan et al. 2010). The rhizomes of *A. calamus* reportedly relieve stomach cramps, dysentery, and asthma, and are used as anthelmintic, insecticides, tonics and stimulants. Alcoholic rhizome extracts of *A. calamus* growing in KwaZulu-Natal, South Africa, were previously found to have anthelmintic and antibacterial activity. Using bioassay-guided fractionation, the phenyl propanoid β–asarone was isolated from the rhizome. This compound was shown to possess anthelmintic and antibacterial activity. It has previously been isolated from *A. calamus*, and a related species, *A.gramineus*. Different varieties of *A. calamus* exhibit different levels of ß-asarone, with the diploid variety containing none of the compounds. Mammalian toxicity and carcinogenicity of asarones have been demonstrated by other researchers, supporting the discouragement of the medicinal use of *A. calamus* by traditional healers (Van 2002).

2.4. *Asparagus racemosus* Willd.

A. racemosus plant with a woody stem sends runners out, has needle-like leaves with small white flowers (Sharma et al. 2000, Room 2010), reported to be growing throughout Himalayas. The root is employed in diarrhea as well as in cases of chronic colic and dysentery. Root boiled with some bland oil, is used in various skin diseases. Root boiled in milk is the most important herb in ayurvedic medicine for curing women's fertility problems. It is taken internally in the treatment of infertility, loss of libido, threatened miscarriage, menopausal problems, hyperacidity, stomach ulcer, and bronchial infection. Externally it is used to treat stiffness in the joints (Brown 1995). The root is used fresh in the treatment of dysentery. It is harvested in the autumn and dried for use in treating other complaints. The whole plant is used in the treatment of diarrhea, rheumatism, diabetes and brain complaints. It is also used in the management of the behavioral disorder and minimal brain dysfunction (Sheth 1991). The rhizome is a soothing tonic that acts mainly on the circulatory, digestive, respiratory and female reproductive organs. The root is alterative, antispasmodic, aphrodisiac, demulcent, diuretic, galactagogue and refrigerants (Chopra et al. 1986).

Table 1: Nutraceutical values and chemical compounds of wild plants

Botanical names	Family	Vernacular names	Chemical compounds	References
Acorus calamus L.	Acoraceae	Bach, vacha, gorabach, safedbach, Ugragandha, shadgrantha, Shataparva, sweet root, sweet flag, and calamus	calamene, a-pinene, a-asarone, eugenol, c-asarone, p-cymene, a-caryophyllene, 1,8-cineole, b-pinene, methyl ether, camphene, terpinolene, camphor and hydrocarbons	(Balakumbahan et al. 2010, Nigam et al. 1990, Srivastava et al. 1997)
Allium sativum L.	Amaryllidaceae	Garlic	allicin	(Singh & Singh 2008)
Ananascomosus (L.) Merr.	Bromeliaceae	Pine apple	bromalin	(Pulliainen et al. 2010)
Asparagus racemosus Willd.	Asparagaceae	satmul, Bojhidan, shatavar, shatamuli, shatavari, satuli, vrishya and Spiry Asparagus	hexanal, furfural, decanoic acid, camphor, 2-asparagine, folic acid, tyrosine, arginine, several flavonoids (quercetin, rutin, and hyperosides)	(Chaudhuryet al. 1957)
Asparagus abyssinicus Hochst. ex A.Rich	Asparagaceae	Shatavari	Carnitine or L-Carnitine	(Dray1992)
Brassica oleracea L.	Brassicaceae	Broccolli	betaine	(Block 1985)
Capsicum annuum L.	Solanaceae	Red pepper	capsaicin or trans-8-methyl-N-vanillyl-5 non enamide	(Chen, et al. 2002)
Cinnamomumcamphora (L.) J.Presl	Lauraceae	Dalchinikapoor	camphor	(Rabelo et al.2004)
Curcuma longa var. vanaharidra Velay., Pandrav., J.K.George & Varapr.	Zingiberaceae	Turmeric	curcumin	(Okui et al. 1963)
Dioscoreadeltoidea Wall. ex Griseb.	Dioscoreaceae	Singli-mingli, Varahik and andwild yam.	diosgenin, cortisone, smila genone, carbohydrates,	(Lahiri et al. 1964)

Contd.

			glycosides,alkanoides, flavonoids, saponin, tannin, unsaturated triterpinoides, sterol, and resin.	
Emblica officinalis Gaertn.	Phyllanthaceae	Amla, olive plant	Vitamin E, C, ellagic acid, etc.	(Bhattacharya et al. 1999, Vormisto et al. 1997, Santos et al. 1999)
Fragaria abnormis Tratt.	Rosaceae	Strawberry and raspberry	ellagic Acid	(Steiber et al. 2004)
Justicia adhatoda (L.) Huth	Acanthaceae	basak, vasa, vasaka, adulsa,adhatoda and Malabar nut.	quinazoline alkaloid, vasicine, vasicol, fatty acids, adhatonine, vasicinone,vasicinol, vasicinolone (Chowdhury, 1987), vitamin C, saponins, flavonoids and steroids, terpene, ketone and phenolic ether	(Mohanta et al. 2001, Chakrabarty et al. 2001, Shrivastava et al. 2006)
Linumusitatissimum L.	Linaceae	Linseed	Omega 3 Fatty Acids	(Machowetz et al. 2007)
Malus domestica Baumg.	Rosaceae	*Apple*	phloratine	(Holman 1998)
Moringa oleifera Lam.	Moringaceae	*Drumstick tree*	Flavonoidaisothiacyanates	(Kou et al.2018)
Ocimumtenuiflorum L.	Lamiaceae	*Holy basil, Tulsi*	Vitamin C and A , proteins	(Pattanayak et al. 2010)
Olea europaea L.	Oleaceae	Brown olive, Indian olive	olive oil	(Chang et al. 2008)
Paris polyphylla Sm.	Melanthiaceae	Singpan, Satwa and Satuwa.	przewalskinone B, polyphyllin C, polyphyllin D, polyphyllins.	(Zang, et al. 2010)
Ricinus communis L.	Euphorbiaceae	Castor	ricinoleic acid	(Vattem et al. 2005)
Rubusellipticus Sm.	Rosaceae	akkhe, acchuye, heere, hisalu, hinsar, golden Himalayan raspberry and yellow Himalayan raspberry	magnesium, calcium, potassium, and phosphorus	(Ward 1962)

Contd.

Secale cereale L. *Glycine max* (L.) Merr., and *Brassica oleracea* L.	Poaceae, Brassicaceae	rye, soybeanand brocaoli	betaine, aspartic acid, glycine steric acid, riboflavin, and thiamine	(Mohan et al 2000, Tidke et al 2015)
Tagetes patula L.	Asteraceae	Marigold	lutein or lutein ester	(Prasad 2005, Mohan et al 2000)
Tinospora cordifolia (Willd.) Miers	Menispermaceae	amrita bali,gilloy, guduchi, amruthvalli, gulancha, and heart-leaved moonseed.	Carbohydrates, protein calcium and vitamin c	(Kumari et al 2012)
Trichosanthestri cuspidata Lour.	Cucurbitaceae	Indrayan, LalIndrayan, Koundal, Redballsnakegourd	bryonolic acid, D-glucose, glyceryl monostearate, cucurbitacin-B, isocucurbitacin-B, isocucurbitacin-D,	(Saboo et al.2012)
Vitis vinifera L.	Vitaceae	Grapes	proanthocyanins	(Donald et al. 1952, Valverde et al. 2011)

2.5. *Justicia adhatoda* (L.) Huth

J. adhatoda is an evergreen shrub, commonly known as vasaka occurs throughout India. Cattle do not eat this plant as the leaves emit an unpleasant smell. The leaves, roots and flowers are extensively used in indigenous medicine as a remedy for cold, cough, bronchitis and asthma. For relief in asthma, the dried leaves should be smoked. In Ayurveda, a preparation made from vasaka flowers, known as gulkand is used to treat tuberculosis. A few fresh petals of vasaka flowers should be bruised and put in a pot of chill clay. Even the juice from its leaves is useful in treating tuberculosis. About 30 ml of the juice is taken thrice a day with honey. It relieves the irritable cough by its soothing action on the nerve and by liquefying the sputum, which makes expectoration easier. For coughs, 7 leaves of the plant are boiled in water, strained and mixed with 24 grams of honey. Its leaves, bark, the root-bark, the fruit, and flowers are useful in the removal of intestinal parasites. The decoction of its root and bark in doses of 30 grams twice or thrice a day for 3 days can be given for this purpose. The juice of its fresh leaves can also be used in doses of a teaspoon thrice a day for 3 days. The juice from its leaves should be given in doses of 2 to 4 grams in treating diarrhea and dysentery. A poultice of its leaves can be applied with beneficial results over fresh wounds, rheumatic joints, and inflammatory swellings. A warm decoction of its leaves is useful in treating scabies and other skin diseases (Sam path et al. 2010).

2.6. *Dioscorea deltoidea* Wall. ex Griseb

D. deltoidea is a climbing herb with rhizomatous rootstock (Ali 2012, Dangwal and Chowhan 2015). It mainly occurs in the tropics and subtropics region of the world, mainly in West Africa, parts of Central America and the Caribbean, Pacific Islands and Southeast Asia (Anand 2011). In Asia, the plant is found mainly in Cambodia, Bhutan, Afghanistan, China, Pakistan, India (Western Himalaya), Nepal, Vietnam, and Thailand. It is used to get relief from the intestinal worms. It acts as a vermifuge (worm repellent) for children especially. Its roots are found being active against uterine sedative. It is diuretic expectorant in major regions of Himalaya. It is also used as bio-poison for fishes (Rashid et al. 2010). Its tubers are employed to treat bilious colic and to kill lice. Its roots contain a good amount of diosgenin used as starting material for hormone preparation. It is also used in soap making due to its saponin content (Muslim & Sikander 2010). Studies indicate that it was used to wash clothes as a detergent because of saponins present in this species. Traditionally *D. deltoidea* is found to be anti-rheumatic and treat ophthalmic conditions (Jain 1975, Kumari et al. 1975) reported that the powder from the rhizome of the plant is taken orally to cure dysentery, abdominal pain, and piles.

2.7. *Tinospora cordifolia* (Willd.) Miers

T.cordifolia is a large, glabrous, deciduous, climbing shrub. (Spandana et al. 2013). The species is widely distributed in India, extending from the Himalayas down to the southern part of Peninsular India. It is also found in neighboring countries like Bangladesh, Pakistan, and Srilanka. The plant is reported from southeast Asian continent such as Malaysia, Indonesia, and Tamilnadu. The plant possesses anti-oxidant, anti-hyperglycemic, anti-neoplastic, anti-stress, anti-dote, anti-spasmodic, anti-pyretic, antiallergic, anti-leprotic, antiinflammatory, anti-hyperlypidaemia, Immunomodulatory properties. Various parts of the plant contain immense medicinal property (Spandana 2013).

2.8. *Paris polyphylla* Sm.

Paris polyphylla, locally known as Singpan by Manipuris, Satwa by Garhwalis of Uttarakhand, Satuwa by Nepalese of Nepal is an important member of this genus. It is a glabrous herb, perennial in nature growing on light sandy and moist humus-rich soil with full or partial shade. The rhizomes of *Paris polyphylla* have been used as an anti-helminthic and vermifuge in folk communities of Nepal (Devkota 2005). The powdered roots of this plant are used as ethnopediatrics for diarrhoea in Garhwal Himalaya, Uttarakhand, India (Tiwari et al. 2010, June 1989).

2.9. *Allium sativum* L.

Garlic, *Allium sativum* L. is plant member of the Liliaceae family, has been widely recognized as a valuable spice and a popular remedy for various ailments and physiological disorders. Garlic is one of the important bulb crops grown and used as a spice or a condiment throughout India. According to the Unani and Ayurvedic systems, as practiced in India, garlic is carminative and is a gastric stimulant and thus help in digestion and absorption of food. Allicin present in aqueous extract of garlic reduces cholesterol concentration in human blood. Garlic extract has antimicrobial activity against many genera of bacteria, fungi, and viruses. Cultivated practically throughout the world, garlic appears to have originated in central Asia and then spread to China, the Near East, and the Mediterranean region before moving west to Central and Southern Europe, Northern Africa and Mexico (Singh & Singh 2008). In herbal medicine, garlic has been atherosclerosis traditionally used for asthma, deafness leprosy, bronchial congestion, arteriosclerosis i.e. hardening of arteries, fevers, worms and liver gall bladder trouble. Garlic is good for the heart, a portion of food for the hair and a stimulant to appetite. In recent times, experiments have confirmed several ancient beliefs about the healing value of this herb. These experiments have in fact proven much greater power of garlic than known previously. The unpleasant odour in garlic is due to its sulfur content. Garlic has proved effective in certain diseases of the chest. It reduces the stinking of the breath in pulmonary gangrene. Garlic is also useful in the treatment of tuberculosis. In Ayurveda, a decoction of garlic boiled in milk is considered a wonderful drug for tuberculosis. One gram of garlic, 250ml of milk and a liter of water are boiled together till it reduces to one-fourth of the decoction. It should be taken thrice a day. Taken in sufficient quantities, it is a marvelous remedy for pneumonia (Breithaupt 1997). Three cloves of garlic boiled in milk can be used every night with excellent results in asthma. A pod of garlic is peeled, crushed and boiled in 120ml pure malt vinegar. It is strained after cooling and an equal quantity of honey is mixed and preserve in a clean bottle. One or two teaspoons of this syrup can be taken with fenugreek decoction in the evening and before retiring. This has been found effective in reducing the severity of asthmatic attacks. Garlic is one of the most effective remedies for lowering blood pressure. Pressure and tension are reduced because it has the power to ease the spasm of the small arteries. It also modifies the pulse and modifies the heart rhythm. In Russia garlic is used extensively in the treatment of rheumatism and associated diseases. Garlic helps to break up cholesterol in the blood vessels, thereby preventing any hardening of arteries which leads to high blood pressure and heart attack. If a patient takes garlic after a heart attack, the cholesterol level comes down. Though the earlier damage may not be repaired, the chances of new attacks are reduced. It has been found to help remove toxins revitalize the blood, stimulate circulation and promote intestinal flora, or colony of bacteria that prevent infection by harmful bacteria. Garlic has been used successfully for a variety of skin disorders. Pimples disappear without a scar when rubbed with raw garlic several times a day. Even very persistence forms of acne in some adults have been healed with garlic. Garlic rubbed over ringworm, gives quick relief. The area is burnt by the strong garlic and later the skin peels off and the ringworm is cured.

2.10. *Brassica oleracea* L.

Broccoli, *Brassica oleracea*, variety Italica, belongs to the family Brassicaceae, is a fast-growing annual plant that grows 60-90 cm tall. They are native to the eastern Mediterranean and Asia and were later introduced to England and America in the 1700s. Plants in general, are known to be extremely rich in a variety of secondary metabolites that are found to possess anti-microbial properties. Many anti-microbial agents that are derived from traditional medicinal plants are available for treating many diseases caused by micro-organisms. The groups of anti-microbial phytochemicals are of several categories that include alkaloids, flavonoids, tannins, polyphenols, essential oils, phenolics and polypeptides (Perumals & Gopalakrishnakone 2010). They are also known to contain a high content of flavonoids, vitamins and mineral nutrients. Vitamin C, insoluble complexes, is a good adjuvant in iron therapy but can interfere with the metabolism of some drugs and antineoplastic agents. The presence of these contents has shown that broccoli provides immense benefits in protecting humans against cancer, and also assures to reduce the risk of specific cancers.

3. Conclusion

Indian herbs are well known for their nutraceutical values. Most part of our India, basically Himalayan belt is enriched with such type of medicinal plants equipped with nutraceutical values. The information of the plants can fulfill nutritional deficiency as well as can be used as medicines. These plants could be easily explored and utilized as drugs after isolation, semi-synthesis or synthesis of active constituents present in it. We have done a small effort for the documentation of some Indian origin plants to mitigate the nutritional deficiency as well as medicinal purpose against various ailments. As prospects of this chapter can be fruitful for the readers to carry out research on such type of plants and solve various human health issues.

Abbreviation used

No abbreviation was used in this manuscript

Conflict of Interest

There is no conflict of interest.

Acknowledgements

Authors are thankful to CSIR IIIM Jammu for the help to formulate this chapter.

References

Afolayan AJ & Jimoh FO (2009). Nutritional quality of some wild leafy vegetables in South Africa. *International Journal of Food Sciences and Nutrition* 60(5): 424-431; DOI:10.1080/09637480701777928.

Ali M (2012). Plant Growth-A Note on the vegetative growth in *Dioscorea deltoidea. Journal of Aromatic and Medicinal Plants* 1(8): 2-6; DOI:10.4172/2167-0412.1000116.

Anderson AM, Mitchell MS&Mohan RS(2000). Isolation of curcumin from turmeric. *Journal of Chemical Education* 77(3): 359; DOI:10.1021/ed077p359.

Balakumbahan R, Rajamani K & Kumanan K (2010). *Acorus calamus*, An overview. *Journal of Medicinal Plant and Research* 4 : 2740-2745; ISSN 1996-0875

Bernal J, Mendiola JA, Ibanez E & Cifuentes A (2011). Advanced analysis of nutraceuticals. *Journal of Pharmaceutical and Biomedical Analysis* 55(4): 758-774;DOI:10.1016/j.jpba.2010.11.033.

Bhandari S, Dobhal U, Sajwan M & Bisht NS (2008). *Trichosanthes tricuspidata*: A medicinally important plant. *Trees for Life Journal* 3: 5.

Bhattacharjya DK & Borah PC (2008). Medicinal weeds of crop fields and role of women in rural health and hygiene in Nalbari district, Assam. *Indian Journal of Traditional Knowledge* 7(3): 501-504.

Bhattacharya A, Chatterjee A, Ghosal S & Bhattacharya SK (1999). Antioxidant activity of active tannoid principles of *Emblica officinalis* (amla). *Indian Journal of Experimental Biology* 37(7):676-80;PMID: 10522157.

Bickford PC, Tan J, Shytle RD, Sanberg CD, El-Badri N & Sanberg PR (2006). Nutraceuticals synergistically promote proliferation of human stem cells. *Stem Cells and Development* 15(1): 118-123;DOI: 10.1089/scd.2006.15.118.

Bown D (1995). Encyclopedia of Herbs & their uses. London. 124. Published by Dorling Kindersley 424 pages.

Breithaupt GK, Michael L, Harisios B & Gustav GB (1997).Protective effect of chronic garlic intake on elastic properties of aorta in the elderly. *Circulation* 96(8): 2649-2655; DOI: 10.1161/01.CIR.96.8.2649.

BV Shetty & V Singh (2010). Flora of Rajasthan. pp.756-1001. 1st Edition, Merut publishers and Distributors, Meerut. Vol 1.

Caddick LR, Wilkin P, Rudall PJ, Hedderson TAJ & Chase MW (2002).Phylogenetics of Dioscoreales based on combined analyses of morphological and molecular data. *Botanical Journal of the Linnean Society* 138: 123–144; DOI: 10.1046/j.1095-8339.2002.138002123.x.

Chakrabarty A & Brantner AH (2001). Study of alkaloids from Adhatoda vasicanees on their anti-inflammatory activity. *Phytotherapy Reseasrch* 15: 532-534; PMID: 11536385.

Chang YY, Liu JS, Lai SL, Wu HS & Lan MY (2008). Cerebellar hemorrhage provoked bycombined use of nattokinase and aspirin in a patient with cerebral microbleeds. *Internal Medicine* 47 (5): 467–9;DOI: 10.2169/internal medicine. 47.0620.

Chaudhury SS, Gautam SK & Handa KL (1957). Composition of calamus oil from calamus roots growing in Jammu and Kashmir. *Indian Journal of Pharmacy* 9: 183–186.

Chen PF, Chang ST & Wu HH (2002).Antimite activity of essential oils and their components from *Cinnamomum osmophloeum*. *Quarterly Journal of Chinese Forester* 35: 397–404.

Chopra RN, Nayar SL, Chopra IC, Asolkar LV & Kakkar KK (1986). Glossary of Indian medicinal plants. (Including the supplement), New Delhi: Council of Scientific & Industrial Research, 1956-92.

Chopra RN, Nayar SN & Chopra IC (1956). Glossary of Indian Medicinal Plants. New Delhi: Council of Scientific and Industrial Research. 247.

Dangwal LR & Chauhan AS (2015). *Dioscorea deltoidea* Wall. Ex Griseb. A highly threatened himalayan medicinal plant: An Overview. *International Journal of Pharmacology and BioSciences* 6(1): 452-460.

Devkota KP (2005). Bioprospecting studies on Sarcococca hookeriana Bail, Sonchus wightianus DC, Parispolyphylla Smith and related Medicinal Herbs of Nepal: Ph D Thesis, HEJ Research Institute of Chemistry, International Centre for Chemical Science,University of Karachi, Karachi-75270, Pakistan.

Dray A(1992). Mechanism of action of capsaicin like molecules on sensory neurons. *Life Sciences* 51 (23): 1759–65;DOI: 10.1016/0024-3205(92)90045-q.

Eric B (1985). The chemistry of garlic and onions. *Scientific American* 252: 114–9;DOI: 10.1038/scientific american 0385-114.

Gaur RD(1999). Flora of the district Garhwal north west himalaya. Srinagar, Uttarakhand (India): Transmedia Publishers. 811.

Gehm BD, Joanne M Mc A, Chien PY & Larry J (1997). Resveratrol, a poly phenolic com pound found in grapes and wine, is an agonist for the estrogen receptor. *Proceeding National Academy of Sciences* 94(25): 14138–14143; DOI:10.1073/ pnas.94.25.14138.

Hamayun M, Khan MA & Begum S (2003). Marketing of medicinal plants of Utror-Gabral Valleys, Swat, Pakistan. *Ethnobotanical Leaflets* 1:13.

James AD (2000). Herbs of the Bible. Years of Plant MedicinePh.D. 1999, Interweave Press.

Holman RT (1998). The slow discovery of theimportance of omega 3 essential fatty acids in human health. *Journal of Nutrition* 28 (2): 427S–433S; DOI :10.1093/jn/ 128.2.427S.

IhantolaVA, Summanen J, Kankaanranta H, Vuorela H, Asmawi ZM & Moilanen E (1997). Anti-inflammatory activity of extracts from leaves of *Phyllanthus emblica*. *Planta Medica* 63(06): 518-524;DOI: 10.1055/s-2006-957754.

Jain SK (1975). Medicinal Plants. 2nd Revised Edition, National Book Trust, New Delhi, India, pp.1-180.

JunZ (1989). Pure & Application Chemistry 61(3): 457-460,16th International Symposium on the Chemistry of Natural Products, Kyoto, Japan, 29 May–3 June 1988.

Kanchanapoom T, Ryoji K & Yamasaki K (2002). Cucurbitane, hexanorcucurbitane and octanorcucurbitane glycosides from fruits of *Trichosanthestri cuspidata*. *Phytochemistry* 59:215-228;PMID: 11809458

Kou X, Li B, Olayanju J, Drake J & Chen N (2018). Nutraceutical or Pharmacological Potential of *Moringa oleifera* Lam. *Nutrients* 10(3): 343; DOI: 10.3390/nu10030343.

Kumari P, Joshi GC & Tiwari LM (2012). Indigenous uses of threatened ethno medicinal plants used to cure different diseases by ethnic peoples of Almora district of Western Himalaya. *International Journal of Aayurvedic and Herbal Medicine* 2 (4): 661-678.

Lahiri PK & Prahdan SN (1964). Pharmacological investigation of vasicinol-an alkaloid from *Adhatodavasica*Nees. *Indian Journal of Experimental Biology* 2: 219-223.

Lewington A. (1993). Medicinal plants and plant extracts: A review of their importation into Europe. Traffic International, Cambridge, UK.

Mac Donald RE & Bishop CJP (1952). Phloretin an antibacterial substanceobtained from apple leaves. *Canadian Journal of Botany* 30: 486-89; DOI : 10.1139/b52-035.

Machowetz A, Poulsen HE, Gruendel S, Weimann A, Fitz M & Marrugat J (2007). Effect ofolive oils on biomarkers of oxidative DNAstress in Northern and SouthernEuropeans. *FASEB Journal* 21: 45-52 : DOI :10.1096/fj.06-6328com.

Mc Gaw LJ, Jager AK, Van SJ & Eloff JN(2002). Isolation of β-asarone, an antibacterial and anthelmintic compound, from *Acorus calamus* in South Africa. *South African Journal of Botany* 68(1): 31-35;DOI: 10.1016/S0254-6299(16)30450-1.

Misra S, Maikhuri RK, Kala CP, Rao KS & Saxena KG (2008). Wild leafy vegetables: A study of their subsistence dietetic support to the inhabitants of Nanda Devi Biosphere Reserve, India. *Journal of Ethnobiology and Ethnomedicine* 4: 15; DOI: 10.1186/1746-4269-4-15.

Mohanta B, Chakraborty A, Sudarshan M, Dutta R.K & Baruah M (2003). Elemental profile in some common medicinal plants of India, Its correlation with traditional therapeutic usage. *Journal of Radioanalytical and Nuclear Chemistry* 258 (1): 175-179; DOI:10.1023/A:1026291000167.

Mukul SA, Uddin M B, & Tito M R (2007). Medicinal plant diversity and local healthcare among the people living in and around a conservation area of Northern Bangladesh. *International Journal of Forest Usufructs Management* 8(2): 50-63.

Muslim CM&Sikander S (2010). Ethnobotany of medicinal plants of Leepa Valley (AzadKashmir). *Pakistan Journal of Forestry* 60: 2.

Naiara BV, Encarna GP, JoseMLR, Rocio GM & Ana BBO (2011).The extraction ofAnthocyanins and Proanthocyanidinsfrom grapes to wine during fermentative maceration is affected by the enological technique. *Journal of Agriculture and Food Chemistry* 59 (10): 5450–5455; DOI: 10.1021/jf2002188.

Nigam MC, Ahmad A & Misra LN (1990). GC-MS examination of the essential oil of *Acorus calamus, Indian Perfume* 34: 282–285;ISSN : 0019-607X.

Okui S, Uchiyama M, MizugakiM (1963). Metabolism of hydroxy fatty acids. II. Intermediates of the oxidative breakdown ofricinoleic acid by genus Candida. *Journal of Biochemistry* 54: 536–540; DOI: 10.1093/oxfordjournals. jbchem.a127827.

Pandey N, Meena RP, Rai SK & Rai SP (2011). Medicinal plants derived nutraceuticals: A Re-emerging health aid. *International Journal of Pharma and Biosciences* 2(4): 420-421; ISSN. 0975-6299.

Parmar C & Kaushal MK (1982).Wild fruits of the Sub-Himalayan region. Kalyani Publishers, New Delhi, India, pp 136.

Pattanayak P, Behera P & Panda S K (2010). An overview of *Ocimum sanctum* Linn.A reservoir plant for therapeutic application. *Pharmacognosy Review* 4 (7):95-105; DOI : 10.4103/0973-7847.65323.

Perumal SR & Gopalakrishnakone P (2010). Therapeutic potential of plants as antimicrobials for drug discovery. *Evidence-based Complementary and Alternative Medicine* 7(3): 283-294; DOI: 10.1093/ecam/nen036.

PrakashA (2011). Uses of some threatened and potential ethnomedicinal plants among the tribals of Uttar Pradesh and Uttrakhand in India. In National Conference on Forest Biodiversity—Earth's Living Treasure (pp. 93-99).

Prasad K (2005). Hypo cholesterolemic and anti-atherosclerotic effect of flax lignin complex isolated from flaxseed.*Atherosclerosis* 179:269–275;DOI: 10.1016/ j.atherosclerosis. 2004.11.012.

Pulliainen K, Nevalainen H, Vakevainen H, Jutila K & Gummer CL (2010). Ananalytical method for the determination ofbetaine (trimethylglycine) from hair. *International Journal of Cosmetic Science* 32: 135–138; DOI: 10.1111/j.1468-2494.2009.00554.x.

Rabelo APB, Tambourgi EB & Pessoa JA (2004). Bromelain partitioning intwo-phase aqueous systems containing PEO-PPO-PEO block copolymers. *Journal of Chromatography B* 807(1): 61-68; DOI: 10.1016/j.jchromb.2004.03.029.

Razaq A, Rashid A,Ali H, Ahmad H & Islam M (2010). Ethnomedicinal potential of plants of Changa Valley District Shangla, Pakistan. *Pakistan Journal of Botany* 42(5): 3463-3475.

Sampath KP, Bhowmik D, Chiranjib PT & Rakesh K (2010). Indian traditional herbs *Adhatoda vasica* and its medicinal application. *Journal of Chemical and Pharmaceutical Research* 2(1): 240-245; ISSN No: 0975-7384.

Saboo SS, Priyanka T, Tapadiya GG & Khadabadi SS (2012). Distribution and ancient-recent medicinal uses of Trichosanthes species. *International Journal of Phytopharmacy* 2(4): 91-97;DOI: https://doi.org/10.7439/ijpp.v2i4.692.

Samant SS & Dhar U (1997). Diversity, endemism and economic potential of wild edible plants of Indian Himalaya. *The International Journal of Sustainable Development & World Ecology* 4(3): 179-191; DOI: 10.1080/13504509709469953.

Santos AR, De Campos RO, Miguel OG, Cechinel-Filho V, Yunes RA & Calixto JB (1999). The involvement of K^+ channels and Gi/o protein in the antinociceptive action of the gallic acid ethyl ester. *European Journal of Pharmacology* 379(1): 7-17; DOI: 10.1016/s0014-2999(99)00490-2.

Sheth SC, Tibrewala NS, Pal PM, Dube BL & Desai RG (1991).Management of behavioural disorders and minimal brain dysfunction. *Probe* 30(3): 222-226.

ShrivastavaN, Srivastava A, Banerjee A & Nivsarkar M (2006). Anti-ulcer activity of *Adhatodavasica*Nees. *Journal of Herbal Pharmacotherapy* 6(2): 43-49; DOI :10.1080/J157v06n02_04.

SinghVK & Singh DK (2008). Pharmacological effects of Garlic (*Allium sativum* L.). *Annual Review of Biomedical Sciences* 10: 6-26;DOI: 10.5016/1806-8774.2008.v10p6.

Spandana U, Ali SL, Nirmala T, Santhi M & Babu SS (2013). A Review on *Tinospora cordifolia*. *International Journal of Current Pharmaceutical Review and Research* 4(2): 61-68.

Srivastava VK, Singh BM, Negi KS, Pant KC & Suneja P (1997). Gas chromatographic examination of some aromatic plants of Uttar Pradesh hills. *Indian Perfumer* 41(4): 129-139.

Steiber A, Kerner J & Hoppel CL (2004). Carnitine: a nutritional, biosynthetic, and functional perspective. *Molecular Aspects of Medicine* 25(5-6): 455-473; DOI: 10.1016/j.mam.2004.06.006.

TidkeSA, Ramakrishna D, Kiran S, Kosturkova G & Ravishankar GA (2015). Nutraceutical potential of soybean. *Asian Journal of Clinical Nutrition* 7: 22-32;DOI: 10.3923/ajcn.2015.22.32.

Tiwari JK, Ballabha R & Tiwari P (2010). Ethnopaediatrics in Garhwal Himalaya, Uttarakhand, India (Psychomedicine and Medicine). *New York Science Journal* 3(4): 123-126.

Vattem DA & Shetty K (2005). Biological functionality of ellagic acid: a review. *Journal of Food Biochemistry* 29(3): 234-266;DOI:https://doi.org/10.1111/j.1745-4514.2005.00031.x.

Ward GM & Johnston FB (1960). Chemical methods of plant analysis, Canada, Department of Agriculture Publication 1064: 59.

Yadav P, Kumar S & Siwach P (2006). Folk medicine used in gynecological and other related problemsby rural population of Haryana. *Indian Journal of Traditional Knowledge* 5(3): 323-326.

Zhang T, Liu H, Liu XT, Xu DR, Chen XQ & Wang Q (2010). Qualitative and quantitative analysis of steroidal saponins in crude extracts from *Paris polyphylla* var. *yunnanensis* and *P. polyphylla* var. chinensis by high performance liquid chromatography coupled with mass spectrometry. *Journal of Pharmaceutical and Biomedical Analysis* 51(1): 114-124;DOI: 10.1016/j.jpba.2009.08.020. Epub 2009 Aug 25.

16

Nature's Anti-diabetic Pharmacy: A Note on Six Valuable Plants

Anupam Changmai and Mohan Lal

Abstract

The percentage of people with obesity and diabetes has increased globally both in developed and developing countries but highest burden is on the indigenous people who are also socially disadvantaged. The modern medical treatment of diabetes is based generally on oral hypoglycaemic agents and insulin. So, there has always been a search for new anti-diabetic agents which are of lower cost and can cure the disease without harming other body functions. Plants being readily available in the Indian-subcontinent have been an important source of drug in the traditional system of medicine in India since time immemorial. The utilization of plants for alternative treatment to combat diabetes using herbal remedies has also been substantiated by World Health Organization. In this communication, we tried to list some scientifically proven anti-diabetic plants commonly available in the Indian subcontinent which can be used in the treatment and management of diabetes.

Keywords: Diabetes, socially disadvantaged, herbal remedies, Indian subcontinent, management of diabetes.

Anupam Changmai and Mohan Lal (✉)

Medicinal Aromatic and Economic Plants Group, BSTD CSIR-North East Institute of Science and Technology, Jorhat-785006, Assam, India

✉*Corresponding author email: drmohanlal80@gmail.com*

Plants for Novel Drug Molecules: Ethnobotany to Ethnopharmacology
Bikarma Singh & Yash Pal Sharma (eds.), (pp. 379-395)

Email: *info@nipabooks.com* Web: *www.nipabooks.com*

1. Introduction

Diabetes is a challenge to the world today. A disease resembling diabetes mellitus was first reported in 1550 B. C. written in papyrus from Egypt. The sweet test of urine was reported by Hindu manuscripts between fifth to sixth centuries. Diabetes mellitus is considered as a health hazard, although diabetes is reported to occur in 2.4 per cent of the population, it is responsible for 8.3 per cent of the National Health budget, accounting for at least £2 billion per year (Health of the Nation 1995).The actual number may vary and can probably much higher. Studies in the USA shows the prevalence of 6.8 per cent in the population aged over than 19 years and 12.3 percent in the 40–75 years age group (Harris et al. 1998).

Changes in lifestyle, human behaviour, environment and globalization have resulted in tremendous change in the health of an individual. The percentage of people with obesity and diabetes has increased globally both in developed and developing countries but the highest burden is on the indigenous people who are also socially disadvantaged (Chen et al. 2012). Diabetes has been a severe problem of India for decades. There has been a tremendous increase in the prevalence of type 2 diabetes in urban Indian adults from $< 3\%$ in the 1970s to $> 12\%$ in 2000 (Ramchandran et al. 2001). The modern medical treatment of diabetes is based generally on oral hypoglycaemic agents and insulin. However, it can also be treated by Indian traditional medicine using anti-diabetic medicinal plants (Chattopadhyay et al. 1993, Ponnachan and Panikkhar 1993, Subramonium et al. 1996).The modern oral hypoglycaemic agents and insulin used in modern medicines may sometimes comes with serious side effects (Prout 1974, Holman and Tuener 1991). Therefore, there has always been a search for new anti-diabetic drug which can cure the disease without harming other body functions. It has become a major health and economic problem worldwide. The adverse effects of conventional medicinal treatments, has led us to search and explore traditional therapeutic agents which are naturally occurring and not harmful as well (Hammeso et al. 2019). Study of traditional medicine and uses of plants for curing diabetes between ethnic groups and the folk of different countries and regions, or brings out valuable information for us to be explored (Jain and Srivastava 2005).Traditionally, the plants are used but a proper study on the doses and the compounds which are responsible is still going on. A proper study towards these plants can avail a life changing drug that could be change the whole scenario of medical science.

2. Anti-Diabetic Plants and Applications

Some traditionally important plants having anti-diabetic properties (Fig.1).

Fig. 1: Antidiabetic plants; **(a)** *Aloe vera;* **(b)** *Andographis paniculata;* **(c)** *Allium sativum;* **d.** *Achyranthes aspera;* **(e)** *Syzygium cumini;* **(f)** *Tinospora cordiafolia.*

2.1. *Aloe vera*

In folk medicine, *Aloe* has been used as a remedy for numerous diseases. Extensive studies are being carried out for curing diabetes using *A.vera* (Manjunath et al. 2016). Aloe gel a known component which is obtained from *A.vera* promotes vasodilatation and promotes homeostasis in tissues surrounding the vascular endotheliums as reported by Heggers et al. 1993. Extracts of *A. vera* showed that it revitalizes pancreatic islet cells by increasing islet mass and increases levels of insulin thus improving glucose metabolism in rats which were diabetic induced (Noor et al. 2017). *A. vera* gel when administered orally negated behavioral deficits at same level as insulin therapy in diabetes induced mice (Tabatabaei et al. 2017). Treating patients suffering from diabetes with *A.vera* may activate immunosystem and may relief complications related to vascular system. Some of these results of scientific papers published gives a scientific basis for the use of *A.vera* for managing diabetes and complications associated with it.

Table 1: *Aloe vera* anti-diabetic property in in-vivo experimental trials

Aloe vera preparation	Administered protocol	Mode Human or Animal	Efficacy	Reference
Aloe vera juice	Orally	Twice a day for two weeks in patients with diabetes	Triglyceride levels and blood sugar in the treated group fell; the levels of cholesterol was undisturbed	Yongchaiyudha et al.1996
High molecular weigh fractions MW: 1000 kDa	Orally	Two tablespoonful (0.05 g) of high molecular weight fraction for 12 weeks three times daily	Significant decrease in blood glucose level which sustained for 6 weeks of the start of the study. The decrease in triglycerides was observed 4 weeks after treatment and continued thereafter. Liver and kidney functions were normal.	Yagi et al.2009
Leaf latex	Orally	Streptozoin induced diabetic mice	Significant antihyperglycemic activities in Streptozoin induced diabetic mice. Other improved parameters incuded areantioxidant activity, lipid profile, body weight, glucose, hypoglycemic and oral tolerance.	Hammeso et al. 2019

2.2. *Andrographis paniculata*

Andrographis paniculata (Burm.f.) Nees is a traditional medicine of South East Asia originated in India and grows widely in Southeast Asian countries(Nugroho et al. 2009). It attains height of 60-70cm and is erect and branched having lanceolate green leaves. The aerial parts and leaves are used mostly for curing ailments. It contains compounds such as 14-deoxyandrographolide, 14-deoxy-11,12-di-dehydroandrographolide,but the major principle component of the plant is andrographolide (Subramanium et al. 2008). The ethanolic extract of *A. paniculata* , showed anti-diabetic effect and hypotriglyceridemic effect which may be accredited to increased glucose metabolism.(Zhang & Tan 2000). Treating obese rats with *A. paniculata* 200mg/kg BW restores irregular metabolic profile of obese rats to standard levels. *A. paniculata* (Burm.f.) Nees showed anti-diabetic effect both in type 1 Diabetic mellitus rats and type 2 diabetis mellitus rats when the purified extract and its active compound andrographolide were used against antidiabetic effects and antihyperlipidemic in high-fructose-fat-fed rat model (Nugroho et al. 2009). When a non-genetic out-bred Sprague-Dawley rat model induced with type 2 diabetes mellitus was treated with leaf water extract of *A.paniculata* the unsual metabolic profiles of the diabeted induced rats restored to normal condition (Akhtar et al. 2016). Augustine et al. (2014) shown that *A. paniculata* effects at molecular levels of insulin secretion significant affecting antihyperlipidemic and antidiabetic sucrose-induced diabetic rat. Blood sugar levels in both glucose and alloxain induced diabetic rat were reduced when treated with *A.paniculata* aqueous extracts. Blood sugar reducing propert was shown by both the extracts of *A.paniculata* i.e ethanolic and aqueous(Hossain et al.2016). The researchers demonstrate that *A.paniculata* has an antidiabetic activity on metabolic profiles of obese rats indicating it can be developed as a natural antidiabetic agent.

2.3. *Allium sativum*

Allium sativum L. is a common spicy flavoring agent also known as garlic has been used since ancient times. It is been cultivated in all over India for its aroma and characteristic flavor and medicinal properties (Zargari 1997). As we know garlic has been used for time immemorial, but recently there has been increase in scientific support of its pharmacological and therapeutics properties. Commercially available garlic preparations in the form of garlic powder, garlic oil and pills are widely used for medicinal purposes, which includes lowering of blood pressure and improvement of lipid profile (Elkayam et al. 2003). One of the researches found that the garlic extract administration significantly decreased cholesterol and serum triglycerides in diabetic experimental rats (Eididi et al. 2006). In accordance with the above data, other workers showed that fresh garlic juice

Table 2: *Andrographis paniculata* anti-diabetic property in in-vivo experimental trials

A.paniculata Preparation	Administered protocol	Mode Human or Animal	Efficacy	Reference
Aqueous and ethanolic extract	Intraperitonial	Alloxain induced diabetic rat	Significant blood sugar lowering effect was exhibited. Moreover, the lowering of blood glucose levels by the aqueous and ethanolic extracts is also comparable.	Hossain et al. 2016
Purified extract and rographolite ($P<0.05$)	The rats received purified extract of *Andrographis paniculata* dose 434.6 mg/kg BW orally, twice daily.	Hyperglycemic rat	The levels of triglyceride, LDL and blood glucose were decreased as compared to controls	Nugroho et al. 2012
Powder (500 mg/kg body weight	Water dissolved plant powder through oral administration	Type II induced diabetic rat model	The induced impairment of glucose transporter 2 protein and insulin receptor were successfully reinstated. The results also showed that the administered powder did not cause a hypoglycemic effect in normal rat thus confirming its activity and potential application in hyperglycemic state	Augustine et al. 2014

Table 3: *Allium sativum* anti-diabetic property in in-vivo experimental trials

Allium sativum preparation	Administration protocol	Animal model	Efficacy	References
Alcoholic extract of garlic	0.25 or 0.5 g/kg bw/day for 14 days through orogastric tube	In rats (70 mg/kg bw, ip)	Increase basal insulin concentration	Eidi et al. 2006
Garlic oil	100 mg/kg bw/2 days for 16 weeks through orogastric tube100 mg/kg bw/2 days for 3 weeks through orogastric tube80 mg/kg bw/2 days for 3 weeks through orogastric tube	Rats (65 mg/kg bw, iv)	1. Improvement in Basal and postglucose loading insulin concentration; 2. Improved insulin 3. Sensitivity and glycogen synthesis in skeletal muscle	Liu et al.2006 Liu et al.2005
Crude garlic extract	100 mg/kg bw/day for 15 weeks through orogastric tube	Streptozotocin induced rats (70 mg/kg bw, ip)	Decreased Fasting blood sugar	Patumraj et al. 2000
Homogenized garlic	Preventive administration with 6.25% w/w in diet for 12 days	Streptozotocin induced mice (200 mg/kg bw, ip)	No effect on Fasting blood sugar; no effect on blood insulin level	Swanston-Falt et al.1990
Garlic juice	A bolus of 25 g garlic/kg bw via oral route	Normal rabbits, concomitant administration with 1g glucose/kg bw	Improved Oral Glucose Tolerence Test	Jain & Vyas 1974
Garlic bulb prepared by decoction	A bolus of 4 mL/kg bw through orogastric tube	Normal rabbits, concomitant administration with 50% dextrose (4 mL/kg bw, sc)	Improved glucose tolerance	Roman-Ramos et al. 1995

Contd.

Allicin	250 mg/kg bw for 1 – 4 h via oral route	Alloxan induced rabbits. Mild diabetes (Fasting Blood Sugar) A 350 mg/dL)	Decreased Fasting blood sugar; increased serum insulin-like activity; improved Oral glucose tolerence; increased liver glycogen concentration, No effect on Fasting Blood Sugar	Mathew et al. 1973

administration or etheric garlic extracts showed improved lipid profile and reduction of cholesterol levels in serum (Knipschild & Ter-Riet 1989). Experimental studies on primary hepatocyte cultures, when used to study the anticholesterogenic properties of garlic, confirms the lipid and cholesterol lowering effect of garlic (Yeh & Yeh 1994). Although garlic was recommended and assumed safe and for many common diseases since time immemoral, too much utilization of garlic must also be avoided. Infrequent allergic response and garlic breathe are predictable adverse effects. Gastrointestinal side effects like diarrhoea and nausea has also been reported to be associated with garlic powder and raw garlic preparation (Sieger et al. 1992, Desai et al.1990)

2.4. *Achyranthes aspera*

Achyranthes aspera is an perennial, procumbent and erect annual herb, with height reaching upto 1-2 metres belonging to the family Amaranthaceae. Commonly found as a weed of the waste places and way sides throughout Indian sub-continent found upto an altitude of 2100 meters including Andaman Islands (Kumar et al. 2011). Administration of *A. aspera* powder orally showed a significant dose-related hypoglycemic effect in diabetic as well as normal in rabbits. The methanol and water extracts also showed decrease in blood glucose levels in normal as well as in alloxan induced diabetic rabbits. Dosages up to 8 g/kg orally did not exhibit any adverse side effects. It was assumed that plant acts by providing certain necessary elements like magnesium, calcium, zinc, copper and manganese to the beta-cells of the pancreas (Mukesh et al. 2013, Akhtar et al.1991). *A. aspera* ethanolic extract showed anti-diabetic activity in experimental rats (Kumar et al. 2011). Similarly, Geetha 2016 showed anti-diabetic action of *A. aspera* with experimental diabetic mice, noticeably by reducing the blood glucose level.

Table 4: *Achyranthes aspera* anti-diabetic property in *in vivo* experimental trials

Achyranthes aspera	Experimental	Effect	References
Powder	Diabetic and Normal rabbit	Produces hyperglycemic effect	Sahu et al.2014
Water extract	Diabetic and Normal rabbit	Decreases blood glucose level	Sahu et al.2014
Methanol extract	Diabetic and Normal rabbit	Decreases blood glucose level	Sahu et al.2014
The whole plant	Antidiabetic agents	Ca^{+2} , Zn^{+2} , Mg^{+2}, Mn^{+2} , Cu^{+2} to the beta-cells	Rawat et al.2013, Akhtar et al.1991

2.5. *Syzygium cumini*

Syzygium cumini also known as Eugenia jambolana found throughout India extending from Himalaya to South India its parts bark, seeds, leaves, fruits is being used for treatment of human disorders in Ayurvedic medicine from very early days(Nadkarni 1954).Aqueous suspension of seed kernel of *S.cumini* when screened for antidiabetic activity showed hypoglycaemic effect when diabetic induced rabbit were medicated with *S.cumini*. It also showed decrease in blood sugar levels in alloxan induced diabetic rats (Nair & Santhakumari 1986). Perera et al.2017 showed the presence of antidiabetic compounds such as umbelliferone, ellagic acid and gallic acid to be present in the decoction of *S. cumini* which can be consumed as a ready to serve drink also mediators such as phenolic compounds were present which showed antiglycation and antioxidant activities indicating its role in mediating its effects in antidiabetic activity and suggesting it to be used in herbal formulations. A study conducted by Prabhakaran et al. 2017 showed that *S.cumini* methanolic extracts possesses phytochemicals of high therapeutic value such as flavonoids, steroids, saponins, phenols, raisins, terpenoids, cardiac glycosides etc. Their study indicated that seed extract of *S.cumini* has potent á-amylase inhibitor activity with high degree of inhibition. Their study suggest the regional pharmaceutical importance of *S.cumini* seed as traditional medicinal plant for the management of diabetes and as a probable candidate for the development of pharmacologically active drug.Vihan et al.2017 showed that *S. cumini* extract had no effect on the blood glucose levels of euglycemic animals but *it* can reduce levels of blood glucose in high fructose diet induced diabetic rats, in a time dependent and dose dependent manner.

2.6. *Tinospora cordifolia*

Tinospora cordifolia has been used in ayurvedic system of medicine since time immemorial (Grover et al. 2002). Shoba et al. 2015 demonstrated hypoglycemic (antidiabetic) activity when alcoholic extract of *T. cordiafolia* was intraperitoneally injected in alloxan treated albino rats. Significant decrease in sugar levels in blood was observed when alcoholic extract was injected at concentration of 20ml/kg bodyweight of test rats. Studies conducted by Puranik et al. 2010 showed that *T. cordifolia* had efficacy of 40% to 80% compared to insulin on diabetic animals. They showed that it decreases glycogenphosphorylase activity and increases hepatic glycogen synthase activity. Patel et al. 2011demonstrated that three alkaloids in *T. cordifolia* viz jatrorrhizine, magnoflorine and palmatine possesses insulin-mimicking activity. Their results showed that these alkaloids possessed antihyperglycemic activity.

Table 5: *Syzygium cumini* anti-diabetic property in in-vivo experimental trials

Syzygium cumini Preparation	Administered protocol	Mode Human or Animal	Efficacy	Reference
Aqueous suspension	Orally	Rabbit ,rats 1g,2g,4g and 6g/kg body weight	Maximum hypoglycemic activity of 42.64% was exhibited at dose levels of 4g/kg in rabbits. In rats it showed hypoglycemic activity of 17.04% in alloxan diabetic rats.	Nair & Santhakumari 1986
Suspension	Orally	Albino rats Dosage of 200mg/kg, 400mg/kg and 800mg/Kg respectively.	All the concentration of (200,400 and 800mg/kg) showed a significant reduction in the blood glucose level.	Vihan & Brashier (2017)

Table 6: *Tinospora cordifolia* anti-diabetic property in in-vivo experimental trials

***Tinospora cordifolia* Preparation**	**Administered protocol**	**Mode Human or Animal**	**Efficacy**	**Reference**
Alcoholic extract	Intraperitoneal	Albino rats180mg/kg body weight	Visible decrease in the blood sugar level	Shoba et al. 2015
Aqueous extract	Orally	Albino Rats different dosages (200 and 400 mg/kg b.w.)	Possesses anti-diabetic activity comparable to 40-80% of insulin	Puranik et al. 2010
Extract in DMSO	In-vitro	In vitro cell culture	Possesses insulin mimicking and insulin releasing activity	Patel & Mishra 2011

3. Discussion

India known as the botanical garden of the world is believed to produce a large amount of medicinal herbs. Out of 21,000 species listed by the World Health Organization that can be used for medicinal purpose around 2500 species are found in India (Seth et al. 2004). There exist a vast number of plants in India having anti-diabetic potential as per Ayurveda (Rizvil and Mishra 2013). A lot of plans are yet to be discovered and only a few has been scientifically proven which can be used for the successful treatment and management of diabetes. Research on ethnomedicinal plants are being performed extensively in the Indian Subcontinent(Grover et al.2002, Mukherjee et al. 2006). There are no reports of complete cure of diabetes and a person recovering totally from it. Therefore, trying to cure diabetes with regular anti-diabetic drugs that makes a patient prone to side effects or are either too expensive for the patient(Dey et al. 2002). The above mentioned medicinal plants may prove to act as better alternative owing to its lower cost and lesser side effects (Modak et al. 2007). The utilization of plants for alternative treatment to combat diabetes using herbal remedies has also been substantiated by WHO (Mandik et al. 2008).

4. Conclusion

Diabetes is one of the greatest problems of mankind and India being the diabetic capital of the world consequently suffers the greatest. Curing diabetes with regular anti-diabetic drugs makes a patient prone to side effects or are either too expensive. Treatment with medicinal plants is considered safe with no or minimal side effect and India being a rich repository, can go a long way in overcoming this problem. It is evident from researches that the biodynamic compounds in these medicinal plants has therapeutic value. As the knowledge about the therapeutic application increases there is a revival in the application of these low cost and less side effective plants to cure diabetes. Therefore, combining traditional knowledge and scientific experiments on phytochemical, pharmalogical, ecological and toxicological studies is the need of the hour India being a biodiversity hotspot with a huge number of ethnic groups with their traditional knowledge may act as the Nature's Anti Diabetic Pharmacy in the years to come.

Abbreviation used

Mw-molecular weight, kDa-kilodalton, BW-bodyweight, LDL-low density lipoprotein, DMSO-Dimethyl sulfoxide.

Conflict of Interest

We declare no conflict of interest.

Acknowledgements

The authors would like to thank Director CSIR-NEIST, Jorhat for encouraging and providing necessary support and CSIR-Phytopharmaceutical Mission for granting the fund in the form of CSIR network project HCP0010.

References

Akhtar MS & Iqbal J (1991). Evaluation of the hypoglycaemic effect of *Achyranthes aspera* in normal and alloxan-diabetic rabbits. *Journal of Ethnopharmacology* 31: 49-57.

Akhtar MT, Sarib BM, Ismail MS, Abas IS, Ismail F, Lajis A & Shaari, K(2016). Anti-Diabetic Activity and Metabolic Changes Induced by *Andrographis paniculata* Plant Extract in Obese Diabetic Rats. *Molecules*. 21(8). DOI: 10.3390/molecules21081026.

Augustine AW, Narasimhan A, Vishwanathan M & Karundevi B (2014).Evaluation of antidiabetic property of *Andrographis paniculata* powder in high fat and sucrose-induced type-2 diabetic adult male rat. *Asian Pacific Journal of Tropical Disease* 4(1): 140-147.

Chattopadhyay RR, Medd CS, Das S, Basu TK & Podder G(1993). Hypoglycaemic and anti-hyperglycaemic effect of Gymnema sylvestre leaf extract in rats. *Fitoterapia* 64: 450–454.

Chen L, Magliano DJ & Zimmet PZ(2012). The worldwide epidemiology of type 2 diabetes mellitus—present and future perspectives. *Nature Reviews Endocrinology* **8:** 228–36.

Desai HG, Kalro RH & Choksi AP (1990). Effect of ginger and garlic on DNA content of gastric aspirate. *Indian Journal Medical Research* 92: 139-141.

Dey L, Attele AS, Yuan CS (2002).Alternative therapies for type 2 diabetes. *Alternative Medicine Review* 7: 45-58.

Eidi A, Eididi M & Esmaeili E (2006).Antidiabetic effect of garlic (Allium sativum L.) in normal and streptozotocin-induced diabetic rats. *Phytomedicine* 13:624–629.

Elkayam A, Mirelman D & Peleg E (2003). The effects of allicin on weight in fructose-induced hyperinsulinemic, hyperlipidemic, hypertensive rats. *American Journal of Hypertension* 16: 1053–1056.

Geetha K (2016). Antidiabetic activity of Achyranthes aspera L. with alloxanised mice for the estimation of level of glucose and cholesterol. *Asian Journal of Plant Science and Research* 6(2): 18-23.

Grover JK, Yadav S & Vats V (2002). Medicinal plants of India with anti-diabetic potential. *Journal Ethnopharmacology* 81(1): 81-100.

Hammeso WW, Emiru YK, Getahun KA & Kahaliw W (2019). Antidiabetic and Antihyperlipidemic Activities of the Leaf Latex Extract of Aloe megalacantha Baker (Aloaceae) in Streptozotocin-Induced Diabetic Model. *Evidence-based Complementary and Alternative Medicine* 1:1-9.

Hasaim MS, Vrbi Z, Sule A & Rahmen KMH (2014) Andrographis paniculata (Burm. f.) Wall.ex Nees: A Review of Ethnobotany, Phytochemistry and Pharmacology. *The scientific World Journal*. Doi.ong/10.1155/2014/274905

Harris M, Flegal KM, Cowie CC, Eberhardt MS, Goldstein DE & Little RR (1998). Prevalence of diabetes, impaired fasting glucose, and impaired glucose tolerance on US adults. The third nation and nutrition examination survey. *Diabetes Care* 21: 518-526.

Health of the Nation: A progress Report London HMSO; HC 656 Session 1995-96 Published 14 August 1996.

Heggers JP, Pelley RP & Robson MC (1993). Beneficial effects of aloe in wound healing.

Holman RR & Turner RC (1991). Oral agents and insulin in the treatment of NIDDM, pp. 467–469. In: Pickup J & Williams G. *Text Book of Diabetes*. Blackwell, Oxford.

International Diabetes Federation (2009).*IDF Diabetes Atlas*. Brussels: 4th Edition.

Jain RC & Vyas CR (1974). Letter: Hypoglycaemia action of onion on rabbits. *British Medical Journal*. DOI:10.1136/bmj.2.5921.730-b.

Jain SK & Srivastava S (2005). Traditional uses of some Indian plants among islanders of the Indian Ocean. *Indian Journal of Traditional Knowledge* 4(4): 345-357.

Knipschild JK & Ter-Riet G (1989). Garlic, onions and cardiovascular risk factors. A review of the evidence from human experiments. Emphasis on commercially available preparations. *British Journal of Clinical pharmacology* 28: 535–544.

Kumar SA, Gnananath K, Gande S, Goud RE, Rajesh P , Nagarjuna S (2011). Anti diabetic activity of ethanolic extract of Achyranthes aspera leaves in Streptozotocin induced diabetic rats. *Journal of pharmacy Research* 4(7): 3124-3125.

Liu CT, Hse H, Lii CK, Chen PS & Sheen LY (2005). Effects of Garlic Oil and Diallyl trisulfide on glycemic control in rats with Streptozotocin-Induced diabetes. *European Journal of Pharmacology* 516:165 – 173.

Liu CT, Wong PL, Lii CK, Hse H & Sheen LY (2006). Antidiabetic effect of garlic oil but not diallyl disulfide in rats with streptozotocin-induced diabetes. *Food and Chemical Toxicology* 44: 1377 – 1384.

Mandlik RV, Desai SK, Naik SR, Sharma G & Kohli RK (2008). Antidiabetic activity of a polyherbal formulation (DRF/AY/5001).*Indian Journal of Experimental Biology* 46:599-606.

Manjunath K , Bhanu PG, Subash KR, Tadvi NA, Manikanta M & Umamaheswara RK. (2016). Effect of Aloe vera leaf extract on blood glucose levels in alloxan induced diabetic rats. *National Journal of Physiology, Pharmacy and Pharmacology* 6(5): 471-474.

Mathew PT & Augusti KT(1937). Studies on the effect of allicin (diallyl disulphide-oxide) on alloxan diabetes. I. Hypoglycaemic action and enhancement of serum insulin effect and glycogen synthesis. *Indian Journal of Biochemistry Biophysics* 10: 209–121.

Modak M, Dixit P, Londhe J, Ghaskadbi S & Devasagayam TP (2007) Indian herbs and herbal drugs used for the treatment of diabetes. *Journal of Clinical Biochemistry and Nutrition* 40(3): 163–173.

Mukherjee PK, Maiti K, Mukherjee K & Houghton PJ (2006). Leads from Indian medicinal plants with hypoglycemic potentials. *Journal of Ethnopharmacology* 106(1): 1-28.

Nadkarni KM (1954). *Indian Materia Medica* Vol. 1, Popular Book Depot, Bombay 7.

Nair R.B & Santhakumari.G (1986). Anti – diabetic activity of the seed kernel of *Syzygium cumini* linn. *Ancient Science of Life* 6(2): 80–84.

Noor A , Gunasekaran S & Vijayalakshmi MA (2017). Improvement of insulin secretion and pancreatic â-cell function in streptozotocin-induced diabetic rats treated with Aloe vera extract. *Pharmacognosy Research* 9(5): 99-104.

Nugroho AE, Andrie M, Warditiani NK, Siswanto E, Pramono S & Lukitaningsih E(2016). Antidiabetic and antihiperlipidemic effect of *Andrographis paniculata* (Burm. f.) Nees and andrographolide in high-fructose-fat-fed rats. *Indian Journal Pharmacology* 44(3): 377-81.

Patel MB & Mishra S (2011). Hypoglycemic activity of alkaloidal fraction of *Tinospora cordifolia*. *Phytomedicine* 18: 1045– 1052.

Patumraj S, Tewit S, Amatyakul S & Jariyapongskul A (2000). Comparative effects of garlic and aspirin on diabetic cardiovascular complications. *Drug Delivery* 7: 91 – 96.

Perera PRD, Ekanayake S & Ranaweera KKDS(2017). Antidiabetic Compounds in Syzygium cumini Decoction and Ready to Serve Herbal Drink Evidence-Based *Complementary and Alternative Medicine*. DOI.org/10.1155/2017/1083589. *Phytotherapy Resarch* 7:48- 52.

Ponnachan TC & Panikkhar KK (1993). Effect of leaf extract of Aeglemarmelos in diabetic rats. *Indian Journal of Experimental Biology* 31: 345–347.

Prabakaran K & Shanmugavel G (2017). Antidiabetic Activity and Phytochemical Constituents of *Syzygium cumini* Seeds in Puducherry Region, South India International *Journal of Pharmacognosy and Phytochemical Research* 9(7):985-989.

Prout TE, Malaisse WJ & Pirart J (1974). *Proceedings VIII Congress of International Diabetes Federation* Published by Excerpta Medica, Amsterdam, 162.

Puranik N, Kammar K.F & Devi.S (2010). Anti-diabetic activity of *Tinospora cordifolia* (Willd.) instreptozotocin diabetic rats; does it act like sulfonylureas? *Turkish Journal of Medical Science* 40 (2): 265-270.

Ramachandran A, Snehalatha C, Kapur A, Vijay V, Mohan V, Das AK, Rao PV, Yajnik C S, Kumar Prasanna KM & Nair JD (2001). High prevalence of diabetes and impaired glucose tolerance in India: National Urban Diabetes Survey. *Diabetologia* 9: 1094–1101.

Rammohan SR, Asmawi MZ, & Sadikun A (2008).In vitro á-glucosidase and á-amylase enzyme inhibitory effects of *Andrographis paniculata* extract and andrographolide. *Acta Biochimica Polonica* 55(2) 391–398.

Rawat M & Parmar N (2013). Medicinal Plants with Antidiabetic Potential - A Review. *American- Eurasian Journal Agriculture and Environmental Science* 13 (1): 81-94.

Rizvil S I & Mishra N (2013). Review Traditional Indian Medicines Used for the Management of Diabetes Mellitus . *Journal of Diabetes Research*. DOI: 10.1155/2013/712092.

Sahu U, Tiwari SP & Roy A (2015). Comprehensive Notes on Antidiabetic Potential of Medicinal plants and Polyherbal Formulation. *UK Journal of Pharmaceutical and Biosciences* 3(3):57-64.

Seth SD & Sharma B(2004). Medicinal plants of India. *Indian Journal of Medical Research* 120: 9–11.

Shobha BK and Gopal PK (2015). Antidiabetic activity of *Tinospora cordifolia* (fam: menispermaceae) in alloxan treated albino rats. *Applied Research Journal* 1(5) 316-319.

Siegers CP (1992). *Allium sativum*. In: Adverse effects of herbal drugs (DeSmet PAGM, Keller K, Hansell R and Chandler R F eds). Published by Springer-Verlag, Berlin, Germany,73-77.

Subramonium A, Pushangadan P & Rajasekaran S (1996). Effects of Artemisia pallens wall on blood glucose levels in normal and alloxaninduced diabetic rats. *Journal of Ethnopharmacology* 50:13–17.

Swanston-Flatt SK, Day C, Bailey CJ & Flatt PR (1990). Traditional plant treatments for diabetes. Studies in normal and streptozotocin diabetic mice. *Diabetologia* 33:462 – 464.

Tabatabaei SRF, Ghaderi S, Bahrami-TM, Farbood Y & Rashno M (2017). Aloe vera gel improves behavioral deficits and oxidative status in streptozotocin-induced diabetic rats. *Biomedicine and Pharmacotherapy* 96:279-290.

Vihan & Brashier (2017). A study to evaluate the antidiabetic effect of *Syzqium* cumini Linn. Seed extract in high fructose diet induced diabetes in Albino rats. *International Journal of Basic and Clinical Pharmacology*. 6(6): 1363-1366.

Yagi A, Hegazy S, Kabbash A & Wahab E A E (2009). Possible hypoglycemic effect of *Aloe vera* L. high molecular weight fractions on type 2 diabetic patients. *Saudi Pharmaceutical Journal* 17(3): 209–215.

Yeh YY & Yeh SM(1994). Garlic reduces plasma lipids by inhibiting hepatic cholesterol and triacyglycerol synthesis. *Lipids* 29: 189–193.

Yongchaiyudha S, Rungpitarangsi V, Bunyapraphatsara N, Chokechaijaroenporn O (1996). Antidiabetic activity of Aloe vera L. juice. I. Clinical trial in new cases of diabetes mellitus. *Phytomedicine* 3(3): 241-243.

Zargari A (1997). *Medicinal Plant*. Published by Tehran University Press, Iran,3, 319–329.

Zhang XF &Tan BK (2000). Anti-diabetic property of ethanolic extract of *Andrographis paniculata* in streptozotocin-diabetic rats. *Acta Pharmacologica Sinica* 12: 1157-64.

17

Blumea lacera (Asteraceae), A Potential Herb of Medicinal Value in Modern Aspects

Tarkeshwar Dubey, Shreyans Kumar Jain and Siva Hemalatha

Abstract

Blumea lacera *(Burm f.) DC. belonging to family Asteraceae is an annual herb with a strong odour. This herb is distributed throughout the plains of North-west India upto an elevation of 2,900 m and is also reported from Tropical Africa, South East Asia, Bangladesh, Nepal, Indonesia, Malaysia, Philippines, Thailand and Vietnam.* Blumea lacera *is a plant with camphor-like odour, and is highly variable, grows in grassland, field, roadside and forest edge. Traditionally, it is used internally and externally as a hemostatic and anti-inflammatory agent. The juice of leaves along with black pepper (Piper nigrum) is given to cure bleeding piles. This herb is also offered as anthelmintic (especially for threadworm). Leaves and roots are used as astringent, diuretic and febrifuge and even in cuts and wounds. Bruises and ulcers can also be treated by applying fresh juice or extract. For treating piles (Hemorrhoids), the leaves of B. lacera has been used in ethnomedicinal practices.The phytoconstituents present in leaves are mainly terpenoids, glycosides, volatile oils, some glycoalkaloids and many more yet to be determined. B. lacera has been used as an ingredient in many pharmaceutical preparations. The chapter gives a brief idea about the ethnomedicinal usage, phytochemical profile, pharmacological usage and the future perspective of B. lacera. and would help the researchers for better understanding of this plant as well as developing plant derived molecules in drug discovery process.*

Keywords: Herbal medicine, terpenoids, pharmacology, ethnomedicine phytoconstituents, tropical drug.

Tarkeshwar Dubey, Shreyans Kumar Jain and Siva Hemalatha(✉)

Department of Pharmaceutical Engineering and Technology, Indian Institute of Technology Banaras Hindu University, Varanasi-221005, Uttar Pradesh, India

✉*Corresponding author email: shemalatha.phe@itbhu.ac.in*

Plants for Novel Drug Molecules: Ethnobotany to Ethnopharmacology
Bikarma Singh & Yash Pal Sharma (eds.), (pp. 397-410)

1. Introduction

From the time immemorial herbs have been used in all cultures throughout history to cure diseases (Doughari 2012). The evidence of the use of herbal remedies some sixty thousand years ago by Neanderthal man was revealed in 1960 in a cave in Northern Iraq (Solecki 1975). As per WHO, 25% of modern medicines are derived from the herbalorigin, and almost 80% of world's population is dependent on herbal medicine for their primary care treatment more abundantly in developing countries like India. Though, herbal drugs are of great interest still the complete systematic information of all the herbs used as medicine is not yet documented. Thus, herbal drugs lack wide acceptance.

Blumea lacera (Burm.f.) DC.is an annual herb with a strong odor, distributed throughout the plains of Northwest India, up to an altitude of 2,000m (Khare 2004). A herbaceous plant, tomentose, capitula in axillary and terminal dense to lax panicles, achene, ribbed, pappus whitish, camphor-like odor, highly variable, in grassland, field, roadside and forest edge (Pornpongrungrueng et al. 2016). The stems of this hairy or glandular herb are erect, simple or branched, very leafy and 1-2 ft in height. The leaves are obovate or oblanceolate, 5-12 cm long, 2-6 cm wide, smaller toward the top, stalked, and toothed or (rarely) lobulated at the margins, leaves are eaten as a vegetable. The bright yellow flowering heads are about 8 mm across, borne on short axillary cymes, and collected in the terminal, spike-like panicles. The involucre-bracts are narrow and hairy (Fig. 1). The achenes are not ribbed, are somewhat 4-angled, and are smooth (Tarkeshwar 2019).

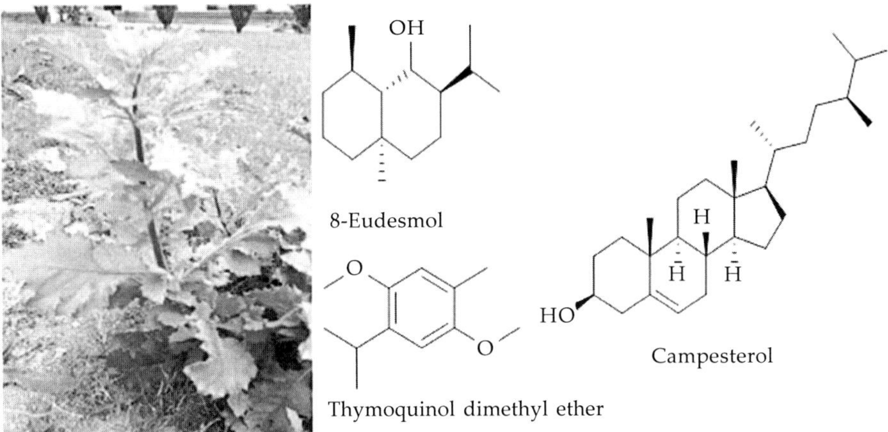

Fig. 1: Habit of *Blumea lacera* and marker compounds.

2. Distribution

Tropical Africa, South East Asia, throughout the plains of India, ascending to 700 m, Bangladesh, Nepal, Indonesia, Malaysia, Philippines, Thailand and Vietnam (Khare 2004).

3. Classification

- Kingdom: Plantae
- Division: Tracheophyta
- Class: Magnoliopsida
- Order: Asterales
- Family: Asteraceae
- Genus: *Blumea* DC.
- Species: *Blumealacera* (Burm f.) DC.

4. Classical and Common Names

Ayurvedic: Kukundara, Kukkuradru, Tamrachuda, Mriducchada; Kukrondaa; Unani: Kakrondhaa; Siddha: Narakkarandai; Hindi: Jangli Muli, Kakronda; Marathi: Bhamurda, Burando; Tamil: Kattumullangi, Narakkarandai; Telugu: Advimulangi, Karupogaku; Bengali: Kukurmuta, Kukursunga; Gujarati: Pilo Kapurio.

5. Synonymous Names

Blumea bodinieri Vaniot, *Blumea chevalieri* Gagnep., *Blumea glandulosa* Benth. & DC, *Blumea glutinosa* DC., *Blumea lacera* (Roxb.) DC., *Blumea lacera* var. *cinerascens* Hook.f., *Blumea lacera* var. *glandulosa* Hook.f., *Blumea mollis* (D.Don) Merr., *Blumea subcapitata* DC.; *Blumea thyrsoidea* Sch.; *Blumea velutina* (H.Lév. & Vaniot), *Conyza dentata* Blanco; *Conyza dentata* Willd.

6. Medicinal Use in Traditional System

According to Bhaavaprakaasha, the herb can cure fever, bronchial infections, vitiated blood (Khare 2004). Traditionally, it is used internally and externally as a styptic and anti-inflammatory agent. The juice of leaves is mixed with black pepper (*Piper nigrum*) and given to cure bleeding piles. This herb is also given as anthelmintic, particularly for threadworm. Leaves and roots are used as astringent, diuretic and febrifuge (Vijayakumar et al. 2007). Bruises and ulcers can also be treated by applying Fresh juice or extract (Umberto 2012). Some major ethnomedicinal uses are provided in Table-1.

7. Phytochemistry

Several phytoconstituents have been reported from *Blumea lacera* especially essential oils. The isolation and characterization of constituents was done mainly from the leaves and aerial parts of this plant using mainly alcoholic, hydroalcoholic, hexane or petroleum ether extracts (Fig. 2). β- Bourbonene, Thymol methyl ether, E-β-Bergamotene, β -Caryophyllene, E-β- Farnesene, β-Humulene, β–Himachalene, Cadinene, Thymoquinol dimethyl ether (33.9%), β- Caryophyllene epoxide, β- recocene I, 8-Eudesmol, thymol hydroquinone-dimethylether, caryophyllene oxide, Neophytadiene & Cetane have been reported in *Blumea lacera* through GC-MS studies. Some other constituents like Gneol, fenchone, coniferyl alcohol derivatives, campesterol, flavonoids, lupeol, hentriacontane, hentriacontanol, alpha-amyrin, beta-sitosterol. Triterpenes have also been reported in this herb. Apart from this, a few glycosides (19α- hydroxyurs-12-ene-24,28-dioate 3-O-β-D-xylopyranoside, 2-isoprenyl-5-isopropylphenol 4-O-fl-D-xylopyranoside, β-pinene-7 β-O-β-D-2,6-diacetylglucopyranoside), and flavonoids (5-Hydroxy 3,6,7,3',4' pentarnethoxyflavone, 5,3',4'-trihydroxy-3,6,7 trimethoxyflavone, 5,4'-dihydroxy-6,7,3'-trimethoxyflavone, 3,5,4'-trihydroxy-6,7,3'-trimethoxyflavone) have been isolated from *Blumea lacera* (Jha et al. , Rao et al. 1977, Agarwal et al. 1995, Mukherjee 2006, Ragasa et al. 2007, Chen et al. 2009, Tiwari et al 2012; Satyal et al. 2015).

Table 1: Ethnomedicinal uses of *Blumea lacera*

Ethnomedicinal uses	Parts used	Region/Tribe	Method of use	References
Diuretic/edema, GI disorders, respiratory tract infection, insect repellent	-	Bagerhat & Rampal, Bangladesh	-	(Mollik, Hossan et al. 2010)
Insecticide	Leaves	Vasubihar village, Bangladesh	Dried leaves are spread around rooms	(Rahmatullah, Islam et al. 2010)
Cuts & bleeding wounds	Leaves	Thane, Maharashtra, India	Juice from squeezed leaves	(Natarajan and Paulsen 2000)
Bruises of toe, cuts & wounds	Leaves	Gorakhpur, U.P. India	Leaf juice (externally)	(Pandey and Tripathi 2011)
Piles	Leaves	Bhoxas tribes, U.P. India	Fresh ground leaves with black pepper	(Harish 1988)
Antihelminthic, febrifuge, diuretic, antiscorbutic, antimicrobial	-	Western Ghats, India	-	(Rao RR, Kavitha et al. 2006)
Fever	Root	Myagdi district, Nepal	Juice of root	(Manandhar 1995)
i) Piles & choleraii)	Root	Trayambakeshwar hill,	-	(Kakulte, Gaikwad
Mouth diseases	Leaves	Nashik, Maharashtra, India		et al.)
Filariasis	Leaves	Musohor tribe, Birganj, Bangladesh	Burned ash of wasp's nest, hornet's nest & *Bambusa glaucescens* with juice of leaves	(Rahman and Rahmatullah 2015)
Indigestion	Leaves	North Bengal plain region, India	-	(Mitra and Mukherjee 2010)
Bleeding piles	-	Chandigarh region, India	-	(Singh 2005)
Cuts, antimicrobial	Leaves	Bihar, India	-	(Upadhyay, Kumar et al. 1998)
Cuts	Leaves	Bantar tribe. Nepal	-	(Acharya and Pokhrel 2006)

Bronchitis, bleeding piles, burning sensation	Leaves	Bhilada tribe, Maharashrastra, India	Extract given orally	(Korpenwar 2012)
Diarrhoea	Root	Tribals of Bargarh district, Orissa, India	Root paste with honey three times a day	(Sen and Behera 2008)
Dog bite	Root	Minority group of Christians, Dinajpur district, Bangladesh	Crushed roots	(Rahmatullah, Kabir et al. 2010)
Blood purifierii) Urinarycomplications	Roots Leaves	Tirunelveli distt. Tamil Nadu, India	-	(Vanila, Ghanthikumar et al. 2008)
Earache, fever, killing the worms	Leaves	Hosangabaddistt. M.P. India	Leaf juice	(Quamar and Bera 2014)
Cuts & Wounds	Leaves	All over India	-	(Kumar, Vijayakumar et al. 2007)
Body ache, intestinal worms, malarial fever, haemostatic, piles (topically), antifungal	Leaves	Chitrakoot, Satna, M.P. India	Decoction crushed leaves	(Beg 2015)
Body ache, intestinal worms, malarial fever, haemostatic, piles (topically), antifungal	Leaves	Eastern U.P. India	Decoction crushed leaves	(Beg 2015)
Cutaneous infection	Root	Meche people, Jhapa district, Eastern Nepal	Root paste	(Rai 2004)
Muscular pain	Whole plant	Jalpaiguri district, W.B. India	-	(Bose, Roy et al. 2015)
Anthelmintic, febrifuge, astringent, diuretic, bleeding piles, cholera	Leaf, Root	Naogaon district, Bangladesh	Leaf juice mixed with black pepper & root juice with black pepper	(Rahman 2013)
Anthelmintic, febrifuge, astringent, diuretic, bleeding piles, cholera	Leaf, Root	Rajshahidivision, Bangladesh	Leaf juice mixed with black pepper & root juice with black pepper	(Rahman 2013)
i) Dog biteii) Leucorrhoea	Leaves Seed	Sahjahanpur district, U.P. India	Leaf juice with black pepper & seed powder	(Sharma, Painuli et al. 2010)

Piles	-	Tribal areas of M.P. India	Dried powdered material is taken as smoke	(Choubey, Dubey et al. 2015)
Blood dysentery in children	Root	Candaka tribals of Orissa, India	Fresh root juice with honey	(Prasad, Lawania et al. 2009)
Fever, conjunctivitis, piles	Leaves	Kalinjar hillock, Banda, U.P. India	-	(Prusti 1998, Mishra 2015)
i) Cuts & Wounds ii) Febrifuge, diuretic, antiscorbutic	Leaf Whole plant	Southern district of West Bengal, India	-	(Pattanayak, Dutta et al. 2012)

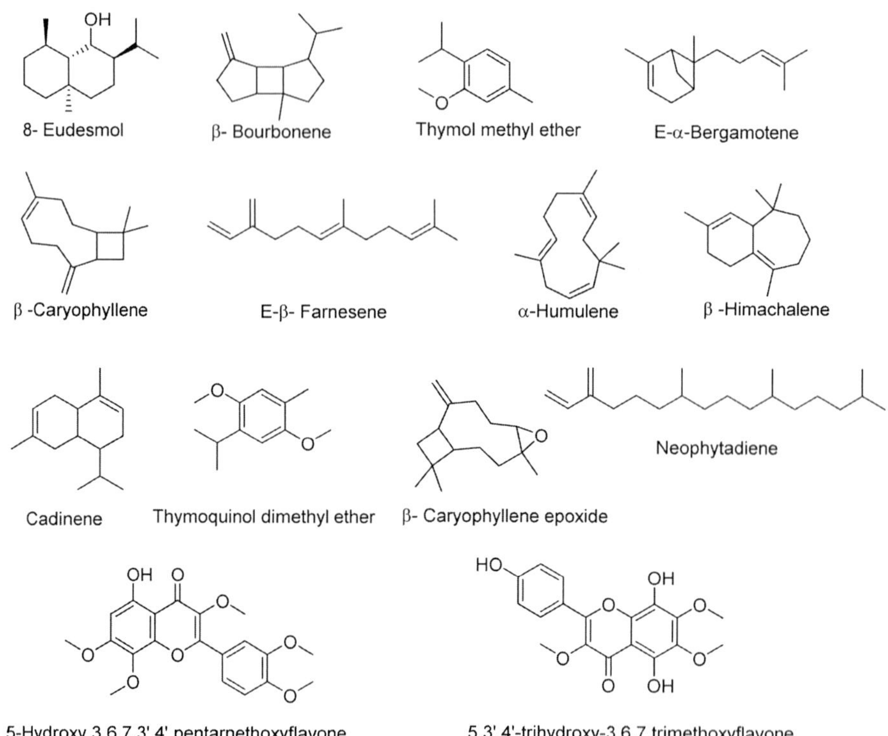

Fig. 2: Major phytoconstituents reported/isolated from *Blumea lacera*

8. Biological Aspects

8.1. Antimicrobial Activity

Whole plant of *Blumea lacera* has been used as antimicrobial agent. Methanolic extract was evaluated against Gram-positive and Gram-negative bacteria and fungi using ciprofloxacin (30μg) and fluconazole (50μg) as reference drugs (Khair et al. 2014). Shahwar et al. (2010) tested three succesive extracts (n-hexane, chloroform and methanol) of *Blumea lacera* against eight bacterial strains (*Streptococcus thermophilus, Micrococcus leuteus, Bacillus subtilis, Salmonella typhimurium, Proteus mirabilis, Nocardia asteroides, Escherichia coli* and *Bacillus licheniformis*) in which methanolic extract was found to be most effective (62-125 μg mL^{-1}). Aqueous and methanolic extracts of whole aerial part of *Blumea lacera* was found to be effective against three bacterial (*Bacillus subtilis, Staphylococcus aureus* (both gram positive) and *Serratia marcescens* (Gram negative) and one fungal (*Candida albicans*) strain expressing its broad spectrum of activity (Mahajan 2010). Leaves of *Blumea lacera* after extraction with ethyl acetate were found to be effective against *Staphylococcus aureus, Pseudomonas aeruginosa,* and *Salmonella typhi* (Khandekar et al. 2013).

8.2. Gastrointestinal (GI) Disorders

Blumea lacera has been reported to be used in various GI disorders mainly in diarrhoea, helminthiasis and some other inflammatory diseases like enterocolitis. In castor oil induced diarrhoea, methanolic as well as ethanolic extract significantly reduced the incidence and defecation rate in dose dependant manner (Pratap and Parthasarathy 2012, Khair et al. 2014).Alcoholic and aqueous extracts were further assessed for anthelmintic activity which showed ameliorative effect in dose dependant manner. Alcoholic extract was found to be superior to the aqueous extract (Pattewar et al. 2012). Pre-treatment of male wistar rats for seven days with ethanolic extract of *Blumea lacera* (aerial parts) exhibited protective effect against indomethacin in induced enterocolitis (Basnet et al. 2015). Anti-motility effect was observed in the leaves of *Blumea lacera* using three different extracts *viz.* methanol, ethanol and chloroform in which all the three extracts reduced the motility of charcoal mixed food in dose dependant manner (Fancy et al. 2015).

8.3. Antidiabetic Activity

Leaves of *Blumea lacera* exerted the potential antidiabetic effect when its methanolic extract was evaluated *in vivo* as well as *in vitro*. Significant reduction in the blood glucose level was evident in oral glucose tolerance (*in vivo*) test while a dose dependant reduction in the starch breakdown *in vitro* was observed in alpha-amylase inhibitory activity (Khair et al. 2014, Hasan et al. 2015). Methanolic whole plant extract was found to be effective in alloxan induced diabetic rats by restoring the blood glucose levels to normal at a dose of 500 mg/kg of rat body weight (Shahwar et al. 2011).

8.4. Antipyretic Activity

It has been observed that methanolic leaf extract of *Blumea lacera* lowered the Brewer's yeast-induced pyrexia in male Wistar rats (200 & 400 mg/kg dose) and in swiss albino mice (400 mg/kg) whereas chloroform and ethanolic extracts exhibited antipyretic activity at doses 200 mg/kg in the same model when tested on swiss albino mice (Verma et al. 2012, Fancy et al. 2015). In another study, methanolic extract of *Blumea lacera* (whole plant)when fractionated with chloroform and given to wistar rats, lowered the normal body temperature causing hypothermia as well as Brewer's yeast-induced pyrexia at doses 200 and 400 mg/kg body weight (Roy 2012).

8.5. Central Nervous System (CNS) Activity

Blumea lacera is a good anxiolytic and CNS depressant as proved by various researchers. Essential oil from leaf extract has shown marked CNS depressant activity in dose dependant manner in the experimental animals (Dixit and Varma 1976). Further, the alcoholic and chloroform extract of the leaf has shown promising results in the open field and hole board tests proving its anxiolytic effects (Khair et al. 2014, Fancy et al. 2015). Surprisingly, the alcoholic and chloroform extract showed anti-depressant effect in contrast to the essential oils when evaluated through forced swimming test (Fancy et al. 2015).

8.6. Miscellaneous Activities

Apart from the above,various other activities have been reported for this plant *viz.* anti-atherothrombosis, membrane stabilising, cytotoxic activity and mosquito repellent activity (Uddin et al. 2011, Khair, et al. 2014, Singh and Mittal 2014, Akter et al. 2015). A new cytotoxic glycoalkaloid have also been isolated from the methanolic extract and has been evaluated against healthy (NIH3T3 and VERO) and cancerous (AGS, HT-29, MCF-7 and MDA-MB-231) human cell lines via MTT assay and has shown marked cytotoxic effect (Akter et al. 2015).

9. Conclusion and Future Perspective

It is evident from the above survey of literature that *Blumea lacera* is a rich source of many pharmacologically and medicinally important phytoconstituents such as thymoquinol dimethyl ether, campesterol, eudesmol, phytosterols, flavonoids and many others. From antimicrobial, anti-inflammatory, antidiabetic, to cytotoxic potential of *Blumea lacera* has been discussed in this chapter. Despite of being highly exploited by ethnomedicinal practitioners in the treatment of various diseases and some other researchers by the means of different animal models, the isolation and characterization of lead molecule with specific mechanism is still lacking. Based on the knowledge of the medicinal properties of *Blumea lacera,* adequate standardisation for lead identification and optimization should be done for this plant. Though, some phytochemicals have been isolated from this plant mainly flavonoids, their usage evidencing mode of action is yet to be determined by the future researchers.

Abbreviation used

μg- microgram, μg mL^{-1}- microgram per ml, viz- namely, etc.- et cetera.

Conflict of Interest

The authors report no conflict of interest in preparing this manuscript.

Acknowledgements

IIT (BHU) is highly acknowledged for providing teaching assistantship to Mr. Tarkeshwar Dubey.

References

Acharya E & Pokhrel B (2006). Ethno-medicinal plants used by Bantar of Bhaudaha, Morang, Nepal. *Our Nature* 4(1): 96-103.

Agarwal R, Singh R, Siddiqui I & Singh J (1995). Triterpenoid and prenylated phenol glycosides from *Blumea lacera. Phytochemistry* 38(4): 935-938.

Akter R, Uddin SJ, Tiralongo J, Grice ID & Tiralongo E (2015). A New Cytotoxic Steroidal Glycoalkaloid from the methanol extract of *Blumea lacera* leaves. *Journal of Pharmacy & Pharmaceutical Sciences* 18(4): 616-633.

Anonymous. (2010). The Plant List. Version 1. Retrieved 1st June, 2019, from http://www.theplantlist.org/.

Basnet S, Adhikari A, Sachidananda VK, Thippeswamy BS & Veerapur VP (2015). Protective effect of *Blumea lacera* DC aerial parts in indomethacin-induced enterocolitis in rats. *Inflammopharmacology* 23(6): 355-363.

Beg MJ (2015). Winter crop weeds employed for the treatment of various diseases in Eastern U.P. *Indian Journal of Applied & Pure Biology* 30(2): 179-185.

Bose D, Roy JG, Mahapatra SD, Datta T, Mahapatra SD & Biswas H (2015). Medicinal plants used by tribals in Jalpaiguri district, West Bengal, India. *Journal of Medicinal Plants* 3(3): 15-21.

Chen M, Jin HZ, Zhang WD, Yan SK & Shen YH (2009). Chemical constituents of plants from the genus Blumea. *Chemistry & Biodiversity* 6(6): 809-817.

Choubey V, Dubey N & Dubey P (2015). Report of some medicinal plants used in folk medicinein tribal areas of mp. *Indian Journal of Life Sciences* 5(1): 21-23.

Dixit V & Varma K (1976). Effect of essential oil of leaves of *Blumea lacera* DC on central nervous system. *Indian Journal of Pharmacology* 8(1): 7.

Doughari JH (2012). Phytochemicals: extraction methods, basic structures and mode of action as potential chemotherapeutic agents. *Phytochemicals-A global perspective of their role in nutrition and health*. InTechOpen.

Dubey T, Kancharla B & Hemalatha S (2019). Quality control standardization of *Blumea lacera* (Burm. f.) DC leaves. *Indian journal of natural products* 33(1): 50-56.

Fancy FA, Shahriar M, Islam MR & Bhuiyan MA (2015). In-vivo anti-pyretic, anti-nociceptive, neuropharmacological activities and acute toxicity investigations of *Blumea lacera. International Journal of Pharmacy and Pharmaceutical Sciences* 7(1): 472-477.

Harish S (1988). Ethno biological treatment of piles by Bhoxas of Uttar Pradesh. *Ancient science of Life* 8(2): 167.

Hasan MN, Rahman MH, Guo R & Hirashima A (2015). Hypoglycemic activity of methanolic leaf extract of *Blumea lacera* in Swiss-albino mice. *Asian Pacific Journal of Tropical Disease* 5(3): 195-198.

Kakulte V, Gaikwad K N & S J Diversity of ethnobotanical plants used by rural community of tryambakeshwar hill of nashik district, maharashtra, India. *Life* 50: 31.

Khair A, Ibrahim M, Ahsan Q, Homa Z, Kuddus MR, Rashid RB & Rashid MA (2014). Pharmacological Activities of *Blumea lacera* (Burm. f) DC: A Medicinal Plant of Bangladesh. *British Journal of Pharmaceutical Research* 4(13): 1677.

Khandekar U, Tippat S & Hongade R (2013). Inverstigation on antioxidant, antimicrobial and phytochemical profile of *Blumea lacera* leaf. *International Journal of Biological & Pharmaceutical Research* 4(11): 756-761.

Khare CP (2004). Indian herbal remedies: rational Western therapy, ayurvedic, and other traditional usage, Botany. Published by Springer science & business mediapages.

Korpenwar A (2012). Ethnomedicinal plants used by bhilala tribals in buldhana district (MS). *DAV International Journal of Science* 1: 60-65.

Kumar B, Vijayakumar M, Govindarajan R & Pushpangadan P (2007). Ethnopharmacological approaches to wound healingexploring medicinal plants of India. *Journal of Ethnopharmacology* 114(2): 103-113.

Kumar S, Jha A, Sahaya L & Pandit N (2016). The Study of GC-MS of *Blumea lacera* (Burm. F.) DC (Family: Asteraceae), Bhagalpur, Bihar.*Journal of Chemistry and Chemical Sciences* 6(4): 392-396.

Mahajan R (2010). Ethnobotanical Significance and Antimicrobial Activity of *Blumea lacera* (Roxb.) DC. *International Journal of Pharmaceutical & Biological Archive* 1(3).

Manandhar NP (1995). An inventory of some herbal drugs of Myagdi district, Nepal. *Economic Botany* 49(4): 371-379.

Mishra A (2015). Study on some ethnomedicinal plants of Kalinjar hillock, Banda district (UP) India. *International Journal of Advanced Research in Engineering and Applied Sciences* 4(7): 1-9.

Mitra S & Mukherjee SK (2010). Ethnomedicinal usages of some wild plants of North Bengal plain for gastro-intestinal problems. *Indian Journal of Traditional Knowledge* 9(4): 705-712.

Mollik MAH, Hossan MS, Paul AK, Taufiq-Ur-Rahman M, Jahan R & Rahmatullah M (2010). A comparative analysis of medicinal plants used by folk medicinal healers in three districts of Bangladesh and inquiry as to mode of selection of medicinal plants. *Ethnobotany Research and Applications* 8: 195-218.

Mukherjee SK (2006). Medicinal plants of asteraceae in india and their uses. Prceeding of National Seminar.

Natarajan B & Paulsen BS (2000). An ethnopharmacological study from Thane district, Maharashtra, India: Traditional knowledge compared with modern biological science. *Pharmaceutical Biology* 38(2): 139-151.

Pandey AK & Tripathi N (2011). Aromatic Plants of Gorakhpur Division: Their Antimycotic Properties and Medicinal Value. *Extraction* 7(2): 1-26.

Pattanayak S, Dutta MK, Debnath PK, Bandyopadhyay SK, Saha B & Maity D (2012). A study on ethno-medicinal use of some commonly available plants for wound healing and related activities in three southern districts of West Bengal, India. *Exploratory Animal and Medical Research* 2(2): 97-110.

Pattewar AM, Dawalbajea A, Gundalea D, Pawarb P, Kavtikwara P, Yerawara P, Pandharkara T & Patawara V (2012). Phytochemistryical & anthelmintic studies on *Blumea lacera*. *Indo Global Journal of Pharmaceutical Sciences* 2(4): 390-396.

Pornpongrungrueng P, Gustafsson MH, Borchsenius F, Koyama H & Chantaranothai P (2016). Blumea (Compositae: Inuleae) in continental Southeast Asia. *Kew Bulletin* 71(1): 1.

Prasad R, Lawania RD & Gupta R (2009). Role of herbs in the management of asthma. *Pharmacognosy Reviews* 3(6): 247.

Pratap SU & Parthasarathy R (2012). Comparative antidiarrhoel activity of ethanolic extract of root of *Blumea lacera* var lacera and Blumea eriantha DC on experimental animals. *Journal of Pharmaceutical and Biomedical Sciences (JPBMS)* 17(17): 1-4.

Prusti AB (1998). Some less known folk claims from candaka tribals of Orissa. *Ancient science of Life* 18(2): 106.

Quamar M & Bera S (2014). Ethno-medico-botanical studies of plant resources of Hoshangabad district, Madhya Pradesh, India: retrospect and prospects. *Jornal of Plant Sciences and Research* 1(1): 1-11.

Ragasa CY, Wong J & Rideout JA (2007). Monoterpene glycoside and flavonoids from *Blumea lacera. Journal of Natural Medicines* 61(4): 474-475.

Rahman A (2013). Medico-botanical study of the plants found in the Rajshahi district of Bangladesh. *Prudence Journal of Medicinal Plants Research* 1(1): 1-8.

Rahman F & Rahmatullah M (2015). Medicinal plant formulations of the Musohor tribe of Birganj in Dinajpur district, Bangladesh. *World Journal of Pharmacy and Pharmaceutical Sciences* 4(11): 177-185.

Rahmatullah M, Islam MR, Kabir MZ, Harun-or-Rashid M, Jahan R, Begum R, Seraj S, Khatun MA & Chowdhury AR (2010). Folk medicinal practices in Vasu Bihar village, Bogra district, Bangladesh. *American-Eurasian Journal of Sustainable Agriculture* 4(1): 86-93.

Rahmatullah M, Kabir A, Rahman MM, Hossan MS, Khatun Z, Khatun MA & Jahan R (2010). Ethnomedicinal practices among a minority group of Christians residing in Mirzapur village of Dinajpur District, Bangladesh. *Advances in Natural and Applied Sciences* 4(1): 45-51.

Rai S (2004). Medicinal plants used by Meche people of Jhapa district, eastern Nepal. *Our Nature* 2(1): 27-32.

Rao CB, Rao TN & Muralikrishna B (1977). Flavonoids from *Blumea lacera*. *Planta medica* 31(03): 235-237.

Rao RR, Kavitha S & K S (2006). Wild aromatic plant species of western ghats: Diversity, conservation and utilization. International Seminar on Multidisciplinary Approaches in Angiosperm Systematics.

Roy A (2012). Antipyretic activity of *Blumea lacera* (Burm. f) DC, A folklore medicine from Chhattisgarh India. *Research Journal of Pharmacognosy and Phytochemistry* 4(1): 1-3.

Satyal P, Chhetri BK, Dosoky NS, Shrestha S, Poudel A & Setzer WN (2015). Chemical composition of *Blumea lacera* essential oil from Nepal, biological activities of the essential oil and (Z)-Lachnophyllum Ester. *Natural product communications* 10(10): 1749-1750

Sen SK & Behera LM (2008). Ethnomedicinal plants used by the tribals of Bargarh district to cure diarrhoea and dysentery. *Indian Journal of Traditional Knowledge* 7(3): 425-428.

Shahwar D, Ullah S, Ahmad M, Ullah S & Ahmad N (2011). Anti-diabetic activity of the methanolic extract of *Blumea lacera* DC (Asteraceae) in alloxan-induced diabetic rats. *Asian Journal of Chemistry* 23(12): 5403.

Sharma J, Painuli R & Gaur R (2010). Plants used by the rural communities of district Shahjahanpur, Uttar Pradesh. *Indian Journal of Traditional Knowledge* 9(4): 798-803.

Singh AP (2005). Ethnobotanical Studies of Chandigarh Region. *Ethnobotanical Leaflets* 2005(1): 33.

Singh S & Mittal P (2014). Mosquito repellent action of *Blumea lacera* (Asteraceae) against Anopheles stephensi and Culex quinquefasciatus. *International Journal of Mosquito Research* 1(1): 10-13.

Singh SK & Beg MJ (2015). Ethnomedicinal plants of Asteraceae from Chitrakoot area of Satna District (M.P.). *Indian Journal of Applied & Pure Biology* 30(1): 55-60.

Solecki RS (1975). Shanidar IV, a Neanderthal flower burial in northern Iraq. *Science* 190: 880-881.

Tiwari P, Saluja G, Pandey AS & Naveen S (2012). Isolation and biological evaluation of some novel phytoconstituents from *Blumea lacera* (Burn f.) DC. *International Journal of Pharmacy and Pharmaceutical Sciences* 4(4): 148-150.

Uddin SJ, Grice ID & Tiralongo E (2011). Cytotoxic effects of Bangladeshi medicinal plant extracts. *Evidence-Based Complementary and Alternative Medicine*.DOI: http://dx.doi.org/10.1093/ecam/ nep111

Umberto Q (2012). CRC world dictionary of medicinal and poisonous plants: common names, scientific names, eponyms, synonyms, and etymology (5 Volume Set). Published by CRC Press.

Upadhyay O, Kumar K & Tiwari R (1998). Ethnobotanical study of skin treatment uses of medicinal plants of Bihar. *Pharmaceutical biology* 36(3): 167-172.

Vanila D, Ghanthikumar S & Manickam V (2008). Ethnomedicinal Uses of Plants in the Plains Area of the Tirunelveli-District, Tamilnanu, India. *Ethnobotanical Leaflets* 2008(1): 159.

Verma LK, Singh AK, Pachade VR, Koley K & Vadlamudi V (2012). Antipyretic activity of *Blumea lacera* leaves in albino rats. *Exploratory Animal and Medical Research* 2(1): 56-59.

18

Phytochemistry and Pharmacological Activities of *Pterocarpus santalinus* an Ayurvedic Crude Drug

Bashir Lone, Venugopal Singamaneni, Arushi Gupta
Upasana Sharma and Prasoon Gupta

Abstract

Recently there has been increasing interest in plants and plant-derived compounds as medicinal agents. In Ayurveda, a wide range of medicinal uses of Pterocarpus santalinus is described. Many important bioactive phytocompounds have been extracted and identified from the heartwood of P. santalinus. The major bioactive compounds present in the heartwood of P. santalinus are santalin A and B, savinin, calocedrin, pterolinus (K,L) and pterostilbenes. These phyto-compounds present only in small amounts and have more effect than nutrients. The heartwood of the plant is used traditionally for the treatment of inflammation, diabetes, headache, skin diseases, jaundice, and in wound-healing. A wide range of pharmacological activities and health benefits of P. santalinus have been reported, including antioxidative, antidiabetic, antimicrobial, anticancer, and anti-inflammatory properties, and protective effects on liver, gastric mucosa and nervous system. All these protective effects were attributed to bioactive compounds present in P. santalinus. Most of the secondary metabolites reported

Bashir Lone, Venugopal Singamaneni, Arushi Gupta, Upasana Sharma and Prasoon Gupta(✉)

Natural Product Chemistry Division, CSIR-Indian Institute of Integrative Medicine Jammu-180001, Jammu & Kashmir, India

Upasana Sharma

Department of Applied Chemistry, Mahant Bachittar Singh College of Engineering and Technology Babliana, Jammu-181101, Jammu and Kashmir, India

✉*Corresponding author email: guptap@iiim.ac.in*

Plants for Novel Drug Molecules: Ethnobotany to Ethnopharmacology
Bikarma Singh & Yash Pal Sharma (eds.), (pp. 411-425)

Email: *info@nipabooks.com* Web: *www.nipabooks.com*

from this plant are terpenes, phenolics, flavones and sterols. The present communication gives the phytochemistry and pharmacological activities of P. santalinus on health with up-to-date information.

Keywords: *Pterocarpus santalinus*, Red Sandalwood, Phytochemistry, Pharmacological activity, Drug Discovery

1. Introduction

As the modern drugs have side effects on health, the World Health Organization (WHO) promotes the evaluation of the potential benefits of plants as effective therapeutic agents (Halim and Misra 2011). About 80% of drugs used in most of countries are derived from medicinal plants (Ramawat and Goyal 2008, Meena et al. 2009). India has a rich biodiversity on medicinal plants which are locally used by tribal population and traditional healers for the treatment of several diseases. *Pterocarpus santalinus* L. is a medium–sized deciduous tree belonging to the Fabaceae family, also named "red sanders" or "red sandalwood", is a rare and commercial tree (Palanisamy et al. 2007). This species is distributed exclusively in well defined forest tracts of Andhra Pradesh in Southern India (Prakash et al. 2006). It is a valuable source of various phyto-compounds. *P. santalinus* has been used as a folk medicine for treatment of inflammation, such as in chronic bronchitis and chronic cystitis, fever, headaches, mental aberrations, ulcers, cancer etc. (Cho et al. 2001, Kwon et al. 2006). Bioactive compounds present in the plant's heartwood show a wide range of biological activities, which represents its potential for the treatment of various diseases (Manjunatha 2006, Arunakumara et al. 2011). Pharmacological studies showed that the heartwood and bark have exhibited antidiabetic, antimicrobial, anti-inflammatory, and hepatoprotective properties (Kwon et al. 2006, Manjunatha 2006, Kondeti et al. 2010). This review summarizes the up to date information on phytochemistry and pharmacological activities of this well known medicinal plant.

2. Diversity, Distribution and Morphology

There are 35 species of *Pterocarpus* currently distributed throughout the world. *Pterocarpus santalinus* is endemic to the hills of Andhra Pradesh, some parts of Karnataka and Tamil Nadu. It is a medium sized deciduous tree growing in dry and rocky ground at 150-900m (Jain and Sastry 1980). Stems are with hard, dark purple heartwood. Bark is blackish brown in colour with rectangular plates, leaves are foliated and flowers are bisexual. Pods are orbicular, seeds are in kidney shaped and reddish brown in colour.

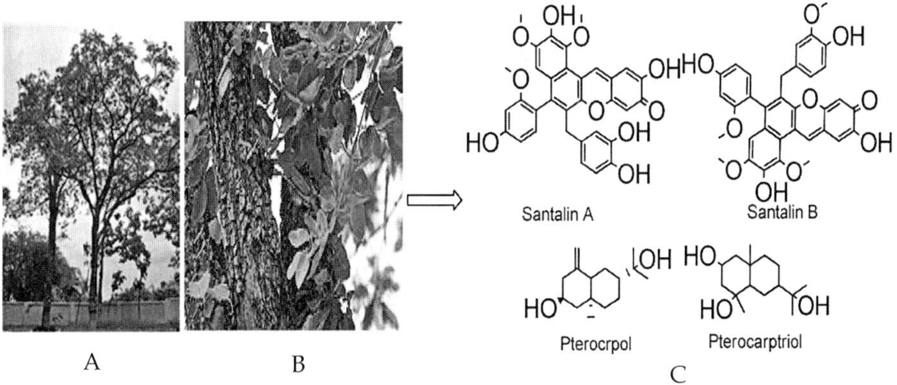

Fig. 1: A. Habit of *Pterocarpus santalinus* in wild B. Part used C. marker compounds

3. Traditional Medicinal Uses

The red sandalwood has natural dye i.e santalin, which is used as coloring agent in pharmaceutical preparations and foodstuffs. In the traditional system of medicine, the decoction from the heartwood is useful in various medicinal properties. It is used in ulcers, eye diseases, inducing vomiting and mental aberrations. The heartwood is known to have anti hyperglycaemic activity, antipyretic, antiinflammatory, anthelmintic, tonic, hemorrhage, dysentery, aphrodisiac, diaphoretic activities and also used as a cooling agent. It has been reported that wood in combination with other drugs is prescribed for snake bites and scorpion stings. A red pigment, santalin, is extracted from the timber and is also widely used domestically and internationally as a colorant in foods. In India it has multiple uses in traditional medicine (Green 1995, Oldfield 1998). It is also used for treating skin diseases, bone fracture, leprosy, spider poisoning, scorpion sting, hiccough, ulcers, general debility and metal aberrations (Chopra et al. 1956, Arokiyaraj et al. 2008). Wood paste applied on boils and other skin eruptions, infections, inflammation and on forehead to relieve headache. Decoction of fruits is used to cure chronic dysentery and to check dermatological conditions including psoriasis. Wood and bark brew taken orally relieves chronic dysentery, worms, blood vomiting, weak vision and hallucination. Wood powder is used to control hemorrhage, bleeding piles and inflammation. The antibacterial, anticancer, hepatoprotective and wound healing properties of this drug have been established recently.

Pterocarpus santalinus is useful in treating bilious affections, skin diseases such as antihelmintic, aphrodisiac and alexiteric as well as vomiting, thirst, eye diseases, ulcers and diseases of the blood (Kirtikar and Basu 2001, Lateef et al. 2008). Infusion of the decoction of fruit is used as astringent tonic in chronic dysentery (Kondeti et al. 2010). Stem bark powder with soft porridge has been used in treating diarrhea and the paste of the wood has been considered as a cooling agent for external application treating inflammations and headache, mental aberrations, and ulcers (Krishnaveni and Rao 2000a).

4. Phytoconstituents

Phytochemical studies of this plant indicate that it contains substances such as alkaloids, phenols, saponins, glycosides, flavonoides, triterpenoides, sterols and tannins. Pterolinuses A-L (**1-5**, **8-13**, **30**, **31**), dehydromelanoxin (**6**), melanoxin (**7**), melanoxoin (**14**), *S*-3'-hydroxy-4,4'-dimethoxydalbergione (**15**), melannein (**16**), dalbergin (**32**) and cearoin (**33**) were isolated from methanol extract of *Pterocarpus santalinus* heartwood (Wu et al. 2011). A group of six closely related sesquiterpenes namely isopterocarpolone (**17**), pterocarptriol (**18**) and pterocarpdiolone (**19**) along with β-eudesmol (**20**), pterocarpol (**21**) and cryptomeridiol (**22**) have been isolated from the heartwood of *Pterocarpus santalinus* (Kumar et al. 1974). Two lignans, savinin (**23**) and calocedrin (**24**), were isolated from the heartwood of *Pterocarpus santalinus* by activity-guided fractionation (Cho et al. 2001). Three flavonoids 6-hydroxy,7,2',4',5'-tetramethoxyisoflavone (**25**), liquiritigenin (**26**) and isoliquiritigenin (**27**) have been isolated from the heartwood of *Pterocarpus santalinus* (Krishnaveni and Rao 2000). Two aurone glycosides, 6-hydroxy-5-methyl 3',4',5' trimethoxy aurone 4-*O*-α-L-rhamnopyranoside (**28**) and 6,4' dihydroxy aurone 4-*O*-rutinoside (**29**) have been isolated from the ethanolic extract of the wood of *Pterocarpus santalinus* (Kesari et al. 2004). An isoflavone glucoside, 4',5-dihydroxy-7-*O*-methoxy isoflavone 3'-O-β-D-glucoside (**34**) together with the santal (**35**) has been isolated from the heartwood (Krishnaveni and Rao 2000). The heartwood of *Pterocarpus santalinus* contain two major red pigments, santalin A (**36**) and santalin B (**37**) (Gurudutt and Seshadri 1974). A pentacyclic triterpene, 3-ketooleanane (**38**) was isolated from the callus induced from the stem cuttings of *P. santalinus* (Krishnaveni and Rao 2000). Triterpenes, acetyl oleanolicaldehyde (**39**), acetyl oleanolic acid (**40**), along with pterostilbene (**41**) have been reported (Kumar et al. 1974). The bark of *P.santalinus* has been found to contain β-amyrone (**42**), lupenone (**43**), epi-lupeol (**44**), lupeol (**45**), β-sitosterol (**46**) and lupenediol (**47**), p-hydroxy benzoic acid (**48**), Gentisic acid (**49**), β-resorcylic acid (**50**), α-resorcylic acid (**51**), vanillic acid (**52**) (Khadem and Marles 2010). Yellow pigments Santalin Y (**53**) and santalin AC (**54**) obtained from *P. santalinus* (Kinjo et al. 1995).

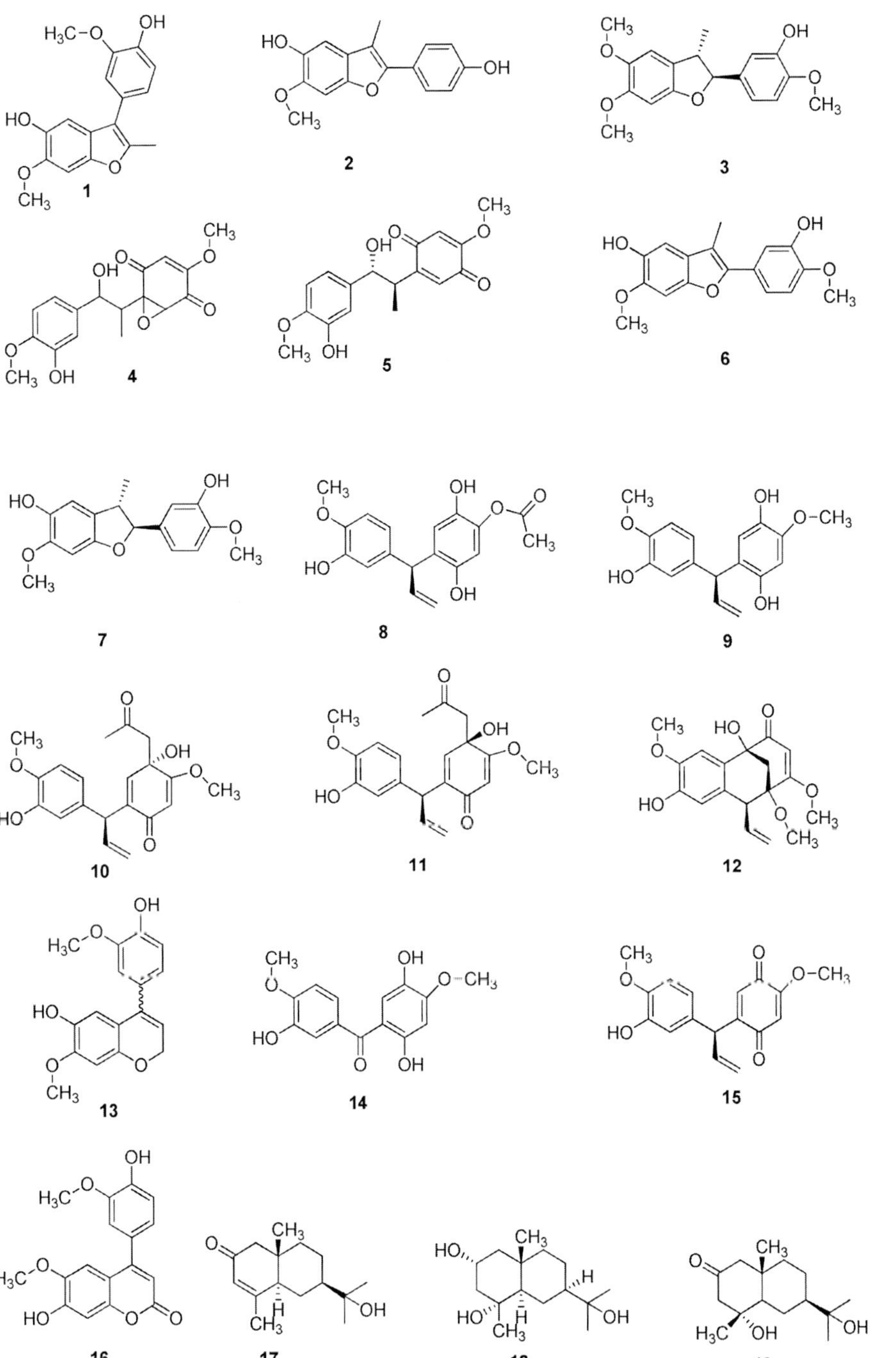

Fig. 1A: Chemical constituents of *Pterocarpus santalinus*

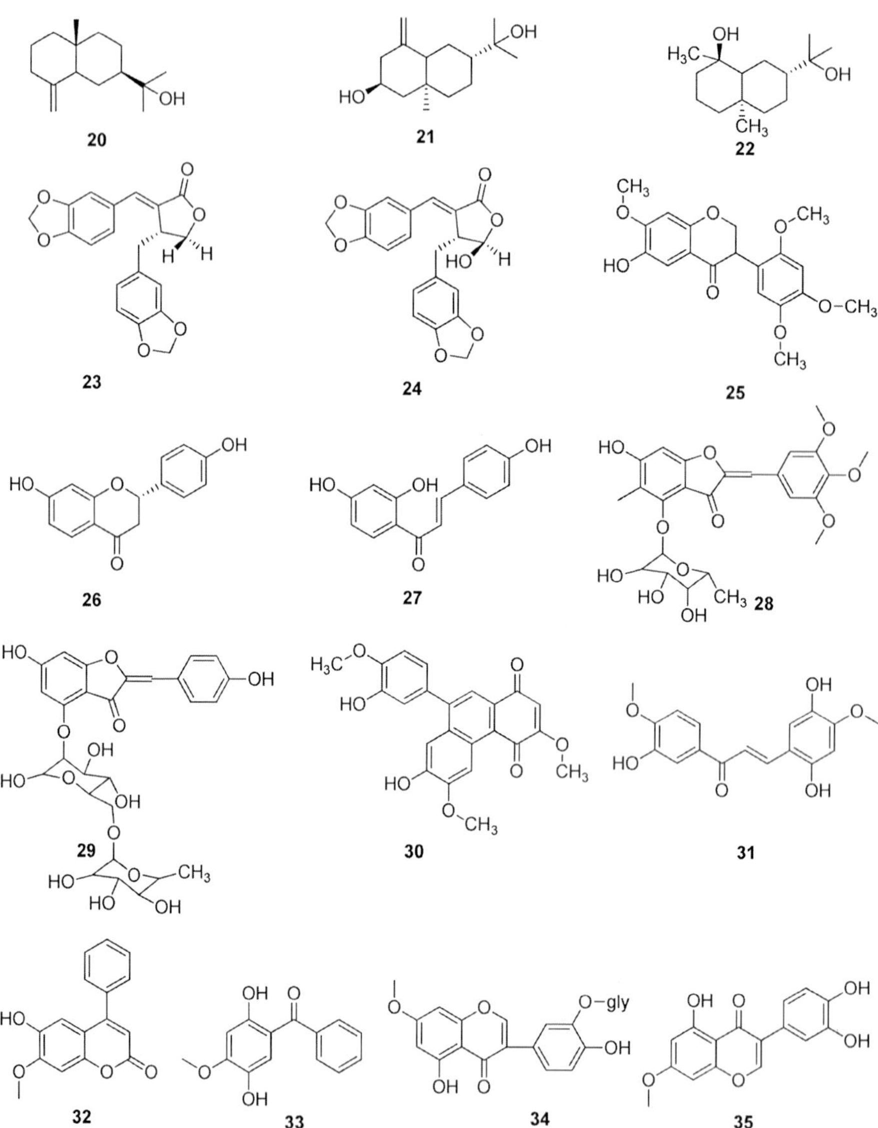

Fig. 1B: Chemical constituents of *Pterocarpus santalinus*

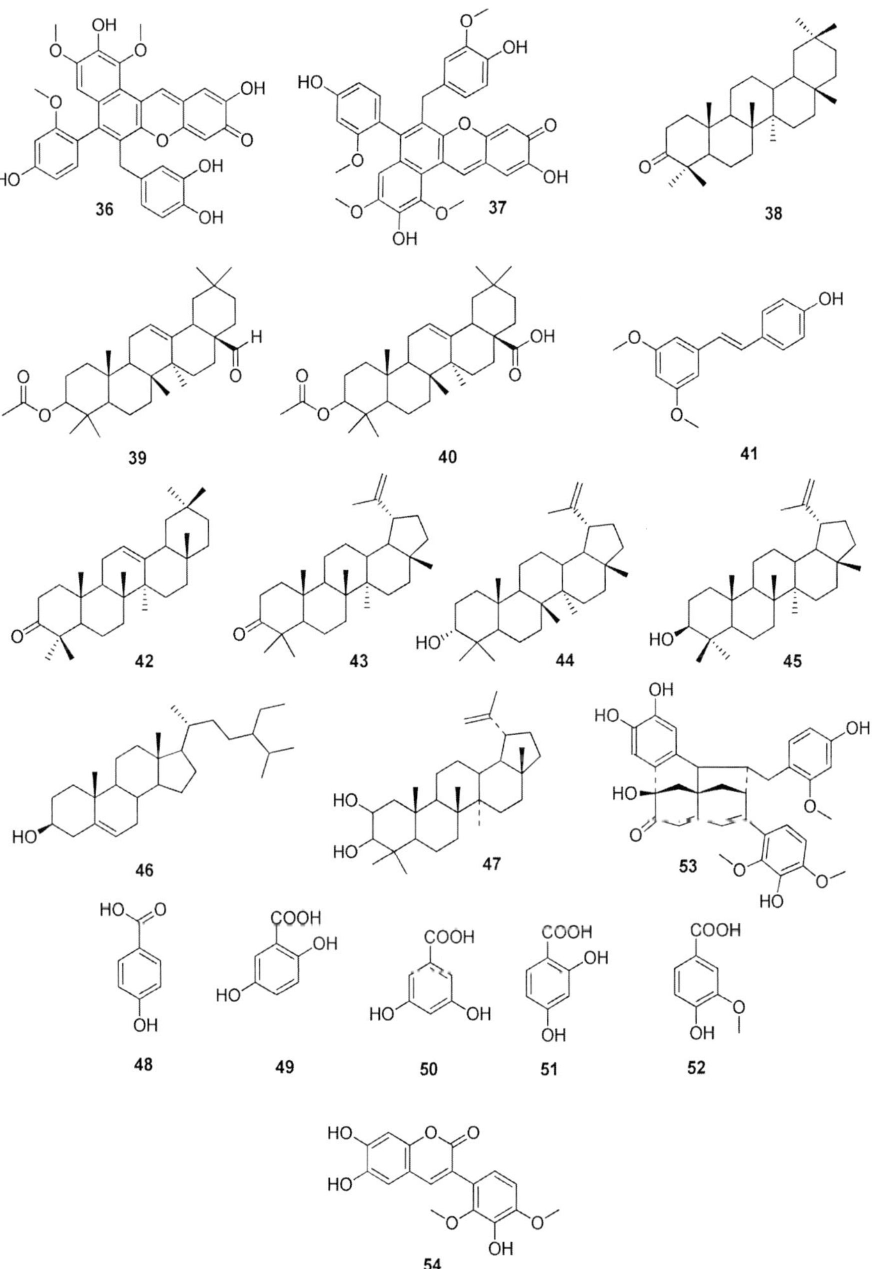

Fig. 1C: Chemical constituents of *Pterocarpus santalinus*

5. Pharmacological Activities

5.1. Radical Scavenging and Antioxidant Activity

The methanolic extract of the leaves exhibited radical-scavenging activity for 1,1-diphenyl-2-picrylhydrazyl (DPPH), nitric oxide, and hydrogen peroxide (Arokiyaraj et al. 2008, Stella et al. 2011). The methanolic extract of heartwood showed Fe^{3+}-reducing capacity and DPPH radical-scavenging activity (Kumar 2011). Pterostilbene exhibited strong *in vitro* antioxidant activity against free radicals such as DPPH, ABTS, hydroxyl, superoxide and hydrogen peroxide (Acharya and Ghaskadbi 2013).

5.2. Antihyperglycemic Activity

Traditionally made heartwood cups for drinking water have been used as a treatment for diabetes (Nagaraju and Rao 1989). Ethanolic fraction from the bark of *Pterocarpus santalinus* at a dose of 0.25 g/kg body weight showed maximum antihyperglycemic activity than that of glibenclamide. 95% ethanol extract produced a maximum glucose lowering effect at the 3rd hour in the fasted, fed, diabetic model and marked suppression of rise in blood sugar at ½ hr in the glucose loaded model (Nagaraju et al. 1991, Badri et al. 2001, Kondeti et al. 2010). Co-administration (250 mg/kg body weight) of aqueous extract of heartwood along with vitamin E to streptozotocin-induced diabetic rats caused significant lowering of blood sugar (Halim and Misra 2011).

5.3. Gastroprotective Activity

The ethanol extract (150 mg/kg body weight/day) of *P. santalinus* showed a significant reduction in gastric lesions, increase in activities of antioxidant enzymes, and decrease in membrane-bound adenylpyrophosphatase (ATPase) activities, against ibuprofen-induced ulcers in albino rats (Narayan et al. 2005). The ethanolic extract (200 mg/kg body weight per day) of heartwood restored the activities of tricarboxylic acid cycle (TCA) enzymes, mitochondrial membrane integrity, and reversed the damage caused by ulcerogens (Narayan et al. 2007). In the presence and absence of *P. santalinus* extract anti *H. pylori* activity was tested on gastric epithelial cells. A reduction in the activity of urease, normal appearance of gastric epithelial cells and decrease in lipid peroxidation (LPO) and lactate dehydrogenase (LDH) activities suggested antipyloric effect (Narayan et al. 2007).

5.4. Angiogenesis and Wound Healing Activity

Acetone extract of *P. santalinus* (1 mg/mL) stimulated angiogenesis more significantly than alcohol and benzene extracts (Jadhav et al. 2012). The wound-healing activity of plant was demonstrated in normal and diabetic wound rat models, and concluded that the ointment made from the plant is effective in treating acute wounds (Biswas et al. 2004).

5.5. Anti-inflammatory Activity

The alcoholic extract showed prominent (55.5±2.8) TNF-α inhibitory activity in LPS stimulated RAW 264.7 cells. Savinin, the lignan from the heartwood inhibit TNF-α production (31.9 μM) and T-cell proliferation (Cho et al. 1999, 2001). Specific lignans, namely, savinin, calocedrin, and eudesmin were found to inhibit TNF-α and also showed antiproliferative effect with an IC_{50} value at 40 μg/mL (Kwon et al. 2006). The heartwood extract (500 μg/mL) showed significant inhibitory activity against carrageenan-induced inflammation in paw edema (Kumar 2011). Benzofurans, neoflavonoids, and pterolinuses from heartwood, including pterolinus B, showed potent anti-inflammatory activity with an IC_{50} value of 0.19 μg/mL (Wu et al. 2011). Pterolinus K and pterolinus L showed significant inhibition of superoxide anion generation and elastase release by human neutrophils in response to formyl-L-methionyl-L-leucyl-L-phenylalanine (FMLP)/cytochalasin B (CB) with IC_{50} values of 0.99 μM and 0.94 μM, respectively (Wu et al. 2011).

5.6. Hepatoprotective Activity

Aqueous (45 mg/mL) and ethanol (30 mg/mL) bark extracts restored CCl_4-induced liver damage in rats (Manjunatha et al. 2006). The heartwood extract (400 mg/kg body weight/day) was tested against D-galactosamine–induced liver damage in albino rats (Palanisamy et al. 2007). *In silico* docking studies revealed that pterocarpol and cryptomeridiol, present in heartwood, targeted the HBx proteins of hepatitis B virus, and were thus reported as potent molecules (Manjunatha et al. 2010). Hepatic fibrosis in chronic liver injury was suppressed by extract of *P. santalinus* (Itoh et al. 2009, 2010). Methanolic extract (250 mg/kg b.wt/day) was given to male albino Wistar rats which were treated with 20% alcohol (5 g/ kg b.wt/day) for 60 days. Alcohol administration significantly increase in the levels of plasma transaminases (ALT and AST), alkaline phosphatase (ALP), lactate dehydrogenase (LDH) and gamma glutamyl transferase (gGT) and these levels were significantly brought back close to normal level by administration of methanolic extract. Alcohol administration also decreased the content of reduced glutathione (GSH) and activities of glutathione peroxidase (GPx), glutathione-s transferase (GST), glutathione reductase (GR), superoxide dismutase (SOD) and catalase (CAT) in liver, which were significantly enhanced by administration of methanolic extract. The active compounds pterostilbene, lignan and lupeols which were present in methanolic extract might be responsible for this activity (Bulle et al. 2016).

5.7. Antibacterial Activity

The antibacterial activity of leaf and bark extract of *P. santalinus* was tested against different bacterial strains. The bark extract showed greater inhibition activity than the leaf extract (Manjunatha 2006, Stella et al. 2011, Mishra and Padhy 2013). The methanolic extract has shown greater inhibitory activity than the aqueous extract (Dey et al. 2014). Silver nanoparticles showed antibacterial activity prepared from the leaf extract of *P. santalinus* (Gopinath et al. 2013). Lignans present in *P. santalinus* have shown antibacterial properties (Yamauchi et al. 2005).

5.8. Anticancer Activity

Treatment of cervical adenocarcinoma (HeLa) cells with methanolic extract resulted in growth inhibition and induction of apoptosis with the IC_{50} value 40 μg/mL. The methanolic extract induced apoptosis through the mitochondrial pathway by the activation of caspase-9 and caspase-3, and degradation of peroxisome-activated receptor protein (PARP). Savinin showed significant inhibition on T cell proliferation from the methylene chloride extract. (Kwon et al. 2006). Benzofuran compounds from heartwood showed cytotoxicity against Ca9-22 cancer cells with an IC_{50} value 0.46 μg/mL (Wu et al. 2011). The anticancer activity of pterostilbene was tested against breast cancer, lung cancer, colon cancer, prostate cancer, and pancreatic cancer (Mena et al. 2012). Pterolinus K and pterolinus L have shown significant anticancer activity with IC_{50} values at 10.86 μM, 9.81 μM, and 8.2 μM, against cancer cell lines HepG2, Hep3B, and A549, respectively (Wu et al. 2011).

5.9. Hypolipidemic Activity

Pterostilbene showed hypolipidemic activity with increased LDL and reduced blood pressure in hypercholesterolemic human volunteers (Rimando et al. 2005, Riche et al. 2014). 3-hydroxybenzoic acid, gentisic acid, α- and β-resorcylic acid, and vanillic acid which are phytoconstituents of *P. santalinus* have potential inhibitory properties against LDL oxidation (Ashidate et al. 2005).

5.10. Other Pharmacological Activities

The aurone glycosides isolated from heartwood have been reported to exhibit antiplasmodial activity (Kayser et al. 2001) and has been used as a potential antileishmanial drug (Kayser et al. 1998). Himoliv, a polyherbal ayurvedic formulation containing *Pterocarpus santalinus* as one of the ingredients has been reported to possess hepatoprotective activity (Bhattacharya et al. 2003). The stem bark extract was shown to contain maximum activity against *Enterobacter aerogenes, Alcaligenes faecalis, Escherichia coli, Pseudomonas aeruginosa, Proteus vulgaris, Bacillus cereus, Bacillus subtilis* and *Staphylococcus aureus* (Manjunatha 2006). Ethanolic stem bark extract is known to possess antihyperglycemic activity (Badri et al. 2001). The leaf extract also showed maximum activity against *E. coli, A. faecalis, E. aerogenes,* and *P. Aeruginosa* (Manjunatha 2006).

6. Conclusion and Future Perspectives

India has a rich biodiversity on medicinal plants which are locally used by tribal population and traditional healers for the treatment of several diseases. The phytoconstituents present in the heartwood of *P. santalinus* are very complex compounds. Further research is necessary to provide a detailed characterization of the novelty of compounds. Bioactive compounds present in the plant's heartwood shows a wide range of biological activities, which represents its potential for the treatment of various diseases. Further studies on molecular biological mechanisms and physiological effects on this plant species needed for these compounds. Most of the pharmacological activities of the plant extracts have been studied *in vitro* or in animal models. Clinical and human intervention studies of the isolated compounds are very limited; therefore focussing on this area is worth investigation.

Abbreviations Used

WHO: World Health Organization; DPPH: 1,1-diphenyl-2-picrylhydrazyl; ABTS: 2,2'-azino-bis(3-ethylbenzothiazoline-6-sulphonic acid); ATPase: adenosine triphosphate; TCA: Tricarboxylic acid cycle; LPO: Lipid peroxidation; LDH: Lactate dehydrogenase; TNF: Tumor necrosis factor; FMLP: formyl-L-methionyl-L-leucyl-L-phenylalanine; CB: cytochalasin B; ALT: Alanine aminotransferase; AST: Aspartate aminotransferase; ALP: Alkaline phosphatase; GT: glutamyl transferase; GSH: reduced glutathione; GPx: glutathione peroxidise; GST: glutathione-s transferase; GR: glutathione reductase; SOD: superoxide dismutase; CAT: catalase and PARP: peroxisome-activated receptor protein.

Conflict of interest

Authors declare no conflict of interest

Acknowledgements

Authors are thankful for the financial assistance from Science and Engineering Research Board, Department of Science and Technology, Government of India (EMR/2016/002584) and CSIR Sickle Cell Mission Project (HCP0008).

References

Achary JD & Ghaskadbi SS (2013). Protective effect of pterostilbene against free radical mediated oxidative damage. *BMC Complementary and Alternative Medicine* 13:238.

Arokiyaraj S, Martin S, Perinbam K, Marie AP & Beatrice V (2008). Free radical scavenging activity and HPTLC finger print *Pterocarpus santalinus* L.-an *in vitro* study. *Indian Journal of Science and Technology* 1:1-3.

Arunakumara KK, Walpola BC, Subasinghe S & Yoon M (2011). *Pterocarpus santalinus* Linn. f. (Roth handgun): A review of its botany, uses, phytochemistry and pharmacology. *Journal of Korean Society for Applied Biological Chemistry* 54: 495-500.

Ashidate K, Kawamura M, Mimura D, Tohda H, Miyazaki S, Teramoto T, Yamamoto Y & Hirata Y (2005). Gentisic acid, an aspirin metabolite, inhibits oxidation of low-density lipoprotein and the formation of cholesterol ester hydroperoxides in human plasma. *European Journal of Pharmacology* 513: 173-179.

Badri KR, Giri R, Kesavulu MM & Rao CA (2001). Effect of oral administration of bark extracts of *Pterocarpus santalinus* L. on blood glucose level in experimental animals. *Journal of Ethnopharmacology* 74: 69-74.

Bhattacharya D, Mukherjee R, Pandit S, Das N & Sur TK (2003). Prevention of carbon tetra chloride induced hepatotoxicity in rats by Himoliv, a polyherbal formulation. *Indian Journal of Pharmacology* 35: 183-185.

Biswas TK, Maity LN & Mukherjee B (2004). Wound healing potential of *Pterocarpus santalinus* Linn: A Pharmacological evaluation. *The International Journal of Lower Extremity Wounds* 3: 143-150.

Bulle S, Reddy VD, Padmavathi P, Maturu P & Varadacharyulu NCh (2016). Modulatory role of *Pterocarpus santalinus* against alcohol-induced liver oxidative/ nitrosative damage in rats. *Biomedicine and Pharmacotherapy* 83: 1057-1063.

Cho JY, Park J, Kim PS, Chae SH, Yoo ES, Baik KU, Lee J & Park MH (1999). Inhibitory effect of oriental herbal medicines on tumor necrosis factor-a production in lipopolysaccharide-stimulated RAW264.7 cells. *Natural Product Sciences* 5: 12-19.

Cho JY, Park J, Kim PS, Yoo ES, Baik KU & Park MH (2001). Savinin, a lignan from *Pterocarpus santalinus* inhibits tumor necrosis factor-a production and T cell proliferation. *Biological and Pharmaceutical Bulletin* 24: 167-171.

Chopra RN, Nayar SL & Chopra IC (1956). Glossary of Indian medicinal plants. Council of Scientific and Industrial Research,India.

Dey SK, Chattopadhyay S & Masanta NC (2014). Antimicrobial activities of some medicinal plants of red and laterite zone of West Bengal, India. *World Journal of Pharmacy and Pharmaceutical Sciences* 3: 719–34.

Gopinath K, Gowri S & Arumugam A (2013). Phytosynthesis of silver nanoparticles using *Pterocarpus santalinus* leaf extract and their antibacterial properties. *Journal of Nanostructure in Chemistry* 3: 68-85.

Green CL (1995). Natural colorants and dye stuff. FAO, Rome, Italy.

Gurudutt KN & Seshadri TR (1974). Constitution of the santalin pigments A and B. *Phytochemistry* 13: 2845-2847

Halim ME & Misra A (2011). The effect of aqueous extract of *Pterocarus santalinus* heartwood and vitamin E supplementation in streptozotocin-induced diabetic rats. *Journal of Medicinal Plant Research* 5: 398-409.

International Legume Database & Information Service (ILDIS) (2005). Genus *Pterocarpus*. 10:01.

Itoh A, Isoda K, Kondoh M, Kawase M, Kobayashi M, Tamesada M & Yagi K (2009). Hepatoprotective effect of syringic acid and vanillic acid on concanavalin a-induced liver injury. *Biological and Pharmaceutical Bulletin* 32: 1215-1219.

Itoh A, Isoda K, Kondoh M, Kawase M, Watari A, Kobayashi M, Tamesada M & Yagi K (2010). Hepatoprotective effect of syringic acid and vanillic acid on CCl4-induced liver injury. *Biological and Pharmaceutical Bulletin* 33: 983-987.

Jadhav J, Mane A & Kanase A (2012). Stimulatory effect of *Pterocarpus santalinus* on vasculogensis in chick chorioallantoic membrane (CAM). *Journal of Pharmacy Research* 5: 208-211.

Jain SK & Sastry ARK (1980). Threatened Plants of India. A State-of-the Art Report, 42.

Kayser O, Kiderlen AF & Brun B (2001). *In vitro* activity of aurones against *Plasmodium falciparum* strains K1 and NF54. *Planta Medica* 67: 718-721.

Kayser O, Kiderlen AF, Folkens U & Kolodziej H (1998). *In vitro* leishmanicidal activity of aurones. *Planta Medica* 65: 316-319.

Kesari AN, Gupta RK & Watal G (2004). Two aurone glycosides from heartwood of *Pterocarpus santalinus*. *Phytochemistry* 65: 3125-3129.

Khadem S & Marles RJ (2010). Monocyclic phenolic acids; Hydroxy and polyhydroxy benzoic acids: Occurence and recent bioactivity studies. *Molecules* 15: 7985-8005.

Kinjo J, Uemura H & Nohara T (1995). Novel yellow pigment from *Pterocarpus santalinus*: Biogenetic hypothesis for santalin analogs. *Tetrahedron Letters* 36: 5599-5602.

Kirtikar KR & Basu BD (2001). Indian Medicinal Plants with Illustrations, (2nd ed.), Oriental enterprises, Dehradun, India.

Kondeti VK, Badri KR, Maddirala DR, Thur SK, Fatima SS, Kasetti RB & Rao CA (2010). Effect of *Pterocarpus santalinus* bark on blood glucose, serum lipids, plasma insulin and hepatic carbohydrate metabolic enzymes in streptozotocin-induced diabetic rats. *Food and Chemical Toxicology* 48: 1281-1287.

Krishnamoorthy P, Stella J, Mohamed AJ & Anand M (2011). Free Radical scavenging and antibacterial evaluation of *Pterocarpus santalinus* leaf *in-vitro* study. *International Journal of Pharmaceutical Sciences and Research* 2:1204-1208.

Krishnaveni KS & Rao JVS (2000). A new triterpene from callus of *Pterocarpus santalinus*. *Fitoterapia* 71: 10-13.

Krishnaveni KS & Rao JVS (2000). An isoflavone from *Pterocarpus santalinus*. *Phytochemistry* 53: 605-606.

Krishnaveni KS & Rao JVS (2000a). A new isoflavone glucoside from *Pterocarpus santalinus*. *Journal of Asian Natural Products Research* 2: 219-223.

Kumar D (2011). Anti-inflammatory, analgesic, and antioxidant activities of methanolic wood extract of *Pterocarpus santalinus* L. *Journal of Pharmacology and Pharmacotherapeutics* 2: 200-202.

Kumar N, Ravindranath B & Seshadri TR (1974). Terpenoids of *Pterocarpus santalinus* heartwood. *Phytochemistry* 13: 633-636.

Kwon HJ, Hong YK, Kim KH, Han CH, Cho SH, Choi JS & Kim BW (2006). Methanolic extract of *Pterocarpus santalinus* induces apoptosis in Hela cells. *Journal of Ethnopharmacology* 105: 229-234.

Latheef SA, Prasad B, Bavaji M & Subramanyam G (2008). A database on endemic plants at Tirumala hills in India. *Bioinformation* 2: 260-262.

Manjunatha BK (2006). Antibacterial activity of *Pterocarpus santalinus*. *Indian Journal of Pharmaceutical Sciences* 68: 115-116.

Manjunatha BK (2006). Hepatoprotective activity of *Pterocarpus santalinus* L. f., an endangered plant. *Indian Journal of Pharmacology* 38: 25-28.

Manjunatha BK, Rupani AR, Priyadarshini P & Paul K (2010). Lead findings from *Pterocarpus santalinus* with hepatoprotective potentials through in silico methods. *International Journal of Pharmaceutical Sciences and Research* 7: 265-270.

Meena AK, Bansal P & Kumar S (2009). Plants-herbal wealth as a potential source of ayurvedic drugs. *Asian Journal of Traditional Medicine*. 4: 152-70.

Mena S, Rodríguez ML, Ponsoda X, Estrela JM, Jäättela M & Ortega AL (2012). Pterostilbene-induced tumor cytotoxicity: A lysosomal membrane permeabilization-dependent mechanism. *PLoSOne* 7: e44524.

Mishra MP & Padhy RN (2013). *In vitro* antibacterial efficacy of 21 Indian timber-yielding plants against multidrug-resistant bacteria causing urinary tract infection. *Osong Public Health Res Perspect* 4: 347-357.

Nagaraju N & Rao KN (1989). Folk-medicine for diabetes from Rayalaseema of Andhra Pradesh. *Ancient Science of Life* 9: 31-35.

Nagaraju N, Prasad M, Gopalakrishna G & Rao KN (1991). Blood sugar lowering effect of *Pterocarpus santalinus* (Red Sanders) wood extract in different rat models. *International Journal of Pharmacognosy* 29: 141-144.

Narayan S, Devi RS & Devi CS (2007). Role of *Pterocarpus santalinus* against mitochondrial dysfunction and membrane lipid changes induced by ulcerogens in rat gastric mucosa. *Chemico-Biological Interactions* 170: 67-75

Narayan S, Devi RS, Srinivasan P & Devi CS (2005). *Pterocarpus santalinus*: A traditional herbal drug as a protectant against ibuprofen induced gastric ulcers. *Phytotherapy Research* 19: 958-962.

Narayan S, Veeraraghavan M & Devi CS (2007). *Pterocarpus santalinus*: An *In Vitro* study on its anti-*Helicobacter pylori* effect. *Phytotherapy Research* 21: 190-193.

Oldfield SO, Lusty C & MacKinven A (1998). The World List of Threatened Trees. World Conservation Press, Cambridge, UK.

Palanisamy D, Kannan SE & Bhojraj S (2007). Protective and therapeutic effects of the Indian medicinal plant *Pterocarpus santalinus* on D-galactosamine-induced liver damage. *Asian Journal of Traditional Medicine* 2: 51-57

Parkash E, Khan PSS, Rao TVJ & Meru ESJ (2006). Micropropagation of red sanders (*Pterocarpus santalinus* L.) using mature nodal explants. *Journal of Forest Research* 11: 329-335.

Ramawat KG & Goyal S (2008). The Indian herbal drugs scenario in global perspectives. *Bioactive Molecules and Medicinal Plants* 18: 325-347.

Riche DM, Riche KD, Blackshear CT, McEwen CL, Sherman JJ, Wofford MR & Griswold ME (2014). Pterostilbene on metabolic parameters: A randomized, double-blind, and placebo-controlled trial. *Evidence-Based Complementary and Alternative Medicine* 2014: 459165.

Rimando AM, Nagmani R, Feller DR & Yokoyama W (2005). Pterostilbene, a new agonist for the peroxisome proliferator-activated receptor alpha-isoform, lowers plasma lipoproteins and cholesterol in hypercholesterolemic hamsters. *Journal of Agricultural and Food Chemistry* 53: 3403-3407.

Wu SF, Chang FR, Wang SY, Hwang TL, Lee CL, Chen SL, Wu CC & Wu YC (2011). Anti-inflammatory and cytotoxic neoflavonoids and benzofurans from *Pterocarpus santalinus*. *Journal of Natural Products* 74: 989-996

Wu SF, Hwang TL, Chen SL, Wu CC, Ohkoshi E, Lee KH, Chang FR & Wu YC (2011). Bioactive components from the heartwood of *Pterocarpus santalinus*. *Bioorganic and Medicinal Chemistry Letters* 21: 5630-5632.

Yamauchi S, Hayashi Y, Nakashima Y, Kirikihira T, Yamada K & Masuda T (2005). Effect of benzylic oxygen on the antioxidant activity of phenolic lignans. *Journal of Natural Products* 68: 1459-1470.

19

Ethnomedicinal Importance of *Arisaema jacquemontii* and Its Futuristic Role in Natural Therapeutics

Rasleen Sudan and Madhulika Bhagat

Abstract

Plants have been used both in the prevention and cure of various diseases throughout the centuries. An important prerequisite for proper utilization of plants as natural resources is that we should have full knowledge regarding their occurrence, frequency, distribution and phenology. Arisaema jacquemontii *Blume (Araceae) is a perennial herb that has been reported in various literatures to be used traditionally in various countries for the treatment of different diseases. Scientific evaluations of this plant have revealed that it harbors various phytoconstituents like alkaloids, phenols, terpenoids, flavonoids, glycosides, tannins and several other chemical constituents which have pharmacological implications with as antioxidant, antifungal, antibacterial, anticancer and immunomodulatory activities. This chapter collaborates the data with respect to the ethnobotanical and ethno medicinal importance of* Arisaema jacquemontii *and its role as a natural source of phytoconstituents that have therapeutic implications.*

Keywords: *Arisaema jacquemontii*, Ethnomedicine, Pharmacology

Rasleen Sudan and Madhulika Bhagat (✉)

School of Biotechnology, University of Jammu, Jammu-180006, J&K, India

✉*Corresponding author email: madhulikasbt@gmail.com*

Plants for Novel Drug Molecules: Ethnobotany to Ethnopharmacology
Bikarma Singh & Yash Pal Sharma (eds.), (pp. 427-443)

Email: *info@nipabooks.com* Web: *www.nipabooks.com*

1. Introduction

Study of indigenous knowledge and practices is increasingly being recognized as a useful multidisciplinary tool to assist in achieving sustainable resources used in many poor rural communities while securing the resource base. Indigenous knowledge has gained international recognition through documents such as the World Conservation Strategy (IUCN 1980) and the Brundtland Commission's "Our Common Future" (WCED 1987), which emphasize the importance of the environmental expertise of local people in the management of natural resources. India had also enacted a number of legislations, in compliance with CBD and WTO, in order to prevent the unfair exploitation of biological wealth of the nation. These legislations, *inter alia,* enjoin for immediate chronicling of the country's biodiversity and the associated indigenous knowledge (Khan et al. 2004). Various other documentations are available that highlight the need to promote greater awareness and a wider application of indigenous knowledge for sustainable biological resource management.

Ethno-botanical knowledge of indigenous communities in curing various diseases, reflects the information they have accumulated with long experience and practices to solve the various problems. The primitive humans explored various remedies from nature for their well-being. Their observations, experiences and beliefs led to the accumulation of knowledge and with the passage of time, this information was transferred from generations to generations throughout centuries in the form of texts. The ancient texts written in different geographical locations led to formulation of their own *Materia Medica* such as the Greek, Indian, Chinese, Japanese and Arabian traditional medicinal systems. These systems have elaborated a vast array of medicinal plants for strengthening innate systems of body, healing and treating a myriad of ailments and health disorders. Apart from the benefits one can generate by exploiting these secret treasures of knowledge, it can also lead to the continuous erosion in the traditional knowledge of many valuable plants being used for the ethno-medicines without any valid documentation and preservation. Thus, there is surging need for extensively exploring the natural sources of medicines and authenticating their implications.

Also, discovering new products from natural sources is encouraging because it is known that out of the 2,50,000 medicinal plant species, only 6% have been evaluated for their bioactive potential. Besides, only around less than 1% of medicinal plant compounds have been assessed in clinical trials (Soosarei et al. 2017). The northern part of India harbors a great diversity of medicinal plants because of the majestic Himalayan range. So far, about 8000 species of angiosperms reported in the Himalayas, out of which 1745 species are medicinal plants (Ramakrishnappa 2013). One of such plant families is Araceae which comprises of 125 genera and about 3750 species of flowering plants (Petruzello 2018). Aroid is the common name for members of the Araceae family of plants, sometimes also known as the Philodendron or Arum family (Christenhusz & Byng 2016). The Araceae are group of monocotyledonous blooming plants in which blooms are borne on a kind of inflorescence called a spadix. The spadix is normally joined by and in some cases mostly encased in, a spathe or leaf-like bract. Species in this family are cormous or rhizomatous or tuberous herbs, root climbers in damp forests, usually with calcium oxalate crystals or raphides and commonly with milky or watery sap (latex).The subject of present study is a plant belonging to Aroid family and genus *Arisaema* i.e. *Arisaema jacquemontii.* It is a high altitude plant (2300-4300 m) which grows in abundance along the shady slopes of North-western Himalayan forest covers, South western Ghats and North-eastern India. One unusal trait shared by all *Arisaema* species, and not those of other genera within the Araceae is the ability of plants to change sex during their lifetime. *Arisaema* plants are typically male when small, and female or hermaphroditic when large, with a single plant capable of changing sex depending on its nutrition and genetics and perhaps changing sex several times during its long life (20 years or more) (Verma et al. 2012). Literature surveys and information from the folklore has revealed that *Arisaema jacquemontii*is used traditionally for treatment of various diseases by the indigenous people of India, China, Pakistan, Nepal and Bhutan. Also, the research work done on this plant reveals that it has tremendous pharmacological potential which needs to be unearthed through extensive and elaborative scientific experimentation.

Keeping the above facts in view, this chapter is an effort tto combine the traditional documentation of *Arisaema jacquemontii* based on the ethnobotanical and ethnomedicinal uses by the indigenous communities of the world, scientific validation of its pharmacological potential and immense scope of this plant in terms of future drug discovery from natural resources.

2. Biological Aspects

2.1. Geographical Distribution

The genus *Arisaema* comprised up of more than 250 herbaceous species of flowering plants from Araceae family who propagate throughout the temperate and tropical areas. Based on the unique look of their flowers, they are better known as Cobra lilies in the east and Jack-in-the-pulpit in the west. *Arisaema jacquemontii,* popularly known as Jacquemont's Cobra Lily is found growing its natural habitat along the rocky and shady slopes in upper forest and lower alpine zones in the drier areas of the Himalayas, 2400-4000 metres (Polunin, 1984). It prefers the cervices between stones under the shade of oak and deodar trees on loamy, peaty and well drained but moist soil full of humus and rocky substrates, shrubberies, sub-alpine areas. It is native to Afghanistan, China, India, Nepal and Pakistan. It also occurs in the Nilgiri Hills in southern India, and the Khasi Hills region of north-east India (Verma et al. 2012).

2.2. Botanical Classification

- Kingdom: Plantae
- Phylum: Magnoliophyta
- Class:Angiospermae
- Order: Alismatales
- Family: Araceae
- Genus: Arisaema
- Species: *Arisaema jacquemontii*

2.3. Synonyms

Arisaema brevispathum Buchet, *Arisaema cornutum* Schott, *Arisaema cylindraceum*Wall., *Arisaema exile* Schott , *Arisaema wightii* Schott

Table 1: Names of *A.jacquemontii* in different regions

Regions	Local names	References
English	Cobra lily, Jack in the Pulpit, Sheathed green dragon, Cobra plant, Snake lily, Jacquemont's Cobra lily, Cobra plant, Snake lily.	(Quattrochi 2012, Crook and Bachman 2013, Wani et al. 2006, Pandey 2006).
India	Sap-kukdi/Hapatmakai/ Surp/Hapat-Brand,Khaprya/Saperimausi/ Meen, Basair, Haput, Gogej, Jinjok, Khaprya, Ki kukri, Kirala, Saperimausi, Kirala, Bankh, Khaprya, Saperimausi, Khyan bank, Sarpabheda, Basair	(Pandey 2006, Khan et al. 2004, Singh and Rawat 2011, Khan et al. 2009, Kala 2015, Bhatt and Negi 2006, Ratha et al. 2015, Lone and Pandit 2007, Akbar et al. 2010, Bala et al. 2019)
Pakistan	Sap kibooti, Hathphees,WaraMarjarai, ZahurButay, Marjarai, Sapmak.	(Tanveer 2013, Shaheen et al. 2012, Butt et al. 2015, Ozturk et al. 2015, Ali and Qaiser 2009, Khan et al. 2007, Baba and Malik 2015)
Nepal	Timju, Baanko, Sarpakomakai	(Pandey 2006)
Bhutan	Dav-ba(dagala)	(Wangchuk 2016)
Tibet	Dahpa	Pandey 2006
China	Zang nan lü nan xing	Li et al. 2010

2.4. Morphological Description

Like all members of the genus *Arisaema,* Jacquemont's cobra lily is a tuberous perennial herb. It grows to about 70 cm tall and dies back to ground level in the winter or dry season. It usually has two palmate (hand-shaped), elliptic-ovate leaves which digitate with 5-9 narrow or long pointed leaflets. The spathe (sheathing bract) is green on the outside and pale-green inside, with faint white stripes, and arches over the spadix (unbranched inflorescence), its tip elongated and coiled. The spadix is unisexual and has a white or purple, tail-like tip. The appendage is rather stout, curved, base dilated, truncate (in male) or cuneate (in female).The fruits are globose berries (4-4.5mm broad), which are red when mature. *A. jacquemontii* plant is typically male when small, and female or hermaphroditic when large, with a single plant capable of changing sex depending on its nutrition and genetics and perhaps changing sex several times during its long life (Verma et al. 2012).

A research work was conducted on *Arisaema serratum* (Thunb.) in Japan regarding the labile sex expression that depends on the size of the plant and alternates between male and female throughout the plant's life (Rumphia 1836). It was found that in the lower part of the spathe tube, male flowers have an exit hole for small midges that is absent from female ûowers. This sexual dimorphism in the spathe tube is the basis of a pitfall-trap ûower pollination system. The unusual life history and mating systems of *Arisaema sp.* present challenging questions in plant reproductive ecology with respect to pollination biology, population genetics, and diversiûcation between *Arisaema* species (Nishizawa et al. 2011, Ved et al. 2016).

Fig. 1: *Arisaemajacquemontii*- **(A)** & **(B)** Whole plant, **(C)** Tubers and **(D)** Ripened fruit

2.5. Ethnobotanical Importance

The folklore of Jammu and Kashmir have been using rhizomes (tubers) of *A. jacquemontii* mixed with edible oil to form a paste, which is used for massage in order to regain muscular strength and to treat skin problems such as blisters and pimples (Verma et al. 2012). The indigenous people of India, Pakistan, Nepal and China use different species of *Arisaema* as anti-nematodal, anti-inflammatory (Jung et al. 1996, Agoreyo et al. 2012, Balekar et al. 2010, Choudhary et al. 2008, Chunxia et al. 2011, Bibi et al. 2010, Kunwar et al. 2010). Chinese Pharmacopiea (2005) elaborates the use of various *Arisaema* sp. as analgesic, antitumor and pesticide agents and for treating dementia, neurological symptoms, heat-phlegm in the lung of children, tic of limbs. The Chinese believe the *Arisaema* sp. to be bitter, acrid in flavor and warm, toxic in nature. Vital functions are eliminating dampness and phlegm, expelling wind to relieve convulsion, dissipating binds, and reduce the swelling (Yang et al. 2007). Table 2 elaborates the ethnobotanical and ethnomedicinal uses of *A. jacquemontii*in various regions.

Table 2: Ethnobotanical and ethnomedicinal uses of *A. jacqemontii*

Region	Plant part	Ethnobotanical/Ethnomedicinal uses	References
Kumaon and Garhwal region (Uttarakhand, India)	Rhizomes	Treatment of ringworm infections and skin diseases	Pande et al. 1999
Garhwal region (Uttarakhand, India)	Fruits	Jaunsari tribe uses fruit decoction is used as snakebite antidote	Bhatt and Negi 2006
Manali wildlife sanctuary (North-western Himalaya, India)	Rhizomes	Treatment of ringworm infections and skin diseases	Rana and Samant 2011
Kedarnath wildlife sanctuary (North-western Himalaya, India)	Rhizomes	Tuber is used in cough, kidney, skin diseases especially as ringworm-killer, antidote to snakebite.	Singh and Rawat 2011
Kashmir (Himalayas, Jammu and Kashmir, India)	Rhizomes	Local tribals use the rhizomes after chopping and mixing with ghee or oil to form a paste. This paste is used on chronic boils as a remedy and to regain muscular strength. The water extract of bulbs is used to get rid of skin eruptions such as blisters or pimples and other skin problems due to cold temperature.	Parvaiz et al. 2006, Malik et al. 2011, Khan et al. 2004, Tantray et al. 2009
Shawat Valley (Swat, Pakistan)	Rhizomes	Cough and respiratory tract infection in cows and buffaloes	Khan et al. 2007
Kaghan Valley (Mansehra District, Pakistan)	Whole plants	Camouflage for snakes in coniferous temperate zone	Awan et al. 2011
Chitral Valley (Pakistan)	Fruits and Rhizomes	Fruits and rhizomes are poisonous and cause sedation. Very small quantity is used during meal for relieving body pain. Also used in small quantities in various preparations by "Hakims" for psychic and nervous disorders	Ali and Qaiser 2009.
KotManzaray Baba Valley (Malakand Agency, Pakistan)	Rhizomes	Antidote for snake bites	Zabihullah et al. 2006.

Contd.

Upper Mustang, Nepal	Rhizomes and flowers	Tibetan therapy system (amchi) uses rhizomes and flowers for treating fever, stomach problems, swelling, toothache, scabies, chest infection, uterus and menstrual disorders, anthelmintic and throat problems and also used as vegetables	Pandey 2006
China	Rhizomes/ Leaves and flowers	Primary *Arisaema* uses and indications include cough caused by stagnated phlegm, vertigo due to wind-phlegm, stroke because of obstructed phlegm, drooping mouth and eye, hemiplegia, epilepsy, convulsions, and tetanus. Externally it treats carbuncles, snake bites, and insect bites. Recommended prepared dosage is from 3 to 9 grams in decoction. For external use, the raw one should be ground into powder and applied to the affected area with vinegar or wine.	Yang et al. 2007

2.6. Cultivation

It prefers a cool peaty soil in the bog garden, woodland garden or a sheltered border in semi-shade. Prefers a loamy or peaty soil and will tolerate a sunny position if the soil is moist but not water-logged and the position is not too hot or exposed. This is probably the hardiest of the Himalayan species and should succeed outdoors in a suitable position in many parts of the country. Only plant out full sized tubers and mulch them with organic matter in the winter. Stored seed remains viable for at least a year and can be sown in spring in the greenhouse but it will probably require a period of cold stratification. Germination usually takes place in 1-6 months at 15° C (Huxley 1992). Plants need protection from slugs. Most species in this genus are dioecious, but they are sometimes monoecious and can also change sex from year to year (Thompson and Morgan 1988, Chittendon 1956).

3. Phytochemistry

Various phytochemicals have been identified in *A. jacquemontii* plant parts. Kletter and Kriechbaum (2001) reported the presence of amino acids alanine, arginine, aspartic acid, leucine, lysine, serine, threonine, tyrosine and valine in *A. jacquemontii* tuber extracts. A novel compound, arisaeminone, a benzophenone derivativehas been isolated from the methanol extract of whole plant and another compound *i.e.*13-phenyltridecanoic acid was found in the seeds (Kletter and Kriechbaum, (2001). The methanol and chloroform extracts of tubers have demonstrated the presence of phenolic compounds, flavonoids, triterpenoids 30-nor-lanost-5- ene-3beta-ol and 30-nor-lanost-5-ene-3-one (Baba et al. 2015, Jeelani et al. 2010). Methanol and chloroform extracts of *A. jacquemontii* leaves have shown the presence of glycosides, terpenoids, coumarins, quinines, saponins, tannins, alkaloids, anthraquinones, flavonoids and phenols (Sudan et al., 2014).Tanveer et al., 2014 reported the presence of terpenes, saponins and glycosides in the methanol and chloroform extracts of fruits. Tanveer et al. 2013 have also reported a triterpenoid 2- hydroxyl diplopterol from the chloroform extract of whole plant.

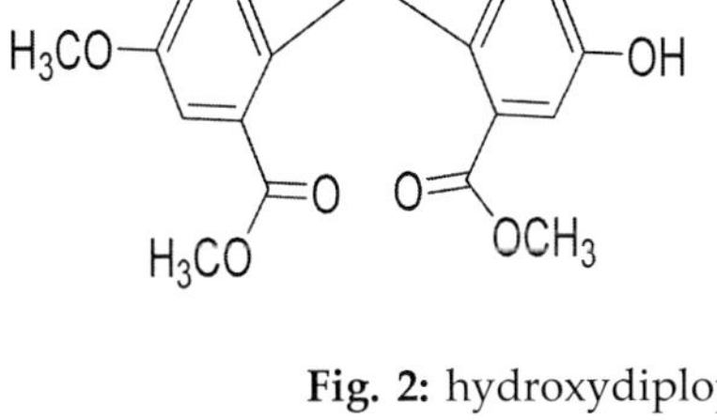

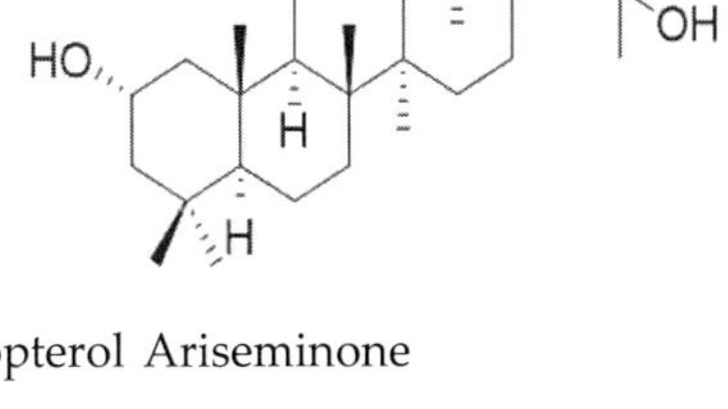

Fig. 2: hydroxydiplopterol Ariseminone

4. Pharmacological Studies

During the recent years, there have been substantial scientific reports authenticating the pharmacological potential of *A. jacquemontii*, as enlisted below.These studies strongly support the future herbal drug formulations from this plant with their possible use in immunomodulation for curing chronic diseases.

4.1. Anti-proliferative Activity

Pharmacology with respect to anti-proliferative effects of *A.jacquemontii* on human cancer cell lines has been studied with some positive results. An effective novel anticancer compound ariseminone was isolated from *A.jacquemontii* (Pandey 2006). Lectins was isolated from tubers and were found capable of inhibiting the growth ofHCT-15, HOP-62, SW-620, HT29, IMR-32, SKOV-3, Colo-205, PC-3, HEP-2, A-549 human cancer cell lines (Kaur et al., 2006), also anothercytotoxic compound was isolated viz.,triterpenoid, 2-hydroxydiplopterol from *A. jacquemontii*as reported by Tanveer et al. (2013).

4.2. Immunomodulation

Methanol and hexane extracts of *A. jacquemontii* leaves demonstrated stimulating effect on humoral response and cytotoxic responses in immune suppressed mice as reported by Sudan et al. (2014), supporting its role as herbal drug formulations with possible use in immunomodulation for curing chronic diseases.

4.3. Chelation Capacity

Sudan et al. (2014) reported considerable chelating capacity of *A. jacquemontii*leaf extracts on ferrous ions.

4.4. Antimicrobial Activity

Baba and Malik (2015) have reported the antibacterial activity of methanol extract of A. *jacquemontii* tubers. Similar studies in 2019 by Bala and co-workers on A. *jacquemontii* highlights the antibacterial and antioxidant properties reporting significant antibacterial activity against *S. aureus, P. aerugenosa*and *S. dysenteriae* proving its ethnomedical uses for dermatological disorders and microbial infections, Bala et al. (2019).

4.5. Antimalarial Effects

The tubers extracts of *A. jacquemontii* evaluated for antimalarial properties against *Plasmodium berghei* (NK-65) demonstrated 70% growth inhibition, Banyal (2014).

4.6. Anticonvulsant Effect

Jeelani et al., 2010 suggest the anticonvulsant potential of chloroform extract of rhizomes and effects on platelet aggregation by *A. jacquemontii* and they have isolated triterpenoids from its tubers.

4.7. Anti-leishmanial Property

In another report by Tanveer et al. (2014), anti-leishmanial, phytotoxic and immunomodulatory properties have been demonstrated by the methanol and chloroform extracts of *A. jacquemontii* leaves and tubers.

4.8. Anti-insect Potential

Lectins isolated from the tubers of *A. jacquemontii*reporting anti-insect potential against the growth of *Bactrocera cucurbitae* (Coquillett) larvae, Kaur et al. (2006).

Scientific validation of various other *Arisaema* species has also been done during the past few years, which is suggestive of the wide pharamacological potential of this genus. For example, the neuroprotectiverole of*A.cumbile*by demonstrating Amyloid α1-40-induced neuronal cell (PC-12) death in case of neurodegenerative diseases like Alzhiemer's disease. It also significantly decrease reactive oxygen species (ROS) production, caspase-3 activity, and DNA fragmen-tation (Ki et al., 2012). Anti-insecticidal and anti-cancer effects of lectins isolated from *A. helleborifolium* have also been reported (Kaur et al., 2006). Ducky et al., 1995 reported the isolation of a phenolic compound from *A. erubescens*, Paeonol, which possesses anti-mutagenic, anti-convulsant and anti-inflammatory activities. Methanol extract of *A. tortuosum* tubers have been reported to exhibit anti inflammatory and anti-proliferative activities against HeLa cancer cell line (Nile and Park, 2014). Chinese Pharmacopoiea describes the formulation of capsules, made from *Arisaema*, ginseng and *Typhonium* rhizome that were used to treatacute gingivitis suggesting the anti-inflammatory potential of *Arisaema sp*. (Jiu et al. 1986). Bioactive metabolites from endophytic fungus of *A. erubescens* have also been reported by Wang et al. (2012).

5. Toxicity

Arisaema bulbs are a strong irritant to the skin and mucous membrane because they contain calcium oxylate crystals. Fruits are poisonous and cause sedation, very small quantity is used during meal for relieving body pain (Ali et al. 2009). These cause an extremely unpleasant sensation similar to needles being stuck into the mouth and tongue if they are eaten raw. It may lead to tongue, pharynx, oral numbness and swelling, mucosal erosion, hoarseness, difficulty in opening mouth, or even slow breathing, choking, etc. Skin contact can cause allergic itching. It was also reported that long-term use of it can cause mental retardation (Zhong 1987). The bulbs are easily neutralized by thoroughly drying or cooking the plant or by steeping it in water before eaten as food. If it is mistakenly taken orally, the regular remedies are to administer fresh ginger juice or perform gastric lavage with dilute vinegar, tannin, tea, and egg whites, and others. Oral erosions can be treated by rinsing the mouth with hydrogen peroxide and boric acid solution, coating the mouth with gentian violet, and giving oxygen or tracheostomy if necessary. In external applications, the allergenic reaction on the skin can be alleviated by washing with water, diluted vinegar, or tannic acid (Zhong 1986).

6. Discussion and Future Prospects

Ethnobotany is an art of collecting useful plants, understanding and description of their uses through anthropogenic methods. It is a discipline that deals with people-plant interactions in a multidisciplinary manner, involving collection and documentation of indigenous uses of plant as well as ecology, economy, pharmacology and public health. However,the ethnobotanical uses of medicinal plants as described in ancient pharmacopoeias need to be scientifically validated in order to authenticate their use as efficient and safe therapeutics.

Recently, the use of medicinal plants has also found its place in developed countries that used to earlier rely more on synthetic drugs for curing diseases. The National Health Statistics (NHS) report released in 2016, stated that about 38% adults and 12% children traditional medicine used in 2008 and this trend has been increasing ever since to about 82 % adults and 52% children (NHIS, 2012), Richard et al. (2016). This trending shift in the use of medicinal plants as alternative treatment is taking place because inspite of tremendous advancements in the field of allopathic medicines, they are still associated with long term and serious side effects, some of which cannot be reversed Karimi et al. 2015. In addition, there has been a steep increase in mortality rate due to deadly diseases like ischemic heart disease and stroke, diabetes, dementia and serious injuries as reported by The Global Burden of Disease (GBD-2015) study, Forouzanfar et al.(2016). According to the BCC Reports (Botanical and Plant-Derived Drugs: Global Markets 2015) the demand for plant derived drugs is expected to increase from $25.6 billion in 2015 to nearly $35.4 billion in 2020, with the annual growth rate of 6.6%, Lawson (2015). Therefore, hazards associated with allopathic drugs, continuous upsurge of serious diseases and the rising demand for natural products has warranted the search for alternative treatments using traditional medicinal plants due to easy accessibility, affordability and lesser side effects (Pan et al. 2014, Tripathi 2015)

The ethnomedicinal knowledge about *Arisaema jacquemontii* along with the scientific validations of its biological potentials are suggestive of the immense scope of further exploration of this plant as source of more potent natural compounds. Plant derivatives possess enormous potential for inhibiting cancer proliferation,because of their ability to delay the cell cycle. About 50% of anticancer drugs developed so far are either direct natural compounds or their derivatives (Cragg & Newmann 2013, Cancer Reasearch UK 2014). Thus, the field of anticancer research warrants the search for new lead compounds from natural sources. Also, a recent review provides useful information about various synthetic immunomodulators, explaining their adverse side effect profiles and generalized effects which pose a major limitation to their use and emphasizes the need for search of more effective and safer immunomodulatory agents. As *A. jacquemontii* has been reported to possess remarkable anticancer as well as immunostimulatory activity. Therefore, this plant may displays the interplay of both the activities and it can be further subjected to series of research experimentations to understand its role as immunotherapeutic in cancer progression. In addition, the immunosuppressive effects of *A. jacquemontii* fractions on mitogen induced lymphocytic proliferation are suggestive of their use in various autoimmune disorders where suppression in order to restore a state of well-balanced immune system. However, detailed studies are required to authenticate their use as immunosuppressants and recognition of the mechanisms of action. Also, the antibacterial studies done on this plant confirm its ethnomedicinal use for curing skin diseases and other infections of the body. The future experimentation can be designed to isolate the antibacterial compounds and understand their mechanisms of action.

In conclusion, *Arisaema jacquemontii* (Himalayan Cobra Lily) widely used by indigenous people of various regions to cure multiple diseases and infections. Some scientific reports available authenticates its pharmacological potential. However,furtherdetailed scientific research on this plant regarding their chemistry and pharmacology along with validation of efficacy, toxicity and safety may eventually pave way for the discovery of new drugswith increased bioavailability and lesser side effects.

Conflict of Interest

No conflict of interest for this research paper

Acknowledgements

This work was a part of the major project funded by University Grant Commission (UGC No. F.No. 41-587/2012 (SR) dated 18-07-2012), Government of India. Authors would also like to recognize facility created by School of Biotechnology with help of funds supported by UGC-SAP, RUSA, DST-FIST, grants and Center of Bioinformatic to carry out our research work. Authors highly acknowledge the help rendered by the elderly people of Bhaderwah (Sh. R RBhagat, Sh. MadanLal and Sh. Omkar Singh) in providing the information about the identification, traditional uses and in collection of the plant samples.

References

Agoreyo BO, Okoro NC &Choudhary MI (2012). Preliminary phytochemical analyses of two varieties of *Adenialobata* (jacq) and the antioxidant activity of their various solvent fractions. *Bayero Journal of Pure and Applied Sciences* 5(1): 182-186.

Akbar M, Seema A, Masood A &Ganai B (2010). Medicinal and aromatic plants from Kashmir Himalayas, Germany, Books on Demand.

Ali H &Qaiser M (2009). The ethnobotany of Chitral valley, Pakistan with particular reference to medicinal plants. *Pakistan Journal of Botany* 41: 2009-2041.

Awan MR, Iqbal Z, Shah SM, Jamal Z, Jan G, Afzal M, et al. (2011). Studies on traditional knowledge of economically important plants of Kaghan Valley, Mansehra District, Pakistan. *Journal of Medicinal Plants Research*. 5:3958-3967.

Baba I, Dubey S, Alia A, Saxena R, Itoo A &Powar K (2012). Ethnobotanical survey of medicinal plants used by the people of District Ganderbal Jammu and Kashmir. *Research Journal of Pharmaceutical, Biological and Chemical Sciences*. 3:549-56.

Baba SA & Malik SA (2015). Determination of total phenolic and flavonoid content, antimicrobial and antioxidant activity of a root extract of *Arisaema jacquemontii* Blume. *Journal of Taibah University for Science*. 9: 449454.

Bala K, Rana J &Sagar A (2019). Antibacterial and Antioxidant Potential *of Arisaema jacquemontii* Blume from Manali, Himachal Pradesh. *Bulletin of Pure and Applied Sciences*. 38(1): 23-33.

Balekar N, Bodhankar SL & Jain DK (2010). Evaluation of Immunomodulatory Activity of Medicinal Plants: Ellagic Acid and Mangiferin. Lap Lambert Academic Publishing.

Banyal H, Tandon A &Nainta M (2014). Antimalarial effects of extracts of *Ariseama jaquemontii* Bl. on Plasmodium berghei Vinke and Lips. *Asian Journal of Biological Sciences*. 7: 131-134.

Bhatt V &Negi G (2006). Ethnomedicinal plant resources of Jaunsari tribe of Garhwal Himalaya, Uttaranchal. *Indian Journal of Traditional Knowledge* 3: 331-335.

Bibi Y, Nisa S, Waheed A, Zia M, Sarwar S, Ahmed S & Chaudhary MF (2010). Evaluation of Viburnum foetens for anticancer and antibacterial potential and phytochemical analysis. *African Journal of Biotechnology* 9(34): 5611.

Butt MA, Ahmad M, Fatima A, Sultana S, Zafar M, Yaseen G, et al. (2015). Ethnomedicinal uses of plants for the treatment of snake and scorpion bite in Northern Pakistan. *Journal of Ethnopharmacology* 168:164-181.

Cancer Research UK (2014) what is cancer? Available at: http://www.cancerresearchuk.org/about-cancer/what-is-cancer

Chittendon F (1956). RHS Dictionary of Plants plus Supplement.

Choudhary K, Singh M & Pillai U (2008). Ethnobotanical survey of Rajasthan-An update. *American-Eurasian Journal of Botany* 1(2): 38-45.

Christenhusz MJM & Byng JW (2016). The number of known plants species in the world and its annual increase. *Phytotaxa*. Magnolia Press. 261 (3): 201-217. doi:10.11646/phytotaxa.261.3.1

Chunxia C, Peng Z, Huifang P, Hanli R, Zehua H & Jizhou W (2011). Extracts of *Arisaemarhizomatum* attenuate inflammatory response on collagen-induced arthritis in BALB/c mice. *Journal of Ethnopharmacology* 133(2): 573-582.

Cragg GM & Newman DJ (2013). Natural products: a continuing source of novel drug leads. *Biochimicaet.Biophysica. Acta (BBA)-General Subjects* 1830(6): 3670-3695.

Crook V & Bachman S (2013). *Arisaema jacquemontii*. The IUCN Red List of Threatened Species. e.T44393284A44482120. http://dx.doi.org/10.2305/IUCN.UK.2013

Ducki S, Hadfield JA, Lawrence NJ, Zhang X & McGown AT (1995). Isolation of paeonol from *Arisaemaerubescens*. *Planta Medica* 61(06): 586-587.

Forouzanfar MH, Afshin A, Alexander LT, Anderson HR, Bhutta ZA, BiryukovS, ...& Cohen AJ (2016). Global, regional, and national comparative risk assessment of 79 behavioural, environmental and occupational, and metabolic risks or clusters of risks, 1990-2015: a systematic analysis for the Global Burden of Disease Study 2015. *The Lancet* 388 (10053):1659.

Huxley A (1992). The New RHS Dictionary of Gardening. MacMillan Press ISBN 0-333-47494-5.

Jeelani S, Khuroo MA, Razadan T (2010). New triterpenoids from *Arisaemajacquemontii. Journal of Asian Natural Products Research*. 12:157-161.

Jiu, ZCYY (1986). Research of Chinese Patent Medicine. 6: 21.

Jung JH, Lee CO, Kim YC & Kang SS (1996). New bioactive cerebrosides from *Arisaema amurense*. *Journal of Natural Products* 59(3): 319-322.

Kala CP (2015). Medicinal and aromatic plants of Tons watershed in Uttarakhand Himalaya. *Journal of Pharmacognosy Phytochemistry and Environmental Sciences*. 3:16-21.

Karimi A, Majlesi M & Rafieian-Kopaei M (2015).Herbal versus synthetic drugs; beliefs and facts. *Journal of Nephropharmacology* 4(1): 27.

Kaur M, Singh K, Rup JP, Saxena AK, Khan RZ, Ashraf MT, Kamboj SS & Singh J (2006). A tuber lectin from *Arisaemahelleborifolium*, Schott. with anti-insect activity against melon fruit Fly, *Bactroceracucurbitae*(Coquillett) and anti-cancer effect on human cancer cell lines. *Archives of Biochemistry and Biophysics*. 445:156–165.

Kaur M, Singh K, Rup PJ, Kamboj SS, Saxena AK, Sharma M et al. (2006). A tuber lectin from *Arisaemajacquemontii*Blume with anti-insect and antiproliferative properties. *Journal of Biochemical and Molecular Biology*. 39:432-440.

Khan M, Kumar S & Hamal IA (2009). Medicinal plants of sewa river catchment area in the Northwest Himalaya and its implication for conservation. *Ethnobotanical Leaflets* 13:5.

Khan SM, Ahmad H, Ramzan M & Jan MM (2007). Ethnomedicinal plant resources of Shawat Valley. *Pakistan Journal of Biological Sciences* 10:1743-1746.

Khan ZA, Khuroo AA & Dar GH (2004). Ethnomedicinal survey of Uri, Kashmir Himalaya. *Indian Journal of Traditional Knowledge* 3(4):351-357.

Ki H, Xian YF & Lin ZX (2012). Neuroprotective Effects of *Radix Rehmanniae* and *ArisaemaCum Bile* on Amyloid-Beta-Induced Toxicity in PC12 Cells. *University of Toronto Journal of Undergraduate Life Sciences*. 6(1).

Kletter C & Kriechbaum M (2001). Tibetan medicinal plants; South Africa, Med. Pharm. Publications.

Kunwar RM, Shrestha KP & Bussmann RW (2010). Traditional herbal medicine in Far-west Nepal: a pharmacological appraisal. *Journal of Ethnobiology and Ethnomedicine* 6(1): 35

Lawson K (2015). Botanical and Plant-Derived Drugs: Global Markets-2015.www.bccresearch.com/market-research/biotechnology/botanical-plant-derived-drugs-report-bio022g.html

Li H, Zhu G, Boyce PC & Murata J (2010). Flora of China. Family Araceae, Science Press (Beijing) & Missouri Botanical Garden (St. Louis), 23.

Lone HA &Pandit AK (2007). Medicinal plant wealth of Langate Forest Division in Kashmir Himalaya. *Journal of Himalayan Ecology & Sustainable Development* 2: 73-78.

Malik A, Siddique M, Sofi P & Butola J (2011). Ethnomedicinal practices and conservation status of medicinal plants of North Kashmir Himalayas. *Research Journal of Medicinal Plant* 5:515-530.

Nile SH & Park SW (2014). HPTLC analysis, antioxidant, anti-inflammatory and antiproliferative activities of *Arisaema tortuosum* tuber extract. *Pharmaceutical Biology* 52(2), 221–227.

Nishizawa T, Watano Y, Kinoshita E, Kawahara T & Nakajima N (2011). Development and characterization of a novel set of microsatellite markers for *Arisaema serratum* (araceae) *American Journal of Botany* e378–e381.

Öztürk M, Hakeem K, Faridah-Hanum I &Efe R (2015). Climate change impacts on high-altitude ecosystems. Switzerland, Springer International Publishing.

Pan SY, Litscher G, Gao SH, Zhou SF, Yu ZL, Chen HQ, Zhang SF, Tang MK, Sun JN & Ko KM (2014). Historical Perspective of Traditional Indigenous Medical Practices: The Current Renaissance and Conservation of Herbal Resources. *Evidence-Based Complementary and Alternative Medicine* 20.

Pande P C, Pokharia DS & Bhatt JC (1999). Ethnobotany of Kumaun Himalaya. Published by Scientific Publishers, Jodhpur, India.

Pandey MR (2006). Use of medicinal plants in traditional Tibetan therapy system in upper Mustang, Nepal. *Our Nature* 4:69-82.

Petruzello M (2018). List of plants in the family Araceae. Encyclopedia Britannica. Published by Encyclopedia Britannica inc.https://www.britannica.com/topic/list-of-plants-in-the-family-Araceae-2075376

Polunin O, Stainton A (1984). Flowers of the Himalayas. Published by Oxford University Press. Oxford–New York, 580 pages.

Quattrocchi U (2012). CRC World Dictionary of Medicinal and Poisonous Plants: Common Names, Scientific Names, Eponyms, Synonyms, and Etymology. Boca Raton, Florida, CRC Press Taylor and Francis Group.

Ramakrishnappa K (2003). Impact of cultivation and gathering of medicinal plants on biodiversity: case studies from India.

Rana SM & Samant SS (2011). Diversity, indigenous uses and conservation status of medicinal plants in Manali wildlife sanctuary, North western Himalaya. *Indian journal of traditional knowledge* 10(3): 439-459.

Ratha KK, Joshi GC, Rungsung W & Hazra J (2015). Use pattern of high altitude medicinal plants by Bhotiya tribe of Niti valley, Uttarakhand. *World Journal of Pharmacy and Pharmaceutical Sciences* 4:1042-1061.

Richard L, Nahin M & Barnes MA (2016). Nation Expenditures on Complementary Health Approaches: United States, 2012 (National Health Statistics Report), 95.

Rumphia (1836). *Arisaema jacquemontii* Blume., *Flora of British India* 6: 505.

Shaheen H, Shinwari ZK, Qureshi RA &Ullah Z (2012). Indigenous plant resources and their utilization practices in village populations of Kashmir Himalayas. *Pakistan Journal of Botany* 44:739-745.

Singh G &Rawat G (2011). Ethnomedicinal survey of Kedarnath wildlife sanctuary in Western Himalaya, India. *Indian Journal of Fundamental and Applied Life Sciences* 1:35-46.

Soosaraei M, Fakhar M, Teshnizi HS, Hezarjaribi HZ & Banimostafavi ES (2017).Medicinal plants with promising antileishmanial activity in Iran: a systematic review and meta-analysis. *Annals of Medical Surgery* (London) (21): 63–80.

Sudan R, Bhagat M, Gupta S, Singh J &Koul A (2014). Iron (FeII) chelation, ferric reducing antioxidant power, and immune modulating potential of Arisaemajacquemontii (Himalayan Cobra Lily). *BioMed Research International* 7.

Tantray MA, Tariq KA, Mir MM, Bhat MA & Shawl AS (2009). Ethnomedicinal survey of shopian, Kashmir (J&K), India. *Asian Journal of Traditional Medicines,* 4 (1).

Tanveer M, Habib-Ur-Rehman, Mesaik M & Choudhary M (2014). Immunomodulatory, antileishmanial and phytotoxicity of *Arisaema jacquemontii* Blume plant extracts. *Archives of Applied Science Research.* 6:12-17.

Tanveer M, Sims J, Choudhary MI & Hamann MT (2013). First ever isolation of cytotoxic triterpenoid 2 hydroxydiplopterol from plant source. *Journal of Medicinal Plants Research.* 7: 2040-2042.

Thompson & Morgan (1988). Growing from Seed, Volume 2.

Tripathy JP (2015). Can naturopathy provide answers to the escalating health care costs in India?. *Journal of Traditional and Complementary Medicine.* 5(2): 63-65.

Ved DK, Sureshchandra ST, Barve V, Srinivas V, Sangeetha S, Ravikumar K, Kartikeyan R., Kulkarni V, Kumar AS, Venugopal SN, Somashekhar BS, Sumanth MV, Begum N, Rani S, Surekha KV &Desale N (2016). FRLHT's ENVIS Centre on Medicinal Plants, Bengaluru http://envis.frlht.org/plant_ details. php?disp_id=3337

Verma H, Lal VK, Pant KK &Soni N (2012). A review on *Arisaema jacquemontii. Journal of Pharmacology Research* 5:1480-1482.

Wang LW, Xu BG, Wang JY, Su ZZ, Lin FC, Zhang CL &Kubicek CP (2012). Bioactive metabolites from Phoma species, an endophytic fungus from the Chinese medicinal plant *Arisaemaerubescens*. *Applied microbiology and Biotechnology* 93(3): 1231-1239.

Wangchuk P, Namgay K, Gayleg K &Dorji Y (2016). Medicinal plants of Dagala region in Bhutan: their diversity, distribution, uses and economic potential. *Journal of Ethnobiology and Ethnomedicine* 12:28.

Wani PA, Dar A, Mohi-Ud-Din G, Ganaie KA, Nawchoo I, Wafai B (2006). Treasure and tragedy of the Kashmir Himalaya. *International Journal of Botany* 2: 402-408.

Yang ZH, Yi JY, Wei ZR & Yan WQ (2007). Effect of extract of RhizomaArisaematis on languish and mechanism of SMMC-7721 cell of human liver cancer lines. *Chinese Medicine of Old People* 27: 142-144.

Zabihullah Q, Rashid A & Akhtar N (2006). Ethnobotanical survey in KotManzarayBaba valley Malakand agency, Pakistan. *Pakistan Journal of Plant Science* 12: 115-121.

20

Indian Snake Root and Devil Root as Distinctive Medicinal Plant for Curing Human Disease: Biology, Chemistry and Cultivation Practices of *Rauwolfia serpentina* and *Rauwolfia tetraphylla*

Rajendra Bhanwaria, Bikarma Singh and Rajendra Gochar

Abstract

India is known as hub of medicinal plants in recent years due to availability of variety of species in different regions which is used in treatment of frequently occurred local diseases. Family a pocynaceae is one of frequently used angiosperm members as medicine in different Indian system of medicine. At present, the genus Rauvolfia is represented by 77 species distributed across the globe, known to contains various unique active phytochemicals in the form of flavonoids, phytosterols, oleoresins, steroids, tannins and alkaloids. R. serpentina (also known as Indian snake root or sarpaghandha) and R. tetraphylla (also known as devil root or barachandrika) are two very important distinctive medicinal shrubs growing in India having wide application in traditional system of medicine for curing human disease and various herbal formulations available in market using these plants. In the present communication, we have discussed the biology, chemistry, pharmacology, cultivation practices

Rajendra Bhanwaria (✉) and Rajendra Gochar

Genetic Resources and Agrotechnology Division,CSIR-Indian Institute of Integrative Medicine Canal Road, Jammu-180001, Jammu & Kashmir, India

Bikarma Singh (✉)

Botanical Garden, CSIR-National Botanical Research Institute, Lucknow-226001 Uttar Pradesh, India

✉*Corresponding author(s) email: rbhanwaria@iiim.res.in; drbikarma.singh@nbri.res.in*

Plants for Novel Drug Molecules: Ethnobotany to Ethnopharmacology
Bikarma Singh & Yash Pal Sharma (eds.), (pp. 445-465)

and future perspectives of R. serpentina and R. tetraphylla for conservation and as a potential plants for future application in herbal medicine discovery.

Keywords: *R. serpentina, R. tetraphylla,* ethnobotany, chemical constituents pharmacology

1. Introduction

Indian traditional medicine system such as Ayurveda, Siddha and Unani are growing at a faster rate in recent years (Pandey et al. 2013). These form of herbal medicines and formulations are not only used by the rural people for their primary health care in developing countries, such types of medicine practice are also used in developed countries such as America and Japan where modern medicines dominate (Ballabh and Chaurasia 2007, Kala 2005). About 80% of Indian rural population uses herbs as medicine (Mukherjee and Wahile 2006). Approximately 960 types of plants are used at the Indian herbal industries for drug development, whose estimated turnover is more than INR 80 billion (Sahoo and Manchikanti 2013).

The genus *Rauvolfia* was first described by Carl von Linnaeus in 1753 in 'Species Plantarum' is a peculiar group of plants under growth-form as evergreen shrubs. Usually the flowers are small and white or greenish white in colour in majority of the species. When the genus was first described was known by the name devil peppers of family apocynaceae and it was named to honor Leonhard Rauwolf, a German physician, Botanist, and traveller (born 21 June 1535-died 15 September 1596). There are 77 species as accepted name in TPL (www.theplantlist.org) under this genus, whose species are reported from tropical regions of America, Africa, Asia, and Oceanic Islands. *R. serpentina* was selected for investigation because of long tapering, snake-like roots (Lobay 2015). Historically, it is of record that Mahatma Gandhi, the famous Indian politician was known to use *Rauwolfia* root to make a herbal tea to relax. Evidence proves that the roots contain an alkaloid called reserpine, which was first reported from Indian *R. serpentina*. Reserpine, drug derived from the roots are used in the treatment of high blood pressure and as a tranquilizer and had been used to cure reptile bites, insomnia, and insanity. Reserpine, isolated in 1952, was the first active alkaloids found in the crude drug. This alkaloid is occasionally also used in treating hypertension.

In this communication, two important species of the genus *Rauwolfia, R. serpentina* and *R. tetraphylla* were selected to review their morphological characters, medicinal value, chemistry, pharmacological aspects and cultivation practices coupled with conservation aspects for future during programme.

2. History of Discovery and Distribution

2.1. *Rauwolfia serpentina*

Rauwolfia serpentina Benth. ex Kurz is commonly called as 'Sarpaghanda' is shrub community plant and it is believed that this plant has been known in Indian system of medicine for approximately 4,000 years (Endress et al. 2000). This plant was first mentioned by Sushruta, an ancient Indian physician, known as "Father of Indian Medicine" and "Father of Plastic Surgery" for inventing and discovering surgical procedure. It is of record that the roots of this species have been used since prevedic period as a drug to cure snakebites and high fever (Kiron et al 2018). The plant also finds its record in ancient literatures (Charaka-1000 to 800 BC) where it was used as insect stings (Panday 1984). As the root of the plant has been used in India for hundreds of years and even the English publication of a clinical report suggest that this plant was used as therapy in 50 cases of essential hypertension. As time passes, this species find its application as medicinal plant in Indian system (ayurveda, siddha, unani) and Western system of medicines (Benjamin et al 1993, Ajayi et al. 2011). Besides India *R. serpentina* recorded as the native plant for the moist deciduous forests of Southeast Asian countries such as Bangladesh, Bhutan, Malaysia, Myanmar and Sri Lanka. In India, this species grows between the elevation of 1300 to 1400 m in the foot hills of Himalaya, Punjab, Assam, Meghalaya, Sikkim, J&K, Uttarakhand, Western Ghats and Andaman's (Vakil 1955, US Dept of Agriculture 2003).

2.2. *Rauwolfia tetraphylla*

Rauvolfia tetraphylla L. was first described by Carl von Linnaeus in 1753 in 'Species Plantarum' as type species of the genus *Rauvolfia*. This species is a critically endangered species and is known by names such as wild snake root, four-leaved devil pepper in english, barachandrika in hindi, papataku in telugu, pampukaalaachchedi in tamil, patalagarudi in oriya, vanasarpagandha in sanskrit and dodda chandrike in kannada. The plant is native plant of West Indies and as time passes gets naturalized in many countries such as India, Pakistan, Sri Lanka, Bangladesh, Nepal, and Myanmar and often cultivated in gardens (Mahalakshmi et al. 2019). In India, the plant is distributed in various states such as Karnataka, Madhya Pradesh, Orissa, West Bengal, Bihar, Andhra Pradesh, Kerala and Tamil Nadu. *R. tetraphylla* is one of the well-known plants being widely used in traditional medicine. The plant contains alkaloids such as reserpine, rauvolscine, ajmalicine, ajmaline, canescine, pseudoyohimbine and yohimbine (Mahalakshmi et al. 2019).

3. Morphological Distinguishing Characters

3.1. Morphology of *Rauwolfia serpentina*

R. serpentina (Fig 1, left) grows between 40 to 80 cm tall. Leaves are elliptical or lanceolate shaped, usually occurred in whorls of 3 to 5 leaflets together, pale green and 7-10 cm long and 3.5-5.0 cm wide (Lobay 2015). The plant look attractive due to shining nature of all parts, and fruits are round and 0.5-0.6 cm in diameter. It also has small pink or white flowers. It also has prominent tubers, soft taproot that reaches a length of 30 to 50 cm and diameter between 1.2 to 2.5 cm (Brijesh 2014).

3.2. Morphology of *Rauwolfia tetraphylla*

Rauvolfia tetraphylla (Fig 1, right) is a pubescent, ever-green shrub with woody stem and reaching a height of 4-6 feet. Leaves are unequal, 5-9 x 3-4 cm, elliptic-ovate, acute at apex, pubescent and usually found in whorls of 4. Flowers are cream colored, about 5mm across, found in terminal corymbose cymes. Calyx lobes are short, ciliate and round. Corolla is white in color, approximately 3 mm long, lobes and tubes are short. Drupes are ovoid, 2-seeded, 5-10 mm across, smooth, jointed to the top, purple when ripe. Flowering occurs throughout the year (Kamble et al. 2013; Bhat 2014; Patil 2015). *Rauvolfia tetraphylla* differs from *R. serpentina* in having leaves in whorls of 4 and short corolla tube.

4. Medicinal Aspects

4.1. General Uses

Since 3000 years ago, *Rauvolfi*a plants got attention as a traditional herb (Ruppert & Stockigt 2005). The roots contain several alkaloids, and the principle use of the drug is sedative and helpful in reducing blood pressure. The roots and drug are now mainly used in diseases of bowels and in fever, hypertension and high blood pressure also used in as a treatment for autistic children, delirium tremens in alcohol and drug addicted patients and improving pruritic and psychogenic dermatosis. The sedative action of the drug is slow, and therefore, the drug is not useful in acute cases; it is more suitable for mild anxiety cases or patients of chronic mental illness. The drug has tranquillizing effect. This drug should not be given to persons suffering from bronchitis, asthma or gastric ulcers. *R. serpentina* is used to cure arthritis, skin cancer, burns, eczema, psoriasis, digestive problems, high blood pressure, sedative and diabetes. Leaves, seeds, roots and fruits of *R. serpentina* are used in treatment of various ailments (Pant & Joshi 2008). Reported that it contain more than 50 indole alkaloids from roots. The ayurvedic formulations available are sarpagandha ghanavati, sarpagandha yoga, sarpagandha churna and maheshvari vati. They are of immense medicinal value and have steady demand in both domestic and international markets. Under ground parts are used for treating various CNS disorders.

R. tetraphylla possesses a range of applications in various countries across the world. The latex from the plant is reported to be cathartic, emetic, and expectorant besides its use in the treatment of dropsy. The juice prepared from the fruit is used as a substitute for ink. Nandhini and Bai (2014b) were done to evaluate the cardiac protective activity in *R. tetraphylla*. An effect of *R. tetraphylla* aqueous leaf extract was done on frog heart in situ preparation. The aqueous leaf extract produced significant positive ionotropic effects (Thinakaran et al. 2009, Jakaria et al. 2016, Patyal et al 2013, Kavitha et al. 2012).

4.2. Traditional System of Medicine

Rauwolfia root extracts are advantageous in gastro intestinal disorders such as cholera, colic and diarrhea, dysentery, which are also used to treat hypertension and breast cancer (Lohani et al. 2011; Stanford et al.1986). *R serpentina* was used in folk medicine in India for centuries to cure a wide variety of infections including bites caused by snake and insects. Juice prepared from fresh fruit can act as a substitute of ink. Juice extracted from leaves used in eye troubles and decoction prepared from leaves is used for toothache. Paste prepared from roots is useful in stomach pain and snake bite (Khare 2007, Caamal et al. 2011, Quattrocchi 2012, Choudhury et al. 2013). In folk medicine, the root is used during delivery to stimulate uterine contractions and promote the expulsion of the fetus. Crying babies are put to sleep by working mothers by making them to suck the breasts, which are smeared by the root-paste. It is also a valuable remedy in the treatment of painful affections of the bowels.

R. tetraphylla is one among the various ethnomedicinal plants being used by Amerindian communities of Caribbean basin (Torres et al. 2015). Table 1 and Table 2 are presented with ethnobotanical uses of *R. serpentina* and *R. tetraphylla*.

5. Phytochemical Constituents

Plants are reported to have a rich repository of phytochemicals and secondary metabolites (alkaloids, terpenes and polyphenolic compounds) exert multifold effects on the health of human beings besides conferring resistance to plants that produce them against insects, pathogens and herbivores. *Rauwolfia* contains many different phytochemicals, including alcohols, alkaloids, glycosides, fatty acids, flavonoids, oleoresins,

Fig. 1: Indian snake root (*Rauvolfia serpentina* and devil root (*R. tetraphylla*) Courtesy: B. Singh).

phytosterols, sugars, steroids and tannins. The most important alkaloids found in the plant are indole alkaloids (>50 types of alkaloids) having been isolated (Verma and Verma 2010). Indole alkaloids are a group of nitrogenous compounds that are derived from the amino acid tryptophan. They share a common 5 and 6 carbon heterocyclic ring structure with 1 nitrogen molecule (Leete 1960). The leaves and stems also contain indole alkaloids, but more concentration recorded from bark of root (Ruyter et al. 1991). The identified indole alkaloids include ajmalidine, ajmaline, ajmalinine, ajmalicine, aricine, canescine, coryanthine, deserpidine, isoajmaline, isoserine, isoserpiline, lankanescine, neoajmaline, papaverine, raubasine, raucaffricine, rauhimbine, rauwolfinine, recanescine, rescinnamine, reserpiline, reserpine, reserpinine, sarpagine, serpentine, serpentinine, thebaine, yohimbine and yohimbinine (Woodsone et al. 1957, Panwar and Guru 2011).

6. Marker Compounds

Reserpine and rescinnamine (Fig. 2 and 3) are two major marker compound in Rauwolfia species. Reserpine is one of the major alkaloids of the plant, whose content has been found to be highest in the root and lower in the stems and leaves (Ruyter et al. 1991, Lobay 2015). The concentration of reserpine in the plant has been found to vary from 0.03% to 0.14% of the dry weight of the plant (Kumar et al. 2018, Lobay 2015, Haresh et al 2010).

Table 1: Ethnobotanical uses of *Rauwolfia serpentina* in India

Disease category	Part used	Mode of Use	State	References
Scabies	Root	Fruits of datura and roots of *R. serpentina* coupled with lemon juice and boric acid are mixed to prepare paste and used externally on infected parts.	Karnataka	(Parinitha et al. 2004)
Itches and Eczema	Root paste	Paste prepared from root of *R. serpentina* and *A. paniculata* is used by Kandhas tribe to cure disease.	Orissa (Kandhamal district)	(Behera et al. 2006).
Antidote against snakebite	Fruits and Seeds	Fruits and seeds of *R. serpentina* is used by local people to cure snake bite.	Tamil Nadu, Rajasthan and Madhya Pradesh	(Rajendran and Agarwal 2007)
Snakebite and scorpion sting	Fruits and Seeds	*R. serpentina* plant used traditionally to treat snakebite and scorpion sting in the form decoction.	Himalayas Tamil and Nadu Orissa	(Prakasha et al. 2010, Behera et al. 2007).
Dog bite	Roots, Fruits and Seeds	Dog bites are treated with *R. serpentina* continuous 7 days.	South India	(Rahamatullah et al. 2010).
High blood pressure, Mental agitation, Insomnia, Sedative and as Hypnotic	Root	Powered roots taken to get relief from blood pressure, mental agitation, insomnia, sedative and as hypnotic.	Kanyakumari	(Meena et al. 2009; Jeeva et al. 2006)
Anxiety, Epilepsy and Nervous disorders	Whole plant	Used to treat high blood pressure by Jaunsari tribe of UP.	(Sonebhadra) Uttar Pradesh and (Garhwal) Uttaranchal	(Singh et al. 2010, Bhatt and Negi. 2006).
Mental disorder	Root power	Root power is taken either with milk or honey in early morning to cure mental disorder by Kandhas tribe.	(Kandhamal) Orissa	(Behera et al. 2006).

Contd.

Malaria	Leaves,root	Leave extract of *R. serpentina* along with *Andrographis paniculata* and *Azadirachta indica* mixed with honey is used to treat malaria fever by Khamptis tribe. Dried powdered root mixed with roots of *Cissampelos pareira* in equal quantities with water and taken to cure malaria by Kol tribe in Similipal biosphere reserve.	Arunachal Pradesh and Orissa.	(Sen et al. 2008, Rout and Thatoi 2009)
Dysentery and Fever	Whole plant	Dysentery and fever are cured by *R. serpentina.* Root extract is taken to cure intestinal disorders by Malamalasar tribe of Parambikulam.	(Kalahandi) Orissa and Kerala.	(Nayak et al. 2004, Yesodharan and Sujana 2007).
Opacities of the eye cornea	Leaves	Rural people usually take decoction of roots during labor pain in woman and juice of leaves to remove opacities of eye cornea.	Kanyakumari	(Raj and Sukumaran, 2008).
Stomach pain	Leaves	Juice extracted from tender leaves is taken empty stomach. Root powder is mixed with black pepper to cure stomach disorder.	(Mayurbhanj) Orissa	(Rout et al. 2009).

Table 2: Ethnobotanical uses of *Rauwolfia tetraphylla* in India

Disease category	Part used	Mode of Use	State	References
Head sore	Root	Paste	(Chittoor) Andhra Pradesh.	(Ganesh and Sudarsanam 2013)
Stomach pain	Root	Extract	(Anuppur) Madhya Pradesh.	(Malaiya 2016)
Blood pressure	Root	Decoction	(Vizianagaram) Andhra Pradesh.	(Babu et al. 2010)
	Root bark	Decoction	(Eastern Ghats) Andhra Pradesh.	(Rao et al. 2016)
Loose motion	Root	Extract	(Paschim Medinipur) West Bengal.	(Manna and Manna 2016)
Reptile bites (Snake), insect bite (Scorpion)	Root	Paste	(Koraput) Odisha.	(Kumar et al. 2016)
	Root bark	Decoction	(Madumalai) Tamil Nadu.	(Rani et al. 2016)
	Whole plant	Paste	(Salem) Tamil Nadu.	(Alagesaboopathi et al. 2018, Alagesaboopathi 2013)
	Root	Paste	(Sundargarh) Orissa.	(Prusti and Behera 2007)
	Root	Powder, paste	(Sundargarh) Orissa.	(Girach et al. 1998)
	Whole plant	Paste	(Salem) Tamil Nadu.	(Alagesaboopathi 2011)
	Root	Paste	(South Western Ghats) Kerala.	(Sulochana et al. 2015)
Skin diseases	Whole plant	Paste	(Ranga Reddy) Telangana	(Ramakrishna et al. 2017)
	Whole plant	Paste	(Kancheepuram) Tamil Nadu.	(Muthu et al. 2006)
	Root	Powder	(Coimbatore and Ooty) Tamil Nadu.	(Kumar et al. 2018)
	Whole plant	Juice	(Bhadrak distric) Odisha.	(Panda et al. 2016)
Chronic wound	Root	Powder, paste	(Latehar) Jharkhand.	(Marandi and Britto 2014)
Intestinal worms	Fruit	Roasted fruits	(Kanyakumari) Tamil Nadu.	(Rani and Jeeva 2017)
Diarrhea and dysentery	Root	Water extract	(Chittagong) Bangladesh.	(Morshed and Nandni 2012)
Malaria	Root	Paste	Odisha.	(Singh et al. 2018)
Diabetes mellitus	Root	Juice	(West Rarr) West Bengal.	(Ghosh 2008)
	Root	Juice	(Birbhum) West Bengal.	(Sarkar et al. 2016)
Piles, sterility	Root, leaf	Juice	(Eastern Ghats) Tamil Nadu.	(Vaidyanathan et al. 2013)

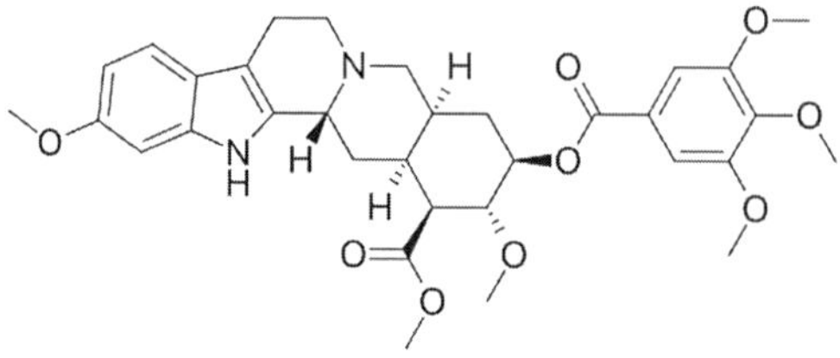

Fig. 2: Reserpine

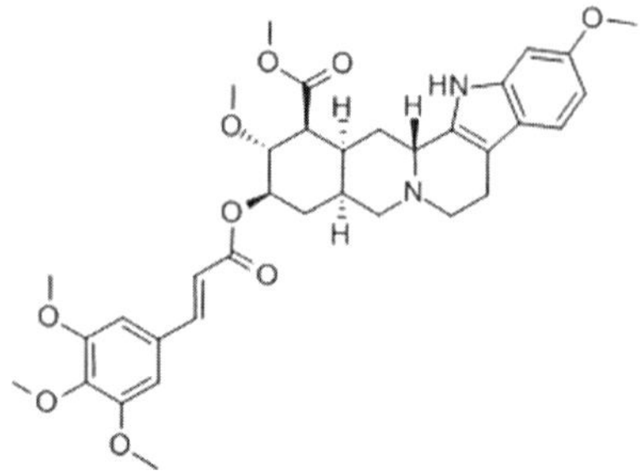

Fig. 3: Rescinnamine

7. Pharmacological Aspects

R. serpentina contains variety active constituents with confirmed antioxidant capacity. This relates to prevention of cardiovascular disease, cancer, diabetes and hypertension. Leaf extract find its application in vivo efficacy studies. (Hansraj et al. 2009). Rauwolfia compound commonly used in Asian medicine, which includes traditional Ayurvedic medicine native to India. Some well known biological activities reported is given below in subheads:

7.1. Antimicrobial Activities

Deshmukh et al. (2012) reported antibacterial activity from the crude extracts determined in accordance with agar-well diffusion method. The bacteria isolated were first grown in a nutrient broth for 18 hrs before use. Two hundred micro liter of the standardized cell suspensions were spread on a Mueller-Hinton agar (MHA). Nath et al. (2012) observed the antimicrobial assay using the crude extract on Mueller–Hinton agar (MHA) medium for test bacteria and PDA for test fungus; where 1 ml of test organism's inoculum was swapped on molten agar plates. Using the sterile cork borer 7 mm wells were made in the plates in which 100 ìl of crude ethanol extract of endophytic fungi was loaded. After incubation for 24 hrs at 37°C for bacteria and 4–5 days at 25°C for fungus, the plates were observed for zone of inhibition. Earlier worker (Khan et al. 2016, Lingaraju et al. 2015, Harisaranraj et al. 2009) reported in antibacterial activities of *R. serpentina* alkaloids. Several studies conducted on antifungal potential of *R. tetraphylla* revealed its effectiveness against a panel of human and phytopathogenic fungi including dermatophytes and seed-borne fungi. The study carried out by Kumaran and Kannabiran (Kumaran and Kannabiran 2003) revealed mycotoxic effect of ethanol extract obtained from roots of *R. tetraphylla* against the growth of *Colletotrichum capsici*. Aqueous leaf extract of *R. tetraphylla* was effective in causing antifungal activity against *Fusarium indicus* and *Aspergillus flavus* dose dependently

(Thinakaran et al. 2009). Whole plant and sometime plant parts such as root, leaves and fruit of *R. tetraphylla* were shown to be effective antibacterial agents. Ethanol extract from *R. tetraphylla* was shown to reveal marked inhibition of gram positive (+ve) and gram negative (-ve) bacteria (Suresh et al. 2008). Aqueous extract of *R. tetraphylla* leaves exhibited concentration dependent inhibition of test bacteria viz. Escherichia coli and Klebsiella pneumoniae (Thinakaran et al.2009). Ethanol extract of leaves were effective in causing concentration dependent inhibition of bacteria. Salmonella typhimurium and Micrococcus luteusinhibited to highest and least extent, respectively (Nandhini and Bai 2014a).

Reserpine in leaves of *R. tetraphylla* display inhibitory properties on gram +ve and gram -ve bacteria (Abubacker and Vasantha 2011). The extract of leaves displayed inhibition of *Staphylococcus aureus* and *Enterobacter faecalis* with marked activity against *S. aureus* (Senthilmurugan et al. 2013). The leaf extract of methanol found to be effective on *S. aureus* and *K. pneumoniae* in a dose dependent manner (Archana and Jeyamanikandan 2015). Compounds viz. 10-Methoxytetrahydroalstonine, reserpiline, isoreserpiline, demethoxyreserpiline, serpentine and α-yohimbine from *R. tetraphylla* were effective in causing inhibition of nalidixic acid sensitive and resistant strains of *Escherichia coli*. Further, these compounds were capable of performing synergistically with nalidixic acid (Dwivedi et al. 2015, Kalaiarasi et al. 2013).

7.2. Antioxidant Activities

The antioxidant activity of the R. serpentina extracts was described by Brand-Williams et al. (1995) via the 2,2 –diphenylpicrylhydrazyl radical neutralization assay. Vinay et al. (2016) was obtained total antioxidant capacity to be high in methanol extract of *R. tetraphylla* leaves at 50 μg and at 100 μg, dichloromethane leaf extract of *R. tetraphylla* (32.2 μg), leaf of n-hexane crude extract (21.8 μg) followed by R. tetraphylla fruit methanol crude extract (35 μg of ascorbic acid/mg of extract). Leaves of all reported five species of Rauwolfia (*R. beddomei, R. densiflora, R. micrantha, R. serpentina* and *R. tetraphylla*) from Southern Western Ghats were chosen to investigate their antioxidant activities, phytochemicals and nutrient composition. The five species were then screened for their antioxidant potentials using various in-vitro models such as total antioxidant capacity, 1,1-diphenyl-2- picrylhydrazyl (DPPH) radical scavenging activity, reducing power and superoxide anion scavenging activity at various concentration. *R. tetraphylla* revealed the highest concentration of β-carotene and Lycopene. (Vadakkemuriyil et al. 2012; Gupta and Gupta 2015; Jakaria et al. 2016). Various parts *viz.* leaf, root and fruit of

R. tetraphylla are shown to exhibit antioxidant activity. Methanol extract obtained from the fruits of *R. tetraphylla* was effective in scavenging DPPH radicals in a dose dependent manner (Oza and Solanki 2016, Jakaria et al 2016).

7.3. Anti-inflammatory Activity

In vitro anti-inflammatory activity of orally administered different extracts of R. tetraphylla root bark in Carrageenan induced acute inflammation in rats was evaluated and showed good activity (Ganga et al. 2012). Rauvolfian RS, a pectic polysaccharide was recovered from the dried callus of *R. serpentina* by extraction with 0.7% aqueous ammonium oxalate, and purified by using membrane ultra-filtration to got the purified rauvolfian RSP in addition to glucan as admixture from the callus, with molecular weights 300 and 100300 kD, respectively. A peroral pretreatment of mice with the crude and purified samples of rauvolfian (RS and RSP) was found to decrease colonic macroscopic scores, the total area of damage, and tissue myeloperoxidase activity in colons as compared with a colitis group. RS and RSP were shown to stimulate production of mucus by colons of the colitis mice. RSP appeared to be an active constituent of the parent RS. The glucan failed to possess antiinflammatory activity (Jakaria et al. 2016; Popov et al. 2007). Rao et al. 2012 investigated anti-inflammatory activity of various solvent extracts of root bark of *R. tetraphylla* by carrageenan induced rat paw edema model. Among extracts, hydro-alcoholic and methanol extract displayed significant reduction in paw edema when compared to hexane and ethyl acetate groups.

8. Toxicological Studies

Contrastly side effects of major compound reserpine include abdominal cramping, bronchospasm, depression, gastric ulceration, hypotension, itching, lethargy, nausea, sedation, psychiatric, vomiting, nightmares, bradycardia, angina-like symptoms, breast enlargement, dysfunction, galactorrhea, skin rash, sexual and withdrawal psychosis. (Lobay 2015). Adequate doses of reserpine that produce decreased blood pressure will not cause reserpine-induced gastric ulcerations (Lobay 2015). Reserpine has been observed to cause a slight edema in some patients (Krogsgaard 1957). Nausea, vomiting, hypotension, sedation, and coma have been reported by patients.

9. Cultivation Practices

Its cultivation needs to be greatly expanded. The plant can be cultivated almost anywhere in the plains, in evergreen forests and sub-Himalayan tracts. The plant are best raised from root cuttings, they are also raised from seed and stem cuttings.

9.1. Land Preparation

The land is plugging through the cultivator one or two times and field is leveled properly by the land leveler. The well drained and deep, rich in organic content, slightly acidic to slightly alkaline clay-loam to silt-loam soil in nature and which pH ranging from 6.5 to 8 are suitable for good growth and commercial cultivation. *Rauwolfiae* is easily propagated by two major methods (I) Seed (II) Cutting (Stem and Root)

9.2. Propagation and Plantation

9.2.1. Seed

Seed germination in Rauvolfia is highly variable. About 4-5 kg of fresh collected seeds is require to rise for one hectare area and the suitable time for nursery rising is first half of May to first half of June. Seed germination was recorded 70-80% in freshly collected heavy seed lot. The nursery is prepared by raised beds as per requirement of planting material and easy for irrigation weeding and other intercultural operation with the addition of well decomposed FYM. Approximately 500 sq m. area of seed bed is ample for producing seedlings enough for transplanting on one hectare of field. The seeds placed, 3-5 cm apart in rows in shallow furrows Seedlings are ready after 40-50 days for transplanting in main field. The plant is transplanted at 30 X 45 cm distance within row to row and plant to plant. Hedayatullah (1959) and Nayar (1956) have investigated the propagation and culture techniques of Rauwolfia species by seeds.

9.2.2. Stem-Root Cutting

Badhwar et al. (1956) were reported the methods of propagation and their effect on root production in *R. serpentina*. Approximately 15 to 22 cm long stem cutting and 5 to 7 cm long root cutting are planted during June and July month in well prepared ridged nursery beds under control condition where sufficient moisture is maintained over a month. An average 100-120 kg of fresh stem cuttings are found enough for planting in one hectare land. The cuttings begin to sprout within 3-4 weeks and after the sprouting of buds, branches and rising roots by the plants, these plants are ready for transplanting at recommended spacing in the already prepared main field.

9.2.3. Manures and Fertilizers

The medicinal plants are usually no needs of fertilizers, pesticide and other chemicals so that they easily grown without using of chemical fertilizers because they have own anti insecticide, fungicide properties. The amount of 10-15 tans well decomposed FYM is mixed and rotated in soil before 15 days of plantation. Application of all types of organic manures such as FYM, Vermicompost, poultry manure, Neem cake, Groundnut cake and Green Manure is enhance the vegetative growth and root yield.

9.2.4. Irrigation and Intercultural Operation

A careful irrigation management schedule is essential to get a good crop of *Rauwolfia*. Plant must steady moisture requirement where temperature increase with insufficient precipitation in monsoon season. Normally, 15-20 irrigations are given during the entire crop period. If plants grown in areas where annual rainfall of 150 cm or above well distributed throughout the growing season. It is suggested that crop must be irrigated fortnight interval during hot dry season and once a month interval in winter is sufficient to plant. The field should be kept relatively weed-free in the first one and half month after transplanting. This means interculture is given with hand hoe one month after transplanting. Second weeding is done one month after the first weeding and hoeing. Loosens the soil helps in the setting or bigger and better roots in the soil. 4-5 inter cultural operation enhance crop yield.

9.2.5. Harvesting and Yield

When plant transplanting is done in kharif or monsoon season the uprooting of roots is indicate with the turning yellowish and shedding of plant leaves during next year. At this stage, the roots found maximum concretion of total alkaloids and root may be penetrating quite deep up to 40-50 cm in the soil. Irrigation before the digging, digging folks are required to dig out the roots. The root-bark should not be damaged during harvesting since the bark contains higher alkaloid percentage compared to wood. After digging the roots are cleaned and thoroughly dried under shade condition to protect it from mould. The yields at 18 months duration crop produce maximum root yield. The average, root yield is depending on types of soil, crop management and harvesting techniques. The 1.5 to 2 tans/ha of dry roots are obtained under the irrigated condition.

10. Conclusion and Future Direction

Plants serve as an excellent source of various therapeutic agents. Reserprine of *Rauwolfia* seems to be a safe and dominant treatment for the disorder of hypertension when perception in convenient less doses. One of the major advantages of using plants is that they do not show the deleterious side effects commonly associated with other drugs. This research work explored the antioxidant as well as the antidiabetic activity of the wild and cultivated variety of *Rauwolfia* plant. The study suggested that the wild source of the plant has more potential to exert antioxidant and antidiabetic activity as compared to the cultivated plant. Although the wild variety of *Rauwolfia* plant is more promising, results showed that it can be substituted with cultivated variety for antioxidant and antidiabetic activity in case of need. There is need to conserve *Rauwolfia* species world-wide and cultivation practices promote huge amount production of roots for value addition of medicine research.

Abbreviation Used

cm = centimeter; *R. serpentine* = *Rauwolfiae serpentine;* kg = Kilogram; ha = Hector; FYM=farm yard manure ; μg = Microgram; ml= Milliliter

Conflict of Interest

Authors declare no conflict of interest for this research paper.

Acknowledgements

Authors would like to thanks Director IIIM Jammu for necessary facilities and encouragement.

References

Abubacker MN & Vasantha S (2011). Antibacterial activity of ethanolic leaf extracts of *Rauwolfia serpentina*(Apocyanaceae) and its bioactive compound reserpine. *Drug Invent Today* 3(3): 16-17.

Ajayi IA, Ajibade O & Oderinde RA (2011). Preliminary phytochemical analysis of some plant seeds. *Research Journal of Chemical Sciences* 1(3): 58-62.

Alagesaboopathi C (2009). An investigation on the antibacterial activity of *Rauvolfia tetraphylla* dry fruit extracts. *Ethnobotanical Leaflets* 13: 644-650.

Alagesaboopathi C (2013). Ethnomedicinal plants used for the treatment of snake bites by Malayali tribal's and rural people in Salem district, Tamil Nadu, India. *International Journal of Bioscience* 3(2): 42-53.

Alagesaboopathi C, Prabakaran G, Subramanian G & Vijayakumar RP (2018). Ethnomedicinal studies of plants growing in the Salem district of Tamil Nadu, India. *International Journal of Pharmacy and Biological Sciences* 8(3): 503-511.

Babu CN, Naidu TM &Venkaiah M (2010). Ethnomedicinal plants used by the tribals of Vizianagaram district, Andhra Pradesh. *Journal of Phytology* 2(6): 76–82

Badhwar RL. Karira GV & Ramaswami S (1956). Methods of propagation and their effect on root production in *Rauwolfia serpentina*. *Indian Journal of Pharmacology* 18: 170-175.

Ballabh B & Chaurasia OP (2007). Traditional medicinal plants of cold desert Ladakh-Used in treatment of cold, cough and fever. *Journal of Ethnopharmacology* 112(2): 341–345.

Behera SK, Panda A, Behera SK & Misra MK (2006). Medicinal plants used by the *Kandhas* of Kandhamal district of Orissa. *Indian Journal of Traditional Knowledge* 5(4): 519-528

Behera, KK, Sahoo S & Mohapatra PC (2007). Medicinal plant resources for bioprospecting and drug development in tribal rich district of Orissa, India. *Ethnobotanical Leaflets* 11: 106-112

Benjamin BD, Roja G & Heble MR (1993). *Agrobacterium rhizogenes* mediated tranformation of *Rauvolfia serpentina*: Regeneration and alkaloid synthesis. *Plant Cell Tissue and Organ Culture* 35: 253-257.

Bhat GK (2014). Flora of South Kanara, Akriti Prints, Mangalore, India.

Bhatt VP & Negi GCS (2006). Ethnomedicinal plant resources of Jaunsari tribe of Garhwal Himalaya, Uttaranchal. *Indian Journal of Traditional Knowledge* 5(3): 331-335

Bhattarai S, Chaudhary RP, Taylor RSL & Ghimire SK (2009). Biological activities of some Nepalese medicinal plants used in treating bacterial infections in human beings. *Nepal Journal of Science and Technology* 10: 83-90

Brand-Williams W, Cuvelier ME & Berset C (1995). Use of a free radical method to evaluate antioxidant activity. *Lebensmittel-Wissenschaft & Technologie* 28: 25–30

Brijesh KS (2014). Rauwolfia: cultivation and collection. Biotech Articles Web site.

Caamal FE, TorresTLW, Simá PP, Peraza SSR & Moo PR (2011). Screening of plants used in Mayan traditional medicine to treat cancer-like symptoms. *Journal of Ethnopharmacology* 135: 719-724.

Chauhan S, Kaur A, Vyas M & Khatik GL (2017). Comparison of antidiabetic and antioxidant activity of wild and cultivated variety of *Rauwolfia serpentina*. *Asian Journal of Pharmaceutical and Clinical Research* 10 (1): 404-406

Choudhury S, Rahaman CH, Mandal S & Ghosh A (2013). Folk-lore knowledge on medicinal usage of the tribal belts of Birbhum district, West Bengal, India. *International Journal of Botany and Research* 3(2): 43-50.

De Britto JA, Sebastian SR & Sujin MR (2011). Phytochemical and antibacterial screening of seven Apocynaceae species against human pathogens. *International Journal of Pharmacy and Pharmaceutical Sciences* 3(5): 278-281.

Deshmukh SR, Ashrit DS & Patil BA (2012). Extraction and evaluation of indole alkaloids from *Rauwolfia serpentina* for their antimicrobial and antiproliferative activities. *International Journal of Pharmacy and Pharmaceutical Sciences* 4 (5): 329-334

Douglas L (2015). *Rauwolfia* in the treatment of Hypertension. *Integrative Medicine* 14(3): 40–46.

Dwivedi GR, Gupta S, Maurya A, Tripathi S, Sharma A, Darokar MP & Srivastava SK (2015). Synergy potential of indole alkaloids and its derivative against drug-resistant *Escherichia coli*. *Chemical Biology and Drug Design* 86: 1471-1481.

Endress ME & Bruyns PV (2000). A revised classification of the Apocynaceae. *Botanical Review* 66(1):1-56.

Fabricant DS & Farnsworth NR (2001).The Value of Plants Used in Traditional Medicine for Drug Discovery. *Environmental Health Perspective* 109(1): 69-75

Ganesh P & Sudarsanam G (2013). Ethnomedicinal plants used by Yanadi tribes in Seshachalam biosphere reserve forest of Chittoor district, Andhra Pradesh India. *International Journal of Pharmacy and Life Sciences* 4(11): 3073-3079.

Ganga BR, Umamaheswara PR, Rao ES, Rao TM & Praneeth VSD (2012). Evaluation of in-vitro antibacterial activity and anti-inflammatory activity for different extracts of *Rauwolfia serpentina* L. Root bark. *Asian Pacific Journal of Tropical Biomedicine* 2(10): 818-821.

Ghorbani A, Naghibi F & Mosaddegh M (2006). Ethnobotany, ethnopharmacology and drug discovery. *Iranian Journal of Pharmaceut Science* 2(2): 109-118.

Ghosh A (2008). Ethnomedicinal plants used in West Rarrh region of West Bengal. *Natural Product Radiance* 7(5): 461-465.

Girach RD (1998). Medicinal ethnobotany of Sundargarh, Orissa. India. *Pharmaceutical Biology* 36(1): 20-24.

Gupta J & Gupta A (2015). Isolation and extraction of flavonoid from the leaves of *Rauwolfia serpentina* and evaluation of DPPH-scavenging antioxidant potential. *Oriental Journal of Chemistry* 31: 1-28.

Hareesh KV, Nirmala SS & Rajendra CE (2010). Reserpine content of *Rauwolfia serpentina* in response to geographical variation. *International Journal of Pharma and Bioscience* 1(4): 429–434.

Harisaranraj R, Suresh K, Saravanababu S & Achudhan VV (2009). Phytochemical based strategies for pathogen control and antioxidant capacities of *Rauwolfia serpentina* extracts. *Recent Research in Science and Technology* 1(2): 067–073

Hedayatullah S (1959). Culture and propagation of *Rauwolfia serpentina* Benth. in east Pakistan. *Pakistan Journal of Scientific and Industrial Research* 2: 118-122.

Hutchings A (1989). A survey and analysis of traditional medicinal plants as used by the Zulu, Xhosa and Sotho. *Bothalia* 19(1): 111-123.

Jakaria MD, Tareq SM, Ibrahim M & Bokhtearuddin S (2016). *Rauvolfia tetraphylla* L. (Apocynaceae): a pharmacognostical, phytochemical and pharmacological review. *Journal of Chemical and Pharmaceutical Research* 8(12): 114-120

Jeeva S, Kiruba S, Mishra BP, Venugopal N, Dhas SSM, Regini GS, Kingston C, Kavitha A, Sukumaran S, Raj ADS & Laloo RC (2006). Weeds of Kanyakumari district and their value in rural life. *Indian Journal of Traditional Knowledge* 5(4): 501-509

Kala CP (2005). Current Status of Medicinal Plants used by Traditional Vaidyas in Uttaranchal state of India. *Ethnobotanical Research and Application* 3: 267-278

Kalaiarasi R, Prasannaraj G & Venkatachalam P (2013). A rapid biological synthesis of silver nanoparticles using leaf broth of *Rauvolfia tetraphylla* and their promising antibacterial activity. *Indo American Journal of Pharmcutical Research* 3(10): 8052-8062.

Kamble RB & Hate S (2013). Chaturvedi A. New additions to the flora of Nagpur district, Maharashtra. *Journal on New Biological Reports* 2(1): 9-13.

Kavitha R, Deepa T, Kamalakannan P, Elamathi R & Sridhar S (2012). Evaluation of phytochemical, antibacterial and antifungal activity of *Rauvolfia tetraphylla. International Journal of Pharmacutical Science Review and Research* 12(1): 141-144.

Khana MIH, Md. Sohrabb H, Ronyb SR, Tareqc FS, Hasana CM & Mazid MA (2016). Cytotoxic and antibacterial naphthoquinones from an endophytic fungus, Cladosporium sp. *Toxicology Reports* 3: 861-865

Khare CP (2007). Indian Medicinal Plants. An illustrated dictionary. Springer-Verlag New York.

Kiran K, Priya AJ & Devi RG (2018). Medicinal and therapeutic uses of *Rauwolfia serpentina* - Indian snakeroot. *Drug Invention Today* 10: 2674-2678

Krogsgaard AR (1957). The effect of reserpine on the electrolyte and fluid balance in man. *Acta Medica Scandinavica* 159(2): 127-32.

Kumar SA, Sankaranarayanan S, Bama P, Baskar R & Kanagavalli K (2018). An ethnobotanical study of medicinal plants used by local people and tribals in Topslip (Annamalai hills) and Ooty (Chinchona village) of Coimbatore and Ooty district. *International Journal of Current Research* 10(5): 69304-69308.

Kumar SS, Padhan B, Palita SK & Panda D (2016). Plants used against snakebite by tribal people of Koraput district of Odisha, India. *Journal of Medicinal Plants Studies* 4(6): 38-42.

Kumaran RS & Kannabiran B (2003). Mycotoxic effect of *Abrus precatorius* and *Rauvolfia tetraphylla* root extracts on the growth of *Colletotrichum capsici. Czech Mycology* 55: 51-56.

Leete E (1960). The biogenesis of the *Rauwolfia* alkaloids, I: the incorporation of tryptophan into ajmaline. *Journal of the American Chemical Society* 82(24):6338-6342.

Lingaraju K, Naikal HR, Manjunath K, Nagaraju G, Suresh D & Nagabhushana H (2015). *Rauvolfia serpentina*-mediated green synthesis of CuO Nanoparticles and its multidisciplinary studies. *Acta Metallurgica Sinica* 8(9): 1134-1140

Lobay D (2015). *Rauwolfia* in the treatment of hypertension. *Integrative Medicine* 14(3): 40-46.

Lohani H, Andola HC, Bhandari U & Chauhan N (2011). HPTLC method validation of reserpine in Rauwolfia serpentina- A high value medicinal *Plant Researcher* 3(3): 34-37

Mahalakshmi SN, Achala HG, Ramyashree KR & Prashith KTR (2019). *Rauvolfia tetraphylla* L. (Apocynaceae)- a comprehensive review on its ethnobotanical uses, phytochemistry and pharmacological activities. *International Journal of Pharmacy and Biological Sciences* 9 (2): 664-682

Malaiya PS (2016). Medicinal plants used by tribal population of Anuppur district Madhya Prades,India. *International Journal of Applied Research* 2(1): 418-421.

Manna S & Manna CK (2016). Diversity in the use of phytomedicines by some ethnic people of Paschim Medinipur, West Bengal, India. *Journal of Medicinal Plants Studies* 4(3): 81-86.

Marandi RR & Britto JS (2014). Ethnomedicinal plants used by the Oraon tribals of Latehar district of Jharkhand, India. *Asian Journal of Pharmacology Research* 4(3): 126-133.

Meena AK, Bansal P & Kumar S (2009). Plantsherbal wealth as a potential source of ayurvedic drugs. *Asian Journal of Traditional Medicine* 4(4): 152- 170

Morshed AJM & Nandni NC (2012). Indigenous medicinal plants used by the tribal healers of Chittagong hill tracts to treat diarrhoea and dysentery. *Hamdard Medicus* 55(2): 48-66.

Mukherjee PK & Wahile A (2006). Integrated approaches towards drug development from Ayurveda and other Indian system of medicines. *Journal of Ethnopharmacology* 103: 25-35.

Muthu C, Ayyanar M, Raja N & Ignacimuthu S (2006). Medicinal plants used by traditional healers in Kancheepuram district of Tamil Nadu, India. *Journal of Ethnobiology and Ethnomedicine* 2: 43.

Nandhini VS & Bai VG (2014a). Screening of phyto-chemical constituents, trace metal concentrations and antimicrobial efficiency of *Rauvolfia tetraphylla. International Journal of Pharmaceutical, Chemical and Biological Sciences* 4(1): 47-52.

Nandhini VS & Bai VG (2014b). Secondary metabolites screening from in-vitro cultured *Rauvolfia tetraphylla* by HPTLC-MS: A special emphasises on their antimicrobial applications. *International Journal of Advances in Pharmaceutical Analysis* 4(2): 81-87.

Nath A, Raghunatha P & Joshi SR (2012). Diversity and biological activities of endophytic fungi of Emblica officinalis, an ethnomedicinal plant of India. *Mycobiology* 40: 18-30

Nayak S, Behera SK & Misra MK (2004). Ethno-medicobotanical survey of Kalahandi district of Orissa. *Indian Journal of Traditional Knowledge* 3(1): 72-79

Nayar SL (1956). Experimental propagation and culture of Rauwolfia serpentina benth. by seeds. *Indian Journal of Pharmacology* 18: 125-126.

Oza N & Solanki H (2016). Estimation of antioxidant activity and total flavonoid content of selected medicinaly important plants. *International Journal of Interdisciplinary Research Centre* 2(1): 53-59.

Panda T, Mishra N & Pradhan BK (2016). Folk knowledge on medicinal plants used for the treatment of skin diseases in Bhadrak district of Odisha, India. *Medicinal and Aromatic Plants* 5: 262.

Pandey MM, Rastogi S & Rawat AKS (2013). Indian traditional ayurvedic system of medicine and nutritional supplementation. *Evidence-Based Complementary and Alternative Medicine* 2013: 1-12.

Pant KK & Joshi SD (2008). Rapid mutiplicatoin of *Rauwolfia serpentina* Benth.Ex. Kurz through tissue culture. *Scientific World* 6: 58-62.

Panwar GS & Guru SK (2011). Alkaloid profiling and estimation of reserpine in *Rauwolfia serpentina* plant by TLC, HP-TLC and HPLC. *Asian Journal of Plant Sciences* 10(8): 393-400.

Parinitha M, Harish GU, Vivek NC, Mahesh T & Shivanna MB (2004). Ethnobotanical wealth of Bhadra wildlife sanctuary in Karnataka. *Indian Journal of Traditional Knowledge* 3(1): 37-50

Patil MB (2015). Ethnomedicines of Nandurbar district, Maharashtra. Laxmi book publication, Solapur, Maharashtra, India.

Patyal R, Kumari R, Rajput CS & Sawhney SS (2013). Therapeutic characteristics of *Rauwolfia serpentina. International Journal of Pharmacutical and Chemical Science* 2(10): 38-42

Popov SV, Vinter VG, Patova OA, Markov PA, Nikitina IR, Ovodova RG, Popova GY, Shashkov AS & Ovodov YS (2007). Chemical characterization and anti-inflammatory effect of Rauvolfian, a pectic polysaccharide of rauvolfia callus . *Biochemistry* 72 (7): 955-962

Prakasha HM, Krishnappa M, Krishnamurthy YL & Poornima SV (2010). Folk medicine of NR Pura taluk in Chikmagalur district of Karnataka. *Indian Journal of Traditional Knowledge* 9(1): 55-60

Prusti A & Behera KK (2007). Ethno-Medico Botanical Study of Sundargarh district, Orissa, India. *Ethnobotanical Leaflets* 11: 148-163.

Quattrocchi U (2012). CRC world dictionary of medicinal and poisonous plants. CRC press.

Rahmatullah M, Mollik MAH, Harun-or-Rashid M, Tanzin R, Ghosh KC, Rahman H, Alam J, Faruque, MO, Hasan MM, Jahan R & Khatun MA (2010). A comparative analysis of medicinal plants used by folk medicinal healers in villages adjoining the Ghaghot, Bangali and Padma rivers of Bangladesh. *Eurasian journal of Sustainable Agriculture* 4(1): 70-85

Raj ADS & Sukumaran S (2008). Rare and endemic plants in the sacred groves of Akumari district in Tamil Nadu. *Indian Journal of Forestry* 31: 611-616

Rajendran SM & Agarwal SC (2007). Medicinal plants conservation through sacred forests by ethnic tribals of Virudhunagar district, Tamil Nadu. *Indian Journal of Traditional Knowledge* 6(2): 328-333

Ramakrishna N, Ranjalkar KM & Saidulu C (2017). Medicinal plants biodiversity of Anantagiri hills in Vikarabad, Ranga Reddy district, Telangana state, India. *Bulletin of Environmental, Pharmacology and Life Science* 6: 249-258.

Rani CPJ & Jeeva S (2017). Ethnomedicinal plants used for gastro-intestinal ailments by Kani tribes of Kanyakumari wildlife sanctuary in Tamil Nadu, India. *International Journal of Current Advanced Research* 6(12): 7995-8000.

Rani JP, Tangavelou AC & Karthikeyan S (2016). Ethnomedicinal plants used by the tribals of Mudumalai wildlife sanctuary for poisonous bites. *International Journal of Modern Research and Reviews* 4(8): 1208-1212.

Rao GB, Rao UP, Rao SE, Rao MT & Praneeth VSD (2012). Evaluation of in-vitro antibacterial activity and anti-inflammatory activity for different extracts of *Rauvolfia tetraphylla* L. root bark. *Asian Pacific Journal of Tropical Biomedicin* 2(10): 818-821.

Rao SD, Rao NGM &Murthy PP (2016). Diversity and indigenous uses of some ethnomedicinal plants in Papikondalu wild life sanctuary, Eastern Ghats of Andhra Pradesh, India. *American Journal of Ethnomedicine* 3(1): 6-25.

Rathi P, Kumari R, Rajput CS & Sawhney SS (2013). Therapeutic characterstics of *Rauwolfia serpentina. International Journal of Pharmaceutical and Chemical Sciences* 2(2): 1038-1042

Rout SD & Thatoi HN (2009). Ethnomedicinal practices of Kol tribes in Similipal Biosphere Reserve, Orissa, India. *Ethnobotanical Leaflets* 13: 379-387

Rout SD, Panda T & Mishra N (2009). Ethno-medicinal plants used to cure different diseases by tribals of Mayurbhanj district of North Orissa. *Ethno Medicine* 3: 27-32.

Ruppert M, Ma X & Stockigt J (2005). Alkaloid biosynthesis in *Rauvolfia*-cDNA cloning of major enzymes of the ajmaline pathway. *Current Organic Chemistry* 9: 1431-1444.

Ruyter CM, Akram M, Illahi I & Stockigt J (1991). Investigation of the alkaloid content of *Rauwolfia serpentina* roots from regenerated plants. *Journal of Planta Medica* 57(4): 328-330.

Sahoo N & Manchikanti P (2013). Herbal drug regulation and commercialization: an indian industry perspective. *The Journal of Alternative and Complementary Medicine* 19(12): 957-963.

Sarkar NR, Mondal S & Mandal S (2016). Phytodiversity of Ganpur forest, Birbhum district, West Bengal, India with reference to their medicinal properties. *International Journal of Current Microbiology and Applied Sciences* 5(6): 973-989.

Sen P, Dollo M, Duttachoudhury M & Choudhury D (2008). Documentation of traditional herbal knowledge of *Khamptis* of Arunachal Pradesh. *Indian Journal of Traditional Knowledge* 7(3): 438-442.

Senthilmurugan G, Vasanthe B & Suresh K (2013). Screening and antibacterial activity analysis of some important medicinal plants. *International Journal of Innovation and Applied Studies* 2(2): 146-152.

Singh H, Dhole PA, Krishna G, Saravanan R, Baske PK (2018). Ethnomedicinal plants used in malaria in tribal areas of Odisha, India. *Indian Journal of Natural Product and Resources* 9(2): 160-167.

Singh PK, Kumar V, Tiwari RK, Sharma A, Rao CV & Singh RH (2010). Medico ethnobotany of Chatara block of district Sonebhadra, Uttar Pradesh, India. *Advance in Biological Regulations* 4(1): 65-80

Stanford JL, Martin EJ, Brinton LA &Hoover RN (1986). *Rauwolfia* uses and breast Cancer: A Case Control study. *Journal of the National Cancer Institute* 76: 817-822.

Sulochana AK, Raveendran D, Krishnamma AP & Oommen OV (2015). Ethnomedicinal plants used for snake envenomation by folk traditional practitioners from Kallar forest region of South Western Ghats, Kerala, India. *Jornal of Intercultural Ethnopharmacological* 4(1): 47-51.

Suresh K, Babu SS & Harisaranraj R (2008). Studies on *in vitro* antimicrobial activity of ethanol extract of *Rauvolfia tetraphylla. Ethnobotanical Leaflets* 12: 586-590.

Thinakaran T, Rajendran A & Sivakumari V (2009). Pharmacognostical, phytochemical and pharmacological studies in *Rauvolfia tetraphylla* L. *Asian Journal of Environmental Science* 4(1): 81-85.

Torres AW, Méndez GM, Durán GR, Boulogne I & Germosén RL (2015). Medicinal plant knowledge in Caribbean basin: a comparative study of Afrocaribbean, Amerindian and Mestizo communities. *Journal of Ethnobiology and Ethnomedicine* 11: 18, doi: 10.1186/s13002-015-0008-4.

Vadakkemuriyil DN, Rajaram P & Ragupathi G (2012). Studies on methanolic extract of Rauwolfia species from Southern Western Ghats of India- In vitro antioxidant properties, characterization of nutrients and phytochemicals. *Industrial Crops and Products* 39: 17-25.

Vaidyanathan D, Senthilkumar SMS & Basha GM (2013). Studies on ethnomedicinal plants used by Malayali tribals in Kolli hills of Eastern Ghats, Tamil Nadu, India. *Asian Journal of Plant Sciences and Research* 3(6): 29-45.

Vakil RJ (1955). Rauwolfia serpentina in the treatment of high blood pressure: a review of the literature. US Dept of Agriculture *Circulation* 12(2): 220–229. .

Verma KC & Verma SK (2010). Alkaloids analysis in root and leaf fractions of sarpaghanda (Rauwolfia serpentina). *Agricultural Science Digest* 30(2): 133–135.

Vinay KN, Lakshmi VV, Satyanarayan ND & Anantacharya GR (2016). Antioxidant activity of leaf and fruit extracts of *Rauwolfia serpentina* L. *International journal of pharmaceutical science and research* 7(4): 1705-1709

Woodson RE, Youngken HW, Schlittler E & Schneider JE (1957). *Rauwolfia*: botany, pharmacognosy, chemistry and pharmacology. *Economic Botany* 11(2): 183-184.

Yesodharan K & Sujana KA (2007). Ethnomedicinal knowledge among *Malamalasar* tribe of Parambikulam wildlife sanctuary, Kerala. *Indian Journal of Traditional Knowledge* 6(3): 481- 485.

21

Role of Allyl Isothiocyanate as a Bioprotective Agent

Sakshi Bhushan, Shivali Verma, Rohit Arora, Namrata Sharma and Saroj Arora

Abstract

Natural plant products are known to prevent several deleterious diseases. This potency of plants to act as medicinal therapeutic source is due to the presence of various bioactive compounds including alkaloids, phenols, flavonoids and glucosinolates (GSLs). Among array of these compounds, GSLs are important class of plant secondary metabolites. The mechanical or physical injury to plant leads to the activation of this "GSL-myrosinase" system resulting in the formation bioactive compounds viz., thiocyanates, nitriles, isothiocyanates, and epithionitriles. Out of these compounds, isothiocyanates (ITCs) are well known for their therapeutic properties. Allyl isothiocyanate (AITC) commonly known as "mustard oil" is one of the naturally occurring ITC which is derived from hydrolysis of sinigrin, the major GSL in cruciferous vegetables like horseradish, wasabi, brussels radish etc. Due to high bioavailability in human body, it has been widely studied for its important bioactivities. Therefore, this review highlights the significant biomodulatory effects viz., *antioxidant, antimutagenic, in vitro, in vivo anticancer, antimicrobial, and insecticidal activities of AITC.*

Keywords: Glucosinolates, Isothicyanates, Allyl Isothicyanate, antioxidant, bioprotective

---

Sakshi Bhushan (✉), Shivali Verma and Namrata Sharma

Department of Botany, University of Jammu, Jammu & Kashmir, India

Rohit Arora

Department of Biochemistry, Sri Guru Ram Das Institute of Medical Sciences and Research, Amritsar Punjab, India

Saroj Arora

Department of Botanical and Environmental Sciences, Guru Nanak Dev University, Amritsar Punjab, India

✉*Corresponding author email: bhushan.sakshi8@gmail.com*

Plants for Novel Drug Molecules: Ethnobotany to Ethnopharmacology
Bikarma Singh & Yash Pal Sharma (eds.), (pp. 467-482)

Email: *info@nipabooks.com* Web: *www.nipabooks.com*

1. Introduction

Plants are well known for plethora of organic compounds that are traditionally classified into "primary" and "secondary" metabolites. Among these, primary metabolites are metabolic intermediates/products found in all living systems which are essential for growth as well as development of plants. In contrary, plant produce secondary metabolites such as terpenes, flavonoids, alkaloids, saponins, steroids, glycosides, tannins, volatile oils etc with a prominent function in the interaction of plants with their environment. They are responsible for the protection of plants against predators and microbial pathogens depending upon their toxic nature and repellence to herbivores and microbes. Such phytoconstituents also play a crucial role in the plant's defence against abiotic stress and for the communication of plants with other organisms. On the basis of odor, color or taste, they function as attractants for the pollinators thus eventually helping in seed dispersal. They also serve as crucial agents of symbiotic association during plant-microbe interactions and plant-plant competition. In addition to this, secondary metabolites show very restricted distribution in the plant kingdom i.e., they are often found to be associated with only one plant species or related group of species. Interestingly, the therapeutic efficacy of plants is usually owed to presence of these secondary metabolites (Aggarwal et al. 2006, Huang et al. 2009). While the role of primary metabolites is extremely well documented, it is the secondary metabolites, which are still less explored and are attracting the attention of several workers recently.

Among array of secondary metabolites, glucosinolates (GSLs) is a major class of nitrogen and sulfur-containing ones present mainly in the cruciferae family. The plant injury/chewing/crushing activates "glucosinolate-myrosinase system which in turn generates unstable aglycone moiety (thiohydroximate-O-sulfate) after releasing glucose molecule. Spontaneous rearrangement of these intermediates and non-enzymatic elimination of the sulfate group forms a variety of bioactive components including isothiocyanates, thiocyanates, nitriles, oxazolidine-2-thiones or epithioalkanes (Bell and Wagstaff 2014, Wang et al. 2016). Isothiocyanates (ITCs) have attracted researchers because of their unique biological and pharmacological activities in cell cultures as well as animal models. The naturally available ITCs include allyl isothiocyanate (AITC), benzyl isothiocyanate, phenethyl isothiocyanate, sulphoraphane and indol-3-carbinol. Natural ITCs have displayed abilities to act as bioprotective agents in different biological assays (Satyan et al. 2006, Wu et al. 2009, Vig et al. 2009, Yuan et al. 2016). Interestingly, ITCs have the ability to act as strong electrophile due to its structure having central carbon atom that readily reacts with sulphur, nitrogen and oxygen-based nucleophiles. However, very few reports have supported their activity as direct antioxidant agents (Barillari et al. 2005, Haina et al. 2010). Instead, ITCs are known to act as indirect antioxidative agents. Several reports in literature have supported the indirect antioxidant potential of ITCs *in vitro* as well as *in vivo*.

Among naturally occurring ITCs, AITC (3-Isothiocyanato-1-propene) is one of the most commonly one which is also designated as "mustard oil" due to its predominance in regularly consumed cruciferous vegetables including mustard. Allyl glucosinolate (Sinigrin) has been reported to be the main constituent in Indian mustard (Hwong et al. 2006). AITC is derived from hydrolysis of its parent compound "sinigrin" in the presence of enzyme myrosinase (Fig. 1). Humans are exposed to AITC extensively due to the consumption of cruciferous vegetables and mustard oil (in spices, pickles and Indian curries) which together are rich source of AITC. Due to the characteristic pungency of this compound, it is purposely added to the food products for enhancement of flavour. As compared to other ITCs the bioavailability of AITC is incredibly high and it is eliminated via urine after absorption through mercapturic acid pathway (Ioannou et al. 1984). Many studies support accumulation of AITC and other ITCs more rapidly in the cells that results in the conjugation of AITC to intracellular thiols (Bruggeman et al. 1986). Therefore AITC is considered an important biological compound and hence has been extensively reviewed further in the following sections (Table 1 and Fig. 2).

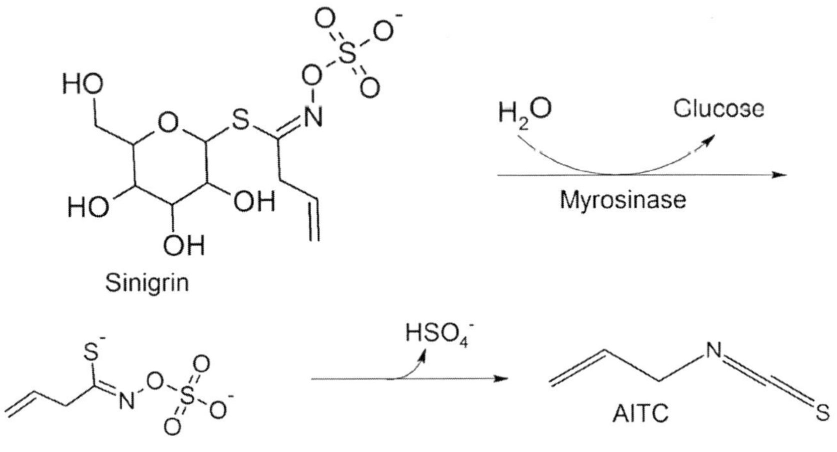

Fig.1: The hydrolytic breakdown of sinigrin that leads to the formation of allyl isothiocyanate

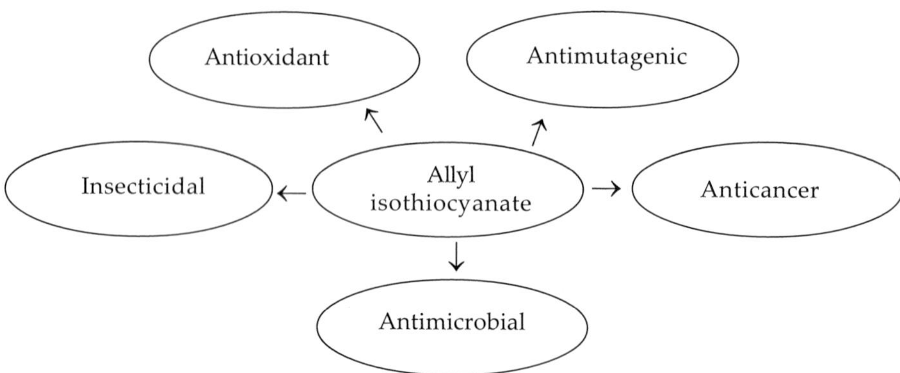

Fig. 2: Schematic representation of important biological properties of AITC

Table 1: The important plant sources of allyl isothiocyanate

Important sources	Reference
Arabidopsis thaliana (L.) Heynh.	(Hofte et al. 1992, Hossain et al. 2013)
Armoracia lapathifolia Gilib.	Zhang , 2010
Brassica oleracea var. *alboglabra*	Wang et al. 2010;Cartea et al. 2008; Vale
Brassica oleracea L.	et al. 2015
Brassica oleracea var. *capitata*	Sarikamis et al. 2008; Rangkadilok et al.
Brassica oleracea var. *acephala*	2002; Bellostas et al. 2007; Rampal et al.
Brassica oleracea var. *gemmifera*	2011
Brassica oleracea var. *italic*	
Brassica campestris var. *pekinensis*	Bennett et al. 2004
Brassica juncea, Coss.	Price et al. 2005; Song et al. 2006;
Brassica juncea italic	Zhao et al. 2007; Rampal et al. 2011
Brassica rapa L. cv. *nakajimana*	Kato et al. 2011; Saka et al. 2017;
Brassica rapa var. *rapa*	Arora et al. 2017
Brassica rapa var. *rapifera*	
Raphanus sativus L. var. *oleiformis*	Sangthong *et al.* 2017
Raphanus sativus var. *caudatus*	
Eruca sativa Mill.	Arora et al. 2017; Bell and Wagstaf, 2014; Bell et al. 2017

2. Protective Activities of Allyl Isothiocyanate

2.1. Antioxidant and Antimutagenic Activities

ITCs are known to modulate the activities of xenobiotic metabolizing enzymes that activate long lasting antioxidant activity. There are very few reports regarding the direct antioxidant activity of AITC. Several studies support the effect of AITC in countering the free radicals generated due to oxidative stress. Cytochrome P450 enzymes result in increase of reactivity of compounds and in this process may lead to formation of reactive moieties. Phase II enzymes in turn enhance the solubility in water and facilitates the excretion of metabolites out of the body. Therefore, suppression of phase I enzymes and exaggeration of phase II enzymes play a crucial role in protecting cells against DNA damage and reactive oxygen species. Many studies have supported the ability of AITC in inducing the phase II enzymes such as glutathione-S-transferase, NAD[P]H:quinone oxidoreductase, or glutamate cysteine in cultured cells *in vitro* as well as in rodent tissues supporting its indirect antioxidant property (Talalay et al. 1995, Ye and Zhang 2001, Tang and Zhang, 2004, Wu et al. 2005, Munday et al. 2006).

It has been observed that AITC and synthetic AITC-NAC conjugate induced quinone reductase (QR) in Hepa1c1c7 murine hepatoma cells. Both the compounds led to the inhibition of cell growth in a concentration-dependent manner by inducing quinone reductase activity and mRNA expression over a range of concentrations (*viz.*, 0.1–2.5 μM). Overall it resulted in 2 to 3 folds increase in quinone reductase (Hwang et al. 2006). The involvement of AITC in enhancement of phase II proteins is linked with activation of nuclear erythroid 2-related factor (Nrf2), which is a chief transcription identity responsible for the expression of phase II enzymes and other genes of significance. The capability of AITC and other natural ITCs to activate the expression of genes associated with Nrf-2 factor suggests its possible role as antioxidant agent in the cellular fibroblasts (Ernst et al. 2011). Also, Bansal et al. (2015) demonstrated the antioxidant, antimutagenic and antimelanoma activities of *Eruca sativa* seed oil and its bio active compounds *viz.*, sulphoraphane, phenylethyl ITC and AITC against B16F10 melanoma cells in C57BL/6 mice model. It was found that ITCs efficiently regulated the different parameters such as glutathione, lipid peroxidation and controlled the tumorous growth efficiently, suggesting the proficient ability of AITC to act as antioxidant agent along with other compounds studied.

The potential of AITC for its antioxidant and anti-tumour ability in Swiss albino mice has been studied at a dose of 25 μg/dose/animal for 5 days which resulted in efficient antioxidant and tumor reducing activity by scavenging hydroxyl and superoxide radicals, reducing lipid peroxidation in liver homogenate (Manesh and Kutten , 2003). Evaluation of six plant derived ITCs including AITC was studied to examine the tissue levels of the phase II detoxifying enzymes *viz.*, glutathione-S-transferase and quinone reductase in the rat tissues. It was observed that AITC (at 40μmol/kg/day) and other ITCs, led to the enhancement of both the phase II enzymes in duodenum, forestomach with pronounced effect in the urinary bladder of all the groups taken into consideration (Mundey and Mundey, 2004)

Several reports have also suggested antimutagenic activity of glucosinolate hydrolytic products (GHPs) (Vig et al. 2009, Arora et al. 2016). Evaluation of oil extract of *Brassica rapa* var. *rapa* seeds exhibited strong inhibition of DNA damage as a result of generation of OH radicals, thus depicting significant antimutagenic property against frameshift and base pair substutution mutations. This activity was attributed to the presence of important GHPs in extracts of *Brassica rapa* (Robin et al. 2015). A study carried out by Rampal et al. (2017) emphasized on the antimutagenicity of ITCs including AITC, benzyl and 3-butenyl ITCs in Ames assay using *Salmonella typhimurium* tester strains against direct acting mutagen sodium azide, 4-nitro-o-phenylenediamine, and indirect acting mutagen 2-aminofluorene. This study suggested the antimutagenic potency of AITC against all the mutagens when given alone and in binary combinations with other ITCs.

2.2. *In vitro* Anticancer Activity

The parent precursor of AITC i.e., "sinigrin" is not known much for its antiproliferative ability however, AITC has shown potency to inhibit proliferation of various human cancer cell lines having 50% inhibition even at lower micro molar concentrations (Zhang et al. 2003). It was observed that treatment with AITC (20 μM) led to the inhibition of human prostate cancer cells PC-3 and LNCaP in a dose dependent manner with an IC_{50} ranging from 15-17 μM. They attributed anti-proliferative effect of AITC due to its accumulation in cells at G2/M phase that further led to the induction of apoptosis (Xiao et al. 2003). Similarly, AITC also studied in brain malignant glioma cells (GBM 8401) in which it significantly decreased the proliferation and viability of abnormal cells in a concentration dependent manner and exhibited IC_{50} of 9.25 μM for 24 h treatment (Chen et al. 2010).

In vitro estimation of anticancer activity of AITC and sulphoraphane derived from *Eruca sativa* seed oil has also been studied against HepG2 human liver carcinoma cells by employing 3-(4,5-dimethylthiazol-2-yl)-2,5-diphenyltetrazolium bromide (MTT) and sulphorhodamine B (SRB) bioassays. It was observed that combination of ITCs resulted in percentage inhibition that ranged from 97.12% to 96.56% when compared to standard drug doxorubicin at concentration of 50 μM (Bansal et al. 2013). Similarly, a study was conducted to evaluate the anti-proliferative effect of AITC in human breast cancer cells UM-UC-3 and rat AY-27 cancer cells. It was found that IC_{50} value of AITC ranged from 2.7 μM to 3.3 μM after the treatment of cells for 72 h. In contrary, the effectiveness of AITC against normal human bladder cells was more than 21 times less as compared to bladder cancer cell line (Bhattacharya et al. 2013). Similar studies were conducted by Savio et al. (2014) who investigated the effect of AITC on bladder cancer cell lines having RT4 (wild type) and T24 (mutated TP53 gene). They evaluated different cellular parameters including morphological changes, cell cycle kinetics and CDK1 gene expression. AITC treatment in both the cell lines suppressed the cell proliferation and led to the induction of morphological changes such as elongated cells and formation of cellular debris.

A study expressed the synergistic effect of AITC when given in combination with cisplatin in human ovarian cancer cells and A549 lung cancer cells. It was observed that treatment with AITC enhanced the apoptosis which was directly co-related to downregulation of antiapoptotic proteins Bcl-2 (Ling et al. 2015). Renal carcinoma is also known to effect human adversely. The treatment of GRC-1 renal carcinoma cell line with AITC reduced the proliferation of cancer cells and also resulted in apoptosis by regulating the discrepancy of Bcl-2/Bax which plays a considerable role in combating the cancerous entities (Jiang et al. 2016). Bo et al. (2016) demonstrated the toxic effect of AITC against MDA-MB-231 and MCF-7 breast cancer cells. It was reported that treatment with AITC effectively stimulated the ROS generation and led to the reduction in mitochondrial membrane potential (MMP) leading to induction of apoptosis in cancerous cells acting via Endo-G signalling pathways and activating caspase-3 and caspase-9 respectively.

2.3. *In vivo* Anticancer Activity

AITC has also been studied extensively in rodent models for its protective properties. Interstingly, the occurrence of tumor incidence was almost decreased by 90% at a dose level of 12.5-100 μmol/kg bw, thus AITC was reported to efficiently inhibit the formation of gastric lesions in Sprague-Dawley rats in a concentration dependent manner in the presence of hydrochloric acid (HCl), ammonia, ethanol, indomethacin and aspirin (Matsuda et al. 2007).

The analysis of mustard seed powder (MSP) comprising of AITC in a rat bladder cancer model, at a dose level of 71.5 mg/kg resulted in inhibition of bladder cancer growth by 34.50 %. They accredited the anticancer ability of mustard seed powder to its active components and their active participation in blocking the muscle invasion by 100%. In addition, AITC-mustard seed powder significantly modulated important therapeutic targets including cyclin B1, caspase-3 along with vascular endothelial growth factor (Bhattacharya et al. 2009; Bhattacharya et al. 2010). In a similar study performed by Bhattacharya et al. (2013) co-treatment of AITC with celecoxib, an important inhibitor of Cox-2, enhanced the tumor inhibition in bladder cancer model in rats, as compared to both the compounds when administered alone. This combination even showed effectiveness in combating the microvessel formation and stimulating microvessel maturation in tissues showing tumor development. Ibrahim et al. (2015) conducted an experiment using Fischer rats in which presence of AITC reduced the lesions to 60% in protective group. However, in curative group with treatment period of 10 weeks, 80% rats showed lesions as compared to 100% lesions in bladder cancer group. Therefore, this study also supported anticancer activity of AITC in suppressing all the stages of the neoplastic phenomenon.

Ahn et al. (2016) evaluated hepatoprotective activity of AITC against the liver toxicity induced in Sprague Dawley rats in the presence of carbon tetrachloride. It was seen that AITC at 5 and 50 mg/kg bw for three days led to the amelioration of liver toxicity by activating kupffer cells along with macrophages. In addition, AITC also regulates oxidative parameters supporting its antioxidant activity. An investigation studying the function of AITC on glycoproteins in DMBA induced mammary carcinogenesis in female Sprague-Dawley rats reported increased levels of these components suggesting the potential of AITC to inhibit the irregular glycosylation that favours neoplastic transformation (Rajkumar et al. 2017).

2.4. Antimicrobial Activity

As compared to sinigrin, AITC has displayed tendency to act against array of pathogenic bacteria such as *Staphylococcus aureus, Salmonella typhimurium, Bacillus cereus, Helicobacter pylori, Penicillium notatum* and *Vibrio parahaemolyticus* with minimum inhibitory concentrations (MIC) ranging from 3.8 μM to 16.7 μM (Shin et al. 2004, Tunc et al. 2007, Wang et al. 2010, Borges et al. 2015).

Luciano and holley, (2009) estimated the effect of AITC against *Escherichia coli* O157:H7. They reported that at lower pH ranging from 4-6, AITC displayed minimum inhibitory concentration (MIC) of 25 μl/L when compared to MIC of 500 μl/L at pH 8.5. In addition to this, AITC showed the capability to interact with "thioredoxin reductase" and "acetate kinase" and it inhibited both the enzymes ultimately effecting *E. coli* resistance. Similarly, AITC was successful in reducing the minimum inhibitory concentration of erythromycin against *Streptococcus pyogenes* (Palaniappan & Holley 2010), and it displayed a synergistic effect along with streptomycin against Gram-negative bacteria including *Pseudomonas aeruginosa* (Saavedra et al. 2010)

The antibacterial mode of action of selected GHPs including AITC against *Pseudomonas aeruginosa, Staphylococcus aureus, Escherichia coli,* and *Listeria monocytogenes* has also been reported (Borges et al. 2014). The MIC for AITC was 100 µg/mL for all the bacteria studied and presence of AITC altered the membrane properties of bacteria resulting in decrease of their surface charge, ultimately effecting cytoplasmic membrane. Furthermore, AITC treatment also modulated the synergistic activity of "nisin" a commercial antibiotic against food-borne pathogenic bacteria (Zou et al. 2013).

In addition to the bactericidal activity, AITC has also been known to exhibit antifungal activity against plethora of fungi including *Pichia anomala, Penicillium polonicum Penicillium discolour, Endomyces fibuliger, Aspergillus flavus* (Zhang, 2010, Baskar et al. 2016). ITCs such as phenethyl ITC, AITC and butyl ITC were observed to be active against the three strains of *Gibberella moniliformis* and further reduced the mycelial size upto 89.7%. Moreover, all the three ITCs also significantly reduced FB2 mycotoxin upto 73 % (Azaiez et al. 2013).

A recent study done by Manyes et al. (2015) reported fungicidal activity of AITC against two fungi *Penicillium expnsum* and *Aspergillus parasiticus* known for producing patulin (PAT) and aflatoxins (AFs). It was observed that AITC was able to inhibit both the strains at a dose level of 5 mg (*A. parasiticus*) and 50 mg (*P. expnsum*). The analysis of risk assessment associated with uptake of AITC also supported the consumption of this compound. The use of AITC in gaseous form inhibited the production of mycotoxins produced by *Fusarium poae* and *Aspergillus parasiticus*. Treatment of 10 µl/L AITC suppressed the generation of aflatoxin, beauvercin and enniatin for 30 days (Nazareth et al. 2016). AITC was also estimated for its antifungal activity against *Candida albicans* alone and along with the presence of fluconazole (FLC). Results from this study revealed its potency of acting against growth and virulence factors of *C. albicans*. In addition to this, co-incubation with AITC and FLC exerted a strong anti-biofilm activity against this pathogen (Raut et al. 2017).

2.5. Insecticidal Activity

AITC has also acted as fumigant against four main insect species of stored products *viz., Rhizopertha dominica* (F.), *Sitophilus zeamais* (Motsch.), *Liposcelis entomophila* (Enderlein) and *Tribolium ferrugineum* (F.). It was observed that fumes of AITC (3 µg mL^{-1}) showed strong toxicity to stored-product pests as indicated by absolute mortality of adults in all the species after treatment for 72 h (Wu et al. 2009). Similarly, the susceptibility of adults of 18 populations of the red flour beetle *Tribolium castaneum* (Herbst.) (Coleoptera: Tenebronidae) to AITC fumes was also checked by Santos et al. (2011). It was observed that the population of adults was affected by AITC with negligible variation. Two selected populations were further used to evaluate the toxicity of AITC against the larvae and pupae of *T. castaneum*. Overall these results indicated the efficacy of AITC as an alternative fumigant against these pests.

Eltayeb et al. (2010) evaluated seven formulations of AITC along with pure AITC and in combination with chlorpyrifos and cypermethrin by using the fumigation and spray application method against *Plutella xylostella* (L.), *Spodoptera litura* (Fab.) and *Pieris rapae* (L.). AITC overall exhibited pronounced activity and can act as an alternative to old fumigant i.e., phosphine. In a similar study, the insecticidal potency of AITC was estimated employing dry and wet application on the insect pests such as *Sitophilus oryzae* (L.), *Tribolium confusum* (Jacquelin Du Val) and *Plodia interpunctella* (Hubner). The range of LC_{50} was found to be from 2.0 to 4.7 $\mu l\ L^{-1}$ respectively. The absolute morality (100%) of insects was recorded after 72-h treatment to AITC fumes at a dose level of 11.5 $\mu l\ L^{-1}$ air. These results were analogous to the observations found in insects treated with 5 mg L^{-1} dose of phosphine. Interestingly, the effect of AITC and phosphine on the midgut and integument of 3rd instar larvae of *T. confusum* was similar, which pointed towards the anti-insect activity of AITC (Mansour et al. 2012).

AITC results in the vacuolization of mitochondrial matrix of AITC-treated adult *Sitophilus zeamais* (Motsch) (Wu et al. 2014). The levels of enzymes including catalase, glutathione-S-transferase, acetylcholinesterase and cytochrome c oxidase were also regulated in the presence of AITC supporting its anti-insect activity. Carbone et al. (2016) studied the effect of GSLs on *Mamestra brassicae* (Noctuidae) and *Pieris rapae* (Pieridae). Total of six genotypes of *Brassica oleracea* var. *acephala* having the presence of sinigrin (parent precursor of AITC) and other GSLs resulted in regulating the growth and development parameters of these two pests.

3. Conclusion

It is well known fact that GSLs are important secondary metabolites present in many plant families. In the presence of myrosinase enzyme these metabolites are broken down into hydrolytic products including ITCs. Interestingly, ITCs have known to possess wide array of bioactivities. They are important to plants due to their critical role in plant defence system. Therefore, nowadays exploration of natural products such as AITC is in demand, for the overall beneficial effects for humans. AITC has shown tremendous benefit in agriculture and medicine as evident from various studies conducted using *in vitro* and *in vivo* approaches. On the whole, there is a need to further unravel and commercialize the use of AITC for the benefit of humans and plant health.

Abbreviation Used

AITC	Allyl isothiocyanate
Bcl-2	B-cell lymphoma 2
DMBA	7,12-Dimethylbenz[a]anthracene
GHP	Glucosinolate Hydrolytic Product
GSL	Glucosinolate
HCl	Hydrochloric acid
HeLA	Human cervical cancer cell line
HepG2	Human liver-cancer cell line
HIG	Histological Index Grading
HL-60	Human leukemia cell line- 60
HSD	Honest Significant Difference
HT-29	Human colon cancer cell line
IC50	Half maximal inhibitory concentration
ITC	Isothiocyanate
MCF-7	Human Breast Cancer Cells
MIC	Minimum inhibitory concentrations
QR	Quinone reductase

Conflict of Interest

There are no conflicts of interest.

Acknowledgements

The authors thank Department of Botanical and Environmental Sciences, Guru Nanak Dev University, Amritsar and Department of Botany, University of Jammu for the compilation of the present review.

References

Aggarwal B B & Shishodia S (2006). Molecular targets of dietary agents for prevention and therapy of cancer. *Biochemical Pharmacology* 71: 1397–1421.

Ahn M, Kim J, Bang H, Moon J, Kim G O & Shin T (2016). Hepatoprotective effects of allyl isothiocyanate against carbon tetrachloride-induced hepatotoxicity in rat. *Chemico-Biological Interactions*. 254: 102-108.

Arora R, Arora S & Vig A P (2017). Development of validated high-temperature reverse-phase UHPLC-PDA analytical method for simultaneous analysis of five natural isothiocyanates in cruciferous vegetables. *Food Chemistry* https://doi.org/10.1016/j.foodchem 2017.07.059.

Arora R, Singh B, Vig A P & Arora, S (2016). Conventional and modified hydrodistillation method for the extraction of glucosinolate hydrolytic products: a comparative account. *Springer Plus* 5: 479.

Azaiez I, Meca G, Manyes L & Fernández-Franzón M (2013). Antifungal activity of gaseous allyl, benzyl and phenyl isothiocyanate *in vitro* and their use for fumonisins reduction in bread. *Food Control* 32: 428-434.

Bansal P, Medhe S, Ganesh N & Srivastava M M (2013). Isothiocyanates: naturally occurring HEPG2 human liver carcinoma inhibitor. *International Journal of Bioassays* 2: 325-328.

Bansal P, Medhe S, Ganesh N & Srivastava M M (2015). Antimelanoma potential of *Eruca sativa* seed oil and its bioactive principles. *Indian Journal of Pharmaceutical Sciences* 77: 208.

Barillari J, Cervellati R, Paolini M, Tatibouet A, Rollin P & Iori, R (2005). Isolation of 4-methylthio-3-butenyl glucosinolate from *Raphanus sativus* sprouts (kaiware daikon) and its redox properties. *Journal of Agriculture and Food Chemistry* 53: 9890–9896.

Baskar V, Park SW & Nile SH (2016). An update on potential perspectives of glucosinolates on protection against microbial pathogens and endocrine dysfunctions in humans. *Critical reviews in Food Science and Nutrition* 56: 2231-2249.

Bell L, Yahya H N, Oloyede OO, Methven L & Wagstaff C (2017). Changes in rocket salad phytochemicals within the commercial supply chain: glucosinolates, isothiocyanates, amino acids and bacterial load increase significantly after processing. *Food Chemistry* 221: 521-534.

Bell Luke & Wagstaff C (2014). Glucosinolates, myrosinase hydrolysis products, and flavonols found in rocket (*Eruca sativa* and *Diplotaxis tenuifolia*). *Journal of Agricultural and Food Chemistry* 62: 4481-4492.

Bellostas N, Sorensen J C & Sorensen H (2007). Profiling glucosinolates in vegetative and reproductive tissues of four *Brassica* species of the U triangle for their biofumigation potential. *Journal of the Science of Food and Agriculture* 87: 1586-1594.

Bennett R N, Mellon F A & Kroon P A (2004). Screening crucifer seeds as sources of specific intact glucosinolates using ion-pair high-performance liquid chromatography negative ion electrospray mass spectrometry. *Journal of Agricultural and Food Chemistry* 52: 428-438.

Bhattacharya A, Li Y , Shi Y & Zhang, Y (2013). Enhanced inhibition of urinary bladder cancer growth and muscle invasion by allyl isothiocyanate and celecoxib in combination. *Carcinogenesis* 34: 2593-2599.

Bhattacharya A, Li Y, Wade KL, Paonessa JD, Fahey JW & Zhang Y (2010). Allyl isothiocyanate-rich mustard seed powder inhibits bladder cancer growth and muscle invasion. *Carcinogenesis* 31: 2105-2110.

Bhattacharya A, Tang L, Li Y, Geng F, Paonessa JD, Chen SC, Wong MKK & Zhang Y (2009). Inhibition of bladder cancer development by allyl isothiocyanate. *Carcinogenesis* 31: 281-286.

Bo P, Lien JC, Chen YY, Yuz FS, Lu H F, Yu CS, Chou YC, Yu CC & Chung JG (2016). Allyl isothiocyanate induces cell toxicity by multiple pathways in human breast cancer cells. *The American Journal of Chinese Medicine* 44: 415-437.

Borges A, Abreu AC, Ferreira C, Saavedra MJ, Simoes LC & Simoes M (2015). Antibacterial activity and mode of action of selected glucosinolate hydrolysis products against bacterial pathogens. *Journal of Food Science and Technology* 52: 4737-4748.

Borges A, Simoes LC, Saavedra, MJ & Simoes M (2014). The action of selected isothiocyanates on bacterial biofilm prevention and control. *International Biodeterioration & Biodegradation* 86: 25-33.

Bruggeman IM, Temmink JH & Van Bladeren PJ (1986). Glutathione and cysteine-mediated cytotoxicity of allyl and benzyl isothiocyanate. *Toxicology Applied Pharmacology* 83: 349–359.

Carbone S S, Sotelo T, Velasco P and Cartea EM (2016). Antibiotic properties of the glucosinolates of *Brassica oleracea* var. acephala similarly affect generalist and specialist larvae of two lepidopteran pests. *Journal of Pest Science* 89: 195-206.

Cartea ME & Velasco P (2008). Glucosinolates in *brassica* foods: bioavailability in food and significance for human health. *Phytochemistry Reviews* 7: 213-229.

Chen NG, Chen KT, Lu CC, Lan Y H, Lai CH, Chung YT, Yang JS & Lin YC (2010). Allyl isothiocyanate triggers G2/M phase arrest and apoptosis in human brain malignant glioma GBM 8401 cells through a mitochondria-dependent pathway. *Oncology Reports*. 24: 449-455.

Eltayeb EM, Zhang GA, Xie J G & Mi FY (2010). Study on the effect of allyl isothiocyanate formulations on three lepidopterous insect larvae, the diamond back moth, *Plutella xylostella* (L.), the small cabbage white butterfly, *Pieris rapae* (L.) and tobacco cut worm, *Spodoptera litura* (Fab.). *American Journal of Environmental Sciences* 6: 168–176.

Ernst I M, Wagner A E, Schuemann C, Storm N, Höppner W, Döring F, Stocker A & Rimbach G (2011). Allyl, butyl-and phenylethyl-isothiocyanate activate Nrf2 in cultured fibroblasts. *Pharmacological research* 63: 233-240.

Haina Y, Shanjing Y, Yuru Y, Gongnian X & Qi Y (2010). Antioxidant activity of isothiocyanate extracts from broccoli. *Chinese Journal of Chemical Engineering* 18: 312-321.

Hofte H, Hubbard L, Reizer J, Ludevid D, Herman E M & Chrispeels M J (1992). Vegetative and seed-specific forms of tonoplast intrinsic protein in the vacuolar membrane of *Arabidopsis thaliana*. *Plant Physiology* 99: 561-570.

Hossain MS, Ye W, Hossain MA, Okuma E, Uraji M, Nakamura Y & Murata Y (2013). Glucosinolate degradation products, isothiocyanates, nitriles, and thiocyanates, induce stomatal closure accompanied by peroxidase-mediated reactive oxygen species production in *Arabidopsis thaliana*. *Bioscience, Biotechnology and Biochemistry* 77: 977-983.

Huang WY, Cai YZ & Zhang Y (2009). Natural phenolic compounds from medicinal herbs and dietary plants: potential use for cancer prevention. *Nutrition and Cancer* 62: 1-20.

Hwang ES & Lee HJ (2006). Allyl isothiocyanate and its N-acetylcysteine conjugate suppress metastasis via inhibition of invasion, migration, and matrix metalloproteinase-2/-9 activities in SK-Hep1 human hepatoma cells. *Experimental Biology and Medicine* 231: 421–430.

Ibrahim HA, Gabr MN, El-kholy WM, Sami W & Abd-Elkhalek MMZ (2015). Protective and curative role of allyl isothiocyanate against bladder carcinogenesis in rats. *International Journal* 3: 544-554.

Ioannou YM, Burka LT & Matthews HB (1984). Allyl isothiocyanate: comparative disposition in rats and mice. *Toxicology and Applied Pharmacology* 75: 173-181.

Jiang Z, Liu X, Chang K, Liu X & Xiong J (2016). Allyl isothiocyanate inhibits the proliferation of renal carcinoma cell line GRC-1 by inducing an imbalance between bcl2 and bax. *Medical science monitor: International Medical Journal of Experimental and Clinical Research* 22: 4283-4288.

Kato M, Imayoshi Y, Iwabuchi H & Shimomura K (2011). Kinetic changes in glucosinolate-derived volatiles by heat-treatment and myrosinase activity in Nakajimana (*Brassica rapa* L. cv. *nakajimana*). *Journal of Agricultural and Food Chemistry* 59: 11034-11039.

Ling X, Westover D, Cao F, Cao S, He X, Kim HR, Zhang Y & Fengzhi CF (2015). Synergistic effect of allyl isothiocyanate (AITC) on cisplatin *efficacy in vitro* and *in vivo*. *American Journal of Cancer Research* 5: 2516-2530.

Luciano FB & Holley RA (2009). Enzymatic inhibition by allyl isothiocyanate and factors affecting its antimicrobial action against *Escherichia coli* O157: H7. *International Journal of Food Microbiology* 131: 240-245.

Manesh C & Kutten G (2003). Anti-tumour and anti-oxidant activity of naturally occurring isothiocyanates. *Journal of Experimental and Clinical Cancer Research* 22:193-9.

Mansour, EE, Mi F, Zhang G, Jiugao X, Wang Y & Kargbo A (2012). Effect of ally lisothiocyanate on *Sitophilus oryzae, Tribolium confusum* and *Plodia interpunctella*: Toxicity and effect on insect mitochondria. *Crop protection.* **33:** 40-51.

Manyes L, Luciano F B, Manes J & Meca G (2015). *In vitro* antifungal activity of allyl isothiocyanate (AITC) against *Aspergillus parasiticus* and *Penicillium expansum* and evaluation of the AITC estimated daily intake. *Food and Chemical Toxicology* 83: 293-299.

Matsuda H, Ochi M, Nagatomo A & Yoshikawa M (2007). Effects of allyl isothiocyanate from horseradish on several experimental gastric lesions in rats. *European Journal of Pharmacology* 561: 172-181

Munday R & Munday CM (2004). Induction of phase II detoxification enzymes in rats by plant-derived isothiocyanates: comparison of allyl isothiocyanate with sulforaphane and related compounds. *Journal of Agricultural and Food Chemistry* 52: 1867-1871.

Munday R, Zhang Y, Fahey JW, Jobson HE, Munday CM, Li J & Stephenson KK (2006). Evaluation of isothiocyanates as potent inducers of carcinogen-detoxifying enzymes in the urinary bladder: critical nature of *in vivo* bioassay. *Nutrition and Cancer* 54: 223-231.

Nazareth T M, Bordin K, Manyes L, Meca G, Mañes J & Luciano F B (2016). Gaseous allyl isothiocyanate to inhibit the production of aflatoxins, beauvericin and enniatins by *Aspergillus parasiticus* and *Fusarium poae* in wheat flour. *Food Control* 62: 317-321.

Palaniappan K & Holley RA (2010). Use of natural antimicrobials to increase antibiotic susceptibility of drug resistant bacteria. *International Journal of Food Microbiology* 140: 164-168.

Price AJ, Charron CS, Saxton A M & Sams CE (2005). Allyl isothiocyanate and carbon dioxide produced during degradation of *Brassica juncea* tissue in different soil conditions. *HortScience* 40: 1734-1739.

Rajakumar T, Pugalendhi P, Thilagavathi S, Ananthakrishnan D & Gunasekaran K (2017). Allyl isothiocyanate, a potent chemopreventive agent targets AhR/ Nrf2 signaling pathway in chemically induced mammary carcinogenesis. *Molecular and Cellular Biochemistry* 1-12. DOI 10.1007/s12282-017-0783-y.

Rampal G & Vig AP (2011). Evaluation of antifungal and antioxidative potential of hydrolytic products of glucosinolates from some members of Brassicaceae family. *Journal of Plant Breeding and Crop Science*. 3: 218-228.

Rampal G, Thind TS, Arora R, Vig AP & Arora S (2017). Synergistic antimutagenic effect of isothiocyanates against varied mutagens. *Food and Chemical Toxicology*. https://doi.org/10.1016/j.fct.2017.05.017

Rangkadilok N, Nicolas ME, Bennett RN, Premier RR, Eagling DR & Taylor PW (2002). Determination of sinigrin and glucoraphanin in *Brassica* species using a simple extraction method combined with ion-pair HPLC analysis. *Scientia Horticulturae*. 96: 27-41.

Raut JS, Bansode BS, Jadhav AK & Karuppayil SM (2017). Activity of allyl isothiocyanate and its synergy with fluconazole against *Candida albicans* Biofilms. *Journal of Microbiology and Biotechnology* 27: 685.

Robin, Arora S & Vig AP (2015). Inhibition of DNA oxidative damage and antimutagenic activity by dichloromethane extract of *Brassica* rapa var. rapa L. seeds. *Industrial Crops and Products* 74: 585-591.

Saavedra MJ, Borges A, Dias C, Aires A, Bennett RN, Rosa ES & Simoes M (2010). Antimicrobial activity of phenolics and glucosinolate hydrolysis products and their synergy with streptomycin against pathogenic bacteria. *Medicinal Chemistry* 6: 174-183.

Saka B, Djouahri A, Djerrad Z, Souhila T, Aberrane S, Sabaou N, Baaliouamer A & Boudarene L (2017). Chemical variability and biological activities of *Brassica rapa* var. rapifera parts essential oils depending on geographic variation and extraction technique. *Chemistry and Biodiversity* 14: 1-20.

Sangthong S, Weerapreeyakul N, Lehtonen M, Leppanen J and Rautio J (2017). High-accuracy mass spectrometry for identification of sulphur-containing bioactive constituents and flavonoids in extracts of *Raphanus sativus* var. caudatus Alef (Thai rat-tailed radish). *Journal of Functional Foods* 31: 237-247.

Santos JC, Faroni LRA, Sousa AH & Guedes RNC (2011). Fumigant toxicity of allyl isothiocyanate to populations of the red flour beetle *Tribolium castaneum. Journal of Stored Products Research* 47: 238-243.

Sarikamis G, Balkaya A & Yanmaz R (2008). Glucosinolates in kale genotypes from the blacksea region of Turkey. *Biotechnology and Biotechnological Equipment* 22: 942-946.

Satyan KS, Swamy N, Dizon DS, Singh R, Granai CO & Brard L (2006). Phenethyl isothiocyanate (PEITC) inhibits growth of ovarian cancer cells by inducing apoptosis: role of caspase and MAPK activation. *Gynecologic Oncology* 103: 261-270.

Savio ALV, Da Silva G N, De Camargo EA & Salvadori D M F (2014). Cell cycle kinetics, apoptosis rates, DNA damage and TP53 gene expression in bladder cancer cells treated with allyl isothiocyanate (mustard essential oil). *Mutation Research/Fundamental and Molecular Mechanisms of Mutagenesis*.762: 40-46.

Shin I S, Masuda H & Naohide K (2004). Bactericidal activity of wasabi (*Wasabia japonica)* against *Helicobacter pylori. International Journal of Food Microbiology* 94: 255-261.

Song L, Iori R &Thornalley PJ (2006). Purification of major glucosinolates from *Brassicaceae* seeds and preparation of isothiocyanate and amine metabolites. *Journal of the Science of Food and Agriculture* 86: 1271-1280.

Talalay P, Fahey JW, Holtzclaw WD, Prestera T & Zhang Y (1995). Chemoprotection against cancer by phase 2 enzyme induction. *Toxicology letters* 82: 173-179.

Tang L & Zhang Y (2004). Dietary isothiocyanates inhibit the growth of human bladder carcinoma cells. *Journal of Nutrition* 134: 2004–2010.

Tunc S, Chollet E, Chalier P, Preziosi-Belloy L & Gontard N (2007). Combined effect of volatile antimicrobial agents on the growth of *Penicillium notatum. International Journal of Food Microbiology* 113: 263-270.

Vale AP, Santos J, Brito NV, Fernandes D, Rosa E & Oliveira MBP (2015). Evaluating the impact of sprouting conditions on the glucosinolate content of Brassica oleracea sprouts. *Phytochemistry.* 115: 252-260.

Vig AP, Rampal G, Thind TS & Arora S (2009). Bio-protective effects of glucosinolates–A review. *LWT-Food Science and Technology* 42: 1561-1572.

Wang K, Zang Y, Wang H, Xia X & Liu T (2010). Influence of three larval diets on susceptibility of selected insecticides and activities of detoxification esterases of *Helicoverpa assulta* (Lepidoptera: Noctuidae). *Pesticide Biochemistry and Physiology* 96: 51–55.

Wang SY, Chen CT and Yin JJ (2010). Effect of allyl isothiocyanate on antioxidants and fruit decay of blueberries *Food Chemistry.* 120: 199-204.

Wang J, Barba FJ, Frandsen H B, Sørensen S, Olsen K, Sørensen J C & Orlien V (2016). The impact of high pressure on glucosinolate profile and myrosinase activity in seedlings from Brussels sprouts. *Innovative Food Science and Emerging Technologies* 38: 342-348.

Wu B, Zhang G, Shuang S, Dong C, Choi MM & Lee AW (2005). A biosensor with myrosinase and glucose oxidase bienzyme system for determination of glucosinolates in seeds of commonly consumed vegetables. *Sensors and Actuators B: Chemical* 106: 700-707

Wu H, Liu XR, Yu DD, Zhang X & Feng JT (2014). Effect of allyl isothiocyanate on ultrastructure and the activities of four enzymes in adult Sitophilus zeamais. *Pesticide Biochemistry and Physiology*109: 12-17.

Wu H, Zhang GA, Zeng S & Lin KC (2009). Extraction of allyl isothiocyanate from horseradish (*Armoracia rusticana*) and its fumigant insecticidal activity on four stored product pests of paddy. *Pest Management Science* 65: 1003-1008.

Xiao D, Srivastava SK, Lew KL, Zeng Y, Hershberger P, Johnson CS, Trump DL & Singh SV (2003). Allyl isothiocyanate, a constituent of cruciferous vegetables, inhibits proliferation of human prostate cancer cells by causing G 2/M arrest and inducing apoptosis. *Carcinogenesis* 24: 891-897.

Ye L & Zhang Y (2001). Total intracellular accumulation levels of dietary isothiocyanates determine their activity in elevation of cellular glutathione and induction of Phase 2 detoxification enzymes. *Carcinogenesis* 22: 1987-1992.

Yuan J M, Stepanov, I, Murphy S E, Wang R, Allen S, Jensen J & Kurzer M S (2016). Clinical trial of 2-phenethyl isothiocyanate as an inhibitor of metabolic activation of a tobacco-specific lung carcinogen in cigarette smokers. *Cancer Prevention Research* 9: 396-405.

Zhang Y (2010). Allyl isothiocyanate as a cancer chemopreventive phytochemical. *Molecular Nutrition and Food Research* 54: 127-135.

Zhang Y & Tang L (2007). Discovery and development of sulforaphane as a cancer chemopreventive phytochemical. *Acta Pharmacologica Sinica* 28: 1343-1354.

Zhang Y, Tang L & Gonzalez V (2003). Selected isothiocyanates rapidly induce growth inhibition of cancer cells. *Molecular Cancer Therapeutics* 2: 1045-1052.

Zhao D, Tang J & Ding X (2007). Analysis of volatile components during potherb mustard (*Brassica juncea*, Coss.) pickle fermentation using SPME–GC-MS. *LWT-Food Science and Technology* 40: 439-447.

Zou Y, Jung LS, Lee SH, Kim S, Cho Y & Ahn J(2013). Enhanced antimicrobial activity of nisin in combination with allyl isothiocyanate against *Listeria monocytogenes, Staphylococcus aureus, Salmonella Typhimurium* and *Shigella boydii*. *International Journal of Food Science and Technology* 48: 324-333.

22

Indian Sarsaparilla, *Hemidesmus indicus*, an Endangered Medicinal Plant of India

Jnanesha AC, Ashish Kumar and Manoj Kumar Singh

Abstract

Medicinal plants are being widely used either as single drug or in combination in health care delivery system. Hemidesmus indicus is a rare and endangered plant of Deccan plateau of India and is a potential source of several active principles of therapeutic value. Its immense medicinal values can bring H. indicus as a royal source of herbal medicine in India. It contains various phytoconstituents belonging to the category glycosides, flavonoids, tannins, sterols and volatile oils. The main pharmacological activity of H. indicus include anti-inflammatory, anti-oxidants, anti cancerous activity, anti acne activity, anti venom activity, antinociceptive, renoprotective, heptoprotective activity, anti arthritis and anti leprotic activity. The root of Hemidesmus indicus was used mainly for the preparation of drink locally called as nannari or sugandhapala. The drink is medicinal, which cools the system, gives good appetite and acts as a blood purifier and roots are used for preparation of chutneys and pickles.

Keywords: *Hemidesmus indicus*, Pharmacological activity, Chemical constituents Conservation

---

Jnanesha AC (✉) and Ashish Kumar

CSIR-Central Institute of Medicinal and Aromatic Plants, Research Centre Boduppal Hyderabad-500092, Telangna, India

Manoj Kumar Singh

Kulbhaskar Ashram Post Graduate College, Prayagraj, Uttar Pradesh-211002, India

✉*Corresponding author(s) email: jnangowda@gmail.com*

Plants for Novel Drug Molecules: Ethnobotany to Ethnopharmacology
Bikarma Singh & Yash Pal Sharma (eds.), (pp. 483-494)

Email: *info@nipabooks.com* Web: *www.nipabooks.com*

1. Introduction

Hemidesmus indicus commonly known as Anantmul or Indian Sarsaparilla belongs to the family of Asclepiadaceae and is synonymously known as *Periploca indica*. *H. indicus* has long been used as a folk medicine and found to be an important ingredient in ayurvedic and unani preparations (Anonymous 1997). It is perennial, diffusely twinning or prostrate semi erect shrub a woody root stock having numerous slender wiry laciferous branches with purplish brown bark. This plant is found throughout India growing under mesophytic to semi dry conditions in the plains and upto an altitude of 600 m. It is quite common in open scrub jungles, hedges, and uncultivated soil. It is found in India, Sri Lanka, Pakistan, Iran, Bangladesh and Molusccas. This plant also found in Upper Gangetic plain, eastwards to Bengal and Sundarbans and from Central provinces to Travancore and South India (Anonymous 2015). *H. indicus* root is sweet, cooling and demulcent. It is used as tonic, diuretic and aphrodisiac. Different species available in the market are *H. Indicus, Decalepis hamiltonii* and *Cryptolepis buchanani* belongs to the family Asclepiadaceae; *Ichnocarpus frutescens* and *Vallaris solanacea* of the family Apocynaceae (Murthy et al. 2006)

Use of this plant against leucorrhoea at Bargarh district in Orissa and Sattordem Village of Goa has been reported (Sen and Behera 2000; Kamat 2001). Antipyretic use of this plant has also been reported (Singh and Kumar 1999). Siddique et al. (2004) have reported the use of *H. indicus* among the local people and herbal practitioners of Barind Tract of Bangladesh against diarrhoea, rheumatism, fever, headache, asthma, eye disease and wounds. Rajasab and Isaq (2004) have reported the use of *H. indicus* among the tribes of North Karnataka. Ayyanar and Ignacimuthu (2005) have reported traditional uses of *H. indicus* among the Kani tribals in Kouthalai of Tirunelveli hills, Tamil Nadu. Uses of *H. indicus* among the Korku tribe of Amravati district of Maharashtra have been reported by Jagtap et al. (2006).

In Southern State, particularly Andhra Pradesh and Telangana State the juice or drink prepared from Hemidesmus indicus root is locally called as nannari or sugandhapala. The drink is medicinal, which cools the system, gives good appetite and acts as a blood purifier and it is also called as poor man's drink and roots are used for preparation of chutneys and pickles. It is a medicine for food indigestion, constipation and gas troubles. The drink nannari is served especially during special occasions to the people in Deccan Plateau region and the roots are sold in important pilgrim centres and local shanties at throw away price. Decoction of leaves of *Hemidesmus indicus* was prescribed by charaka in sallow complexion, loss of voice, cough, menstrual disorders and dysentery whereas entire plant is prescribed for treating asthma, cough, abdominal swelling and aching limbs (Khare 2007).

2. Origin and Distribution

H. indicus is widespread in the Indian Plains and the coast and in the mountains up to a height of 600 m, further beyond not exceeding 1000m. Globally the species is distributed in India, South East Asia, Sri Lanka, Pakistan, Iran, Bangladesh, Afghanistan and Malaysia. In India, it is found in mainly Southern India and Northern India, Sikkim and Peninsular India. This species is globally distributed in Indo-Malaysia. Within India, it is found in moist deciduous forests, scrub jungles, hedges and degraded sites from Upper Gangetic plains of North India to Assam and Peninsular India (Aadhan and Anand 2018).

3. Taxonomical Position and Botanical Description

Hemidesmus indicus is a twining shrub, belonging to the family Asclepiadaceae having chromosome number 2n=22. It is also known as as Sariva false sarsaparilla and Indian sarsaparilla, anantmool is part of the milkweed family which is taxonomically distinct from "true" sarsaparilla (smilax febrifuga, family:smilacaceae) Asides from these names, Anantmool has also been referred to as Sugandhipala (Telugu), Magrabu (Hindi), Ananta(Sanskrit),Nannari (Tamil, Malayalam), Syama (Sanskrit), Namdaberu (Kannada), Sogadaberu (Kannada), and Indian sarsaparilla (English).

Kingdom: Plantae	**Order:** Gentianales
Division: Magnoliophyta	**Family:** Periplocaceae
Sub-Division: Magnoliophytina	**Sub-family:** Asclepiadiaceae
Class: Magnoliopsida	**Genus:** *Hemidesmus*
Sub-class: Magnoliodae	**Species:** *Hemidesmus indicus* (L.) R. Br.

The Indian Sarsaparilla, *H. indicus* locally known as "Anantamul" is an endemic medicinal plant. The Roots of this taxon have been used in folk medicine as well as in ayurvedic and unani preparations. *H. indicus* a perennial twiner, nodes thickened. Leaves are simple, opposite decussate, narrow or broadly oblong, glabrous above, often streaked with white lines. Flowers (about 5 cm. across) are crowded in subsessile cymes. Calyx lobes 5, ovate, acute. Corolla is greenish outside, purple inside with a short tube and spreading lobes. The stem bark is purplish brown. Fruit is a follicle and seeds are with brownish-white hairs. The slender but strong roots of *H. indicus* (diameter less 1 cm) are pleasant to smell and sweetish in taste.

Fig. 1: Plant and Roots of *Hemidesmus indicus*. (Photo Courtesy Mr. Ravi Gollamandala, Andhra Pradesh Urban greening and Beautification Corporation)

4. Chemical Constituents

The roots of *H. indicus* contain hexatriacontane, lupeol, its octacosanoate, β-amyrin, α-amyrin, its acetate and sitosterol. It also contains new coumarino-lignoid-hemidesminine, hemidesmin I and hemidesmin II, six pentacyclic triterpenes including two oleanenes, and three ursenes. The stem contains calogenin acetylcalogenin-3-0-β-D-digitoxopyrannosyl-0-β-D-digitoxopyronsyl-0-β-D-digitoxopyranoside. It also afforded 3-keto-lup-12-en-21 28-olide along with lupanone, 4-methoxy-3-methoxybenzalaldehyde, lupeol-3-β-acetate, hexadecanoic acid, and 3-methoxy-4-5methoxybenzalaldehydglycosides-indicine and hemidine. The leaves contain tannins, flavonoids, hyperoside, rutin and coumarino. Leucodermalignoids such as hemidesminine, hemidesmin I and hemidesmin II are rare group of naturally occurring compounds present in leaves.

Table 1: Molecular formula and molecular weight of chemical constituents of *Hemidesmus indicus*

Name of the Compound	Molecular Formula	Molecular Weight
α-amyrin	$C_{30}H_{50}O$	426
β-amyrin	$C_{30}H_{50}O$	426
Lupeol	$C_{30}H_{50}O$	426
Ursane	$C_{30}H_{52}$	412
Oleanane	$C_{30}H_{52}$	412

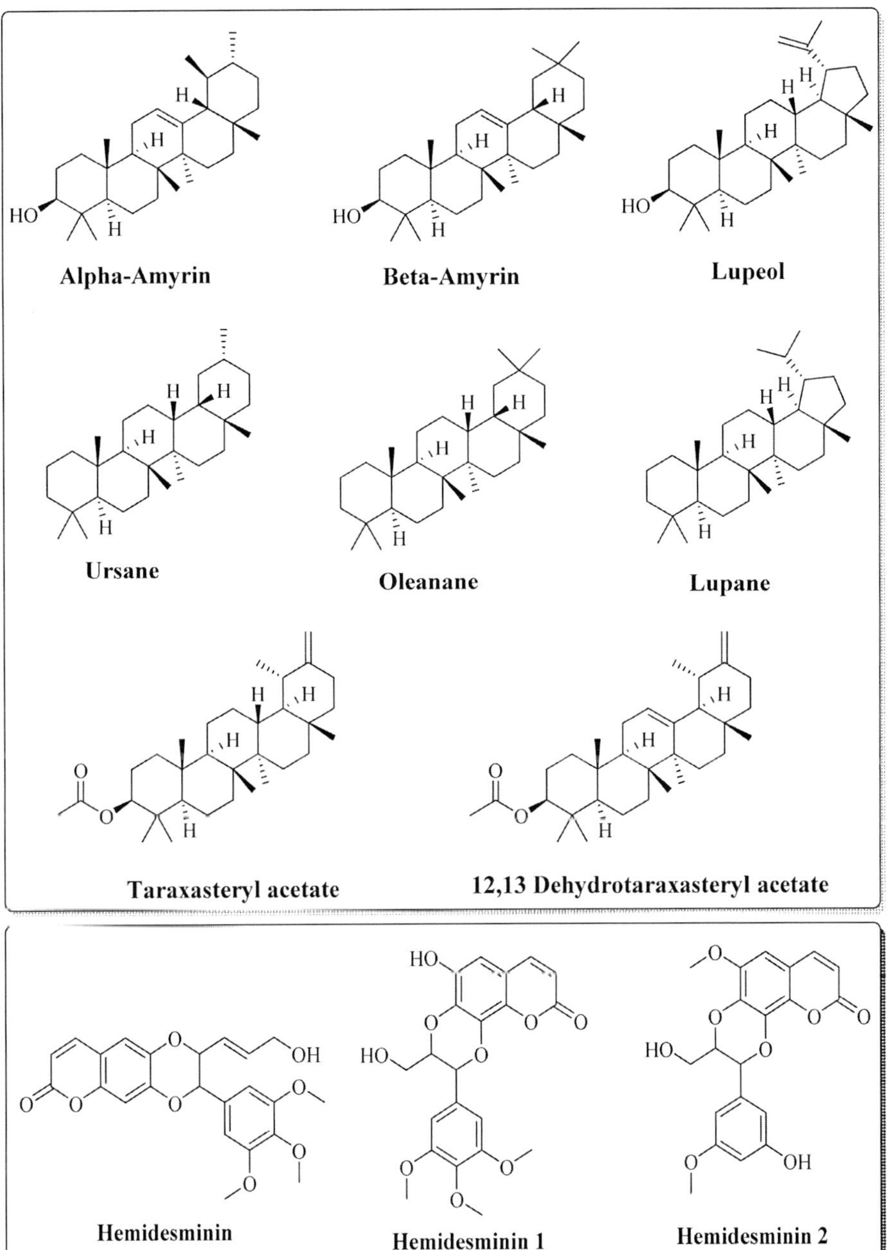

Fig. 1: Chemical structures of constituents of *Hemidesmus indicus*

5. Pharmacological Activity

The pharmacological activities includes antioxidant, analgesic activity, anti arthritic activity, anti cancerous activity, antipyretic activity, hepatoprotective activity, anti-diabetes activity, antimicrobial activity, antileprotic activity, antiacne activity, antithromobotic activity, antihyperlipidaemic activity, antinociceptive activity, wound healing activity, larvicidal activity, anticonvulsant activity, renoprotective activity, anti-ulcer activity, anti-psychotic activity, nootropic activity, antigenotoxic effect and anti-angiogenic activity. Due to the effective usage of this herb in biomedical science, it is essential to optimize the standard protocols for its propagation and enhancement of bioactive molecules (Sethi et al. 2006).

5.1. Antioxidant Activity

The aqueous extract of whole plant of *Hemidesmus indicus* showed significant free radical scavenging activity which shows that the plant extract has potential source of antioxidants and helps to prevent many diseases (Satheesh kumar 2013). Methanolic extract of *Hemidesmus indicus* roots showed a concentration dose dependent inhibition of 1,1-diphenyl-2- picryl hydrazyl (DPPH) radical, superoxide radicals and moderate nitric oxide scavenging activity due to presence of polar components. Ethanolic extract of *Hemidesmus indicus* showed potent antioxidant effect and provided protection against free radical mediated oxidative stress in kidney in ethanol induce nephrotoxicity in rats (Nadana 2007). Administration of *Hemidesmus indicus* extract 500 mg/kg/day for 30 days of experiment significantly reduced the level of serum-urea, uric acid, creatinine and kidney-thiobarbituric acid reacting substances (TBARS), lipid peroxides and conjugated dienes. The extract also increased level of kidney superoxide dismutase (SOD), catalase (CAT), glutathione peroxidises (GPx), and reduced glutathione (GSH). Terpenoidal fraction obtained from successive extraction of *Hemidesmus indicus* roots possesses potent free radical scavenging activity.

5.2. Anticarcinogenic Activity

The root decoction of *H. indicus* showed cytotoxic on HepG2 cells. Treatment of mouse skin with *H. indicus* extract prior to application of cumene hydroxide prevented induction of ornithine decarboxylase activity and DNA synthesis which is considered to be a biochemical marker to evaluate tumor promoting potential of an agent (Thabrew et al. 2005 & Iddamaldeniya *et al.* 2003). The extract showed inhibition of tumor growth in mouse skin and hence can be considered as a potent chemopreventive agent. Crude methanolic extract of roots of *Hemidesmus indicus* showed protective effect against cytotoxicity induced by *Salmonella typhimurium* in human intestinal cell lines. Root Extract of *Hemidesmus indicus* protect microsomal membranes by reducing lipid peroxidation and also protect DNA from radiation induced strand breaks (Paumarthi 2011).

5.3. Anti-inflammatory Activity

The treatment with the hydroalcoholic root extract of *Hemisdesmus indicus* at different doses (100, 200 and 300 mg/kg b.w., p.o.) significantly prevented increase in volume of paw edema and formation of granulation tissue in dose dependent manner and maximal effect was observed at 300 mg/kg b.w which was comparable to phenylbutazone 100 mg/kg b.w., i.p. In carragenan induced paw oedema, methanolic roots extract also exhibited significant reduction in volume between 2 -4 h after treatment (Lakshman et al. 2006). A saponin from the *Hemisdesmus indicus* is found to have anti inflammatory activity against formalin induced edema (Satheesh kumar et al. 2013)

5.4. Hepatoprotective Activity

Methanolic root extract of *Hemidesmus indicus* showed hepatoprotective activity against paracetamol induced liver toxicity and hepatotoxicity in rats and also showed protective against Rifampicin and Isoniazid induced liver toxicity. This activity is might be due to presence of coumarino lignoids *viz.*, hemidesmin-I and hemidesmin-II which has free radical scavenging activity (Mookan et al. 2000). 50% aqueous ethanolic extract of *H. indicus* (400mg/kg, per orally) similar hepatoprotective activity against carbon tetrachloride (CCl4) induced liver damage. These effects were attributed to its free radical scavenging and anti -lipid peroxidative activities. Methanolic extract of roots of *Hemidesmus indicus* showed hepatoprotective effect against carbon tetrachloride (CCl4) (Mohan rao et al. 2005)

5.5. Antileprotic Activity

Delayed in cutaneous hypersensitivity stimulation when aqueous extract of *Hemidesmus indicus* root was orally administered at 2 % in mice infected with *Mycobacterium leprae* (Gupta. 1981)

5.6. Antiacne Activity

The root extract of *Hemidesmus indicus* showed strong inhibitory effect on *Propioni bacterium* acne and *Staphylococcus epidermis*. Minimum inhibitory concentration for *Propioni bacterium* acne and *Staphylococcus epidermis* was found to be 0.051mg/ml and 1.25mg/ml. But high concentrations were required to act as bactericidal agent (Kumar et al. 2008)

5.7. Antinociceptive Activity

Administration of root extract of *Hemidesmus indicus* extract in mice showed dose dependent antinociceptive effect in all the mice models for antinociception and it blocked both the neurogen and inflammatory pain (Verma et al. 2005).

5.8. Renoprotective Activity

The root extract of *Hemidesmus indicus* showed inhibition of Cisplatin induced lipid peroxidation and also ehthanolic extract of *Hemidesmus indicus* roots at different dose levels of 250 and 500 mg/kg showed dose dependent reduction in the elevated blood urea, serum creatinine and increase in the GSH and GST enzyme level in Cisplatin induced renal injury in rats (Kaur et al. 2011).

5.9. Antivenom Activity

The methanol root extract was explored for the first time for neutralization of snake venom (*Vipera russellii*) activity and the extract significantly neutralized the viper venom- induced lethality and hemorrhagic activity in albino rat and mouse Lupeol acetate isolated from the root extract of *Hemidesmus indicus* could significantly neutralize lethality, haemorrhage, defibrinogenation, edema, PLA2 Activity induced by the Daboia russellii venom (Alam et al. 1996).

5.10. Antiarthritis Activity

The *in vitro* study by inhibition of protein denaturation method emphasizes the anti-arthritic effect of *Hemidesmus indicus* root extract to that of the standard drug diclofenac sodium. The anti-arthritic activity may be due to the presence of chemical profile like flavonoids, phenols, polyphenols and steroids. Hydro alcoholic extract and ethyl acetate fraction of *Hemidesmus indicus* showed significantly higher anti-arthritic activity than chloroform and residual fraction (Mehta et al. 2012).

6. Trade and Marketing of *Hemidesmus indicus*

In India, cultivation of *Hemidesmus indicus* is more in Deccan plateau region. The root comes to maturity in about 12-14 months after planting depending upon the soil and climatic conditions. Harvesting may be done in December and January. The harvested roots are washed, dried in shade, and stored in moisture-free packing in cool and dry places. The roots can be easily harvested by digging around the plant leaving the central core of the roots and stem for regeneration. The folk healers generally collect the medicinal herbs from their own localities (Vedavathy. 1997). The forest dwellers collect and sell the roots to traders, contractors or directly to the companies. In more organized areas raw materials is sold to tribal co-operatives. In the USA and Europe Anantamul is receiving renewed attention as an Ayurvedic medicinal and herbal product. It's having wider distribution in the Indian sub-continent, and the supply, mostly from the wild, would have been sufficient to meet the domestic demand for Ayurveda and folk medicine and for preparation of popular beverages like "nannari sherbet" in South India. As the raw material prices increased the profits are largely cornered by the contractors, wholesalers and the marketing companies, with only sketchy accounts of demand and supply to arrive at trade volumes (Austin 2008).

Table 5: *H. indicus* available in pharmaceutical forms in Indian market

S.No.	Pharmaceutical forms in market		
1.	Anantamul capsules	8.	Nannari Milk
2.	Anantamul syrup	9.	Nannari drink
3.	Anantamulvati	10.	Mahani pickle
4.	Anantamul choorn	11.	Sarivakalpa
5.	Anantamul powder	12.	Sarasaparilla extract
6.	Anantamul ghana	13.	SarivaArka
7.	Nannarisherbat	14.	Sarivadibhati

7. Conservation Perspectives

Indian sarsaparilla is threatened with extinction in India due to indiscriminate collection and over-exploitation of natural resources for commercial purposes to meet the requirement of the pharmaceutical industry, coupled with limited cultivation. Collection and conservation of this plant were carried out from south Karnataka and Western Ghats of India. The genetic erosion has affected the species greatly and populations left in India have very poor alkaloid content. It was found to be endangered in Southern Western Ghats of India (Kumar and Jnanesha 2017). It has been categorized as globally endangered. The plant was described as critically endangered in Northeast India. Mainly in rural and tribal areas the people will uproot the plant and collect the root and sold to the local market without giving much attention to the multiplication. Government organization like CSIR-CIMAP and ICAR institute should give prime importance by conserve these species and develop a variety which is suitable to northern region. Forest department and NGO have to create awareness among the rural youths about the importance of these crops. Conservation and cultivation of *Hemidesmus indicus* can be treated as an alternative income generation source for the rural unemployed without hampering their ongoing income generating activities. Compared to other crops, cultivation of *Hemidesmus indicus* requires less attention and expenditure and can be successfully adopted by the cultivators. By doing so, we will not only be able to conserve the precious wealth of medicinal plants but also we will achieve the goal of conserving the rare and endangered species, which are threatened, and at the verge of extinction. In view of this, there is a need in Vitro and in *Vivo* for the conservation of this important threatened plant species.

8. Conclusion

Hemidesmus indicus is commonly found throughout India and widely recognized in Indian system of medicine. Global resurgence in the use of plant-based drug is an opportunity for India to attain self-reliance and boost the export of herbal drug and the naturally derived phytochemicals of medicinal value are the potential source of the new drugs and one such important source is *Hemidesmus indicus*. The demand for *Hemidesmus indicus* plant is escalating phenomenally, nationally and internationally, with the increased realization of their effectiveness in treating various diseases and their applications latent in the Indian society, especially among the forest tribes and village communities and the formulations using this herb have emerged even in the western market especially in the management of cardivasular, renal problems and for chronic skin diseases. The current pharmaceutical knowledge of *Hemidesmus Indicus* has to be potentially tapped in a commercial way for the production of therapeutic products. Urgent need to Develop a varieties suited to northern region and develop a agro technologies and market linkages to farmer in order to extend the area and generate income to the farmer.

Conflict of Interest

Authors have no conflict of interest on this publication.

Acknowledgments

The authors are thanks to the Director, CSIR-Central Institute of Medicinal and Aromatic Plants (CSIR-CIMAP), Lucknow, for facilities.

References

Aadhan K & Anand SP (2018). Ethanomedicinal plants utilized by Pallyars tribe in Sadhuragiri hills, Southern Western Ghats, Tamil Nadu, India. *International Journal of Biological Research* 3(1):7-15.

Alam MI, Auddy B & Gomes A (1996). Viper venom neutralization by Indian medicinal plant (*Hemidesmus indicus* and *Pluchea indica*) root extracts. *Phytotherapy Research*. 10(1):58-61.

Anonymous (1997). The Wealth of India. Raw materials, Vol. III, V and X, CSIR, New Delhi, India.

Anonymous (2005). Quality standards of Indian Medicinal Plants, ICMR, New Delhi; 2:119-128.

Ashish Kumar & AC Jnanesha (2016). Conservation of Rare and Endangered Plant Species for Medicinal Use; *International Journal of Science and Research* 5 (12): 1370-1372.

Ashish Kumar & Jnanesha AC (2017). Genetic diversity and conservation of medicinal plants in Deccan plateau region in India. *Journal of Medicinal Plants Studies* 5(3): 27-30.

Ashish Kumar & Jnanesha AC (2018). Conservation and Uses of Some Medicinal Plants. In. Research Trends in Medicinal Plant Sciences. Eds. Manzoor Hussain. Published by AkiNik Publications, Delhi, India, 385 pages.

Austin A (2008). A review on Indian Sarsaparilla, *Hemidesmus indicus* (L.) R. *Brazilian Journal of Biological Science* 8(1):1-12.

Ayyanar M & Ignacimuthu S (2005). Traditional knowledge of Kani tribals in Kouthalai of Tirunelveli hills, Tamil Nadu, *Indian Journal of Ethnopharmacology* 102: 246-255.

Gupta PN (1981). Antileprotic action of an extract from Anantamul. (*Hemidesmus indicus* R. Br.) *Indian Journal of Leprosy* 1981; 53: 354-359.

Iddamaldeniya SS, Thabrew MI, Wickramasinghe SMDN, Ratnatunge N &Thammitiyagodage MG (2003). Protection against diethylnitrosamine-induced hepatocarcinogenesis by an indigenous medicine comprised of Nigella sativa, *Hemidesmus indicus* and Smilax glabra-a preliminary study. *Journal of Carcinogenesis* (2):1-6.

Jagtap A & Sing NP (1999). Fascicles of Flora of India. Botanical Survey of India, Government of India, India.

Kamat SV (2001). Folk medicines of Sattordem Village of Goa. A note on Ethno botany, USA.

Kaur A, Singh S, Shirwaikar A & Setty MM (2011). Effect of Ethanolic Extract of *Hemidesmus indicus* Roots on Cisplatin Induced Nephrotoxicity in Rats. *Journal of Pharmacy Research* 4(8): 2523- 2525.

Khare CP (2007). Indian Medicinal Plants: An Illustrated Dictionary. Springer.

Kumar G, Jayaveera K, Kumar Ashok C Bharathi T, Umachigi S & Vrushabendra S (2008). Evaluation of antioxidant and antiacne properties of terpenoidal fraction of *Hemidesmus indicus* (Indian sarsaparilla). *The Internet Journal of Aesthetic and Antiaging Medicine* 1 (1).

Lakshman K, Shivaprasad HN, Jaiprakash B & Mohan S (2006). Anti- inflammatory and antipyretic activities of *Hemidesmus indicus* root extract. *African Journal of Traditional complementary and Alternative Medicine* 3(1) : 90 –94.

Mehta A, Sethiya NK, Mehta C & GB Shah (2012). Anti-arthritis activity of roots of *Hemidesmus indicus* R.Br. (Anantmul) in rats. *Asian Pacific Journal of Tropical Medicine*. 130-135.

Mohana Rao GM, Venkateswararao CH, Rawat AKS, Pushpangadan P & Shirwaikar A (2005). Antioxidant and Antihepatotoxic activities of *Hemidesmus indicus* R. Br. *Acta Pharmaceutica Turcica* 47:107– 113.

Mookan P, Rangasamy A & Thiruvengadam D (2000). Protective effect of *Hemidesmus indicus* against rifampicin and isoniazid - induced hepatotoxicity in rats. *Fitoterapia* 71: 55-59.

Murthy CKN, Rajasekaran T, Giridhar P, Raviahankar GA (2006). Antioxidant property of *Hemidesmus indicus* (L.) R. Br. *Indian Journal of Experimental Biology* 44: 832-837.

Nadana S & Namasivayam N (2007). Impact of *Hemidesmus indicus* R.Br. extract on ethanol-mediated oxidative damage in rat kidney. *Redox Report* 12(5): 229-235.

Pasumarthi S, Chimata MK, Chetty CS & Challa S (2011). Screening of phytochemical compounds in selected medicinal plants of Deccan Plateau and their viability effects on $Caco_2$ cells. *Journal of Medicinal Plants Research* 5(32): 6955-6962.

Rajasab AH & Isaq M (2004). Documentation of folk knowledge on edible wild plants of North Karnataka. *Indian Journal of Traditional Knowledge* 3: 419-429.

Satheesh Kumar D, Pooja M, Harika K, Haswitha E, Nagabhushanamma G & Vidyavathi N (2013). In-vitro antioxidant activities, total phenolics and flavonoid contents of whole plant of Hemidesmus indicus (linn.). *Asian Journal of Pharmaceutical and Clinical Research* 6(2):249-251.

Sen SK & Behera LM (2000). Ethno medicinal plants used against leucorrhoea at Bargarh district in Orissa (India). *Neo Botanica* 8:19-22.

Sethi A, Srivastav SS & Srivastav S (2006). Pregnane glycoside from *Hemidesmus indicus*. *Indian Journal of Heterocyclic Chemistry* 16:191-192.

Siddique NA, Bari MA, Naderuzzaman ATM, Khatun N, Rahman MH, Sultana RS, Matin MN, Shahnewaz S & Rahman MM. (2004). Collection of indigenous knowledge and identification of endangered medicninal plants by questionnaire survey in Barind Tract of Bangladesh. *Journal of Biological Sciences* 4: 72-80.

Singh KK & Kumar K (1999). Ethno therapeutics of some medicinal plants used as antipyretic agents among the tribals of *India. Journal of Economic and Taxonomic Botany* 23: 135-141.

Thabrew MI, Mitry RR, Morsy MA & Hughes RD (2005). Cytotoxic effects of a decoction of Nigella sativa, *Hemidesmus indicus* and *Smilax glabra* on human hepatoma HepG2 cells. *Life Sciences* 77(12): 1319-1330.

Vedavathy S, Mrudula, V & Sudhakar (1997). A tribal medicine of Chittoor district, Andhra Pradesh, Herbal Folklore Research centre, Tirupathi, India. pp. 165.

Verma PR, Joharapurkar AA, Chatpalliwar VA & Asnani AJ (2005). Antinociceptive activity of alcoholic extract of *Hemidesmus indicus* R. Br. in mice. *Journal of Ethno pharmacology* 102: 298–301.

23

Exploiting Therapeutic Potential of Kirayat (*Andrographis paniculata*): Identity, Chemical Constituents and Biological Aspects for Future Reference in Drug Discovery

Dipayan Ghosh, Ajit Kumar Shasany and Narendra Kumar

Abstract

Andrographis paniculata (Burm.f) Nees (Family: Acanthaceae) is one of the most popular and valuable medicinal plants with high trade value. It is out mentioned in various systems of medicine since ancient times in Indian, Malay, Chinese and Thai. Andrographis paniculata is native to Southeast Asian countries i.e., India, China, Taiwan and Sri Lanka. In India it is distributed from Uttar Pradesh, West Bengal, Assam to Mizoram and also cultivated throughout the country. The plant being used as Ayurvedic medicine in various diseases. The principal chemical constituents are lactones, diterpenoid and flavonoids. Andrographolide, neoandrographolide are the principle secondary metabolites of Andrographis paniculata with more than 80 compounds reported till now. Modern studies revealed that Andrographis paniculata has an extensive range of pharmacological activities such as hepatoprotective, anti-inflammatory, antimalarial, antimicrobial, antidiarrhoeal, anti-human immunodeficiency virus (HIV), anti-oxidant and anti-pyretic. Therefore, such plants need value addition for product development.

Keywords: *Andrographis paniculata*, Medicinal plant, Chemistry, Pharmacology, Drug

---

Dipayan Ghosh, Ajit Kumar Shasany and Narendra Kumar (✉)

CSIR-Central Institute of Medicinal and Aromatic Plants, Kukrail Picnic Spot Road Lucknow-226015, Uttar Pradesh

✉*Corresponding author email: narendrakumar@cimap.res.in*

Plants for Novel Drug Molecules: Ethnobotany to Ethnopharmacology
Bikarma Singh & Yash Pal Sharma (eds.), (pp. 495-514)

1. Introduction

In the traditional system of medicines, medicinal plants play an important role to cure various ailments and to maintain good health and longevity. In these systems, drugs primarily prepared from medicinal plants are used either as a single drug or as the polyherbal formulations. Medicinal uses of plants are exhaustively reported in the Indian classical texts such as *Rig-Veda* and *Atharva-Veda* (3500-1600 BC). *Charak Samhita* (900 BC) and *Sushruta Samhita* (600 BC), described about 700-800 raw drugs which were used in the form of different preparations for the treatment of several disorders. Since then, the herbal medicines are extensively used by human in their primary healthcare. The World Health Organization (WHO) estimates that more than 80% of the populations in developing countries depending on the conventional system of medicine for their primary health care (Nathiya et al. 2012). However, the herbal medicines are being popular in the developed and developing countries owing to very less or no side effects as compared to modern medicines. The widespread use of plant-based medicines resulted in speedy growth of herbal market. However, India's share in the global herbal market is only 0.5 percent. The minimal export of Indian herbal drugs is due to quality and efficacy issues.

Andrographis paniculata is commonly known as Mahatita or King of Bitters or kalmegh. It is one of the most potential herb possesses high medicinal properties. It is an annual, branched erect, herb, up to 1m height. The plant is mention in traditional systems of medicines all over the world such as Traditional Indian Systems of medicines, Traditional Chinese Medicine (TCM), Traditional Malaya system of Medicine and Traditional Thai Medicine. In Ayurveda, Siddha, Unani and Homeopathy medicines, it is used for the treatment of fever, liver related disorders, intestinal worm infestation, diarrhoea and skin diseases. The aerial parts and root of *Andrographis paniculata*, contains chemical constituents such as diterpene lactones, diterpene glycosides, flavonoids, flavonoid glycosides, quinic acid derivatives and xanthones (Subramanian et al. 2012). *Andrographis paniculata* encompasses four important bioactive secondary metabolites viz., andrographolide, neo-andrographolide, dehydro-andrographolide, iso-andrographolide. These metabolites are mainly responsible for anti-inflammatory, hepatoprotective, astringent, anti-diabetes, anti-diarrhoeal, antimicrobial, anticancer, anti-HIV, antimalarial, immune stimulator activity (Hossain et al. 2014). The crude drug of kalmegh (*Andrographis paniculata*) comprises of the fragmented fresh or dried aerial parts. The aerial part contains leaves, stem, flowers and fruits.

2. Systematic Position, Common Names, Distribution and Taxonomy

2.1. Systematic Classification

- Kingdom: Plantae
- Division: Angiospermae
- Class: Dicotyledonae
- Subclass: Gamopetalae
- Series: Bicarpellatae
- Order: Personales
- Family: Acanthaceae
- Genus: *Andrographis*
- Species: *Andrographis paniculata (Burm.f) Nees.*

Fig. 1: *Andrographis paniculata*

2.2. Common Name

Hindi: Kiryat; English: The King of Bitters; Ayurvedic: Kaalmegha, Bhuunimba; Siddha: Nilavembu; Unani: Kiryaat; Gujarathi: Kariyatu; Kannada: Nelaberu; Malayalam: Kiriyattu; Marathi: Olenkiryata; Oriya: Bhuinimba; Bengali: Kalmegh, Mahatita; Telugu: Nilavembu; Sanskrit: Kalmegha, Bhunimba; Tamil: Nilavembu; Arab: Quasabhuva; Canarese: Kreata; Olikiriyat: Hasada, Kalamegh; Decan: Charayetah; Java: Sadilata; Sadani: Bhuinim; Persian: Naine-havandi; Sinhalese: Hinbinkohomba

2.3. Diversity and Geographical Distribution

The genus *Andrographis* represents about 40 species, distributed throughout the tropical Asian countries (Anonymous, 1948). About 19 species were reported from India, most of them are wild (Anonymous, 1948). *Andrographis paniculata* is native and abundantly distributed to Southeastern Asian countries, i.e., India, Pakistan, Sri Lanka and Indonesia. The species is cultivated in many parts of India, China, Thailand, the East and West Indies, and Mauritius(Subramanian et al. 2012). In India, it is grown throughout the plains and hills, from Uttar Pradesh to Assam, especially in West Bengal and Mizoram. It is also cultivate in the southern part of India. The plant is naturally grown in wastelands, and road-sides. Kalmegh is usually propagates from seeds during the rainy phase of summer season (Kharif) crop in India.

2.4. Morphological and Anatomical Enumeration

2.4.1. Macroscopic Identification

Andrographis paniculata is an erect, annual, branched herb, about 1m height. Leave linear to lanceolate, tip acute or acuminate, margin entire, 3-7cm long, green, glabrous with a short petiole. Stems thin, hard, quadrangular, branched with prominent nodes and internodes, green in colour. Axillary or terminal panicle inflorescence. Flower white with three light-pink spots on the lower lip of corolla. Sepal 5, pubescent, green; petal bilabiate, upper lip two-lobed and lower lip three-lobed, oblong, hairy, whitish in colour; stamen 2, inserted in the corolla throat, ovary 2-celled, superior. Fruit capsule, flattened linear-oblong in shape, 0.3-0.4x2.0-2.5 cm in size, longitudinally furrowed on both the compressed sides. Seeds are small, numerous, sub-quadrate with the rugose surface and yellowish-brown in color.

Fig. 2: Raw drug of *Andrographis paniculata* contain aerial parts

2.4.2 *Microscopic identification*

In the transverse section, leaves show a dorsiventral pattern, prominentepiderm is on both the side of the lamina and triangular midrib. Upper and lower epidermal cells are uniseriate, compactly arranged rectangular with thin cuticle layer. There is a single palisade layer below the upper epidermis that traverse through the midrib. Palisade cells are columnar and densely filled with chloroplasts. Spongy parenchyma cells are 3 to 4 layered, loosely arranged with large intercellular spaces and present below the palisade cells in the lamina region. Lower epidermis sometimes shows glandular trichomes. Midrib possesses an arched vascular bundle in the centre. Radially arranged xylem vessels in rows with about 25–40 μm in diameter and narrow phloem.

In transverse section, stem shows a quadrangular outline with four collenchymatous lobes at the corner. Epidermis single-layered with rectangular cells. The projection of collenchymatous cells forms all four corners with 2–3 layered of cells. Sessile glandular trichomes are present on the epidermis. Cortex is composed of parenchymatous cells; some of these cells possess cystoliths and crystals. A wide vascular bundle is present next to the cortex. Xylem vessels are solitary, small, thin-walled and arranged in radial rows, about 15 μm to 30 μm in diameter. Conspicuous medullary rays are present with vessels rows. In the centre, parenchymatous pith with thin-walled cells of which may occasionally contain acicular crystals.

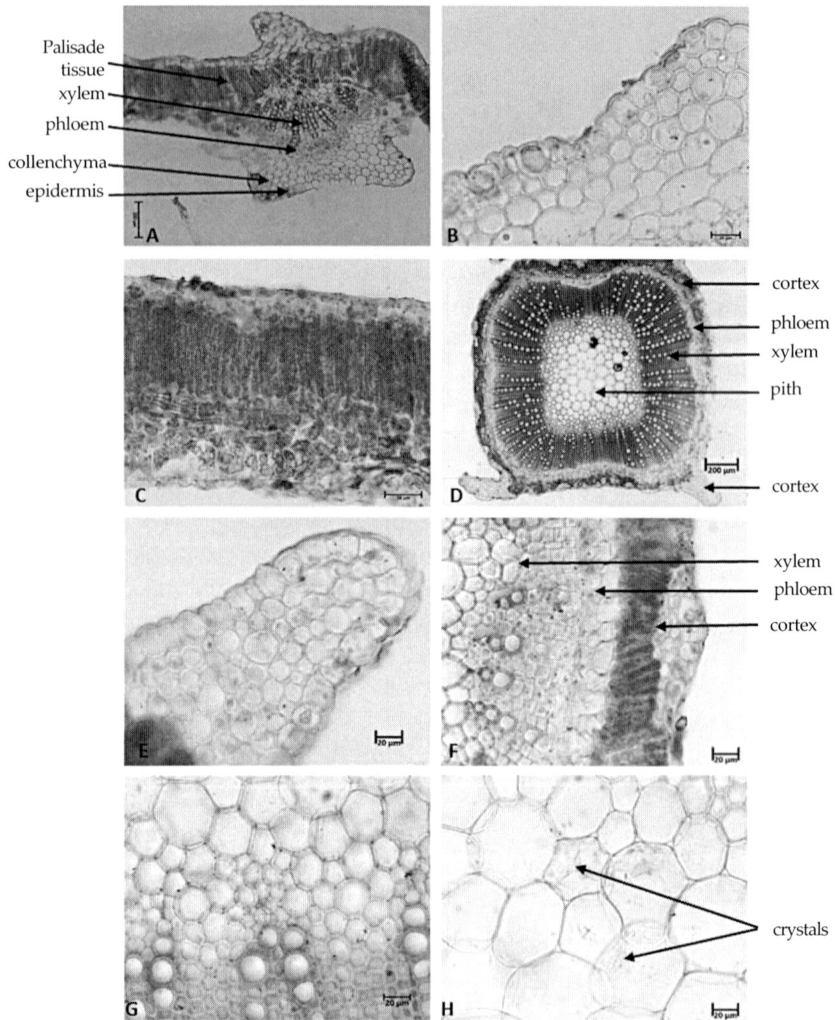

Fig. 3: *Andrographis paniculata.* A-C Transverse section of leaf. A) Midrib; B) Collenchymatous tissue in midrib; C) T.S. of lamina. D-H Transverse section of stem. D) quadrangular stem with lobes; E) collenchymatous tissue in lobes; F) arrangement of cortex and vascular tissues; G) protoxylem and pith region; H) acicular crystals in pith cells.

3. History and Physico-chemical Properties

3.1. Earlier Historical Records

In *Adarsh Nighantu, Andrographis paniculata* mentioned as *Vasadi Varga* with synonyms *Kalpanath, Yavatikta* and *Shankhini*. Since the ancient time it is a well-reputed plant in Indian Systems of Medicines. The plant has been mentioned in Indian classical text, i.e., *Bhavaprakash Nighantu, Charaka*

Samhita, The Indian Meteria Medica for treatment of fever, liver disorders, throat and upper respiratory tract infections, gastrointestinal problems and other contagious and chronic diseases (Nadkarni & Nadkarni 1976; Bapalal 1999). Its therapeutic actions are based on its properties such as *Rasa, Guna, Virya,* and *Vipaka*. According to Ayurvedic Pharmacopeia of India (Anonymous 2011), the properties and actions of *Kalmegh* are listed below

- ***Rasa*** (Taste): Rasa (Taste)—Tikta (Bitter), *Tikta* (Bitter),
- ***Guna*** (Quality)—*Laghu* (Light for digestion), *Ruksha* (Dry),
- ***Virya*** (Potency)—*Sita* (Cold),
- ***Vipaka*** (Metabolic Property)—*Katu* (Transforms into Pungent taste after digestion).
- ***Karma*** (Action)—*Kaphapitta shamaka* (reduces vitiated *Kapha* and *pitta*), *Dipana* (appetizer), *Pachana* (digestive), *Yakrut uttejaka* (stimulates liver), *Jwaraghna* (antipyretic), *Krimighna* (wormicidal), *Raktashodhak* (purifies blood), *Shothahar* (reduces edema), *Svedajanana* (stimulates sweating).

3.2. Physico-chemical Studies

Foreign organic matter: Not more than 1%

Total ash: Not more than 7%

Acid-insoluble ash: Not more than 1.5%

Water-soluble extractive: Not less than 25%

Alcohol-soluble extractive:Not less than 12%

(value in dry weight basis)

3.3 Organoleptic Characters

Odour characteristic; taste intensely bitter; green in colour.

3.4 Adulteration/Substitute

Andrographis echioides is used as an adulterant of *Andrographis paniculata*, where andrographolide is not present (Jani et al. 2017).

4. Medicinal Use, Formulations and Dosages

4.1. Ethnobotanical Usages

Since ancient times, the whole plant of *Andrographis paniculata* has been used ethnobotanically in Southeast Asia, China and India to cure the broad spectrum of health concerns. It is commonly used by the traditional Vaidyas, folk-tribes or folk-communities as a remedy in different ailments, i.e. common cold, fever, malaria, diarrhoea, skin diseases and expulsion of worms (Akbar 2011). The decoction of aerial parts of the herb is recommended in cough, muscular pains, liver disorders, gastric infections, sore throat and upper respiratory tract infections and also given as blood purifier (Akbar 2011). It is also widely used in folklore medicine as an antidote for snake-bite and poisonous scorpion sting (Nazimudeen et al. 1978; Yabesh et al. 2014). The plant is also referred by folk-Vaidyas for the treatment of scabies, skin eruptions, leprosy, boils and gonorrhoea. The leaf decoction of *Andrographis paniculata* mixed with goat or buffalo milk for the treatment of jaundice and enlarged liver by Malayali tribes of Tamil Nadu, India (Alagesaboopathi 2015).

4.2. Formulations and Dosages

It is an essential constituent in several important Ayurvedic formulations. In Ayurveda, *Kalmegha* is used in the form of *Churna Kalpana* (Powder form) 1-3 gms; *Taral Satva* (Liquid extract): 1/2-1 ml; *Swaras* (Juice): 5-10 ml; *Kwath* (Decoction): 20-40 ml (Anonymous 2011).

5. Chemistry Aspects

5.1. Chemical Constituents

Andrographis paniculata contains bioactive chemical constituents, mainly lactones, diterpenoids, diterpene glycosides, quinic acid derivatives, flavonoids and flavonoid glycosides and xanthones class of compounds. From the last few decades, more than fifty diterpene lactones, thirty flavonoids, eight quinic acid derivatives and four xanthones are reported from the aerial part and root of *Andrographis paniculata* (Subramanian et al. 2012). Among them, andrographolide, neoandrographolide, deoxyandrographolide, 14-deoxy-11, 12-didehydroandrographolide are major compounds (Siripong et al. 1992; Chao & Lin 2010). Chakravarti & Chakravarti (1952) revealed that the leaves contain andrographolide and kalmeghin. Further study confirm four lactones andrographolide, deoxyandrographolide, neoandrographolide and 14-deoxy-11, 12-didehydroandrographolide in aerial part (Chan et al. 1968). Additionally some minor compounds are also found in roots and aerial parts contained

5-hydroxy- 7, 20, 60-trimethoxyflavone (flavone), 14-deoxy-15-isopropylidene-11, 12-didehydroandrographolide (23-carbon terpenoid) (Reddy et al. 2003; Rao et al. 2004) and deoxyandrographolide- 19-D-glucoside (diterpene glucoside) (Wiart et al. 2005), 3-deoxy-andrographoside (diterpenoid lactones), 14-deoxy-15-methoxy-andrographolide (Weiming et al. 1982), were isolated from *Andrographis paniculata*. Some minor but important constituent in *Andrographis paniculata* are isoandrographolide,14-epi-andrographolide, 12-epi-14-deoxy-12-methoxyandrographolide,14-deoxy-12-methoxyandrographolide, 14 deoxy -11-hydroxyandrographolide and14-deoxy-12-hydroxyand-rographolide, 6'acetylneoandrographolide and 14-deoxy-11,12-didehydroandrographi-side; and rographolactone (Wang et al. 2009) along with diterpene dimer, bis andrographolides A,B,C,D,andropanolide (labdane type diterpenoid) (Matsuda et al. 1994; Pholphana et al. 2004; Pramanick et al. 2006), 14-deoxy-15-methoxy-andrographolideand 3-deoxy-andrographoside (diterpenoid lactones) (Wang et al. 2017).

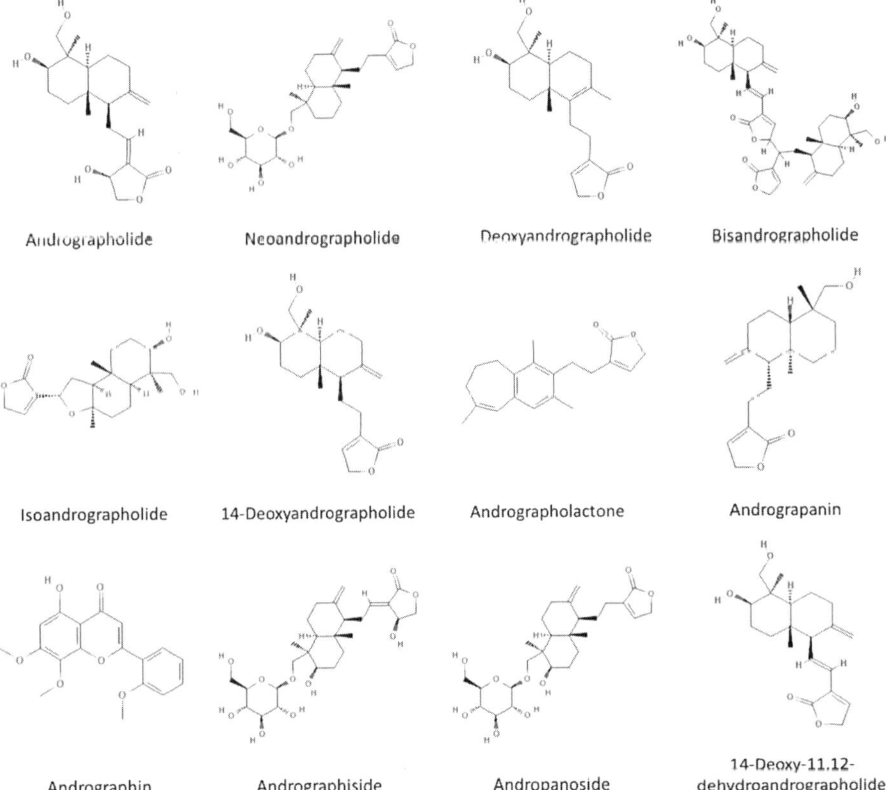

Fig. 3: Major chemical constituents of *Andrographis paniculata*

5.2. Chemical Marker

Andrographolide, Neoandrographolide are the major bioactive markersin the plant. These bioactive markers along with other chemical markers i.e. andrograpanine,14 deoxy-11,12-didehydroandographolide are used as standardized markers for quality evaluation.

6. Biological Properties: Pharmacology Aspects

Andrographis paniculata possesses a broad range of clinical or pharmacological properties, and some of them areenormously beneficial for human health. The biological properties associated with this plant is given in subheads:

6.1. Hepatoprotective

Andrographis paniculata is mainly considered for hepatoprotective and a hepatostimulative activity. Andrographolide, neoandrographolide and andrographiside compared with well-known hepatoprotective agent silymarin, shows protective effects on hepatotoxicity, induced by carbon tetrachloride or tert-butyl-hydroperoxide (tBHP) intoxicationin mice. Andrographolide have lower potential than andrographiside and neoandrographolide (Negi et al. 2008). The aqueous extract also has significant hepatoprotective activity on severe liver damage in Swiss male mice induced by hexachlorocyclohexane (Trivedi & Rawal, 2005). A comparative study between *Andrographis paniculata* and *Swertia chirayita* shows that the ethanolic extract of *Andrographis paniculata* have better hepatoprotective activity (Nagalekshmi et al. 2011)

6.2. Immuno-modulatory

The plant significantly reduces nitric oxide (NO) concentration in avian lymphocyte culture as compared to control with aqueous extract of leaves. There was a marked increase in B cell proliferation in treated cells as compared to control (Thakur et al. 2013). Another study shows that andrographolides containing standardised extract of *Andrographis paniculata* can enhance and modulate immune system during antigen interaction (Naik & Hule 2009).

6.3. Common Cold

The "Kan jang", a tablet made from the dried extract of *Andrographis paniculata,* is uses for the treatment of common colds and upper respiratory tract infections, such as pharyngotonsillitis. Kan Jang shows better activity at doses of 1.2 to 6.0 g/day for four days. It is also studied that Kan Jang can lower the relative risk of catching common cold at the time of weather change (Cacer et al. 1997).

6.4. Anti-inflammatory

Andrographis paniculata has significant inhibitory effects on the production of pro-inflammatory (Nitric oxide, Interleukin-1 beta and interleukin-6), inflammatory (prostaglandin E2 and thromboxane B2) and allergic (leukotriene B4) mediators but no inhibition against histamine release (Chandrasekaran et al. 2010). It also shows inhibitory effect against inflammatory and allergic mediators like reactive oxygen species (ROS) production by phorbol-12-myristate-13-acetate (PMA), as well as N-formyl-methionyl-leucyl-phenylalanine (fMLP) induced rat neutrophils adhesion. Andrographolide prevents reactive oxygen species production and neutrophils adhesion. (Shen et al. 2002, Reyes et al. 2006).

6.5. Anti-diabetes

Extract of aerial parts prepared in ethanol shows potent glucose lowering effect in streptozotocin (STZ)-induced diabetic rats (Zhang et al. 2000). These extract can increase the survival rate, density of endocrine cells, estrous cycle improvement and reduced the 'insulin resistance' phenomenon in streptozotocin (STZ) - induced diabetic rats (Syahrin et al. 2006). In alloxan-induced female diabetic rats, it could restore impaired estrous cycle (Reyes et al. 2006).

6.6. Anti-diarrhoeal

The alcoholic extract of *kalmegh* shows significant antidiarrhoeal activity against enterotoxins of *E. coli*. Four isolated diterpenes from *Andrographis puniculata,* andrographolide, neoandrographolide, andrographiside and deoxyandrographolide. Among them, andrographolide and neoandrographolide represents the superior activity against *E. coli* ST enterotoxin, responsible for epidemics of neonatal diarrhoea (Gupta et al. 1990). Another study displays the alcohol, chloroform, hexane, butanol and aqueous extracts of different parts of *Andrographis paniculata* possesses highly significant antisecretory activity (Thanagkul et al. 1985, Chaichantipyuth et al. 1986, Gupta et al. 1993).

6.7. Antiviral

Ethanolic extract of *Andrographis paniculata* shows anti-herpes activity. Active constituent like andrographolide, neoandrographolide and 14-deoxy-11, 12 didehydroandrographolide possess viricidal activity against herpes simplex virus 1 (HSV-1). Among them, neo-andrographol-ide exhibits the highest rate of antiviral activity (Wiart et al. 2005).

6.8. Antibacterial

The extract of *Andrographis paniculata* has antimicrobial activity against many pathogenic bacterial strains such as (*Escherichia coli* K 12 ROW, *Salmonella typhi* 59, *Shigella sonnei* 2, *Vibrio cholerae* 811, *Vibrio cholerae* 854, *Shigella boydii* 8, *Staphylococcus aureus* ML-59, *Bacillus licheniformis* 10341, *Vibrio alginolyteus, Salmonella typhimurium* NCTC 74 and *Staphylococcus aureus* 29737) (Mishra et al. 2009). Dichloromethan, methanolic and aqueous extract shows significant antibacterial activity against 12 skin disease-causing bacterial strains of seven Gram-positive strains (*Staphylococcus epidermis, Staphylococcus saprophyticus, Staphylococcus aureus, Bacillus anthracis, Streptococcus pyogenes, Enterococcus faecalis* and *Micrococcus luteus*) and five Gram-negative strains (*Proteus mirabilis, Neisseria meningitis, Proteus vulgaris, Klebsiella pneumoniae* and *Pseudomonas aeruginosa*). The minimum inhibitory concentration and minimum bactericidal concentration between 150 to 300 μg/ml and 250 to 400 μg/ml respectively (Singha et al. 2003, Sule et al. 2010, Sule et al. 2011).

6.9. Antimalarial

Chloroform and methanolic extract of *Andrographis paniculata* shows significant antimalarial activity against *Plasmodium berghei* (*in vivo*) and *Plasmodium falciparum* (*in vitro*) parasites. Among them, chloroform extract has better activity than the methanolic extract (Rahman et al. 1999). Andrographolide, neoandrographolide, deoxyandrographolide and andrographiside shows dose-dependent inhibition on *Plasmodium berghei* NK 65 (Misra et al. 1992). The anti-malarial activity of 1,2-dihydroxy-6,8-dimethoxy xanthone were observed on Swiss Albino mice infected with *Plasmodium falciparum* without cytotoxicity (Dua et al. 2004).

6.10. Cardiovascular and Antithrombotic

Andrographolide has a potent cardiovascular and antithrombotic activity (Amroyan et al. 1999). Crude water extract of *Andrographis paniculata*, its semi-purified butanol fraction and aqueous fraction possess cardiovascular activities contribute more to the hypotensive response than a direct action on the heart (Zhang and Tan 1997).

6.11. Anticancer

The methanolic extract of *Andrographis paniculata* (aerial part) shows anticancer activity on mouse myeloid leukaemia cell (Siripong et al. 1992; Talei et al. 2013). Andrographolide inhibits the *in-vitro* proliferation of different tumour cell lines of different types of cancers. The andrographolide has potent cytotoxic activity against KB cells, HT-29 (colon cancer) cells (Kumar et al. 2004) and P388 lymphocytic leukaemia. This chemical compound induces apoptosis in human cancer cells via activation of caspase 8 in the extrinsic death receptor pathway (Zhou et al. 2006). Another study shows that the compound directly arrests at G0/G1 phase of cell-cycle through induction of cell-cycle inhibitory protein p27 and decreased expression of cyclin-dependent kinase 4 (CDK4) (Rajagopal et al. 2003). Besides, andrographolide and some other related compounds also showed potent growth inhibitory activity against M1-tumor cell line (Matsuda et al. 1994).

6.12. Anti-human Immunodeficiency Virus (HIV)

In-vitro study shows that *Andrographis panicuiata* has an inhibitory effect against the human immunodeficiency virus (HIV) (Calabrese 2000). Dehydro-andrographolide succinic acid monoester (DASM) is present in aerial part of the plant. It possesses inhibitory effect to HIV-1 (IIIB) at the average concentrations of 108 pg/ml and the average minimal concentration of 2.0 pg/ml. It has no cytotoxic effect on H9 (cutaneous T lymphocyte cell line) (Chang et al. 1991).

6.13. Antifeedant and Antioviposition

The 14-deoxyandrographolide has a synergistic action in antifeedant properties against the larva of the diamondback moth (DBM) larva *Plutella xylostella* (Lepidoptera: Yponomeutidae) (Hermawan et al. 1993). Furthermore, the antioviposition effect of the acetone extract suppresses feeding activity at the concentrations of 60 ppm, 500 ppm of the 1st, 4th-instar larvae and 125 ppm, of females DBM, respectively (Hermawan et al. 1994).

6.14. Antileishmanial

The capsules prepared from extract of *Andrographis paniculata*are useful in leishmaniasis. Andrographolide and its derivatives have cytostatic activity against *Leishmania donovani* in a hamster model (Sinha et al. 2000).

6.15. Anti-hyperlipidemia

Hyperlipidemia, rise in lipid level in the body, can cause heart attack and other obstruction in the blood channels of the brain (Briel et al. 2009). Andrographolide and neoandrographolide shows a significant hypolipidemic effect without any significant level of liver damage (Yang et al. 2012).

6.16. Antioxidant

Andrographis paniculata has significant anti-oxidant activity on nicotine-induced male Wistar rats. Nicotine can generate an excessive amount of free radicals that causea high level of oxidative stress in the brain. Treatment with andrographolide and aqueous crude extract can invert the changes induced by nicotine (Das et al. 2009). Tert-butanolic extract of *Andrographis paniculata* exhibits high antioxidant activity in an in-vitro assay (Rao and Rathod 2015).

6.17. Analgesic, Antipyretic and Anti-ulcerogenic

Andrographolide, isolated from *Andrographis paniculata* shows significant analgesic activity in acetic acid-induced writhing in mice, antipyretic effect in Brewer's yeast-induced pyrexia in rats and also significant anti-ulcerogenic activity in aspirin-induced ulceration in rats (Madav et al. 1995).

6.18. Anti-dengue

The methanolic extracts of *Andrographispaniculata* possess inhibitoryactivity against dengue virus serotype 1 (DENV-1) infected Vero E6 cells with maximum non-toxic dose in (Tang et al. 2012).

6.19. Immuno-stimulatory

Andrographolide, 14-deoxy-11,12-didehydroandrographolide and 14-deoxyandrographolide shows the immunostimulatory activity by increasing proliferation and interleukin-2 (IL-2) induction in human peripheral blood lymphocytes (HPBLs) (Kumar et al. 2004).

6.20. Anti-snakebite

Andrographis paniculata has potent activity against the snake venom neutralization (Martz 1992, Yabesh et al. 2014, Silambarasan and Ayyanar 2015). Plant extract effectively inhibited the toxic effect of snake venoms and apply externally at the place of snake-bite (Samy et al. 2008, Upasani et al. 2017). Ethanolic extract of the aerial parts is given intraperitoneally (25 g/kg body weight) to cobra venom poisoned mice resulting in the occurrence of respiratory failure and death. The same extract induce contractions in guinea-pig ileum at concentrations of 2 mg/ml (Nazimudeen et al. 1978).

6.21. Anti-scorpion-sting

The extract of *Andrographis paniculata* has potent anti-scorpion sting activity (Yabesh et al. 2014).

6.22. Nematicidal Activity

Andrographolide of *Andrographis paniculata,* shows nematicidal activity (Bandhopadhyayet al. 1986).

7. Contraindications

7.1. Antifertility Effect

Dry powdered extract of *Andrographis paniculata,* administrate orally to pregnant female rats, at 200, 600, and 2000 mg per kg body weight as a dose. The result indicate that *Andrographis paniculata* do not have any effect on pregnancy, level of progesterone secretionand thus, it cannot induce abortion (Panossian et al. 1999).

7.2. Reproductive Toxicity in Male Rats

Andrographis paniculata has antispermatogenic and antiandrogenic activity. The dose of 20 mg dry leaf extract per day administrate orally to male albino rats for 60 days, resulted in reproductive toxicity in male rat like the interruption of spermatogenesis, damage in Leydig cells and degenerative changes in the seminiferous tubules, epididymis, ventral prostate and seminal vesicle (Akbarsha et al. 1990). Another study shows the dried extract of *Andrographis paniculata* is give in male Sprague Dawley rats, at a dose level of 20, 200 and 100 mg/kg for 60 days resulting inno testicular toxicity was found (Burgo et al. 1997).

7.3. Reduction in Metabolic Activity

Andrographolide might reduce the metabolic activity of intestinal enzyme cytochrome P450 3A4 (CYP3A4). It is recommended that the andrographolide given to patients as dose-dependent manner of 0.3–0.6 gm/day,in clinically (Qiu et al. 2012).

7.4. Effect on Sleeping Time

The dose-dependent effect of andrographolide injection significantly shows the reduction of sleeping time within 90 min. The sleeping time was 61.33 min, at a dose of 20 mg/100 kg body weight, which was significantly lower ($p < 0.001$) than the control value 76.33 min (Tripathi and Tripathi 1991).

7.5. Choleratic Action

Andrographolide exhibite an intense choleretic action and induce in increasing bile flow when given intraperitoneally to albino rats (Tripathi and Tripathi 1991).

8. Conclusion

This chapter contains information about identity, chemical constituents and pharmacological activities. There are several isolated compounds reported from various part of *Andrographis paniculata*, but pharmacological action is lacking in more than half of the compounds and have not reported yet. More focus on Research and development, especially on conservation purpose, due to it overexploitation some species have become rare or endangered. Hence, the large scale cultivation and awareness about the plant is needed.

Conflict of Interest

Authors declare no conflict of interest.

Acknowledgement

Authors are thankful to the Director, CSIR-CIMAP, Lucknow for facilities and encouragement. The financial support from the Council of Scientific and Industrial Research (CSIR), New Delhi, India as phytopharmaceutical mission project (HCP-0010) is gratefully acknowledged.

References

Akbar S (2011). *Andrographis paniculata*: a review of pharmacological activities and clinical effects. *Alternative Medicine Review* 16(1) 66–77.

Akbarsha MA, Mannivannan B, Shahul Hamid K & Vijayan B (1990). Antifertility effect of *Andrographis paniculata* (Nees) in male albino rats. *Indian Journal of Experimental Biology* 28:421-426.

Alagesaboopathi C (2015). Medicinal plants used for the treatment of liver diseases by malayali tribes in Shevaroy hills, Salem district, Tamilnadu, India. *World Journal of Pharmaceutical Research* 4:816-826.

Amroyan E, Gabrielian E, Panossian A, Wikman G & Wagner H (1999). Inhibitory effect of andrographolide from *Andrographis paniculata* on PAF-induced platelet aggregation. *Phytomedicine* 6(1):27–31.

Anonymous (1948). The Wealth of India: Raw Materials. I. Publications and Information Directorate, C.S.I.R., New Delhi.

Anonymous (2011). API- Ayurvedic Pharmacopeia of India, Part I, Volume VIII, First Edition, Ministry of H & FW, Government of India, New Delhi.

Bandhopadhyay K, Datta SK & Sukul NC (1986). A spectrophotometric estimation of andrographolide and examination of nematicidal activity of the crude extract of *Andrographis paniculata*. *Indian Drugs* 23:510-512.

Bapalal GV (1999). Nighantu Adarsh Chaukhambha Bharati Academy Varanasi, 2nd edition, Vol 1.

Briel M, Ferreira-Gonzalez I &You JJ (2009). Association between change in high density lipoprotein cholesterol and cardiovascular disease morbidity and mortality: systematic review and metaregression analysis. *British Medical Journal* 338: 692–692.

Burgos RA, Caballero EE, Sfinchez NS, Schroeder RA, Wikman GK & Hancke JL (1997) Testicular toxicity assessment of *Andrographis paniculata* dried extract in rats. *Journal of Ethnopharmacology* 58: 219–224.

Cacer DD, Hancke JL, Burgos RA & Wickman JK (1997). Prevention of common cold with *Andrographis paniculata* Nees. Dried extract. A pilot double blind trial. *Phytomedicine* 4:101-104.

Calabrese C, Berman SH & Babish JG (2000). A phase I trial of andrographolide in HIV positive patients and normal volunteers. *Phytotherapy Research* 14: 333-338.

Chaichantipyuth C & Thanagkul B (1986). *Andrographis paniculata* Nees as antidiarrhoeal and antidysentery drug in Thailand. *Asian Journal of Pharmacy* 6(Suppl.): 59–60.

Chakravarti D & Chakravarti RM (1952). Andrographolide 1. *Journal of Chemical Society* 5: 1657-1700.

Chan WR, Taylor DR, Willis CR & Fehlhaber HW (1968). The structure of neoandrographolide a diterpene glucoside of *Andrographis paniculata* Ness. *Tetrahedron letters* 46: 4803-4806.

Chandrasekaran CV, Gupta A& Agarwal A (2010). Effect of an extract of *Andrographis paniculata* leaves on inflammatory and allergic mediators *in vitro*. *Journal of Ethnopharmacology* 129:203–207.

ang RS, Ding L, Qing CG & Choa PQ (1991). Dehydroandrographolide succinic acid monoester as an inhibitor against human immunodeficiency virus43225 (HIV). *Society for Experimental Biology and Medicine* 197: 59–66.

Chao WW & Lin BF (2010). Isolation and identification of bioactive compounds in *Andrographis paniculata* (Chuanxinlian). *Chinese Medicine* 5(17):1-15.

Das ES, Gautam N, Dey SK, Maiti T & Roy S (2009). Oxidative stress in the brain of nicotine-induced toxicity: protective role of *A. paniculata* Nees. and vitamin. *Applied Physiology, Nutrition, and Metabolism* 34:124–135.

Dua, VK, Ojha VP, Roy R, Joshi BC, Valecha N, Devi CU, et al. (2004). Anti-malarial activity of some xanthones isolated from the roots of *Andrographis paniculata*. Journal of Ethnopharmacology 95(2-3): 247–251.

Gupta S, Ahmad MC, Yadava JNS, Srivastava V & Tandon JS (1990). Antidiarrhoeal activity of diterpenes of *Andrographis paniculata* Nees. (Kal-Megh) against Escherichia coli enterotoxin *in vivo* Models. *Journal of Pharmaceutical Biology* 28:273-283.

Gupta S, Yadava JNS & Tandon JS (1993). Anti-secretory (Antidiarrhoeal) Activity of Indian Medicinal Plants Against *Escherichia coli* Enterotoxin-Induced Secretion in Rabbit and Guinea Pig Ileal Loop Models. *International Journal of Pharmacognosy* 31(3):198-204.

Hermawan W, Kajiyama S, Tsukuda R, Fujisaki K, Kobayashi A & Nakasuji F (1994). Antifeedant and antioviposition activities of the fractions of extract from a tropical plant, *Andrographis paniculata* (Acanthaceae), against the diamondback moth, Plutella xylostella (Lepidoptera: Yponomeutidae). *Applied Entomology and Zoology* 29:533–538.

Hermawan WR, Tsukuda K, Fujisaki A, Kobayashi & Nakasuji F (1993). Influence of crude extracts from a tropical plant, *Andrographis paniculata* (Acanthaceae), on suppression of feeding by the diamondback moth, *Plutella xylostella* (Lepidoptera: Yponomeutidae) and oviposition by the azuki bean weevil, *Callosobruchus chinensis* (Coleoptera: Bruchidae). *Applied Entomology and Zoology* 28(2): 251–254.

Hossain MS, Urbi Z, Sule A, Rahman KMH (2014). *Andrographis paniculata* (Burm. f.) Wall. ex Nees: a review of ethnobotany, phytochemistry, and pharmacology, Scientific World Journal 2014:274905.http://dx.doi.org/10.1155/2014/274905

JaniS, Harishal C, Bhupesh P (2017). Detailed comparative micromorphological and micrometric evaluation of *Andrographis paniculata* and its adultarant (*A. echioides*). *World Journal of Pharmaceutical Research* 6: 825.

Kumar RA, Sridevi K, Kumar NV, Nanduri S & Rajagopal S (2004). Anticancer and immunostimulatory compounds from *Andrographis paniculata, Journal of Ethnopharmacology* 92:291–295.

Madav S, Tripathi HC, Tandan SK & Mishra S.K. 1995. Analgesic, Antipyretic and Antiulcerogenic Effects of Andrographolide. Indian Journal of Pharmaceutical Sciences 57(3): 121- 125.

Matsuda T, Kuroyanagi M, Sugiyama S, Umehara K, Ueno A & Nishi K (1994). Cell differentiation-inducing diterpenes from *Andrographis paniculata* Nees. *Chemical and Pharmaceutical Bulletin* 42(6): 1216-25.

Mishra US, Mishra A, Kumari R, Murthy PN & Naik BS (2009). Antibacterial activity of ethanol extract of *Andrographis paniculata. Indian Journal of Pharmaceutical Sciences* 71(4): 436–438.

Misra P, Pal NL, Guru PY, Katiyar JC, Srivastava V & Tandon JS (1992). Antimalarial activity of *Andrographis paniculata* (Kalmegh) against *Plasmodium berghei* NK 65 in *Mastomys natalensis, International Journal of Pharmacognosy* 30(4): 263–274.

Nadkarni KM & Nadkarni AK (1976). Indian Materia Medica. Bombay, Popular Prakashan.

Nagalekshmi R, Menon A, Chandrasekharan DK, Nair CKK (2011). Hepatoprotective activity of *Andrographis Paniculata* and *Swertia Chirayita. Food and Chemical Toxicology* 49: 3367–3373.

Naik SR & Hule A (2009). Evaluation of immunomodulatory activity of an extract of andrographolides from *Andographis paniculata. Planta Medica* 75:785–791

Nazimudeen SK, Ramaswamy S & Kameswaran L(1978). Effect of *Andrographis paniculata* on snake venom induced death and its mechanism. *Indian Journal of Pharmacological Science* 40: 132–133.

Negi AS, Kumar JK, Luqman S, Shanker K, Gupta MM & Khanuja SPS (2008). Recent Advances in Plant Hepatoprotectives: A Chemical and Biological Profile of Some Important Leads. *Medicinal Research Reviews* 28(5): 746-772.

Panossian A. et al. (1999). Effect of *Andrographis paniculata* extract on progesterone in blood plasma of pregnant rats. *Phytomedicine.* 6:157–161.

Pholphana N, Rangkadilok N, Thongnest S, Ruchirawat S, Ruchirawat M & Satayavivad J (2004). Determination and variation of three active diterpenoids in *Andrographis paniculata* (Burm. f) Nees. *Phytochemical Analysis* 15: 365-371.

Pramanick S, Banerjee S, Achari B, Das B, Sen AK, Mukhopadhyay S, Neuman A & Prangé T (2006). Andropanolide and isoandrographolide, minor diterpenoids from *Andrographis paniculata*: Structure and X-ray crystallographic analysis. *Journal of Natural Products* 69: 403-405.

Qiu F, Hou XL,Takahashi K, Chen LX, Azuma J & KangN (2012). Andrographolide inhibits the expression and metabolic activity of cytochrome P450 3A4 in the modified Caco-2 cells. *Journal of Ethnopharmacology* 141:709–713.

Rahman NN, Furuta T, Kojima S, Takane K & Mohd MA (1999). Antimalarial activity of extracts of Malaysian medicinal plants. *Journal of Ethnopharmacology* 64: 249-254.

Rajagopal S, kumar RA, Deevi DS, Satyanarayana C & Rajagopalan R (2003). Andrographolide, a potential cancer therapeutic agent isolated from *Andrographis paniculata. Journal of Experimental Therapeutics and Oncology* 3: 147–158,

Rao PR & Rathod VK (2015). Rapid extraction of andrographolide from *Andrographis paniculata* Nees by three phase partitioning and determination of its antioxidant activity. *Biocatalysis and Agricultural Biotechnology* 4(4): 586-593.

Rao YK, Vimalamma G, Rao CV, Tzeng YM (2004). Flavonoids and andrographolides from *Andrographis paniculata. Phytochemstry* 65:2317–2321.

Reddy MK, Reddy MVB, Gunasekar D, Murthy MM, Caux C & Bodo B (2003). A flavone and an unusual 23-carbon terpenoid from *Andrographis paniculata. Phytochemistry* 62: 1271–1275.

Reyes BAS, Bautista ND, Tanquilut NC, Anunciado RV, Leung AB, Sanchez GC, Magtoto RL, Castronuevo P, Tsukamura H & Maeda KI (2006). Anti-diabetic potentials of *Momordica charantia* and *Andrographis paniculata* and their effects on estrous cyclicity of alloxan-induced diabetic rats. *Journal of Ethnopharmacology*105:196–200

Shen YC, Chen CF & Chiou WF (2002). Andrographolide prevent oxygen radical production by human neutrophils: possible mechanism(s) involved in its anti-inflammatory effect. *British Journal of Pharmacology* 135: 399-406.

Singha PK, Roy S & Dey S (2003). Antimicrobial activity of *Andrographis paniculata. Fitoterapia* 74(7-8): 692–694.

Sinha J, Mukhopadhyay S, Das N & Basu MK (2000). Targeting of liposomal andrographolide to *L. donovani* infected macrophages *in vivo.* Drug *Delivery* 7: 209-213.

Siripong P, Kongkathip B, Preechanukool K, Picha P, Tunsuwan K & Taylor WC (1992). Cytotoxic diterpenoid constituents from *Andrographis paniculata* leaves. *Journal of the Science Society of Thailand* 18 (4): 187-194.

Subramanian R, Asmawi MZ & Sadikun A (2012). A bitter plant with a sweet future? A comprehensive review of an oriental medicinal plant:*Andrographis paniculata. Phytochemistry Reviews* 11: 39–75.

Sule A, Ahmed QU, Samah OA & Omar M (2010). Screening for antibacterial activity of *Andrographis paniculata* used in Malaysian folkloric medicine: a possible alternative for the treatment of skin infections. *Ethnobotanical Leaflets* 4: 445–456.

Sule A, Ahmed QU, Samah OA & Omar MN (2011). Bacteriostatic and bactericidal activities of *Andrographis paniculata* extracts on skin disease causing pathogenic bacteria. *Journal of Medicinal Plants Research* 5(1): 7–14.

Syahrin A, Amrah S, Chan K, Lim B, Hasenan N, Hasnan J & Mohsin S (2006). Effect of spray dried ethanolic extract of *Andrographis paniculata* (Burms F.) Nees on streptozotocin induced Diabetic Rats. *International Journal of Diabetes in Developing Countries* 26: 163-168.

Talei D, Valdiani A, Maziah M, Sagineedu SR & Saad MS (2013). Analysis of the Anticancer Phytochemicals in *Andrographis paniculata* Nees under Salinity Stress. BioMed Research International 11: http://dx.doi.org/10.1155/2013/319047

Tang LIC, Ling APK, Koh RY, Chye SM & Voon KGL (2012). Screening of anti-dengue activity in methanolic extracts of medicinal plants. *BMC Complementary and Alternative Medicine.* 12:3

Thakur AV, Ambwani TK & Ambwani S (2013). Immunomodulatory Potential of *Andrographis paniculata* in Chicken Lymphocytes Ulture System. *Journal of Immunology and Immunopathology* 15:108-110.

Thamlikitkul V, Theerapong S, Boonroj P, Ekpalakorn W, Taechaiya S, Orn Chom-jan T, Pradipasena S, Timsard S, Dechatiwongse T, Chantrakul c, Punkrut W, Boontaeng N, Petcharoen S, Riewpaiboon W, Riewpaiboon A& Tenambergen E (1991). Efficacy of *Andrographis paniculata* Ness for Pharyngotonsillitis in Adults. *Journal of the Medical Association of Thailand* 74: 437-442.

Thanagkul B & Chaichantipayut C (1985). Double-blind study of *Andrographis paniculata* Nees and tetracycline in acute diarrhoea and bacillary dysentery. *Ramathibodi Medical Journal* 8: 57–61.

Tripathi GS, Tripathi YB (1991). Choleretic Action of Andrographolide Obtained From *Andrographis Paniculata* in Rats. *Phytotherapy Research* 5:176-178.

Trivedi N & Rawal UM (2005). Hepatoprotective and toxicological evaluation of *Andrographis paniculata* on severe liver damage. *Indian Journal of Pharmacology* 32:288-293.

Wang GC, WangY, Ian D, Williams H, Sung HY, Zhang XQ, Zhang DM, Jiang RW, Yao XS, Ye WC (2009). Andrographolactone, a unique diterpene from *Andrographis paniculata*. *Tetrahedron Letters* 50:4824–4826.

Wang GY, Wen T, Liu FF, Tian H, Fan C, Huang X, Ye W&Wang Y (2017). Two new diterpenoid lactones isolated from *Andrographis paniculata*. *Chinese Journal of Natural Medicines* 15(6): 0458- 0462.

Weiming C& Xiaotian L (1982). Deoxyandrographolide-19â-D-Glucoside from the Leaves of *Andrographis paniculata*. *Journal of Medicinal plant Research* 45:245-246.

Wiart C, Kumar K, Yusof MY, Hamimah H, Fauzi ZM & Sulaiman M (2005). Anti-viral properties of ent-labdene diterpenes of *Andrographis paniculata* Nees. *Phytotherapy Research* 19: 1069-1070

Yabesh JEM, Prabhu S, Vijayakumar S (2014). An ethnobotanical study of medicinal plants used by traditional healers in silent valley of Kerala, India. *Journal of Ethnopharmacology* 154: 774–789

Yang T, Shi H, Wang Z & Wang C (2013). Hypolipidemic Effects of Andrographolide and Neoandrographolide in Mice and Rats. *Phytotherapy Research* 27: 618-623

Zhang CY & Tan BKH (1997). Mechanisms of cardiovascular activity of *Andrographis paniculata* in the anaesthetized rat. *Journal of Ethnopharmacology* 56:97-101.

Zhang XF & Tan BK (2000). Antihyperglycaemic and anti-oxidant properties of *Andrographis paniculata* in normaland diabetic rats. *Clinical and Experimental Pharmacology and Physiology* 27:358–363.

Zhou J, Zhang S, Ong CN & Shen HM (2006). Critical role of pro-apoptotic Bcl-2 family members in andrographolide-induced apoptosis in human cancer cells. *Biochemical Pharmacology* 72: 132-144.

24

Diversity, Medicinal Value and Associated Biological Activities of Lianas Growing in Western Himalaya Used for Human Health Care

Uma Bharti and Bikarma Singh

Abstract

Lianas are slightly lesser known angiosperms, grows typically in the form of climbing vines mostly in tropical and subtropical forests. Such plants are characterized by scandent thick stems, woody stem in nature and vary in shape and size. They depend on supports in the form of trees, fence and usually climb upward towards the sunlight as they needed for their survival. Keeping the importance of lianas in forests and household, this study was conducted to document the liana angiosperms growing in Western Himalaya along with their medicinal potential for human and animals. This study reveals a total 64 lianas species belonging to 24 families recorded from Jammu and Kashmir, Ladakh, Uttarakhand, Himachal Pradesh and Punjab. This investigation also provided with medicinal potential of these species, which is used as ethnomedicine by local people living in the interior regions of hills and mountains of Himalaya. In this communication, we highlight lianas as an example of an underestimated forest element that, on the one hand, is a key component in and around forests. This study also helps in conservation of endangered liana species in Himalaya.

Keywords: Lianas, Folklore knowledge, Himalaya, Conservation

Uma Bharti

Department of Botany, University of Jammu, Jammu-180006, Jammu and Kashmir, India

Bikarma Singh(✉)

Botanical Garden, CSIR-National Botanical Research Institute, Lucknow-226001 Uttar Pradesh, India

✉Corresponding author email: *drbikarma.singh@nbri.res.in*

Plants for Novel Drug Molecules: Ethnobotany to Ethnopharmacology
Bikarma Singh & Yash Pal Sharma (eds.), (pp. 515-524)

Email: *info@nipabooks.com* Web: *www.nipabooks.com*

1. Introduction

Himalayan tropical and subtropical forests are considered as an important source of biodiversity and species richness represents the wide variety of plant and animal species that have adapted to the unique habitats (Hilje et al. 2017). Lianas are a structural group of plants that have received growing attention for their role in community structure and ecosystem functioning and are considered as a key component (Sánchez-Azofeifa et al. 2006). These group of angiosperms are an example of an underestimated forest element that, on the one hand, is a key component in and around forests and, on the other, receives very little attention (Bongers et al. 2002). The lack of knowledge about lianas is not common in the general public, but also applies to forest managers, who often advocate extensive liana cutting throughout the forest (Appanah and Putz 1984, Parren and Bongers 2001). In terms of ecology, lianas have a very high canopy versus stem ratio, which results in a higher proportion of photosynthetic biomass than is present in most woody plants. Lianas always remain rooted to the ground throughout their lives and often have special adaptations to attach themselves to their host and climb into the forest canopy. These adaptations include stem twining, clasping tendrils arising from stem, leaf and branch modifications, thorns and spines that attach the liana to its host, downward-pointing adhesive hairs, and adhesive adventitious roots. Therefore, eventually we can say that these vines lose their ability to support their weight and subsequently rely on external supports, such as trees, to aid their ascent to the forest canopy (Gerwing 2006). Most of these plant types found in any tropical forest, but some strategies appear to be better than others at colonizing different succession stages of a forest (Dewalt et al. 2000), or at reaching up to several heights in the forests, leading to vertical stratification of lianas in the forest.

The origin of the high diversity of liana throughout tropical and subtropical forests is probably a result of the repeated independent evolution of the climbing habit. Nearly 60% of all dicotyledonous plant orders have at least one representative climber (Heywood 1993). This repeated but independent evolution could explain why lianas have so many different yet homologous adaptations, such as the varied adaptations for climbing mentioned above. Although lianas have a wide range of pollination syndromes, seed sizes and seed dispersal mechanisms, overall, the relative abundance of these characteristics differs from that of other plant groups. For example, lianas tend to have fairly small, wind-dispersed seeds compared with trees or shrubs (Gentry 1991). Fruits and floral nectar are the main direct food resources birds obtain from lianas; however, only 25% of liana species in the Neotropics have fleshy fruits, thus, birds are not very dependent on lianas (Muller-Landau et al. 2005). The animals that pollinate lianas tend to differ from that of other plant groups, with large bees and beetles disproportionately well represented (Gentry 1991). Although more, detailed studies on the pollination biology of lianas are needed, the variation in liana taxa, species diversity, climbing strategy, seed size and dispersal mechanisms suggested that there are many ways to be a liana. Lianas compose a substantial part of the plant community in forests around the world, and these have the competitive advantage of growing during times of drought due to an efficient vascular structure and resource allocation; as a result, liana abundance is often higher in tropical dry forests than other tropical evergreen forests (Schnitzer 2005). Lianas constitute 25% of the woody species in lowland tropical moist and wet forests; in some forests, such as those on the rim of the Amazon basin, liana diversity can be as high as 44% of the woody species (Perez-Salicrup et al. 2001, Bongers et al. 2002). At a broader scale, the mean abundance, diversity and partly also taxonomic composition of lianas in lowland tropical forests are similar among the continents (Africa, America, Asia), although mean liana abundance appears to be somewhat higher in Africa (Gentry 1991). Cucurbitaceae, Asclepidiaceae, Rubiaceae, Leguminosae, Celastraceae and Apocynaceae are dominant families constituting liana species.

When we talk about ethnobotany which deals with the collection of valuable medicinal plants by a group of people and describes their different uses (Safa et al. 2012). Hence, identification of useful medicinal plants is an excellent policy to understand their properties by indigenous inhabitants. Western Himalaya are encompassed with varied climate, geographical regions, different types of mountains, plains, hills, rivers and lakes, which is considered as a center for accessing valuable species. Therefore, this mountainous belts has been selected for studying the liana diversity.

2. Study Area

Geographically, the Himalayan mountain range straddles a transition zone between Palearctic and Indo-Malayan realms. Species from both are represented in the hotspot. In addition, geological, climatic and altitudinal variations in the region, as well as topographic complexity, contribute to the biological diversity of the mountains along east-west and north-south axes. Western Himalaya regions extend southeast approximately 560 km from river Indus in northwest to river Sutlej in southeast. The upper Indus region separate them from the Karakoram mountain to the north. The part of State Punjab, called Punjab Himalayas constitutes the westernmost section of the vast Himalaya mountain range. It lies mainly in the disputed Kashmir region of the northern Indian subcontinent, including portions administered by Pakistan, the northwestern part is coming under jurisdictions of Himachal Pradesh state, India. The forest types vary from typical Himalayan tropical to alpine meadows, including subtropical and temperate forests. Of the estimated 10,000 species of plants in the Himalaya Hotspot, about 3,160 are endemic. The largest family of flowering plants in the hotspot is the Orchidaceae, with 750 species. Cushion plants have been recorded at more than 6,100 meters, while a high-altitude scree plant in the mustard family, *Ermania himalayensis,* was found at 6,300 meters on the slopes of Mt. Kamet in the northwestern Himalayas (Source: https://www.cepf.net/our-work/biodiversity-hotspots/himalaya/ species).

3. Liana Diversity and Medicinal Applications

A total of 64 liana species belonging to 24 families have been recorded from the study area (Table 1). These have been studied for ethnomedicinal uses, edible, fodder items. In case of medicinal plants, information on fever, stomach-ache, diarrhea, dysentery, rheumatic pains, gout, cut, headache, spermatorrhoea etc. were observed dominant in the study area. Some plant have antiviral, antibacterial, insecticidal, analgesic and anti-inflammatory properties. As far as plant parts are concerned, roots of 11 species, leaves of 24 species, fruits of 14 species, whole plants of 4 species, seeds of 10 species, exclusively used by the tribes residing in valleys and on hills of Himalaya mountains. Yineger et al. (2008) and Estomba et al. (2006) also reported that medicinal plant knowledge and use increases with age when the community had suffered an important erosion of ethnomedicinal plant knowledge. In the present study, the illiterates were having more knowledge of medicinal plants than the literates and a significant negative correlation was found between the number of medicinal plant reports and education level. This may be attributed to the fact that the literate people are more likely to be exposed to modernization (Shankar et al. 2001, Balami 2004, Hoare 2007, Emmanuel and Didier 2011).

Table 1: Medicinal values and associated biological functions of lianas growing in Western Himalaya

Botanical name	Family	Occurrence	Part Used	Mode of Administration	Biological function
Abrus precatorius L.	Fabaceae	W	Se	Indirect through decoction	Used in treatment of diabetes and for abortion of foetus.
Anisomeles indica (L.) Kuntze	Lamiaceae	W	Fr	Indirect through decoction	Taken as nutrient supplement helping in body building tissues.
Antigonon leptopus Hook. & Arn.	Polygonaceae	P	R	Indirect through decoction	Taken in curing stomach problems.
Argyreia capitiformis (Poir.) Ooststr	Convolvulaceae	W	Se	Indirect through decoction	Taken to enhance sexual weakness and in the treatment of animal bites.
Argyreia nervosa (Burm.f.) Bojer	Convolvulaceae	W	Se	Taken through milk	Taken to get relief from body pain, and in curing cold and cough.
Aristolochia bracteolata Lam.	Aristolochiaceae	W	Fr	Taken as vegetable	Taken as energy giving food.
Aristolochia indica L.	Aristolochiaceae	W	Fr	Fruit powder	Taken to cure fever and snake bite.
Aristolochia ringens Vahl	Aristolochiaceae	W		Taken through milk	Used in treatment of rheumatism and scorpion bite.
Basella alba L.	Basellaceae	C	Lv	Taken as cooked vegetable	Taken to cure skin disorders.
Bauhinia vahlii Wight & Arn.	Caesalpiniaceae	W	Se	Taken through milk	Taken along with milk as substitute of cashew nut while preparing sweet dish.
Cajanus scarabaeoides (L.) Thouars	Fabaceae	C	WP	Taken as cooked vegetable	Taken to get relief from body pain and swellings.
Cassytha filiformis L.	Lauraceae	W	WP	Taken as cooked vegetable	Taken in treatment of whooping cough.
Cayratia pedata (Lam.) Gagnep.	Vitaceae	W	Lv	Taken as cooked vegetable	Tender leaves cooked as vegetable.

Contd.

Celastrus paniculatus Willd.	Celastraceae	W	Se	Indirect through decoction	Taken as blood purifier.
Cissampelos pareira L.	Menispermaceae	W	AeP	Taken as cooked vegetable	Taken as anti-spasmodic and anti-dysenteric plant.
Cissus quadrangularis L.	Vitaceae	P	WP	Direct as paste	Applied to treat bone fractures.
Clematis grata Wall.	Ranunculaceae	W	Lv	Indirect through decoction	Taken in treatment of jaundice.
Clitoria ternatea L.	Fabaceae	C	R	Indirect through decoction	Taken to get relief from headache, migraine, fever, malaria and snakebite.
Coccinia grandis (L.) Voigt	Cucurbitaceae	W	WP	Taken as cooked vegetable	Taken as nutrient supplement vegetables.
Cocculus hirsutus (L.) W.Theob.	Menispermaceae	W	Lv	Indirect through decoction	Taken in treatment of spermatorrhoea.
Combretum roxburghii Spreng.	Combretaceae	P	Lv	Direct as paste	Applied to cure skin infections.
Convolvulus arvensis L	Convolvulaceae	W	Lv	Direct as paste	Applied to cure skin infections.
Cryptolepis dubia (Burm.f.) M.R.Almeida	Apocynaceae	W	WP	Indirect through decoction	Used in treatment of rheumatism, and also taken as blood purifier.
Dioscorea bulbifera L.	Dioscoreaceae	W	Tb	Taken as cooked vegetable	Used in treatment of rheumatism.
Dioscorea hispida Dennst.	Dioscoreaceae	W	Tb	Taken as cooked vegetable	Taken as energy giving food.
Dioscorea oppositifolia L.	Dioscoreaceae	W	Tb	Taken as cooked vegetable	Taken as energy giving food.
Dioscorea pentaphylla L.	Dioscoreaceae	W	Tb	Taken as cooked vegetable	Given to patient suffering from spleen disorders.
Dioscorea wallichii Hook.f.	Dioscoreaceae	W	Tb	Indirect through decoction	Taken to expel intestinal worms.
Diplocyclos palmatus (L.) C.Jeffrey	Cucurbitaceae	W	Se	Direct as paste	Taken to increase chances of conception.
Erycibe expansa Wall. ex G.Don	Convolvulaceae	W	Fl	Indirect through decoction	Used in treatment of madness.

Contd.

Gymnema sylvestre (Retz.) Schult.	Asclepiadaceae	C	Lv	Indirect through decoction	Used in treatment of diabetes.
Hemidesmus indicus (L.) R.Br.	Periploceeae	W	R	Indirect through decoction	Used as blood purifier to treat eczema.
Holostemma annulare (Roxb.) K.Schum.	Apocynaceae	W	R	Direct as paste	Used to treat urine infection and chronic constipation.
Ichnocarpus frutescens (L.) W.T.Aiton	Apocynaceae	W	AeP	Indirect through decoction	Taken as tonic.
Indigofera cassioides DC.	Fabaceae	W	Fl	Indirect through decoction	Used to treat throat wound and blisters.
Ipomoea aquatica Forssk.	Convolvulaceae	W	AeP	Indirect through decoction	Taken to keep brain cool and for sound sleeping.
Ipomoea hederifiolia L.	Convolvulaceae	W	Fr	Indirect through decoction	Taken to cure chronic constipation.
Ipomoea pes-tigridis L.	Convolvulaceae	W	Lv	Direct as paste	Used to treat acne.
Jasminum arborescencs Roxb.	Oleaceae	W	Fl	Indirect through decoction	Used in treatment of leucorrhoea.
Jasminum grandiflorum L.	Oleaceae	P	Fr	Indirect through decoction	Used in treatment of diarrhea in children.
Luffa acutangula (L.) Roxb.	Cucurbitaceae	C	Fr	Indirect through decoction	Used to treat piles.
Merremia emarginata (Burm. f.) Hallier f.	Convolvulaceae	W	Lv	Indirect through decoction	Taken in treatment of earache, headache and stomach disorders.
Momordica charantia L.	Cucurbitaceae	C	Lv	Direct as paste	Used in controlling diabetics.
Momordica dioica Roxb. ex Willd.	Cucurbitaceae	C	R	Indirect through decoction	Used in treatment of bowel infections and urinary problems.
Mucuna pruriens (L.) DC.	Fabaceae	W	Se	Indirect through decoction	Used in treatment of weakness, cough, and intestinal worms.
Olax scandens Roxb.	Olacaceae	F	Fr	Taken direct as tonic	Taken to enhance memory power.
Operculina turpethum (L.) Silva Manso	Convolvulaceae	W	R	Taken as powder along with water or milk	Used in treatment of chronic constipation.

Contd.

Passiflora foetida L.	Passifloraceae	P	R	Direct as paste	Used in treatment of bone fracture.
Pergularia daemia (Forssk.) Chiov.	Asclepidiaceae	W	Ltx	Indirect through decoction	Used as tonic to increase milk secretion.
Phaseolus lunatus L.	Fabaceae	C	Lv	Taken as powder	Used in treatment of rheumatism.
Piper betle L.	Piperaceae	C	Lv	Direct	Used in treatment of eye disorders.
Piper longum L.	Piperaceae	C	Fr	Directly taken as fruit	Used in treatment of fever, cough, cold and nose running.
Piper nigrum L.	Piperaceae	W	Fr	Taken as powder	Used in treatment of leucorrhea.
Piper retrofractum Vahl	Piperaceae	W	Fr	Indirect through decoction	Used to get relief from cough and cold.
Pueraria tuberosa (Willd.) DC.	Fabaceae	W	Tb	Indirect through decoction	Used in treatment of fever, nervous system, hot stomach and rheumatism.
Scindapsus officinalis (Roxb.) Schott	Araceae	W	Fl	Direct as paste	Used in treatment of rheumatism.
Smilax lanceifolia Roxb.	Smilacaceae	W	R	Direct as paste	Used to cure gynecological disorders and amenorrhea.
Solena amplexicaulis (Lam.) Gandhi	Solanaceae	W	Fr	Indirect through decoction	Used as tonic.
Symphorema polyandrum Wight	Lamiaceae	W	Se	Direct as paste	Used in treatment of snake bite.
Tinospora sinensis (Lour.) Merr.	Menispermaceae	W	St	Indirect through decoction	Used in treatment of malaria fever and rheumatism.
Tragia involucrata L.	Euphorbiaceae	W	R	Direct as paste	Used on wound after detoxification against asthma.
Trichosanthes tricuspidata Lour	Cucurbitaceae	C	F	Indirect through decoction	Used to remove filthy mucous from nose.
Ventilago denticulate Willd	Rhamnaceae	W	Lv	Indirect through decoction	Used as nutrient supplement.
Ventilago maderaspatana Gaertn	Rhamnaceae	W	Fl	Direct	Used to cure eye infection.

Note-Parts Used: Lv=Leaves, Fl=Flower, R=Root, AeP=Aerial parts, St=Stem, Se=Seed, Fr=Fruits, Tb=Tuber, Ltx=Latex, WP: Whole plant; Occurrence: C=Cultivated, P=Planted, W=Wild

4. Conclusion

The reported 64 lianas growing in Western Himalayas were ethobotanically important as herbal medicine. The recent reports indicated that indigenous knowledge are getting depleted at a faster rate due to less exposure of medicinal plants in the younger generation and also due to shift christianity. Plants are important to humans and therefore, it becomes necessary to collect and document precious knowledge from the tribal and remote areas on plants before complete depletion. There is need to increase awareness among the tribal communities for sustainable use of plant wealth and their conservation. The plants with folklore knowledge should be chemically analyzed so that the biologically active chemical constituents from them can be identified and use for the development of new drugs. Nutraceutical analysis of wild edible vegetables and fruits should also be encouraged for search of new or less known potential edible plants to supplement food requirements of the over growing global population.

Conflict of Interest

The authors have no conflict of interest.

References

Appanah S & Putz FE (1984). Climber abundance in virgin dipterocarp forest and the effect of pre-felling timber cutting on logging damage. *The Malaysian Forester* 47: 335-342.

Balami NP (2004). Ethnomedicinal use of plants among the newer community of constraints and opportunities. The Rainforest Foundation, UK.

Bongers F, Schnitzer SA & Traore D (2002). The importance of lianas and consequences for forest management in West Africa. *Bioterre* 59-70.

DeWalt DA, Berkman ND, Sheridan S, Lohr KN & Pignone MP (2004) Literacy and health outcomes: A systematic review of the literature. *Journal of General Internal Medicine* 19: 1228-1239.

Emmanuel MM & Didier DS (2011). Medicinal plant knowledge of ethnic groups in Douala town, Cameroon. *American Journal of Food and Nutrition* 1(4): 178–184.

Estomba D, Ladio A & Lozada M (2006). Medicinal wild plant knowledge and gathering patterns in Mapuche community from Northwestern Patagonia. *Journal of Ethnopharmacology* 103(1): 109-119.

Gentry AH (1991). The distribution and evolution of climbing plants, pp. 3-49. In: Mooney HA & Putz FE (eds) The biology of vines. Cambridge University Press, Cambridge.

Gerwing JJ (2006). The influence of reproductive traits on liana abundance 10 years after conventional and reduced-impacts logging in the eastern Brazilian Amazon. *Forest Ecology and Management* 221: 83-90

Heywood V (2000). Management and sustainability of the resource base for medicinal plants. In: Honnef S & Melisch R (eds) Medicinal utilization of wild species: challenge for man and nature in the new millennium. WWF Germany/TRAFFIC Europe-Germany EXPO Hannover, Germany.

Hilje B, Stack S & Sanchez-Azofeifa A (2017). Lianas Abundance is Positively Related with the Avian Acoustic Community in Tropical Dry Forests. *Forests* 8(9): 311; https://doi.org/10.3390/f8090311

Hoare AL (2007). The Use of Non Timber Forest Products in the Congo-basin of Dolakha district, Nepal. *Journal of Ethnopharmacology* 86: 81-96.

Muller-Landau HC & Hardesty BD (2005). Seed dispersal of woody plants in tropical forests: Concepts, examples and future directions, pp. 267-309. In: Burslem D, Pinard M & Hartley S (eds) Biotic Interactions in the Tropics: Their Role in the Maintenance of Species Diversity. Cambridge University Press, Cambridge, UK.

Parren MPE & Bongers F (2001). Does climber cuttingreduce felling damage in southern Cameroon? *Forest Ecology and Management* 141: 175-188.

Pérez-Salicrup DR (2001). Effect of liana cutting on tree regeneration in a liana forest in Amazonian. *Bolivia Ecology* 82: 389-396.

Safa O, Soltanipoor MA, Rastegar S, Kazemi M, Nourbakhsh Dehkordi Kh & Ghannadi A (2012). An ýethnobotanical survey on Hormozgan province, Iran. *AJP* 3(1): 64-81.

Sánchez-Azofeifa GA & Castro-Esau K (2006). Canopy observations on the hyperspectral properties of a community of tropical dry forest lianas and their host trees. *International Journal of Remote Sensing* 27: 2101-2109.

Schnitzer SA (2005). A mechanistic explanation for global patterns of liana abundance and distribution. *American Nature* 166: 262-276.

Shankar U, Lama SD & Bawa KS (2001). Ecology and economics of domestication of non timber forest products: an illustration of Broomgram in Darjeeling. Himalaya. *Journal of Tropical Forestry Science* 13 (1): 171-191.

Shrestha PM & Dhillion SS (2003). Medicinal plant diversity and use in the highlands of Dolakha district, Nepal. *Journal of Ethnopharmacology* 86(1): 81-96.

Yineger H, Yewhalaw D & Teketay D (2008). Ethnomedicinal plant knowledge and practice of the Oromo ethnic group in southwestern Ethiopia. *Journal of Ethnobiology and Ethnomedicine* 4(11); https://ethnobiomed. biomedcentral. com/ articles/10.1186/1746-4269-4-11

25

Ethnomedical Applications and Phytochemistry of *Boerhaavia diffusa*

Savita Sharma and Sharada Mallubhotla

Abstract

Medicinal plants in recent years attracted much attention of researchers and become the basis of modern pharmaceuticals because of their active biologically compounds. Boerhavia diffusa commonly known as Punarnava belongs to family Nyctaginaceae is a widely distributed herb that has been naturalized in many areas of the world. It is utilized as a green leafy vegetable in many Asian and African countries besides possessing anticancer, antiestrogenic, immunomodulatory and antiamoebic activities due to the presence of valuable phytochemical compounds. Medicinal value of this plant is because of presence of specific kind of phytoconstituents like alkaloids, flavonoids, triterpenoids, lipids, steroids, lignins, proteins, boeravinone A-F, glycoproteins and carbohydrates which are present in huge quantity in this herb. In view of its therapeutic importance, it is widely used in Ayurvedic, Siddha, Homoeopathy, Unani, Modern and Tribal medicinal system. Whole plant sometime its parts are utilized for curing different disorders by tribal people in India. It is a well known ethnomedicinal plant species mostly found in subtropical and tropical regions of the world.

Keywords: Ethnomedicinal, Nyctaginaceae, Phytoconstituents, *Boerhaavia diffusa*

Savita Sharma and Sharada Mallubhotla (✉)

School of Biotechnology, Faculty of Sciences, Shri Mata Vaishno Devi University, Katra-182320 Jammu and Kashmir, India

✉*Corresponding author: sharda.p@smvdu.ac.in*

Plants for Novel Drug Molecules: Ethnobotany to Ethnopharmacology
Bikarma Singh & Yash Pal Sharma (eds.), (pp. 525-539)

Email: *info@nipabooks.com* Web: *www.nipabooks.com*

1. Introduction

Recent studies on alternative drug development strategies have lead to the discovery of numerous medicinally important molecules of plant origin. These therapeutic agents can be revered irreplaceable in treating different types of illnesses, thereby, contributing to betterment of human health and wellbeing. In developing countries, plants have been used in systematic ways, and are the basis of traditional medicinal systems (Hunt 2000). Inspite of the fact that modern synthetic medicines exist along with traditional medicinal system, however, plant based drugs have still maintained their prevalence for social, historical and economical reasons (Adenubi et al. 2016). Earlier some plant species such as *Catharanthus roseus, Lantana camara, Artemisia annua, Taxus* and *Bacopa* species. were considered as harmful but recent research on pharmacological properties of these species, confirms that they are medicinally useful herbs containing valuable phytochemicals (Dar et al. 2017). This system of medicine is globally accepted in China, India, Sri Lanka, Japan, Pakistan and Thailand. In China around 40 % of total drugs utilized are derived from plant products. A vast majority of medicinal plants around the world are not exploited for their medicinal values, so they have bright future and could be strongly used in present and future studies (Singh 2015). In Indian sub-continent, about 1,700 species out of 7,500 species of plants are considered as medicinal plants and they are mostly used in the Ayurvedic system of medicines for their therapeutic properties by different sections of people (Kala 2006). But before utilizing these plant species for medicinal purposes, full characterization of them must be done to check out their efficiency for production of medicinal products (Saxena 2001). Researcher and pharmaceutical industries are highly interested in them as they have a great potential to treat numerous diseases. Studies on natural product chemistry revealed that most of the new drugs are plant based but the discovery of these useful products is time consuming and difficult (Newman & Cragg 2007).

Beside health care, they also play an important role in Indian economy and further help poor people to sustain their livelihood. India is ranked second globally for the export of these valuable plant species after China. For trade, the plant parts are mostly harvested from their wild habitats. Studies on medicinal plants, continuously lead to the extraction of various phytochemicals or phytoconstituents, which are mainly responsible for the medicinal values of these herbs and protect them from microbial infection and herbivores (Nweze et al. 2004, Amor et al. 2009). The practice of utilizing biologically active agents for treatment is solely based on preexisting traditional knowledge wherein plant products are formulated to produce different types of effective drugs (Rajavel et al. 2012). World

Health Organization (WHO) promotes utilization of these herbal drugs in national health care system because of their easy availability, low cost, lesser side effects and renewable nature.

2. Plant Secondary Metabolites

Plant secondary metabolites have gained much attention in the present era as a result of their positive benefits towards health care and may turn out to be invaluable in treatment of various health disorders like malignancy, diabetes and cardiovascular illnesses. Moreover, they are considered as suitable natural resources for pharmaceuticals, nutraceuticals and phytomedicines (Crozier et al. 2006). Secondary metabolites are complex and have vast structural diversity among different plant species (Dewick 2002). Besides their protective nature, secondary metabolites also help in pollination and seed dispersal (Oksman & Inze 2004). Hydroxylation, glycosylation and methylation pathways are mostly responsible for the synthesis of bioactive compounds from primary metabolites and under stress conditions plants metabolism shifts towards the production of more secondary metabolites for their defense purposes (Lacy & Kennedy 2004).They have a wide range of activity, like they act as antibiotics and restrain microbial growth or have antioxidant activities. But it is difficult to designate a single function to a metabolite as some of them are multifunctional in nature. For instance, salicylic acid act as a signal molecule in both pollination as well as in plant pathogen attack (Heinrich et al. 2004). These secondary metabolites are broadly divided into three main categories namely terpenes, alkaloids and phenolics on the basis of their chemical composition.

2.1. Alkaloids

Alkaloids are nitrogen containing basic alkali like organic compounds mostly found in plant species. Around 3,000 varied types of alkaloids have been reported from more than 4,000 plant species (Raffauf 1996). Plants belonging to family Papaveracea, Ranunculaceae, Maryllidaceae and Solanaceae, etc. are the prominent sources of these specific metabolites. Medicinally they act as narcotics, sedatives, relaxants and stimulants (Agbafor et al. 2011). It has been reported that the higher concentration of alkaloids in *Tamarindus indica* resulted into its antibacterial activity against *Pseudomonas aeruginosa, Staphylococcus aureus* and *Escherichia coli* (Abukakar et al. 2008).

2.2. Phenolics

Phenolics are most widely spread secondary metabolites within the plant kingdom and identified by an aromatic ring which is attached with either one or more hydroxyl groups (Mandal et al. 2010). They are well known for their flavour and taste inducing qualities in fruits and vegetables and are required for plant growth and reproduction (Tomas & Espin 2001). Phenolic compounds vary from simple, low weight catechols to complex lignins and tannins that protect plants from pathogens, cold conditions and harmful UV radiations (Harborne 2000). Their biosynthesis takes place by means of shikimate, mevalonate or polyketide pathways.

2.3. Terpenes

Terpenes comprise the biggest class of bioactive compounds with more than 30,000 familiar structures and are important constituents of essential oils. Most of them are hydrocarbon but some terpene derivatives like aldehyde, ketones or alcohol are not hydrocarbons and are termed as terpenoids. They are identified by number of isoprene unit (five carbon) they contain (Fig. 1). Terpene hydrocarbons have a molecular formula $(C_5H_8)_n$ and n is the number of isoprene units they contain. Essential oils are for the most part comprised of monoterpenes or sesquiterpenes (Carson & Hammer 2010, Savoia 2012). Extracts of various plant parts (bark, stem, leaves, root, etc.) of *Aspilia mossambicensis* and *Taenia asiatica* are reported to have antibacterial activity due to the presence of their terpenoids (Munyendo et al. 2011).

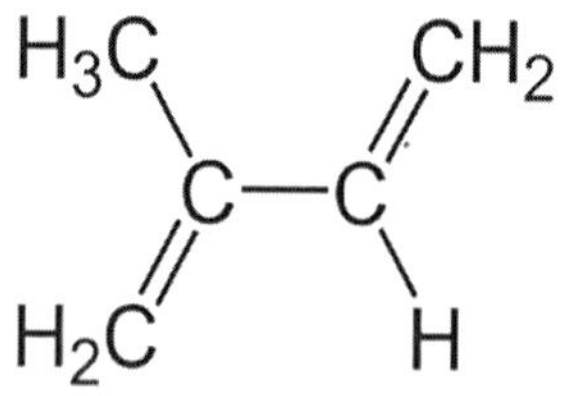

Fig. 1: Isoprene unit structure

3. Flavonoids

Flavonoids are polyphenolic structures, containing two aromatic rings (six carbon each) which are linked by a three carbon heterocyclic ring structure. Epidermal layer of leaf, seeds and skin of fruits are the richest source of these metabolites and they protect plants against disease causing organisms and harmful radiations. Flavonoids are also well known for their antitumor, antiinflammatory and antioxidant activities (Strack & Wray 1992). They are more effective antioxidants as compared to carotenoids, vitamin C and vitamin E, which protect plants against oxidative stress of free radicals. Their antioxidant potential depends upon the number of hydroxyl groups present in them (Dai & Mumper 2010). Flavonoids such as kaempferol, quercetin, and rutin, isolated from *Zingiber officinale* are reported to have a potential in suppressing human breast cancer cell line growth (Ghasemzadeh & Jaafar 2011). Depending upon the chemical nature and substituent position on rings; flavonoids are further sub classified into flavonols, flavan-3-ols, flavones, anthocyanidins, isoflavones and flavanones. Other flavonoids which are presented in minute quantities involve, chalcones, coumarins, aurones, dihydrochalcones and flavan-3,4-diols (Crozier et al. 2006).

4. Non-flavonoid Compounds

These are the phenolic compounds which contribute specific characteristics in wine grapes. Stilbenes and hydroxycinnamates are two important classes of these compounds primarily found in wine grapes and are associated with their antioxidant properties of it. Pharmacologically, non-flavonoid polyphenols are well known for their antiinflammatory properties.

5. Targeted plant *Boerhaavia diffusa* L.

Boerhaavia diffusa L. most popularly known as Red Hogweed or Spreading Hogweed or Tarvine in English, Itsit in Dogri and Punjabi, Raktapunarnava in Hindi, Satodi in Gujarati, Punarnava in Sanskrit, Marathi and Telugu, Mukaratte- Kirai in Tamil and belongs to family Nyctaginaceae which is commonly known as four O' clock family (Sina & Deniz 2019) (Fig. 2). In privilege of Dutch physician Hermannn Boerhaave, this plant was named as *Boerhaavia diffusa* and most popularly known as punarnava or rejuvenator because during summers the plant top dries up and regenerates again in the rainy seasons (Kumar & Dora 2012).

Fig. 2: *Boerhaavia diffusa* **(A)** Natural habitat, **(B)** Aerial part, **(C)** Flowering twig.

5.1. Diversity and Distribution

Genus *Boerhaavia* consists of 40 species and in India it is represented by six species namely *B. diffusa, B. repens, B. erecta, B. chinensis, B. rubicunda* and *B. rependa* (Nayak & Thirunavoukkarasu 2016). Plant belongs to kingdom- Plantae, Division - Angiosperm, Class - Eudicotyledons, Order -Caryophyllales, Family -Nyctaginaceae, Genus - *Boerhaavia* and species - *diffusa*. It prefers to grow in subtropical, tropical and warmer regions of the world and is widely distributed in Africa, America, Australia, Brazil, Ceylon, China, Egypt, India, Iran, Pakistan, Sri Lanka, Sudan and USA. In India it is mostly found in the state of Gujarat, Bihar, Assam, Kerala, Madhya Pradesh, Maharashtra, Punjab, Odisha, Tamil Nadu, Rajasthan and Uttar Pradesh. *Boerhaavia* is a perennial, diffuse herb which grows upward or prostrate in agricultural fields, ditches, wasteland, roadsides, grasslands, residential areas, marshy places and fallow lands (Chopra 1969).

5.2. Botanical Enumeration

Boerhaavia diffusa is a medicinal herb of up to 15 m length and more (Figure 2). Roots are strong, rod-shaped, tuberous, firm with thicked root stock. Stems are cylindrical, woody or succulent, hairy, ascending or prostrate, swallow at the nodes and purplish green. Leaves are in unequal pairs, simple, ovate-cordiform or orbicular ovate, thick, fleshy, hairy, short petioled, white underneath and glabrous above; round at the base with smooth and flat margins. Flowers are pink, small, hermaphrodite, pedunculate, glandular, 5 - ribbed, 4mm long, 4-10 together in terminal or axillary with pink tubular perianth, 3 stamens, peltate stigma and 5 bracts. Flowering took place during April-June. Fruit is small, viscid, curved, 5- ribbed, club or oval shaped anthocarp, hairy and fruiting is throughout the year (Thakur et al. 1989, Patil & Patil 2012).

5.3. Ethnobotanical Uses

Young leaves and shoot parts of *Boerhaavia* are cooked and eaten as vegetable by Purulia tribe in West Bengal and in Assam as it is considered as rich source of minerals, proteins, carbohydrates and vitamins (Cho et al. 2004). People of Garhwal (Uttrakhand) used its roots for curing piles. In district Jhabua of Madhya Pradesh, its root paste is utilized for treatment of dysentery. *Boerhaavia* decoction is utilized for treatment of rheumatism, leucorrhea and stomach ache in Lalitpur district of Uttar Pradesh by Sahariya tribe. This plant in additionally used for elephantiasis by tribes of Ambikapur district. Tribals of terai region of Indo-Nepal Himalayan, collect this plant from its natural habitat and used for treatment of blood pressure and seminal weakness (Mitra & Gupta 1997, Kumar & Dora

2012). It is also used for accelerating child birth as well as for regaining men virility by most of the tribal people. Leaves of *Boerhaavia* are used for treatment of skin diseases, jaundice, night blindness, low blood pressure and also act as antidote against snake bite. Its decoction is used to treat body swelling of women after delivery in Jajpur district of Orissa (Girach et al. 2006). Its fruit paste when taken daily with fruit paste of *Piper nigrum* helps to cure cold within seven days (Lal & Yadav 1983). For curing jaundice, cough, enlargement of spleen, abdominal pain, pulmonary cavitations, hemorrhoids and cancers it is reportedly used (Dhar et al. 1968). Its root powder along with the powder of *Thalictrum foliolosum* is taken to cure eye diseases, night blindness and corneal ulcers (Gupta et al. 1962). To purify blood and to relieve muscular pain it can also be used orally. It is also used to treat hypertension, seminal weakness and renal ailments (Anand 1995, Gupta et al. 2004). In various food preparations its seeds and roots are added and seeds are also used as bird feed. This plant is also well known for enhancing milk yielding capacity of cattles in different regions of West Bengal. Besides India, this plant is also known for its ethnomedicinal uses in various countries which are summarized in Table 1 (Nayak & Thirunavoukkarasu 2016).

Table 1: Ethnomedicinal uses of *Boerhaavia diffusa* by different countries

Country	Ethnomedicinal uses
Iran	Used for treatment of gonorrhea, joint pain, nephritis, diuretic, expectorant, intestinal gas problem and edema.
Nigeria	for asthma, boils, fever, epilepsy, and boils
Brazil	for beri beri, edema, hepatitis, jaundice, hypertension, renal disorders, spleen, kidney and liver disorders.
Guatemala	for guinea worms and erysipelas
Elsewhere	for sterility, enhancing childbirth and guinea worms
Ghana	For treatment of boils and asthma
Philippines	For fever, diuretic, vermifuge and purgative
West Africa	For curing menstrual irregularities and guinea worms

(*Source:* Nayak & Thirunavoukkarasu, 2016)

5.4. Phytoconstituents

In view of its therapeutic importance it is used for curing different ailments which are summarized in Table 2 (Pramanick 2015). Medicinal value of this plant is because of presence of specific kind of phytoconstituents like alkaloids, flavonoids, triterpenoids, lipids, steroids, lignins, proteins, boeravinone A-F, glycoproteins and carbohydrates which are present in huge quantity in this herb (Surange & Pendse 1972). Beside these it also contain other constituents *viz.* palmitic acid, α-2- sitosterol, β- Sitosterol, tetracosanoic, stearic acid, β- sitosterol ester, β- Ecdysone, triacontanol, hentriacontane and palmitic acid (Verma et al. 1979). Whole herb contains fifteen amino acids out of which six are essential ones and alkaloids are mostly present in its root part whereas allantoin and fatty acids are present in its minute seeds.

Table 2: Uses of Punarnava in different medicinal system

Health problem against which *Boerhaavia* is utilized	Medicinal System
Urinary tract infection, cough and cold, liver disease, headache and heart diseases	Homeopathy system
Anemia, tumors, dyspepsia, heart diseases, asthma, enlargement of spleen in children, cough and cold, sprains and abdominal pain.	Ayurvedic system
Diuretic, asthma, jaundice, dropsical swelling, laxative and emetic.	Unani system
Anti-inflammatory, laxative, expectorant, diuretic and hepato-protective	Siddha system
Abdominal pain, asthma, laxative, muscular pain, paralysis, jaundice, stomach pain, dislocated joint, pneumonia, elephantiasis and fistula	Tribal system
Night blindness, antiviral, dyspepsia, cardiac disorder, restoration in virility of man, lowering serum uric acid level and anticonvulsant.	Modern system

Source: Pramanick, 2015

Biological activity of alkaloid named punarnavine, rotenoids (known as boeravinonone) and ursolic acid has been studied in detail (Ahmad & Hossain 1968, Mishra & Tiwari 1971, Aftab et al. 1996). Rotenoids known as boeravinones *viz.* boeravinone A, boeravinone B, boeravinone C, boeravinone D, boeravinone E and boeravinone F have been isolated from its roots and studied in details for their pharmacological potential and among them boeravinone B is most potent and interesting constituent from therapeutic point of view and is known for anticancer and anti inflammatory properties of this plant species (Seth et al. 1986, Jain & Khanna 1989). Other constituents which are isolated from *Boerhaavia* are C methyl flavone, dihydroisofuroxanthone- borhavine, phytosterols and two lignans *viz.* liriodendrin and syringaresinol mono β-D glycoside (Gupta & Ahmed 1984). Herb is rich source of fats and proteins and punarnavoside has been reported from this plant which acts as antifibrolytic agent. Boerhaavic acid and boerhavin has also been reported to present in its green stalk and around 6% potassium nitrate is present in its shoot parts (Kokate et al. 2005). Boerhavisterol, diusarotenoid, boerhadiffusene, boerhavinone A and boerhavilanastenyl are new compounds isolated from its roots and their structures were also elucidated (Gupta & Ali 1998) and Fig. 3 depicted the diversity at structural level of major phytochemicals isolated from *Boerhaavia* till date.

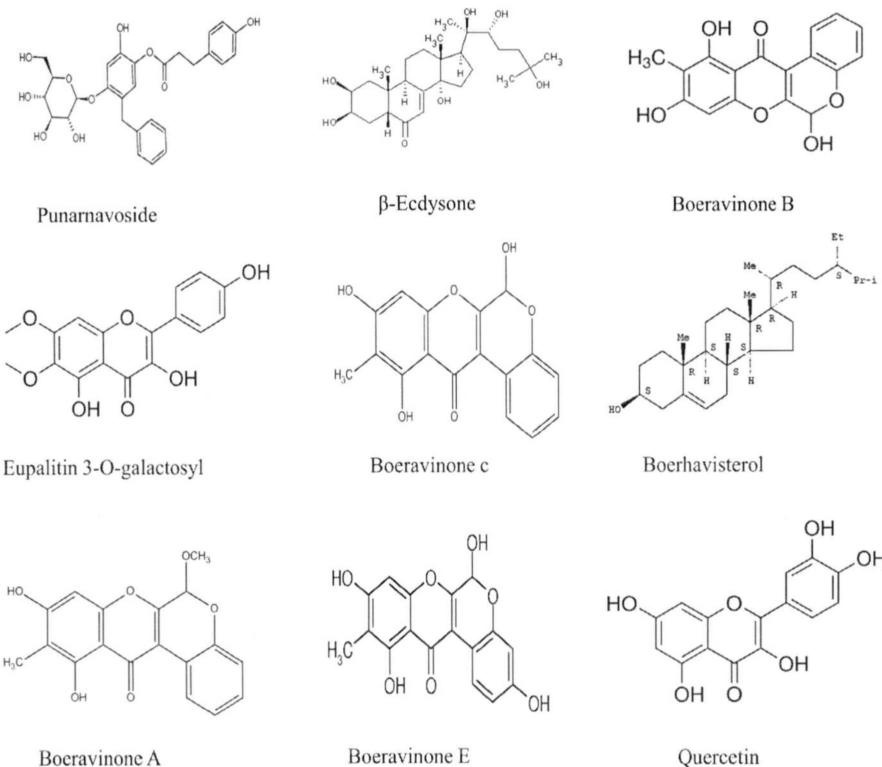

Fig. 3: Major phytochemicals isolated from *Boerhaavia diffusa*

5.5. Medicinal Uses

Its therapeutic properties and chemical constituents have been widely studied and more than 35 different well known formulations contain the herb as a major ingredient (Mishra et al. 2014). Punarnava kshar, punarnava taila and punarnavastaka kvath are some of the Ayurvedic preparation of *Boerhaavia* which are mentioned in old Indian medicinal books like Charaka Samhita and Sushrita Samhita and these formulations are used for curing different ailments and its various well known formulation which are used in different system of medicines are enlisted in Table 3. Its root, shoot part and sometime whole plant are used for its medicinal attributes. Antifibrinolytic agent, punarnavoside is chiefly present in its roots and is used as diuretic and laxative agent in treatment of jaundice, asthma, anemia, dropsy, gonorrhea, ascites, urinary infection and also as blood purifier (Olukoya et al. 1993, Wesely et al. 2010, Reddy et al. 2019). It was utilized to treat blood pressure and seminal weakness and furthermore used for curing cough, stomach ache, cold and its plain juice act as a powerful cure for bites of snake and rodent. The seeds and flowers are utilized for urinary issue and as contraceptive (Chopra et al. 1956).

For curing liver problems its fried leaves are taken with rice by Koch tribe of Meghalaya (Majumdar et al. 2019). Its root juice is also useful for treating asthma and jaundice (Dwivedi et al. 2019). In addition to chemotherapy, it was also utilized in Ayurvedic system of medicine as an adjuvant for curing pulmonary tuberculosis (Kant et al. 2001). Many Ayurvedic formulations like Abana (Heartcare), Evecare (Menstricare), Diabecon (Glucocare), Digyton, Immunol etc. contain *Boerhaavia* as an important constituent and in almost all herbal preparations which are utilized for controlling obesity; punarnava is present as a major ingredient. Its fresh root juice is helpful in treating eye disorders like conjunctivitis and night blindness. Roots of *Boerhaavia* are rich in proteins and act as antiviral agents which protects many useful crops like mung bean, cucumber, watermelon, papaya, beans etc. from harmful mosaic viruses (Awasthi 200, Singh & Awasthi 2002, Awasthi et al. 2003). Recently due to its anti proliferating impact on cancerous cells, it has been used as an adjuvant in anticancer therapy (Milic 2008) and anti tumor effect of this herb is attributed to the presence of punarnavine alkaloid.

Table 3: Punarnava based various formulations used in different medicinal system

Formulation name	Formulation type	Used against
Amavatari kashayam, Arthoven tablets	Unani formulations	Rheumatoid arthritis, Immuno disorders
Punarnavaristha, Punarnavadi kvatha, Punarnava guggula, Punarnava mandura, Sukumara ghrita, Dashamoolarishta and Shothaghna lepa	Ayurvedic formulations	Jaundice, inflammatory disorders, anemia, cardiac disorders, ascites, myxedema, abdominal pain, dyspnoea, piles, fever
Biliarin, Liva, Liv-77, Livomap, Livotrit, Livol, Jaundex syrup, Livin, Liva-16, Acilvan, Hepex, Livodin, Styplon and Neoliv-100.	Allopathic formulations	Sinus infection, pneumonia, bronchitis, urinary tract infection, hepatitis, liver dysfunction, anxiety, seizures, behavioral disorder
Caranai, Mukkiratai	Siddha formulations	Fever, inflammation, vitamin B 12 deficiencies, renal stones, kidney failure, anemia

Source: Pramanick, 2015

5.6. Pharmacological Activity

Different parts of *Boerhaavia diffusa* like leaves, root, or its aerial shoots have been utilized in various regions of India for curing different ailments. Pharmacological studies were conducted on this plant species and it was observed that both its root as well as shoot extracts exhibit wide range of therapeutical properties which will be helpful for wellness of human beings. It has gain lot of importance due to its pharmacological activities like hepatoprotective, antibacterial, antiinflammatory, antioxidant, antifibrinolytic, antidiabetic, antistress and antiproliferating.

5.7. Adulterants and Substitutes

Boerhaavia samples in market are often confused and adulterants with *Trianthema portulacastrum* L. But these two plant species differ in their chemical composition and are sources of two different medicinal formulations, one is Punarnava prepared from *Boerhaavia* other one is Varshabhu from *Trianthema* and these species can be differentiated on the basis of their palisade ratios and stomatal indices and among them Trianthema possess higher values of given aspects (Goyal et al. 2010).

6. Conclusions

Due to the presence of certain kind of phytocompounds, plant based medicines are highly valuable and they are accepted globally as a system of modern pharmaceuticals. Researchers and pharmaceutical industries are highly interested in them as they have a great potential to treat numerous long lasting ailments. *B. diffusa* is also well known therapeutically important plant species used in various traditional medicinal systems which were further supported by clinical and pharmacological studies. From this literature review it is concluded that this plant species must be explored by researchers on toxicological parameter for obtaining valuable market products.

7. Future Prospective

This work serves as a basis for exploring novel secondary metabolites from *Boerhaavia diffusa* which are responsible for its distinctive therapeutic properties with scientific validation. Further, it can assist in selection of high metabolite yielding tissues which can be utilized for enhancement of its secondary metabolite production.

Conflict of interest

Conflict of interest: None.

Acknowledgements

The authors are highly thankful to Head, School of Biotechnology, Shri Mata Vaishno Devi University for providing the necessary facilities of BIF Lab for sourcing the information.

References

Abukakar MG, Ukwuani AN & Shehu RA (2008). Phytochemical screening and anti-bacterial activity of *Tamarindus indica* pulp extract. *Asian Journal of Biochemistry* 3: 134-138.

Adenubi OT, Fasina FO, McGaw LJ & Eloff JN (2016). Plant extracts to control ticks of veterinary and medical importance, a review. *South African Journal of Botany* 105: 178-193.

Aftab K, Usmani SB, Ahmad SI & Usmanghani K (1996). Naturally occurring calcium channel blockers-II. *Hamdard Medicus* 39: 44-54.

Agbafor KN, Akubugwo EI, Ogbashi ME, Ajah PM & Ukwandu CC (2011). Chemical and antimicrobial properties of leaf extracts of *Zapoteca portoricensis. Research Journal of Medicinal Plants* 5: 605-612.

Ahmad K & Hossain A (1968). Isolation, synthesis and biological action of hypoxan-thine-9- larabinofuranoside. *Journal of Agricultural and Biological Sciences* 11: 41-45.

Anand RK (1995). Biodiversity and tribal association of *Boerhaavia diffusa* in India-Nepal Himalayan Terai Region. *Flora and Fauna* 2: 167-170.

Awasthi LP (2000). Protection of crop plants against virus diseases through root extract of *Boerhaavia diffusa. Indian Phytopathology* 54: 508-509.

Awasthi LP, Kumar P & Singh RV (2003). Effect of *Boerhaavia diffusa* inhibitor on the infection and multiplication of cucumber green mottle mosaic virus in musk melon plants. *Indian Phytopatholog* 56: 362-366.

Carson CF & Hammer KA (2010) Chemistry and bioactivity of essential oils. In: Lipids and essential oils as antimicrobial agents. Published by Wiley & Sons, New York, USA, pages 238.

Cho E, Seddom J, Ronser B, Willet W & Hankison S (2004). Prospective study of intake of fruits, vegetables, vitamins and carotenoids and related musclopathy. *Archives of Opthalmology* 122: 883-892.

Chopra GL (1969). Angiosperms: Systematic and life cycle. Published by . S. Nagin and Co, Jalandhar, Punjab, India, 204 pages.

Chopra RN, Nayar SL & Chopra IC (1956). Glossary of Indian medicinal plants. Published by Council of Scientific and Industrial Research (CSIR), New Delhi, India, 330 pages.

Crozier A, Jaganath IB & Clifford MN (2006). Phenols, polyphenols and tannins: An overview pp. 1-24.. In: Plant Secondary Metabolites: Occurrence, Structure and Role in the Human Diet, Crozier A, Clifford MN and Ashihara H (Eds.), Blackwell publishing Ltd, UK.

Dai J & Mumper R (2010). Plant phenolics: Extraction, analysis and their antioxidant and anticancer properties. *Molecules* 15: 7313-7352.

Dar RA, Shahnawaz M & Qazi PH (2017). General overview of medicinal plants: A review. *The Journal of Phytopharmacology* 6: 349-351.

Dewick PM (2002). Medicinal natural products: A biosynthetic approach. Published by Chichester, West Sussex, England, Wiley, 550 pages.

Dhar ML, Dhar MM, Dhawan BN & Mehrotra BN (1968). Screening of Indian plants for biological activity: part I. *Indian Journal of Experimental Biology* 6: 232-247.

Dwivedi T, Kanta C, Singh LR & Prakash I (2019). A list of some important medicinal plants with their medicinal uses from Himalayan state Uttarakhand, India. *Journal of Medicinal Plants Studies* 7: 106-116.

Ghasemzadeh A & Jaafar HZE (2011). Anticancer and antioxidant activities of Malaysian young ginger (*Zingiber officinale* Roscoe) varieties grown under different CO_2 concentrations. *Journal of Medicinal Plant Research* 5: 3247-3255.

Girach RD, Aminuddin & Singh VK. (2006). Notable plants in ethnomedicine of Jajpur district, Orissa, India pp. 495-501. In: Recent progresses in Medicinal Plants, Govil JN, editors, Vol. 14. Biopharmaceuticals.

Goyal BM, Bansal P, Gupta V, Kumar S, Singh R & Maithani M (2010). Pharmacological potential of *Boerhaavia diffusa*: An overview. *International Journal of Pharmaceutical Sciences and Drug Research* 2: 18-23.

Gupta AK, Sharma M, & Tandon N. (2004). Boerhaavia diffusa Linn. (Nyctaginaceae) pp. 204-235. In: Gupta AK, Tendon N, editors, Reviews on Indian Medicinal Plants, S. Narayan and Co. New Delhi, India.

Gupta D & Ahmed B (1984). A new C - methylflavone from *Boerhaavia diffusa* linn. roots. *Indian Journal of Chemistry* 23: 682-684.

Gupta J & Ali M (1998). Chemical constituents of *Boerhaavia diffusa* Linn. roots. *Indian Journal of Chemistry* 37: 912-917.

Gupta RBL, Singh S & Dayal Y (1962). Effect of punarnava on the visual acuity and refractive errors. *Indian Journal of Medicinal Research* 50: 428-434.

Harborne JB (2000). The flavonoids: advances in research since 1992. *Phytochemistry* 55: 481-504.

Heinrich M, Barnes J, Gibbons S & Williamson EM (2004). Fundamentals of pharmacognosy and phytotherpy. Pub;ished by Churchill Livingstone, Elsevier Science Ltd, UK. 360 pages.

Hunt DI (2000). Ecological ethno-botany: Stumbling toward new practices and paradigms. *Model Assisted Statistics and Applications* 16: 1-13.

Jain GK & Khanna NM (1989). Punarnavoside: A new antifibrinolytic agent from *Boerhaavia diffusa* Linn. *Indian Journal of Chemistry* 28: 163-166.

Kant S, Agnihotri MS & Dixit KS (2001). Clinical evaluation of *Boerhaavia diffusa* as an adjuvant in the treatment of pulmonary tuberculosis. *Phytomedica* 2: 89-94.

Kokate CK, Purohit AP & Gokhale SB (2005). Pharmacognosy. Published by Nirali Prakashan, Pune, India, 330 pages.

Kumar A & Dora J (2012). Herb of life punarnava. *International Journal of Research and Reviews in Pharmacy and Applied Science* 2: 347-359.

Lacy AO & Kennedy R (2004). Studies on coumarins and coumarin related compounds to determine their therapeutic role in the treatment of cancer. *Current Pharmaceutical Design* 10: 3797-3811.

Lal SD & Yadav BK (1983). Folk medicines of Kurukshetra district (Haryana), India. *Economy Botany* 37: 299-284,

Majumdar HC, Shyam JM, Chowdhury U, Koch D & Roy N (2019). Traditional hepatoprotective herbal medicine of Koch tribe in the South West Garo hills district, Meghalaya. *Indian Journal of Traditional Knowledge* 18: 312-317.

Mandal SM, Chakraborty D & Dey S (2010). Phenolic acids act as signaling molecules in plant microbe symbioses. *Plant Signaling Behavior* 5: 359-368.

Milic N (2008). Biological and phytochemical studies on *Boerhaavia diffusa*. Thesis, University of Naples, Federico II, Italy.

Mishra AN & Tiwari HP (1971). Constituents of the roots of *Boerhaavia diffusa. Phytochemistry* 10: 3318-3321.

Mishra S, Aeri V, Gaur PV & Jachak SM (2014). Phytochemical, therapeutic and ethnopharmacological overview for a traditionally important herb: *Boerhavia diffusa* Linn. *BioMed Research International* 5: 1-19.

Mitra R & Gupta RC. (1997). Punarnava - An Ayurvedic drug of repute pp. 209-227. In: Applied botany abstracts, Economic Botany Information Service, National Botanical Research Institute, Lucknow, Uttar Pradesh, India.

Munyendo WLL, Orwa JA, Rukunga GM & Bii CC (2011). Bacteriostatic and bactericidal activities of *Aspilia mossambicensis, Ocimum gratissimum* and *Toddalia asiatica* extracts on selected pathogenic bacteria. *Research Journal of Medicinal Plants* 5: 717-727.

Nayak P & Thirunavoukkarasu M (2016). A review of the plant *Boerhaavia diffusa*: its chemistry, pharmacology and therapeutical potential. *The Journal of Phytopharmacology* 5: 83-92.

Newman DJ & Cragg G (2007). Natural products as source of new drugs over the last 25 years. *Journal of Natural Product* 70: 461-477.

Nweze EL, Okafor JL & Njoku O (2004). Antimicrobial activities of methanolic extracts of *Trume guineesis* (Scchumn and Thorn) and *Morinda Lucinda* used in Nigerian herbal medicinal practice. *Journal of Biological Research and Biotechnology* 2: 34-46.

Oksman CKM & Inze D (2004). Plant cell factories in the post-genomic era: New ways to produce designer secondary metabolites. *Trends in Plant Science* 9: 433-440.

Olukoya DK, Tdika N & Odugbemi T (1993). Antibacterial activity of some medicinal plants from Nigeria. *Journal of Ethnopharmacology* 39: 69-72.

Patil SJ & Patil HM (2012). Ethnomedicinal herbal recipes from Satpura hill ranges of Shirpur tehsil, Dhule, Maharashtra, India. *Research Journal of Recent Sciences* 1: 333-366.

Raffauf RF (1996). A guide to their discovery and distribution, Hawkworth Press Inc, New York, pages 234.

Rajavel R, Mallika P, Kumar PK, Moorthy S & Kumar ST (2012). Antinociceptive and anti-inflammatory effects of the methanolic extract of *Oscillatoria annae. Research Journal of Chemical Sciences* 2: 53-61.

Reddy AM, Babu MVS & Rao RR (2018). Ethnobotanical study of traditional herbal plants used by local people of Seshachalam Biosphere reserve in Eastern Ghats, India. *International Journal of Natural Fibres and Medicinal Plants* 65: 40-54.

Savoia D (2012). Plant derived antimicrobial compounds: Alternatives to antibiotics. *Future Microbiology* 7: 979-990.

Saxena PK (2001). Development of plant based medicines: Conservation, efficacy and safety. Kluwer Academic Publishers, Dordrecht, Netherlands. 242 pages.

Seth RK, Khanna M, Chaudhary M, Singh S & Sarin JPS (1986). Estimation of punarnavosides, a new antifibrinolytic compound from *Boerhaavia diffusa. Indian Drugs* 23: 583-584.

Sina MM & Deniz MA (2019). Medicinal plants potential against diabetes mellitus: review article. *International Journal of Pharmacognosy* 6: 39-53.

Singh R (2015). Medicinal plants: a review. *Journal of Plant Sciences* 3: 50-55.

Singh S & Awasthi LP (2002). Prevention of infection and spread of bean common mosaic virus disease of mung bean through botanicals. *Indian Journal of Mycology and Plant Pathology* 32: 141-145.

Strack D & Wray V (1992). Anthocyanins. In: The flavonoids: Advances in research since 1986. Harborne Jb (Eds.), Chapman & Hall, London. 676 pages.

Surange SR & Pendse GS (1972). Pharmacognostic study of roots of *Boerhavia diffusa* (Punarnava). *Journal of Research in Indian Medicine* 7: 1-6.

Thakur RS, Puri HS & Husain A. (1989). Major medicinal plants of India. Published by Central Institute of Medicinal and Aromatic Plants (CIMAP), Lucknow, Uttar Pradesh, India, 585 pages.

Tomas BF & Espin JC (2001). Phenolic compounds and related enzymes as determinants of quality of fruits and vegetables. *Journal of the Science of Food and Agriculture* 81: 853-876.

Verma HN, Awasthi LP & Saxena KC (1979). Isolation of virus inhibitor from the root extract of *Boerhaavia diffusa* inducing systemic resistance in plants. *Canadian Journal of Botany* 57: 1214-1218.

Wesely EG, Johnson M, Regha P & Kavitha MS (2010). *In vitro* propagation of *Boerhaavia diffusa* L. through direct and indirect organogenesis. *Journal of Chemical and Pharmaceutical Research* 2: 339-347.

Index to Scientific Names

A

B

C

D

E

F

G

H

I

M

N

O

P

Q

R

S

T

U

V

W

X

Z

Colour Plates

Chapter 2: Traditional Uses of Plants as Medicine Among Khasi Tribe of Meghalaya, Northeast India

Coix lacryma-jobi *Ilex khasiana* *Houttuynia cordata*

Nepenthes khasiana *Erythrina stricta* *Rubia cordifolia*

Ilex embelioides *Pouzolzia hirta* *Schima khasiana*

Swertia angustifolia *Prunella vulgaris* *Mimosa pudica*

Photo Plate. Medicinal plants of Khasi Hills

Chapter 3: Assessing Ethnobotanical Value and Threat Status of *Hodgsonia heteroclita* (Roxb.) Hook.f. & Thomson, A Lesser Known Liana Species of Sikkim Himalaya

Fig. 1.1 a-c Plant with fruits in its natural habitat **d)** Leaves, **e)** Flowers and **f)** Seed

Chapter 5: Pharmacological Potential of Ethnomedicinal Plants of Asteraceae Family from Arunachal Pradesh Northeast India

Photo Plates 1: *Ageratum conyzoides* 2. *Crassocephalum crepidioides* 3. *Chromolaena odorata* 4. *Spilanthes paniculata*; 5. *Spilanthes acmella*; 6. *Artemisia* sp.

Chapter 6: Analysis of Indigenous Uses of Ethnobotanical Herbaceous Plants and their Diversity Among Nature-Dependent Communities in Atraulia of Burhanpur Tehsil, District Azamgarh, Uttar Pradesh

Fig. 3: Medicinal plant diversity in the study area (A-*Croton bonplandianum*, B-*Euphorbia hirta*, C-*Mallotus philippinensis*, D-*Cyperus rotundus*, E-*Achyranthus aspera*, F-*Rumex dentatus*)

Chapter 7: Ethnobiology of Lichens of North Western Himalayas

Fig. 5: Representative samples of different lichens present in North-west Himalayas (**i.** *Leptogium* sp., **ii.** *Bulbothrix meizospora* (Nyl) Hale, **iii.** *Canomaculina subtinctoria* (Zahlbr.) Elix, **iv.** *Canoparmelia texana* (Tuck.) Elix & Hale, **v.** *Cetrelia collata* (Nyl) Culb. & C.Culb, **vi.** *Evernia mesomorpha* Nyl., **vii.** *Everniastrum cirrhatum* (Fr) Hale, **viii.** *Everniastrum nepalense* (Taylor) Hale ex Sipman, **ix.** *Flavoparmelia caperata* (L.) Hale, **x.** *Flavopunctelia flaventior* (Stirton) Hale, **xi.** *Parmotrema tinctorum* (Despr. ex. Nyl.) Hale, **xii.** *Melanelia infumata* (Nyl.) Essl., **xiii.** *Usnea bailyi* (Strilon) Zahlbr., **xiv.** *Buellia subsororiodes* S. Singh and Awasthi, **xv.** *Heterodermia diademata* (Taylor) Awasthi, xvi. *Ramalina* sp.

Chapter 9: Traditional Usages and Chemical Constituents of High Value Medicinal Plants Recorded from ChitharaVillage Panchayat (Gautam Buddha Nagar) of Uttar Pradesh, India

Fig. 2: High usages medicinal plants of Chithara, **A)** *Bacopa monnieri* (Brahmi), **B)** *Cannabis sativa* (Bhang), **C)** *Cissampelos pareira* (Padha), **D)** *Coccinia grandis* (Kundru), **E)** *Eclipta prostrata* (Bhringraj), **F)** *Euphorbia hirta* (Dudhi, Asthma plant), **G)** *Hyptis suaveolens* (Vilaiti Tulsi), **H)** *Justicia adhatoda* (Arus), **I)** *Solanum americanum* (Makoy), **J)** *Withania somnifera* (Ashwagandha)

Chapter 10: Ethnobotanical Studies on Members of Family Apiaceae from Jammu Division (Jammu and Kashmir) Western Himalaya India

Fig. 2: Some ethnobotanically used plants of family Apiaceae from Jammu province- **(A)** *Ferula jaeschkeana* Vatke. **(B)** *Anethum sowa* Roxb. ex Fleming **(C)** *Bunium persicum* (Boiss.) B.Fedtsch. **(D)** *Hydrocotyle javanica* Thunb. (E) *Chaerophyllum reflexum* var. *acuminatum* (Lindl.) Hedge & Lamond **(F)** *Centella asiatica* (L.) Urb.

Chapter 12: Ethnobotany and Tribals of Ladakh-Need for Scientific Initiatives and Interventions

Fig. 3: *Acantholimon lycopodiodies*

Fig. 4: *Physochlaina praealta*

Fig. 5: *Arnebia guttata*

Fig. 6: *Artemisia* sp.

Fig. 7: *Pedicularis longiflora*

Fig. 8: *Pedicularis pectinata*

Fig. 9: *Myricaria squamosa*

Fig. 10: *Morus alba*

Fig. 11: *Lancea tibetica*

Fig. 12: *Iris lactea*

Fig. 13: *Dracocephalum moldavica*

Fig. 14: *Cirsium arvense*

Plate 1A: Plant diversity of the study area

Fig. 15: *Leontopodium nanum*

Fig. 16: *Taraxacum officinale*

Fig. 17: *Ribes orientale*

Fig. 18: *Tanacetum gracile*

Fig. 19: *Solanum villosum*

Fig. 20: *Rosa ecae*

Fig. 21: *Juniperus indica*

Fig. 22: *Rheum webbianum*

Fig. 23: *Potentilla anserina*

Fig. 24: *Plantago depressa*

Fig. 25: *Echinops cornigerus*

Fig. 26: *Ephedra gerardiana*

Plate 1B: Plant diversity of the study area

Fig. 27: *Oxyria digyna* **Fig. 28:** *Geranium himalayense* **Fig. 29:** *Urica hyperborea*

Fig. 30: *Elaeagnus angustifolia* **Fig. 31:** *Physalis alkekengi* **Fig. 32:** *Vitis vinifera*

Fig. 33: *Dactylorhiza hatagirea* **Fig. 34:** *Hippophae rhamnoides* **Fig. 35:** *Prunus armeniaca*

Fig. 36: *Fagopyrum esculentum* **Fig. 37:** *Mentha longifolia* **Fig. 38:** *Rosa webbiana*

Plate 1C: Plant diversity of the study area

Fig. 39: *Tagetes erecta* **Fig. 40:** *Capparis spinosa* **Fig. 41:** *Melilotus officinalis*

Fig. 42: *Medicago saliva* **Fig. 43:** *Populus nigra* **Fig. 44:** *Juglans regia*

Plate 1D: Plant diversity of the study area

Chapter 14: Eco-Taxonomy, Ethnobotany and Active Chemical Constituents of Ten High Value Plants of Kathua District (J&K) of India in Southeast Asia With Special Reference to Jasrota Wildlife Sanctuary

Image 1: High Value selected medicinal plants growing in Kathua district (Jasrota Wildlife Sanctuary), J&K (India) in Southeast Asia; **(A)** *Aerva sanguinolenta,* **(B)** *Centella asiatica,* **(C)** *Cissampelos pareira,* **(D)** *Colebrookea oppositifolia,* (E) *Tinospora cordifolia,* **(F)** *Holarrhena pubescens,* **(G)** *Justicia adhatoda,* **(H)** *Vitex negundo,* **(I)** *Woodfordia fruticosa,* **(J)** *Cryptolepis dubia.*

Chapter 16: Nature's Anti-diabetic Pharmacy: A Note on Six Valuable Plants

Fig. 1: Antidiabetic plants; **(a)** *Aloe vera*; **(b)** *Andographis paniculata*; **(c)** *Allium sativum*; **d.** *Achyranthes aspera*; **(e)** *Syzygium cumini*; **(f)** *Tinospora cordiafolia.*

Chapter 19: Ethnomedicinal Importance of *Arisaema jacquemontii* and Its Futuristic Role in Natural Therapeutics

Fig. 1: *Arisaemajacquemontii-* **(A)** & **(B)** Whole plant, **(C)** Tubers and **(D)** Ripened fruit

Chapter 20: Indian Snake Root and Devil Root as Distinctive Medicinal Plant for Curing Human Disease: Biology, Chemistry and Cultivation Practices of *Rauwolfia serpentina* and *Rauwolfia tetraphylla*

Fig. 1: Indian snake root (*Rauvolfia serpentina* and devil root (*R. tetraphylla*) Courtesy: B. Singh).

Chapter 22: Indian Sarsaparilla, *Hemidesmus indicus*, an Endangered Medicinal Plant of India

Fig. 1: Plant and Roots of *Hemidesmus indicus*. (Photo Courtesy Mr. Ravi Gollamandala, Andhra Pradesh Urban greening and Beautification Corporation)

Chapter 23: Exploiting Therapeutic Potential of Kirayat (*Andrographis paniculata*): Identity, Chemical Constituents and Biological Aspects for Future Reference in Drug Discovery

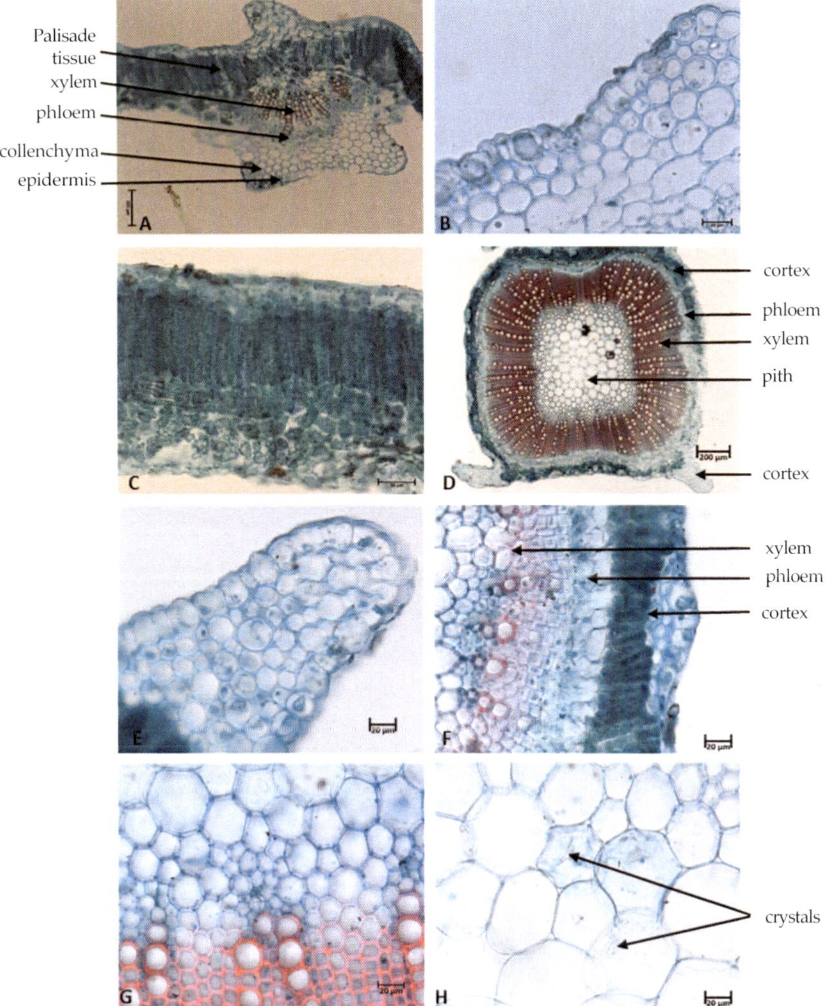

Fig. 3: *Andrographis paniculata*. A-C Transverse section of leaf. A) Midrib; B) Collenchymatous tissue in midrib; C) T.S. of lamina. D-H Transverse section of stem. D) quadrangular stem with lobes; E) collenchymatous tissue in lobes; F) arrangement of cortex and vascular tissues; G) protoxylem and pith region; H) acicular crystals in pith cells.

Chapter 25: Ethnomedical Applications and Phytochemistry of *Boerhaavia diffusa*

Fig. 2: *Boerhaavia diffusa* **(A)** Natural habitat, **(B)** Aerial part, **(C)** Flowering twig.